THE NAKED APE
TRILOGY

THE NAKED APE
TRILOGY

The Naked Ape
The Human Zoo
Intimate Behaviour

DESMOND MORRIS

JONATHAN CAPE
LONDON

First published in this edition 1994

1 3 5 7 9 10 8 6 4 2

The Naked Ape © Desmond Morris 1967
The Human Zoo © Desmond Morris 1969
Intimate Behaviour © Desmond Morris 1971

Desmond Morris has asserted his right
under the Copyright, Designs and Patents Act, 1988
to be identified as the author of this work

Published in the United Kingdom in 1994 by
Jonathan Cape
Random House, 20 Vauxhall Bridge Road, London SWIV 2SA

Random House Australia (Pty) Limited
20 Alfred Street, Milsons Point, Sydney,
New South Wales 2061, Australia

Random House New Zealand Limited
18 Poland Road, Glenfield,
Auckland 10, New Zealand

Random House South Africa (Pty) Limited
PO Box 337, Bergvlei, South Africa

Random House UK Limited Reg. No. 954009

A CIP catalogue record for this book
is available from the British Library

ISBN 0–224–04140–1
Typeset by Pure Tech Corporation, Pondicherry, India
Printed in Great Britain by
Clays Ltd, St. Ives PLC

CONTENTS

ACKNOWLEDGMENTS

The three books that make up this trilogy are intended for a general audience and, because of this, academic authorities are not mentioned in the text. To have done so would have broken the flow of words and is a practice suitable only for more technical works. But many important papers and books were referred to during the assembly of these volumes and it would be wrong not to acknowledge their valuable assistance. I have therefore included Chapter References at the end of each book relating the topics discussed to the major authorities concerned. A Bibliography giving the detailed references can be found at the back.

I would also like to express my debt of gratitude to the countless friends and colleagues who helped me, directly or indirectly, in discussions, correspondence and in numerous other ways. There are so many that it is impossible to list them all here, but in particular I must mention my wife Ramona, whose active support and encouragement was so great that she was virtually a co-author; my publisher and editor, Tom Maschler, whose unflagging enthusiasm inspired the trilogy, and my two brilliant teachers, Peter Medawar and Niko Tinbergen.

D.M.

INTRODUCTION TO THE TRILOGY

The Naked Ape was first published in 1967. Everything I wrote in it seemed obvious enough to me, but it shocked many people.

They were upset for several reasons. The main objection was that I had written about human beings as though they were just another animal species to study. As a zoologist I had already spent twenty years examining the behaviour of a wide variety of creatures, from fish to reptiles and from birds to mammals. My scientific papers on subjects ranging from the courtship behaviour of fish to the pairing of birds, to the food-hoarding of mammals, had been read by a handful of specialists and had caused little or no controversy. When I turned to writing books for a more general audience, about animals such as snakes, apes and pandas, there was again no great fuss. They were read and accepted by a small audience of interested parties. But then, when I produced a similar study of an unusual, naked-skinned primate, everything changed.

Suddenly, every word I had written became the subject of heated debate. The human animal, I discovered, was still finding it hard to come to terms with its biological nature.

I must confess I was surprised to find myself fighting a rearguard action for Charles Darwin. After a century of scientific progress and the discovery of more and more fossils of human ancestors, I had assumed that most people were ready to face the fact that we are an integral part of primate evolution. I thought they were ready to take a close look at their animal qualities and to learn from them. That was the aim of my book, but it soon became clear that I had a greater struggle on my hands.

In some parts of the world *The Naked Ape* was banned and illicit copies were confiscated and burned by the Church, or the idea of human evolution was ridiculed and the book was viewed as a bad joke in appalling taste. I was deluged with religious tracts to encourage me to mend my ways.

The Chicago Tribune pulped an entire issue of its magazine because its owners were offended by a review of the book that appeared on its pages. Why were they so offended? Because the review in question included the word 'penis'.

Sexual honesty, it seemed, was another of the faults to be found in the book. The same newspaper included endless reports of violence and murder. The word 'gun' appeared frequently. As I pointed out at the time, it was strange that they were prepared to mention something that shot death, but not something that shot life. But logic had no place here. By exchanging my fish and birds for men and women, I had uncovered a sleeping giant of human prejudice.

In addition to breaking religious and sexual taboos, I was also accused of 'making man beastly' by insisting that the human species is driven by powerful inborn urges. This flew in the face of a great deal of fashionable psychological theorizing suggesting that everything we do is determined by learning and conditioning.

It was claimed that I was putting forward the dangerous idea that mankind was trapped by brutish animal instincts from which there was no escape. This was another misreading of what I had written. There was no good reason why my suggestion of inborn 'animal impulses' should make man brutish in the derogatory sense that was implied. A brief glance at the various chapters of the book reveals that the inborn patterns I referred to include such features as the powerful urge to form loving pair-bonds, to care for our children, to seek a varied diet, to keep ourselves clean, to settle disputes by display and ritual rather than by bloodshed and, above all, to exhibit playfulness, curiosity and inventiveness. These are our main 'animal urges' and to say that they make us bestial or brutish is wilfully to misrepresent the zoological way of looking at human behaviour.

There was, in addition, a political misunderstanding. It was assumed, again wrongly, that I was portraying the human species as condemned to some primeval status quo. The extremes of the political spectrum viewed this as outrageous. To them, the human animal must be completely pliable, able to adapt to any regime imposed upon it. The idea that, beneath the skin, all humans may be guided by a set of genetic suggestions, inherited from their parents, is repulsive to political tyrants. It means that these leaders will always encounter a deep-seated resistance to their extreme social ideas. And that, as history teaches us, is what has happened time and again. Tyrannies may come, but they also go. Friendly, co-operative human nature eventually reasserts itself.

Finally, there were those who felt that to call human beings 'Naked Apes' was insulting and pessimistic. Nothing could be further from the truth. I used the title simply to emphasize that I was attempting a zoological portrait of our species. Viewed alongside other primates, 'The Naked Ape' is a valid description. To say it is insulting is to insult animals. To say it is pessimistic is to fail to marvel at the extraordinary success story of a modestly designed mammal.

When an illustrated edition of *The Naked Ape* was published in 1986 I was asked to update the text. There was only one alteration I felt the need to make. I had to change a 3 to a 4. In 1967, when the book first appeared, the world population of human beings had stood at 3,000 million. In the intervening years it had risen to 4,000 million. Writing now, in 1994, it has risen yet again to well over 5,000 million. By the year 2000 it will be 6,000 million.

The effects on human life of this massive increase in numbers concerned me. During the millions of years of our evolution we had been thin on the ground, living in small tribes. That tribal life moulded us, but it has left us unequipped for modern urban life. How does the Tribal Ape manage to cope as a City Ape?

This question became the subject of the sequel to *The Naked Ape*. I had often heard it said that 'the city is a concrete jungle', but I knew this to be false. I had studied jungles and they were not like cities. They were not overcrowded. They were organic and changed only very slowly. Cities blossomed almost overnight. In biological terms, Rome *was* built in a day.

When, as a zoologist, I studied the behaviour of city-dwellers, they did remind me of something. Living in their cramped quarters they reminded me not of jungle wildlife, but of captive zoo animals. The city, I decided, was not a concrete jungle, it was a human zoo, and this became the title of my second volume in the Naked Ape trilogy.

In *The Human Zoo* I looked closer at the aggressive, sexual and parental behaviour of our species, as it manifested itself under the stresses and pressures of urban living. What happens when the tribe becomes a super-tribe? What happens when status becomes super-status? How does our family-based sexuality survive when each individual is surrounded by thousands of strangers?

Why, if cities are so stressful, do people flock to them? The answer to this last question adds a pleasant element to a sometimes depressing picture. For the city, despite all its faults, acts as a giant stimulus-centre where our great inventiveness can flourish and develop.

To complete the trilogy, in a volume entitled *Intimate Behaviour*, I addressed the subject of what has happened to our personal relations within this new environment. How has our intensely sexual and loving nature reacted to modern life? In our intimate relations, how much has been lost and how much has been gained?

In many ways, we have remained remarkably faithful to our biological origins. Our genetic programming has proved flexible, but is nevertheless resistant to major changes. Where straightforward loving relations prove impossible for us, we employ our inventiveness to devise substitutes that will see us through. Our ingenuity as a species enables us to enjoy the technical comforts and excitements of modern living while at the same time managing to obey our primeval imperatives.

This has been the secret of our extraordinary success and, if we are lucky, will enable us to continue to walk our increasingly hazardous evolutionary tightrope. Those who picture the future as a ruined, polluted landscape are misguided. They watch the newscasts, recoil at the worst we can do and extend that a thousand times to create their gloom-ridden scenario. They overlook two things. First, the news that is brought to us is nearly always bad news, but for every act of violence or destruction that occurs there are a million acts of peaceful friendliness. We are indeed an amazingly peaceful species, given our population levels, but our widespread peacefulness fails to make the headlines.

Second, when visualizing the future, they usually overlook the possibility of revolutionary new inventions. Every generation has seen startling technical advances and there is no reason to suppose that these will suddenly stop. On the contrary, they will almost certainly increase dramatically. Nothing is impossible. If we can imagine it, sooner or later we will be able to do it. But even when we have made our mainframe computers look as primitive as clay tablets, we ourselves will still be no more than Naked Apes, made of flesh and blood. Even if, in our relentless quest for progress, we have destroyed all our close animal relatives, we will remain biological phenomena, subject to biological rules.

With this in mind I am delighted that my Naked Ape trilogy, originally published between 1967 and 1971, is now reappearing as a single volume. A quarter of a century later, the message remains the same – you are a member of the most extraordinary animal species that has ever lived. Understand your animal nature and accept it.

DESMOND MORRIS
Oxford, 1994

THE NAKED APE

A Zoologist's Study of the Human Animal

CONTENTS

INTRODUCTION

There are one hundred and ninety-three living species of monkeys and apes. One hundred and ninety-two of them are covered with hair. The exception is a naked ape self-named *Homo sapiens*. This unusual and highly successful species spends a great deal of time examining his higher motives and an equal amount of time studiously ignoring his fundamental ones. He is proud that he has the biggest brain of all the primates, but attempts to conceal the fact that he also has the biggest penis, preferring to accord this honour falsely to the mighty gorilla. He is an intensely vocal, acutely exploratory, overcrowded ape, and it is high time we examined his basic behaviour.

I am a zoologist and the naked ape is an animal. He is therefore fair game for my pen and I refuse to avoid him any longer simply because some of his behaviour patterns are rather complex and impressive. My excuse is that, in becoming so erudite, *Homo sapiens* has remained a naked ape nevertheless; in acquiring lofty new motives, he has lost none of the earthy old ones. This is frequently a cause of some embarrassment to him, but his old impulses have been with him for millions of years, his new ones only a few thousand at the most – and there is no hope of quickly shrugging off the accumulated genetic legacy of his whole evolutionary past. He would be a far less worried and more fulfilled animal if only he would face up to this fact. Perhaps this is where the zoologist can help.

One of the strangest features of previous studies of naked-ape behaviour is that they have nearly always avoided the obvious. The earlier anthropologists rushed off to all kinds of unlikely corners of the world in order to unravel the basic truth about our nature, scattering to remote cultural backwaters so atypical and unsuccessful that they are nearly extinct. They then returned with startling facts about the bizarre mating customs, strange kinship systems, or weird ritual procedures of these tribes, and used this material as though it were of central importance to the behaviour of our species as a whole. The work done by these investigators was, of course, extremely interesting and most valuable in showing us what can happen when a group of naked apes becomes side-tracked into a cultural blind alley. It revealed just how far from the normal our behaviour patterns can stray without a complete social collapse. What it did not tell us was anything about the typical behaviour of typical naked apes. This can only be done by examining the common behaviour patterns that are shared by all the ordinary, successful members of the major cultures – the mainstream specimens who together represent the vast majority. Biologically, this is the only sound approach. Against this, the old-style anthropologist would have argued that his technologically simple tribal groups are nearer the heart of the matter than the members of advanced civilizations. I submit that this is not so. The simple tribal groups that are living today are not primitive, they are stultified. Truly primitive tribes have not existed for thousands of years. The naked ape is essentially an exploratory species and any society that has failed to advance has in some sense failed, 'gone wrong'. Something has happened to it to hold it back, something that is working against the natural tendencies of the species to explore and investigate the world around it. The characteristics that the earlier anthropologists studied in these tribes may well be the very features that have interfered with the progress of the groups concerned. It is therefore dangerous to use this information as the basis for any general scheme of our behaviour as a species.

Psychiatrists and psychoanalysts, by contrast, have stayed nearer home and have concentrated on clinical studies of mainstream specimens. Much of their earlier material, although not suffering from the weakness of the anthropological information, also has an unfortunate bias. The individuals on which they have based their pronouncements are, despite their mainstream background, inevitably aberrant or failed specimens in some respect. If they were healthy, successful and therefore typical individuals, they would not have had to seek psychiatric aid and would not have contributed to the psychiatrists' store of information. Again, I do not wish to belittle the value of this research. It has given us an immensely important insight into the way in which our behaviour patterns can break down. I simply feel that in attempting to discuss the fundamental biological nature of our species as a whole, it is unwise to place too great an emphasis on the earlier anthropological and psychiatric findings.

(I should add that the situation in anthropology and psychiatry is changing rapidly. Many modern research workers in these fields are recognizing the limitations of the earlier investigations and are turning more and more to studies of typical, healthy individuals. As one investigator expressed it recently: 'We have put the cart before the horse. We have tackled the abnormals and we are only now beginning, a little late in the day, to concentrate on the normals.')

The approach I propose to use in this book draws its material from three main sources: (1) the information about our past as unearthed by palaeontologists and based on the fossil and other remains of our ancient ancestors; (2) the information available from the animal behaviour studies of the comparative ethologists, based on detailed observations of a wide range of animal species, especially our closest living relatives, the monkeys and apes; and (3) the information that can be assembled by simple, direct observation of the most basic and widely shared behaviour patterns of the successful mainstream specimens from the major contemporary cultures of the naked ape itself.

Because of the size of the task, it will be necessary to oversimplify in some manner. The way I shall do this is largely to ignore the detailed ramifications of technology and verbalization, and concentrate instead on those aspects of our lives that have obvious counterparts in other species: such activities as feeding, grooming, sleeping, fighting, mating and care of the young. When faced with these fundamental problems, how does the naked ape react? How do his reactions compare with those of other monkeys and apes? In which particular respect is he unique, and how do his oddities relate to his special evolutionary story?

In dealing with these problems I realize that I shall run the risk of offending a number of people. There are some who will prefer not to contemplate their animal selves. They may consider that I have degraded our species by discussing it in crude animal terms. I can only assure them that this is not my intention. There are others who will resent any zoological invasion of their specialist arena. But I believe that this approach can be of great value and that, whatever its shortcomings, it will throw new (and in some ways unexpected) light on the complex nature of our extraordinary species.

I

ORIGINS

THERE IS a label on a cage at a certain zoo that states simply, 'This animal is new to science'. Inside the cage there sits a small squirrel. It has black feet and it comes from Africa. No black-footed squirrel has ever been found in that continent before. Nothing is known about it. It has no name.

For the zoologist it presents an immediate challenge. What is it about its way of life that has made it unique? How does it differ from the three hundred and sixty-six other living species of squirrels already known and described? Somehow, at some point in the evolution of the squirrel family, the ancestors of this animal must have split off from the rest and established themselves as an independent breeding population. What was it in the environment that made possible their isolation as a new form of life? The new trend must have started out in a small way, with a group of squirrels in one area becoming slightly changed and better adapted to the particular conditions there. But at this stage they would still be able to inter-breed with their relatives nearby. The new form would be at a slight advantage in its special region, but it would be no more than a race of the basic species and could be swamped out, reabsorbed into the mainstream at any point. If, as time passed, the new squirrels became more and more perfectly tuned-in to their particular environment, the moment would eventually arrive when it would be advantageous for them to become isolated from possible contamination by their neighbours. At this stage their social and sexual behaviour would undergo special modifications, making inter-breeding with other kinds of squirrels unlikely and eventually impossible. At first, their anatomy may have changed and become better at coping with the special food of the district, but later their mating calls and displays would also differ, ensuring that they attracted only mates of the new type. At last, a new species would have evolved, separate and discrete, a unique form of life, a three hundred and sixty-seventh kind of squirrel.

When we look at our unidentified squirrel in its zoo cage, we can only guess about these things. All we can be certain about is that the markings of its fur – its black feet – indicate that it is a new form. But these are only the symptoms, the rash that gives a doctor a clue about his patient's disease. To really understand this new species, we must use these clues only as a starting point, telling us there is something worth pursuing. We might try to guess at the animal's history, but that would be presumptuous and dangerous. Instead we will start humbly by giving it a simple and obvious label: we will call it the African black-footed squirrel. Now we must observe and record every aspect of its behaviour and structure and see how it differs from, or is similar to, other squirrels. Then, little by little, we can piece together its story.

The great advantage we have when studying such animals is that we ourselves are not black-footed squirrels – a fact which forces us into an attitude of humility that is becoming to proper scientific investigation. How different things are, how depressingly different, when we attempt to study the human animal. Even for the zoologist, who is used to calling an animal an animal, it is difficult to avoid the arrogance of subjective involvement. We can try to overcome this to some extent by deliberately and rather coyly approaching the human being as if he were another species, a strange form of life on the dissecting table, awaiting analysis. How can we begin?

As with the new squirrel, we can start by comparing him with other species that appear to be most closely related. From his teeth, his hands, his eyes and various other anatomical features, he is obviously a primate of some sort, but of a very odd kind. Just how odd becomes clear when we lay out in a long row the skins of the one hundred and ninety-two living species of monkeys and apes, and then try to insert a human pelt at a suitable point somewhere in this long series. Wherever we put it, it looks out of place. Eventually we are driven to position it right at one end of the row of skins, next to the hides of the tailless great apes such as the chimpanzee and the gorilla. Even here it is obtrusively different. The legs are too long, the arms are too short and the feet are rather strange. Clearly this species of primate has developed a special kind of locomotion which has modified its basic form. But there is another characteristic that cries out for attention: the skin is virtually naked. Except for conspicuous tufts of hair on the head, in the armpits and around the genitals, the skin surface is completely exposed. When compared with the other primate species, the contrast is dramatic. True, some species of monkeys and apes have small naked patches of skin on their rumps, their faces, or their chests, but nowhere amongst the other one hundred and ninety-two species is there anything even approaching the human condition. At this point and without further investigation, it is justifiable to name this new species the 'naked ape'. It is a simple, descriptive name based on a simple observation, and it makes no special assumptions. Perhaps it will help us to keep a sense of proportion and maintain our objectivity.

Staring at this strange specimen and puzzling over the significance of its unique features, the zoologist now has to start making comparisons. Where else is nudity at a premium? The other primates are no help, so it means looking farther afield. A rapid survey of the whole range of the living mammals soon proves that they are remarkably attached to their protective, furry covering, and that very few of the 4,237 species in existence have seen fit to abandon it. Unlike their reptilian ancestors, mammals have acquired the great physiological advantage of being able to maintain a constant, high body temperature. This keeps the delicate machinery of the body processes tuned in for top performance. It is not a property to be endangered or discarded lightly. The temperature-controlling devices are of vital importance and the possession of a thick, hairy, insulating coat obviously plays a major role in preventing heat loss. In intense sunlight it will also prevent over-heating and damage to the skin from direct exposure to the sun's rays. If the hair has to go, then clearly there must be a very powerful reason for abolishing it. With few exceptions this drastic step has been taken only when mammals have launched themselves into an entirely new medium. The flying mammals, the bats, have been forced to denude their wings, but they have retained their furriness elsewhere and can hardly be counted as naked species. The burrowing mammals have in a few cases – the naked mole rat, the aardvark and the armadillo, for example – reduced their hairy covering. The aquatic mammals such as the whales, dolphins, porpoises, dugongs, manatees and hippopotamuses have also gone naked as part of a general streamlining. But for all the more typical surface-dwelling mammals, whether scampering about on the ground or clambering around in the vegetation, a densely hairy hide is the basic rule. Apart from those abnormally heavy giants, the rhinos and the elephants (which have heating and cooling problems peculiar to themselves), the naked ape stands alone, marked off by his nudity from all the thousands of hairy, shaggy or furry land-dwelling mammalian species.

At this point the zoologist is forced to the conclusion that either he is dealing with a burrowing or an aquatic mammal, or there is something very odd, indeed unique, about the evolutionary history of the naked ape. Before setting out on a field trip to observe the animal in its present-day form, the first thing to do, then, is to dig back into its past and examine as closely as possible its immediate ancestors. Perhaps by examining the fossils and other remains and by taking a look at the closest living relatives, we shall be able to gain some sort of picture of what happened as this new type of primate emerged and diverged from the family stock.

It would take too long to present here all the tiny fragments of evidence that have been painstakingly collected over the past century. Instead, we will assume that this task has been done and simply summarize the conclusions that can be drawn from it, combining the information available from the work of the fossil-hungry palaeontologists with the facts gathered by the patient ape-watching ethologists.

The primate group, to which our naked ape belongs, arose originally from primitive insectivore stock. These early mammals were small, insignificant creatures, scuttling nervously around in the safety of the forests, while the reptile overlords were dominating the animal scene. Between eighty and fifty million years ago, following the collapse of the great age of reptiles, these little insect-eaters began to venture out into new territories. There they spread and grew into many strange shapes. Some became plant-eaters and burrowed under the ground for safety, or grew long, stilt-like legs with which to flee from their enemies. Others became long-clawed, sharp-toothed killers. Although the major reptiles had abdicated and left the scene, the open country was once again a battlefield.

Meanwhile, in the undergrowth, small feet were still clinging to the security of the forest vegetation. Progress was being made here, too. The early insect-eaters began to broaden their diet and conquer the digestive problems of devouring fruits, nuts, berries, buds and leaves. As they evolved into the lowliest forms of primates, their vision improved, the eyes coming forward to the front of the face and the hands developing as food-graspers. With three-dimensional vision, manipulating limbs and slowly enlarging brains, they came more and more to dominate their arboreal world.

Somewhere between twenty-five and thirty-five million years ago, these pre-monkeys had already started to evolve into monkeys proper. They were beginning to develop long, balancing tails and were increasing considerably in body size. Some were on their way to becoming leaf-eating specialists, but most were keeping to a broad, mixed diet. As time passed, some of these monkey-like creatures became bigger and heavier. Instead of scampering and leaping they switched to brachiating – swinging hand over hand along the underside of the branches. Their tails became obsolete. Their size, although making them more cumbersome in the trees, made them less wary of ground-level sorties.

Even so, at this stage – the ape phase – there was much to be said for keeping to the lush comfort and easy pickings of their forest of Eden. Only if the environment gave them a rude shove into the great open spaces would they be likely to move. Unlike the early mammalian explorers, they had become specialized in forest existence. Millions of years of development had gone into perfecting this forest aristocracy, and if they left now they would have to compete with the (by this time) highly advanced ground-living herbivores and killers. And so there they stayed, munching their fruit and quietly minding their own business.

It should be stressed that this ape trend was for some reason taking place only in the Old World. Monkeys had evolved separately as advanced tree-dwellers in both the Old and the New World, but the American branch of the primates never made the ape grade. In the Old World, on the other hand, ancestral apes were spreading over a wide forest area from western Africa, at one extreme, to south-eastern Asia at the other. Today the remnants of this development can be seen in the African chimpanzees and gorillas and the Asian gibbons and orangutans. Between these two extremities the world is now devoid of hairy apes. The lush forests have gone.

What happened to the early apes? We know that the climate began to work against them and that, by a point somewhere around fifteen million years ago, their forest strongholds had become seriously reduced in size. The ancestral apes were forced to do one of two things: either they had to cling on to what was left of their old forest homes, or, in an almost biblical sense, they had to face expulsion from the Garden. The ancestors of the chimpanzees, gorillas, gibbons and orangs stayed put, and their numbers have been slowly dwindling ever since. The ancestors of the only other surviving ape – the naked ape – struck out, left the forests, and threw themselves into competition with the already efficiently adapted ground-dwellers. It was a risky business, but in terms of evolutionary success it paid dividends.

The naked ape's success story from this point on is well known, but a brief summary will help, because it is vital to keep in mind the events which followed if we are to gain an objective understanding of the present-day behaviour of the species.

Faced with a new environment, our ancestors encountered a bleak prospect. They had to become either better killers than the old-time carnivores, or better grazers than the old-time herbivores. We know today that, in a sense, success has been won on both scores; but agriculture is only a few thousand years old, and we are dealing in millions of years. Specialized exploitation of the plant life of the open country was beyond the capacity of our early ancestors and had to await the development of advanced techniques of modern times. The digestive system necessary for a direct conquest of the grassland food supply was lacking. The fruit and nut diet of the forest could be adapted to a root and bulb diet at ground level, but the limitations were severe. Instead of lazily reaching out to the end of the branch for a luscious ripe fruit, the vegetable-seeking ground ape would be forced to scratch and scrape painstakingly in the hard earth for his precious food.

His old forest diet, however, was not all fruit and nut. Animal proteins were undoubtedly of great importance to him. He came originally, after all, from basic insectivore stock, and his ancient arboreal home had always been rich in insect life. Juicy bugs, eggs, young helpless nestlings, tree-frogs and small reptiles were all grist to his mill. What is more, they posed no great problems for his rather generalized digestive system. Down on the ground this source of food supply was by no means absent and there was nothing to stop him increasing this part of his diet. At first, he was no match for the professional killer of the carnivore world. Even a small mongoose, not to mention a big cat, could beat him to the kill. But young animals of all kinds, helpless ones or sick ones, were there for the taking, and the first step on the road to major meat-eating was an easy one. The really big prizes, however, were poised on long, stilt-like legs, ready to flee at a moment's notice at quite impossible speeds. The protein-laden ungulates were beyond his grasp.

This brings us to the last million or so years of the naked ape's ancestral history, and to a series of shattering and increasingly dramatic developments. Several things happened together, and it is important to realize this. All too often, when the story is told, the separate parts of it are spread out as if one major advance led to another, but this is misleading. The ancestral ground-apes already had large and high-quality brains. They had good eyes and efficient grasping hands. They inevitably, as primates, had some degree of social organization. With strong pressure on them to increase their prey-killing prowess, vital changes began to take place. They became more upright – fast, better runners. Their hands became freed from locomotion duties – strong, efficient weapon-holders. Their brains became more complex – brighter, quicker decision-makers. These things did not follow one another in a major, set sequence; they blossomed together, minute advances being made first in one quality and then in another, each urging the other on. A hunting ape, a killer ape, was in the making.

It could be argued that evolution might have favoured the less drastic step of developing a more typical cat- or dog-like killer, a kind of cat-ape or dog-ape, by the simple process of enlarging the teeth and nails into savage fang-like and claw-like weapons. But this would have put the ancestral ground-ape into direct competition with the already highly specialized cat and dog killers. It would have meant competing with them on their own terms, and the outcome would no doubt have been disastrous for the primates in question. (For all we know, this may actually have been tried and failed so badly that the evidence has not been found.) Instead, an entirely new approach was made, using artificial weapons instead of natural ones, and it worked.

From tool-using to tool-making was the next step, and alongside this development went improved hunting techniques, not only in terms of weapons, but also in terms of social co-operation. The hunting apes were pack-hunters, and as their techniques of killing were improved, so were their methods of social organization. Wolves in a pack deploy themselves, but the hunting ape already had a much better brain than a wolf and could turn it to such problems as group communication and co-operation. Increasingly complex manœuvres could be developed. The growth of the brain surged on.

Essentially this was a hunting-group of males. The females were too busy rearing the young to be able to play a major role in chasing and catching prey. As the complexity of the hunt increased and the forays became more prolonged, it became essential for the hunting ape to abandon the meandering, nomadic ways of its ancestors. A home base was necessary, a place to come back to with the spoils, where the females and young would be waiting and could share the food. This step, as we shall see in later chapters, has had profound effects on many aspects of the behaviour of even the most sophisticated naked apes of today.

So the hunting ape became a territorial ape. His whole sexual, parental and social pattern began to be affected. His old wandering, fruit-plucking way of life was fading rapidly. He had now really left his forest of Eden. He was an ape with responsibilities. He began to worry about the prehistoric equivalent of washing machines and refrigerators. He began to develop the home comforts – fire, food storage, artificial shelters. But this is where we must stop for the moment, for we are moving out of the realms of biology and into the realms of culture. The biological basis of these advanced steps lies in the development of a brain large and complex enough to enable the hunting ape to take them, but the exact form they assume is no longer a matter of specific genetic control. The forest ape that became a ground ape that became a hunting ape that became a territorial ape has become a cultural ape, and we must call a temporary halt.

It is worth re-iterating here that, in this book, we are not concerned with the massive cultural explosions that followed, of which the naked ape of today is so proud – the dramatic progression that led him, in a mere half-million years, from making a fire to making a space-craft. It is an exciting story, but the naked ape is in danger of being dazzled by it all and forgetting that beneath the surface gloss he is still very much a primate. ('An ape's an ape, a varlet's a varlet, though they be clad in silk or scarlet.') Even a space ape must urinate.

Only by taking a hard look at the way in which we have originated and then by studying the biological aspects of the way we behave as a

species today can we really acquire a balanced, objective understanding of our extraordinary existence.

If we accept the history of our evolution as it has been outlined here,
then one fact stands out clearly: namely, that we have arisen essentially
as primate predators. Amongst existing monkeys and apes, this makes
us unique, but major conversions of this kind are not unknown in
other groups. The giant panda, for instance, is a perfect case of the
reverse process. Whereas we are vegetarians turned carnivores, the
panda is a carnivore turned vegetarian, and like us it is in many ways
an extraordinary and unique creature. The point is that a major switch
of this sort produces an animal with a dual personality. Once over the
threshold, it plunges into its new role with great evolutionary energy –
so much so that it carries with it many of its old traits. Insufficient time
has passed for it to throw off all its old characteristics while it is
hurriedly donning the new ones. When the ancient fishes first conquered dry land, their new terrestrial qualities raced ahead while they
continued to drag their old watery ones with them. It takes millions of
years to perfect a dramatically new animal model, and the pioneer
forms are usually very odd mixtures indeed. The naked ape is such a
mixture. His whole body, his way of life, was geared to a forest
existence, and then suddenly (suddenly in evolutionary terms) he was
jettisoned into a world where he could survive only if he began to live
like a brainy, weapon-toting wolf. We must examine now exactly how
this affected not only his body, but especially his behaviour, and in
what form we experience the influence of this legacy at the present day.

One way of doing this is to compare the structure and the way of life
of a 'pure' fruit-picking primate with a 'pure' carnivore. Once we have
cleared our minds about the essential differences that relate to their
two contrasted methods of feeding, we can then re-examine the naked
ape situation to see how the mixture has been worked out.

The brightest stars in the carnivore galaxy are, on the one hand, the
wild dogs and wolves, and, on the other, the big cats such as the lions,
tigers and leopards. They are beautifully equipped with delicately
perfected sense organs. Their sense of hearing is acute, and their
external ears can twist this way and that to pick up the slightest rustle
or snort. Their eyes, although poor on static detail and colour, are
incredibly responsive to the tiniest movement. Their sense of smell is
so good that it is difficult for us to comprehend it. They must be able
to experience a virtual landscape of odours. Not only are they capable
of detecting an individual smell with unerring precision, but they are
also able to pick out the separate component odours of a complex
smell. Experiments carried out with dogs in 1953 indicated that their
sense of smell was between a million and a thousand million times as

accurate as ours. These astonishing results have since been queried, and later, more careful tests have not been able to confirm them, but even the most cautious estimates put the dog's sense of smell at about a hundred times better than ours.

In addition to this first-rate sensory equipment, the wild dogs and big cats have a wonderfully athletic physique. The cats have specialized as lightning sprinters, the dogs as long-distance runners of great stamina. At the kill they can bring into action powerful jaws, sharp, savage teeth and, in the case of the big cats, massively muscular front limbs armed with huge, dagger-pointed claws.

For these animals, the act of killing has become a goal in itself, a consummatory act. It is true that they seldom kill wantonly or wastefully, but if, in captivity, one of these carnivores is given ready-killed food, its urge to hunt is far from satisfied. Every time a domestic dog is taken for a walk by its master, or has a stick thrown for it to chase and catch, it is having its basic need to hunt catered for in a way that no amount of canned dog-food will subdue. Even the most over-stuffed domestic cat demands a nocturnal prowl and the chance to leap on an unsuspecting bird.

Their digestive system is geared to accept comparatively long periods of fasting followed by bloating gorges. (A wolf, for instance, can eat one-fifth of its total body weight at one meal – the equivalent of you or me devouring a 30–40 lb. steak at a single sitting.) Their food is of high nutritional value and there is little wastage. Their faeces, however, are messy and smelly and defecation involves special behaviour patterns. In some cases the faeces are actually buried and the site carefully covered over. In others, the act of defecating is always carried out at a considerable distance from the home base. When young cubs foul the den, the faeces are eaten by the mother and the home is kept clean in this way.

Simple food storage is undertaken. Carcasses, or parts of them, may be buried, as with dogs and certain kinds of cats; or they may be carried up into a tree-larder, as with the leopard. The periods of intensive athletic activity during the hunting and killing phases are interspersed with periods of great laziness and relaxation. During social encounters the savage weapons so vital to the kill constitute a potential threat to life and limb in any minor disputes and rivalries. If two wolves or two lions fall out, they are both so heavily armed that fighting could easily, in a matter of seconds, lead to mutilation or death. This could seriously endanger the survival of the species and during the long course of the evolution that gave these species their lethal prey-killing weapons, they have of necessity also developed powerful inhibitions about using their weapons on other members of their own species. These inhibitions appear to have a specific genetic basis: they do not have to be learned. Special submissive postures have been evolved which automatically appease a dominant animal and inhibit its attack. The possession of these signals is a vital part of the way of life of the 'pure' carnivores.

The actual method of hunting varies from species to species. In the leopard it is a matter of solitary stalking or hiding, and a last-minute pounce. For the cheetah it is a careful prowl followed by an all-out sprint. For the lion it is usually a group action, with the prey driven in panic by one lion towards others in hiding. For a pack of wolves it may involve an encircling manœuvre followed by a group kill. For a pack of African hunting dogs it is typically a ruthless drive, with one dog after another going in to the attack until the fleeing prey is weakened from loss of blood.

Recent studies in Africa have revealed that the spotted hyaena is also a savage pack-hunter and not, as has always been thought, primarily a scavenger. The mistake has been made because hyaena packs form only at night and minor scavenging has always been recorded during the day. When dusk falls, the hyaena becomes a ruthless killer, just as efficient as the hunting dog is during the day. Up to thirty animals may hunt together. They easily out-pace the zebras or antelopes they are pursuing, which dare not travel at their full day-time speeds. The hyaenas start tearing at the legs of any prey in reach until one is sufficiently wounded to fall back from the fleeing herd. All the hyaenas then converge on this one, tearing out its soft parts until it drops and is killed. Hyaenas base themselves at communal den-sites. The group or 'clan' using this home base may number between ten and a hundred. Females stick closely to the area around this base, but the males are more mobile and may wander off into other regions. There is considerable aggression between clans if wandering individuals are caught off their own clan territory, but there is little aggression between the members of any one clan.

Food-sharing is known to be practised in a number of species. Of course, at a large kill there is meat enough for the whole hunting group and there need be little squabbling, but in some instances the sharing is taken further than that. African hunting dogs, for instance, are known to re-gurgitate food to one another after a hunt is over. In some cases they have done this to such an extent that they have been referred to as having a 'communal stomach'.

Carnivores with young go to considerable trouble to provide food for their growing offspring. Lionesses will hunt and carry meat back to the den, or they will swallow large hunks of it and then re-gurgitate it for the cubs. Male lions have occasionally been reported to assist in this matter, but it does not appear to be a common practice. Male wolves, on the other hand, have been known to travel up to fifteen miles to obtain food for both the female and her young. Large meaty bones may be carried back for the young to gnaw, or hunks of meat may be swallowed at the kill and then re-gurgitated at the entrance to the den.

These, then, are some of the main features of the specialist carnivores, as they relate to their hunting way of life. How do they compare with those of the typical fruit-picking monkeys and apes?

The sensory equipment of the higher primates is much more dominated by the sense of vision than the sense of smell. In their tree-climbing world, seeing well is far more important than smelling well, and the snout has shrunk considerably, giving the eyes a much better view. In searching for food, the colours of fruits are helpful clues, and, unlike the carnivores, primates have evolved good colour vision. Their eyes are also better at picking out static details. Their food is static, and detecting minute movements is less vital than recognizing subtle differences in shape and texture. Hearing is important, but less so than for the tracking killers, and their external ears are smaller and lack the twisting mobility of those of the carnivores. The sense of taste is more refined. The diet is more varied and highly flavoured – there is more to taste. In particular there is a strong positive response to sweet-tasting objects.

The primate physique is good for climbing and clambering, but it is not built for high-speed sprinting on the ground, or for lengthy endurance feats. This is the agile body of an acrobat rather than the burly frame of a powerful athlete. The hands are good for grasping, but not for tearing or striking. The jaws and teeth are reasonably strong, but nothing like the massive, clamping, crunching apparatus of the carnivores. The occasional killing of small, insignificant prey requires no gargantuan efforts. Killing is not, in fact, a basic part of the primate way of life.

Feeding is spread out through much of the day. Instead of great gorging feasts followed by long fasts, the monkeys and apes keep on and on munching – a life of non-stop snacks. There are, of course, periods of rest, typically in the middle of the day and during the night, but the contrast is nevertheless a striking one. The static food is always there, just waiting to be plucked and eaten. All that is necessary is for the animals to move from one feeding-place to another, as their tastes change, or as the fruits come in and out of season. No food storage takes place except, in a very temporary way, in the bulging cheek pouches of certain monkeys.

The faeces are less smelly than those of the meat-eaters and no special behaviour has developed to deal with their disposal, since they drop down out of the trees and away from the animals. As the group is always on the move, there is little danger of a particular area becoming unduly fouled or smelly. Even the great apes that bed down in special sleeping-nests make a new bed at a new site each night, so that there is little need to worry about nest sanitation. (All the same, it is rather surprising to discover that 99 per cent of abandoned gorilla nests in one area of Africa had gorilla dung inside them, and that in 73 per cent the animals had actually been lying in it. This is bound to constitute a disease risk by increasing the chances of re-infection, and is a remarkable illustration of the basic faecal disinterest of primates.)

Because of the static nature and abundance of the food, there is no need for the primate group to split up to search for it. They can move, flee, rest and sleep together in a close-knit community, with every member keeping an eye on the movements and actions of every other. Each individual of the group will at any one moment have a reasonably good idea of what everyone else is doing. This is a very non-carnivore procedure. Even in those species of primates that do split up from time to time, the smaller unit is never composed of a single individual. A solitary monkey or ape is a vulnerable creature. It lacks the powerful natural weapons of the carnivore and in isolation falls easy prey to the stalking killers.

The co-operative spirit that is present in such pack-hunters as wolves is largely absent from the world of the primate. Competitiveness and dominance is the order of his day. Competition in the social hierarchy is, of course, present in both groups, but it is less tempered by co-operative action in the case of monkeys and apes. Complicated, co-ordinated manœuvres are also unnecessary: sequences of feeding action do not need to be strung together in such a complex way. The primate can live much more from minute to minute, from hand to mouth.

Because the primate's food supply is all around it for the taking, there is little need to cover great distances. Groups of wild gorillas, the largest of the living primates, have been carefully studied and their movements traced, so that we now know that they travel on the average about a third of a mile a day. Sometimes they move only a few hundred feet. Carnivores, by contrast, must frequently travel many miles on a single hunting trip. In some instances they have been known to travel over fifty miles on a hunting journey, taking several days before returning to their home base. This act of returning to a fixed home base is typical of the carnivores, but is far less common amongst the monkeys and apes. True, a group of primates will live in a reasonably clearly defined home range, but at night it will probably bed down wherever it happens to have ended up in its day's meanderings. It will get to know the general region in which it lives because it is always wandering back and forth across it, but it will tend to use the whole area in a much more haphazard way. Also, the interaction between one troop and the next will be less defensive and less aggressive than is the case with carnivores. A territory is, by definition, a defended area, and primates are not therefore, typically, territorial animals.

A small point, but one that is relevant here, is that carnivores have fleas but primates do not. Monkeys and apes are plagued by lice and certain other external parasites but, contrary to popular opinion, they are completely flealess, for one very good reason. To understand this, it is necessary to examine the life-cycle of the flea. This insect lays its eggs, not on the body of its host, but amongst the detritus of its victim's sleeping quarters. The eggs take three days to hatch into small, crawling maggots. These larvae do not feed on blood, but on the waste matter that has accumulated in the dirt of the den or lair. After two weeks they spin a cocoon and pupate. They remain in this dormant condition for approximately two more weeks before emerging as adults, ready to hop on to a suitable host body. So for at least the first month of its life a flea is cut off from its host species. It is clear from this why a nomadic mammal, such as a monkey or ape, is not troubled by fleas. Even if a few stray fleas do happen on to one and mate successfully, their eggs will be left behind as the primate group moves on, and when the pupae hatch there will be no host 'at home' to continue the relationship. Fleas are therefore parasites only of animals with a fixed home base, such as the typical carnivores. The significance of this point will become clear in a moment.

In contrasting the different ways of life of the carnivores and the primates, I have naturally concentrated on the typical open-country hunters on the one hand, and the typical forest-dwelling fruit-pickers on the other. There are certain minor exceptions to the general rules on both sides, but we must concentrate now on the one major exception – the naked ape. To what extent was he able to modify himself, to blend his frugivorous heritage with his newly adopted carnivory? Exactly what kind of an animal did this cause him to become?

To start with, he had the wrong kind of sensory equipment for life on the ground. His nose was too weak and his ears not sharp enough. His physique was hopelessly inadequate for arduous endurance tests and for lightning sprints. In personality he was more competitive than co-operative and no doubt poor on planning and concentration. But fortunately he had an excellent brain, already better in terms of general intelligence than that of his carnivore rivals. By bringing his body up into a vertical position, modifying his hands in one way and his feet in another, and by improving his brain still further and using it as hard as he could, he stood a chance.

This is easy to say, but it took a long time to do, and it had all kinds of repercussions on other aspects of his daily life, as we shall see in later chapters. All we need concern ourselves with for the moment is how it was achieved and how it affected his hunting and feeding behaviour.

As the battle was to be won by brain rather than brawn, some kind of dramatic evolutionary step had to be taken to greatly increase his brain-power. What happened was rather odd: the hunting ape became an infantile ape. This evolutionary trick is not unique; it has happened in a number of quite separate cases. Put very simply, it is a process (called neoteny) by which certain juvenile or infantile characters are retained and prolonged into adult life. (A famous example is the axolotl, a kind of salamander that may remain a tadpole all its life and is capable of breeding in this condition.)

The way in which this process of neoteny helps the primate brain to grow and develop is best understood if we consider the unborn infant of a typical monkey. Before birth the brain of the monkey foetus increases rapidly in size and complexity. When the animal is born its brain has already attained seventy per cent of its final adult size. The remaining thirty per cent of growth is quickly completed in the first six months of life. Even a young chimpanzee completes its brain-growth within twelve months after birth. Our own species, by contrast, has at birth a brain which is only twenty-three per cent of its final adult size. Rapid growth continues for a further six years after birth, and the whole growing process is not complete until about the twenty-third year of life.

For you and me, then, brain-growth continues for about ten years *after* we have attained sexual maturity, but for the chimpanzee it is completed six or seven years *before* the animal becomes reproductively active. This explains very clearly what is meant by saying that we became infantile apes, but it is essential to qualify this statement. We (or rather, our hunting ape ancestors) became infantile in certain ways, but not in others. The rates of development of our various properties got out of phase. While our reproductive systems raced ahead, our brain-growth dawdled behind. And so it was with various other parts of our make-up, some being greatly slowed down, others a little, and still others not at all. In other words, there was a process of differential infantilism. Once the trend was under way, natural selection would favour the slowing down of any parts of the animal's make-up that helped it to survive in its hostile and difficult new environment. The brain was not the only part of the body affected: the body posture was also influenced in the same way. An unborn mammal has the axis of its head at right angles to the axis of its trunk. If it were born in this condition its head would point down at the ground as it moved along on all fours, but before birth occurs the head rotates backwards so that its axis is in line with that of the trunk. Then, when it is born and walking along, its head points forwards in the approved manner. If such an animal began to walk along on its hind legs in a vertical posture, its head would point upwards, looking at the sky. For a vertical animal, like the hunting ape, it is important therefore to retain the foetal angle of the head, keeping it at right angles to the body so that, despite the new locomotion position, the head faces forwards. This is, of course, what has happened and, once again, it is an example of neoteny, the pre-birth stage being retained into the post-birth and adult life.

Many of the other special physical characters of the hunting ape can be accounted for in this way: the long slender neck, the flatness of the face, the small size of the teeth and their late eruption, the absence of heavy brow ridges and the non-rotation of the big toe.

The fact that so many separate embryonic characteristics were potentially valuable to the hunting ape in his new role was the evolutionary breakthrough that he needed. In one neotenous stroke he was able to acquire both the brain he needed and the body to go with it. He could run vertically with his hands free to wield weapons, and at the same time he developed the brain that could develop the weapons. More than that, he not only became brainier at manipulating objects, but he also had a longer childhood during which he could learn from his parents and other adults. Infant monkeys and chimpanzees are playful, exploratory and inventive, but this phase dies quickly. The naked ape's infancy was, in these respects, extended right through into his sexually adult life. There was plenty of time to imitate and learn the special techniques that had been devised by previous generations. His weaknesses as a physical and instinctive hunter could be more than compensated for by his intelligence and his imitative abilities. He could be taught by his parents as no animal had ever been taught before.

But teaching alone was not enough. Genetic assistance was required. Basic biological changes in the nature of the hunting ape had to accompany this process. If one simply took a typical, forest-living, fruit-picking primate of the kind described earlier, and gave it a big brain and a hunting body, it would be difficult for it to become a successful hunting ape without some other modifications. Its basic behaviour patterns would be wrong. It might be able to think things out and plan in a very clever way, but its more fundamental animal urges would be of the wrong type. The teaching would be working *against* its natural tendencies, not only in its feeding behaviour, but also in its general social, aggressive and sexual behaviour, and in all the other basic behavioural aspects of its earlier primate existence. If genetically controlled changes were not wrought here too, then the new education of the young hunting ape would be an impossibly uphill task. Cultural training can achieve a great deal, but no matter how brilliant the machinery of the higher centres of the brain, it needs a considerable degree of support from the lower regions.

If we look back now at the differences between the typical 'pure' carnivore and the typical 'pure' primate, we can see how this probably came about. The advanced carnivore separates the actions of food-seeking (hunting and killing) from the actions of eating. They have become two distinct motivational systems with only partial dependence one on the other. This has come about because the whole sequence is so lengthy and arduous. The act of feeding is too remote, and so the action of killing has to become a reward in itself. Researches with cats have even indicated that the sequence there has become further sub-divided. Catching the prey, killing it, preparing it (plucking it), and eating it, each have their own partially independent motivational systems. If one of these patterns of behaviour is satiated, it does not automatically satiate the others.

For the fruit-picking primate the situation is entirely different. Each feeding sequence, comprising simple food-searching and then immediate eating, is comparatively so brief that no splitting up into separate motivational systems is necessary. This is something that would have to be changed, and changed radically, in the case of the hunting ape. Hunting would have to bring its own reward, it could no longer simply act as an appetitive sequence leading up to the consummatory meal. Perhaps, as in the cat, hunting, killing and preparing the food would each develop their own separate, independent goals, would each become ends in themselves. Each would then have to find expression and one could not be damped down by satisfying another. If we examine – as we shall be doing in a later chapter – the feeding behaviour of present-day naked apes, we shall see that there are plenty of indications that something like this did occur.

In addition to becoming a biological (as opposed to a cultural) killer, the hunting ape also had to modify the timing arrangements of his eating behaviour. Minute-by-minute snacks were out and big, spaced meals were in. Food storage was practised. A basic tendency to return to a fixed home base had to be built in to the behavioural system. Orientation and homing abilities had to be improved. Defecation had to become a spatially organized pattern of behaviour, a private (carnivore) activity instead of a communal (primate) one.

I mentioned earlier that one outcome of using a fixed home base is that it makes parasitization by fleas possible. I also said that carnivores have fleas, but primates do not. If the hunting ape was unique amongst primates in having a fixed base, then we would also expect him to break the primate rule concerning fleas, and this certainly seems to be the case. We know that today our species is parasitized by these insects and that we have our own special kind of flea – one that belongs to a different species from other fleas, one that has evolved with us. If it had sufficient time to develop into a new species, then it must have been with us for a very long while indeed, long enough to have been an unwelcome companion right back in our earliest hunting-ape days.

Socially the hunting ape had to increase his urge to communicate and to co-operate with his fellows. Facial expressions and vocalizations had to become more complicated. With the new weapons to hand, he had to develop powerful signals that would inhibit attacks within the social group. On the other hand, with a fixed home base to defend, he had to develop stronger aggressive responses to members of rival groups.

Because of the demands of his new way of life, he had to reduce his powerful primate urge never to leave the main body of the group.

As part of his new-found co-operativeness and because of the erratic nature of the food supply, he had to begin to share out his food. Like the paternal wolves mentioned earlier, the hunting ape males also had to carry food supplies home for the nursing females and their slowly growing young. Paternal behaviour of this kind had to be a new development, for the general primate rule is that virtually all parental care comes from the mother. (It is only a wise primate, like our hunting ape, that knows its own father.)

Because of the extremely long period of dependency of the young and the heavy demands made by them, the females found themselves almost perpetually confined to the home base. In this respect the hunting ape's new way of life threw up a special problem, one that it did not share with the typical 'pure' carnivores: the role of the sexes had to become more distinct. The hunting parties, unlike those of the 'pure' carnivores, had to become all-male groups. If anything was going to go against the primate grain, it was this. For a virile primate male to go off on a feeding trip and leave his females unprotected from the advances of any other males that might happen to come by was unheard of. No amount of cultural training could put this right. This was something that demanded a major shift in social behaviour.

The answer was the development of a pair-bond. Male and female hunting apes had to fall in love and remain faithful to one another. This is a common tendency in many other groups of animals, but is rare amongst primates. It solved three problems in one stroke. It meant that the females remained bonded to their individual males and faithful to them while they were away on the hunt. It meant that serious sexual rivalries between the males were reduced. This aided their developing co-operativeness. If they were to hunt together successfully, the weaker males as well as the stronger ones had to play their part. They had to play a central role and could not be thrust to the periphery of society, as happens in so many primate species. What is more, with his newly developed and deadly artificial weapons, the hunting ape male was under strong pressure to reduce any source of disharmony within the tribe. Thirdly, the development of a one-male-one-female breeding unit meant that the offspring also benefited. The heavy task of rearing and training the slowly developing young demanded a cohesive family unit. In other groups of animals, whether they are fishes, birds or mammals, when there is too big a burden for one parent to bear alone, we see the development of a powerful pair-bond, tying the male and female parents together throughout the breeding season. This, too, is what occurred in the case of the hunting ape.

In this way, the females were sure of their males' support and were able to devote themselves to their maternal duties. The males were sure of their females' loyalty, were prepared to leave them for hunting, and avoided fighting over them. And the offspring were provided with the maximum of care and attention. This certainly sounds like an ideal solution, but it involved a major change in primate socio-sexual behaviour and, as we shall see later, the process was never really perfected. It is clear from the behaviour of our species today that the trend was only partially completed and that our earlier primate urges keep on reappearing in minor forms.

This is the manner, then, in which the hunting ape took on the role of a lethal carnivore and changed his primate ways accordingly. I have suggested that they were basic biological changes rather than mere cultural ones, and that the new species changed genetically in this way. You may consider this an unjustified assumption. You may feel – such is the power of cultural indoctrination – that the modifications could easily have been made by training and the development of new traditions. I doubt this. One only has to look at the behaviour of our species at the present day to see that this is not so. Cultural developments have given us more and more impressive technological advances, but wherever these clash with our basic biological properties they meet strong resistance. The fundamental patterns of behaviour laid down in our early days as hunting apes still shine through all our affairs, no matter how lofty they may be. If the organization of our earthier activities – our feeding, our fear, our aggression, our sex, our parental care – had been developed solely by cultural means, there can be little doubt that we would have got it under better control by now, and twisted it this way and that to suit the increasingly extraordinary demands put upon it by our technological advances. But we have not done so. We have repeatedly bowed our heads before our animal nature and tacitly admitted the existence of the complex beast that stirs within us. If we are honest, we will confess that it will take millions of years, and the same genetic process of natural selection that put it there, to change it. In the meantime, our unbelievably complicated civilizations will be able to prosper only if we design them in such a way that they do not clash with or tend to suppress our basic animal demands. Unfortunately our thinking brain is not always in harmony with our feeling brain. There are many examples showing where things have gone astray, and human societies have crashed or become stultified.

In the chapters that follow we will try to see how this has happened, but first there is one question that must be answered – the question that was asked at the beginning of this chapter. When we first encountered this strange species we noted that it had one feature that stood out immediately from the rest, when it was placed as a specimen in a long row of primates. This feature was its naked skin, which led me as a zoologist to name the creature 'the naked ape'. We have since seen that it could have been given any number of suitable names: the vertical ape, the tool-making ape, the brainy ape, the territorial ape, and so on. But these were not the first things we noticed. Regarded simply as a zoological specimen in a museum, it is the nakedness that has the immediate impact, and this is the name we will stick to, if only to bring it into line with other zoological studies and remind us that this is the special way in which we are approaching it. But what is the significance of this strange feature? Why on earth should the hunting ape have become a naked ape?

Unfortunately fossils cannot help us when it comes to differences in skin and hair, so that we have no idea as to exactly when the great denudation took place. We can be fairly certain that it did not happen before our ancestors left their forest homes. It is such an odd development that it seems much more likely to have been yet another feature of the great transformation scene on the open plains. But exactly how did it occur, and how did it help the emerging ape to survive?

This problem has puzzled experts for a long time and many imaginative theories have been put forward. One of the most promising ideas is that it was part and parcel of the process of neoteny. If you examine an infant chimpanzee at birth you will find that it has a good head of hair, but that its body is almost naked. If this condition was delayed into the animal's adult life by neoteny, the adult chimpanzee's hair condition would be very much like ours.

It is interesting that in our own species this neotenous suppression of hair growth has not been entirely perfected. The growing foetus starts off on the road towards typical mammalian hairiness, so that between the sixth and eighth months of its life in the womb it becomes almost completely covered in a fine hairy down. This foetal coat is referred to

27

as the lanugo and it is not shed until just before birth. Premature babies sometimes enter the world still wearing their lanugo, much to the horror of their parents, but, except in very rare cases, it soon drops away. There are no more than about thirty recorded instances of families producing offspring that grow up to be fully furred adults.

Even so, all adult members of our species do have a large number of body hairs – more, in fact, than our relatives the chimpanzees. It is not so much that we have lost whole hairs as that we have sprouted only puny ones. (This does not, incidentally, apply to all races – negroes have undergone a real as well as an apparent hair loss). This fact has led certain anatomists to declare that we cannot consider ourselves as a hairless or naked species, and one famous authority went so far as to say that the statement that we are 'the least hairy of all the primates is, therefore, very far from being true: and the numerous quaint theories that have been put forward to account for the imagined loss of hairs are, mercifully, not needed.' This is clearly nonsensical. It is like saying that because a blind man has a pair of eyes he is not blind. Functionally, we are stark naked and our skin is fully exposed to the outside world. This state of affairs still has to be explained, regardless of how many tiny hairs we can count under a magnifying lens.

The neoteny explanation only gives a clue as to how the nakedness could have come about. It does not tell us anything about the value of nudity as a new character that helped the naked ape to survive better in his hostile environment. It might be argued that it had no value, that it was merely a by-product of other, more vital neotenous changes, such as the brain development. But as we have already seen, the process of neoteny is one of differential retarding of developmental processes. Some things slow down more than others – the rates of growth get out of phase. It is hardly likely, therefore, that an infantile trait as potentially dangerous as nakedness was going to be allowed to persist simply because other changes were slowing down. Unless it had some special value to the new species, it would be quickly dealt with by natural selection.

What, then, was the survival value of naked skin? One explanation is that when the hunting ape abandoned its nomadic past and settled down at fixed home bases, its dens became heavily infested with skin parasites. The use of the same sleeping places night after night is thought to have provided abnormally rich breeding-grounds for a variety of ticks, mites, fleas and bugs, to a point where the situation provided a severe disease risk. By casting off his hairy coat, the den-dweller was better able to cope with the problem.

There may be an element of truth in this idea, but it can hardly have been of major importance. Few other den-dwelling mammals – and there are hundreds of species to pick from – have taken this step. Nevertheless, if nakedness was developed in some other connection, it might make it easier to remove troublesome skin parasites, a task which today still occupies a great deal of time for the hairier primates.

Another thought along similar lines is that the hunting ape had such messy feeding habits that a furry coat would soon become clogged and messy and, again, a disease risk. It is pointed out that vultures, who plunge their heads and necks into gory carcasses, have lost the feathers from these members; and that the same development, extended over the whole body, may have occurred among the hunting apes. But the ability to develop tools to kill and skin the prey can hardly have preceded the ability to use other objects to clean the hunters' hair. Even a chimpanzee in the wild will occasionally use leaves as toilet paper when in difficulties with defecation.

A suggestion has even been put forward that it was the development of fire that led to the loss of the hairy coat. It is argued that the hunting ape will have felt cold only at night and that, once he had the luxury of sitting round a camp fire, he was able to dispense with his fur and thus leave himself in a better state for dealing with the heat of the day.

Another, more ingenious theory is that, before he became a hunting ape, the original ground ape that had left the forests went through a long phase as an aquatic ape. He is envisaged as moving to the tropical sea-shores in search of food. There he will have found shellfish and other sea-shore creatures in comparative abundance, a food supply much richer and more attractive than that on the open plains. At first he will have groped around in the rock pools and the shallow water, but gradually he will have started to swim out to greater depths and dive for food. During this process, it is argued, he will have lost his hair like other mammals that have returned to the sea. Only his head, protruding from the surface of the water, would retain the hairy coat to protect him from the direct glare of the sun. Then, later on, when his tools (originally developed for cracking open shells) became sufficiently advanced, he will have spread away from the cradle of the sea-shore and out into the open land spaces as an emerging hunter.

It is held that this theory explains why we are so nimble in the water today, while our closest living relatives, the chimpanzees, are so helpless and quickly drown. It explains our streamlined bodies and even our vertical posture, the latter supposedly having developed as we waded into deeper and deeper water. It clears up a strange feature of our body-hair tracts. Close examination reveals that on our backs the directions of our tiny remnant hairs differ strikingly from those of other apes. In us they point diagonally backwards and inwards towards the spine. This follows the direction of the flow of water passing over a swimming body and indicates that, if the coat of hair was modified before it was lost, then it was modified in exactly the right way to reduce resistance when swimming. It is also pointed out that we are unique amongst all the primates in being the only one to possess a thick layer of sub-cutaneous fat. This is interpreted as the equivalent of the blubber of a whale or seal, a compensatory insulating device. It is stressed that no other explanation has been given for this feature of our anatomy. Even the sensitive nature of our hands is brought into play on the side of the aquatic theory. A reasonably crude hand can, after all, hold a stick or a rock, but it takes a subtle, sensitized hand to feel for food in the water. Perhaps this was the way that the ground ape originally acquired its super-hand, and then passed it on ready-made to the hunting ape. Finally, the aquatic theory needles the traditional fossil-hunters by pointing out that they have been singularly unsuccessful in unearthing the vital missing links in our ancient past, and gives them the hot tip that if they would only take the trouble to search around the areas that constituted the African coastal sea-shores of a million or so years ago, they might find something that would be to their advantage.

Unfortunately this has yet to be done and, despite its most appealing indirect evidence, the aquatic theory lacks solid support. It neatly accounts for a number of special features, but it demands in exchange the acceptance of a hypothetical major evolutionary phase for which there is no direct evidence. (Even if eventually it does turn out to be true, it will not clash seriously with the general picture of the hunting ape's evolution out of a ground ape. It will simply mean that the ground ape went through a rather salutary christening ceremony.)

An argument along entirely different lines has suggested that, instead of developing as a response to the physical environment, the loss of hair was a social trend. In other words it arose, not as a mechanical device, but as a signal. Naked patches of skin can be seen in a number of primate species and in certain instances they appear to act as species recognition marks, enabling one monkey or ape to identify another as belonging to its own kind, or some other. The loss of hair on the part of the hunting ape is regarded simply as an arbitrarily selected characteristic that happened to be adopted as an identity badge by this species. It is of course undeniable that stark nudity must have rendered the naked ape startlingly easy to identify, but there are plenty of other less drastic ways of achieving the same end, without sacrificing a valuable insulating coat.

Another suggestion along the same lines pictures the loss of hair as an extension of sexual signalling. It is claimed that male mammals are generally hairier than their females and that, by extending this sex difference, the female naked ape was able to become more and more sexually attractive to the male. The trend to loss of hair would affect the male, too, but to a lesser extent and with special areas of contrast, such as the beard.

This last idea may well explain the sex differences as regards hairiness but, again, the loss of body insulation would be a high price to pay for a sexy appearance alone, even with subcutaneous fat as a partial compensating device. A modification of this idea is that it was not so much the appearance as the sensitivity to touch that was sexually important. It can be argued that by exposing their naked skins to one another during sexual encounters, both male and female would become more highly sensitized to erotic stimuli. In a species where pair-bonding was evolving, this would heighten the excitement of sexual activities and would tighten the bond between the pair by intensifying copulatory rewards.

Perhaps the most commonly held explanation of the hairless condition is that it evolved as a cooling device. By coming out of the shady forests the hunting ape was exposing himself to much greater temperatures than he had previously experienced, and it is assumed that he took off his hairy coat to prevent himself from becoming over-heated. Superficially this is reasonable enough. We do, after all, take our jackets off on a hot summer's day. But it does not stand up to closer scrutiny. In the first place, none of the other animals (of roughly our size) on the open plains have taken this step. If it was as simple as this we might expect to see some naked lions and naked jackals. Instead they have short but dense coats. Exposure of the naked skin to the air certainly increases the chances of heat loss, but it also increases heat gain at the same time and risks damage from the sun's rays, as any sun-bather will know. Experiments in the desert have shown that the wearing of light clothing may reduce heat loss by curtailing water evaporation, but it also reduces heat gain from the environment to 55 per cent of the figure obtained in a state of total nudity. At really high temperatures, heavier, looser clothing of the type favoured in Arab countries is a better protection than even light clothing. It cuts down the in-coming heat, but at the same time allows air to circulate around the body and aid in the evaporation of cooling sweat.

Clearly the situation is more complicated than it at first appears. A great deal will depend on the exact temperature levels of the environment and on the amount of direct sunshine. Even if we suppose that the climate was suitable for hair loss – that is, moderately hot, but not intensely hot – we still have to explain the striking difference in coat condition between the naked ape and the other open-country carnivores.

There is one way we can do this, and it may give the best answer yet to the whole problem of our nakedness. The essential difference between the hunting ape and his carnivore rivals was that he was not physically equipped to make lightning dashes after his prey or even to undertake long endurance pursuits. But this is nevertheless precisely what he had to do. He succeeded because of his better brain, leading to more intelligent manœuvring and more lethal weapons, but despite this such efforts must have put a huge strain on him in simple physical terms. The chase was so important to him that he would have to put up with this, but in the process he must have experienced considerable over-heating. There would be a strong selection pressure working to reduce this over-heating and any slight improvement would be favoured, even if it meant sacrifices in other directions. His very survival depended on it. This surely was the key factor operating in the conversion of a hairy hunting ape into a naked ape. With neoteny to help the process on its way, and with the added advantages of the minor secondary benefits already mentioned, it would become a viable proposition. By losing the heavy coat of hair and by increasing the number of sweat glands all over the body surface, considerable cooling could be achieved – not for minute-by-minute living, but for the supreme moments of the chase – with the production of a generous film of evaporating liquid over his air-exposed, straining limbs and trunk.

This system would not succeed, of course, if the climate were too intensely hot, because of damage to the exposed skin, but in a moderately hot environment it would be acceptable. It is interesting that the trend was accompanied by the development of a sub-cutaneous fat layer, which indicates that there was a need to keep the body warm at other times. If this appears to counterbalance the loss of the hairy coat, it should be remembered that the fat layer helps to retain the body heat in cold conditions, without hindering the evaporation of sweat when over-heating takes place. The combination of reduced hair, increased sweat glands, and the fatty layer under the skin appears to have given our hard-working ancestors just what they needed, bearing in mind that hunting was one of the most important aspects of their new way of life.

So there he stands, our vertical, hunting, weapon-toting, territorial, neotenous, brainy Naked Ape, a primate by ancestry and a carnivore by adoption, ready to conquer the world. But he is a very new and experimental departure, and new models frequently have imperfections. For him the main troubles will stem from the fact that his culturally operated advances will race ahead of any further genetic ones. His genes will lag behind, and he will be constantly reminded that, for all his environment-moulding achievements, he is still at heart a very naked ape.

At this point we can leave his past behind us and see how we find him faring today. How *does* the modern naked ape behave? How does he tackle the age-old problems of feeding, fighting, mating, and rearing his young? How much has his computer of a brain been able to reorganize his mammalian urges? Perhaps he has had to make more concessions than he likes to admit. We shall see.

2

SEX

S EXUALLY the naked ape finds himself today in a somewhat confusing situation. As a primate he is pulled one way, as a carnivore by adoption he is pulled another, and as a member of an elaborate civilized community he is pulled yet another.

To start with, he owes all his basic sexual qualities to his fruit-picking, forest-ape ancestors. These characteristics were then drastically modified to fit in with his open-country, hunting way of life. This was difficult enough, but then they, in turn, had to be adapted to match the rapid development of an increasingly complex and culturally determined social structure.

The first of these changes, from a sexual fruit-picker to a sexual hunter, was achieved over a comparatively long period of time and with reasonable success. The second change has been less successful. It has happened too quickly and has been forced to depend upon intelligence and the application of learned restraint rather than on biological modifications based on natural selection. It could be said that the advance of civilization has not so much moulded modern sexual behaviour, as that sexual behaviour has moulded the shape of civilization. If this seems to be a rather sweeping statement, let me first put my case and then we can return to the argument at the end of the chapter.

To begin with we must establish precisely how the naked ape does behave today when indulging in sexual behaviour. This is not as easy as it sounds, because of the great variability that exists, both between and within societies. The only solution is to take average results from large samples of the most successful societies. The small, backward, and unsuccessful societies can largely be ignored. They may have fascinating and bizarre sexual customs, but biologically speaking they no longer represent the mainstream of evolution. Indeed, it may very well be that their unusual sexual behaviour has helped to turn them into biological failures as social groups.

Most of the detailed information we have available stems from a number of painstaking studies carried out in recent years in North America and based largely on that culture. Fortunately it is biologically a very large and successful culture and can, without undue fear of distortion, be taken as representative of the modern naked ape.

34

Sexual behaviour in our species goes through three characteristic phases: pair-formation, pre-copulatory activity, and copulation, usually but not always in that order. The pair-formation stage, usually referred to as courtship, is remarkably prolonged by animal standards, frequently lasting for weeks or even months. As with many other species it is characterized by tentative, ambivalent behaviour involving conflicts between fear, aggression and sexual attraction. The nervousness and hesitancy is slowly reduced if the mutual sexual signals are strong enough. These involve complex facial expressions, body postures and vocalizations. The latter involve the highly specialized and symbolized sound signals of speech, but equally importantly they present to the member of the opposite sex a distinctive vocalization tone. A courting couple is often referred to as 'murmuring sweet nothings' and this phrase sums up clearly the significance of the tone of voice as opposed to what is being spoken.

After the initial stages of visual and vocal display, simple body contacts are made. These usually accompany locomotion, which is now considerably increased when the pair are together. Hand-to-hand and arm-to-arm contacts are followed by mouth-to-face and mouth-to-mouth ones. Mutual embracing occurs, both statically and during locomotion. Sudden spontaneous outbursts of running, chasing, jumping and dancing are commonly seen and juvenile play patterns may reappear.

Much of this pair-formation phase may take place in public, but when it passes over into the pre-copulatory phase, privacy is sought and the subsequent patterns of behaviour are performed in isolation from other members of the species as far as is possible. With the pre-copulatory stage there is a striking increase in the adoption of a horizontal posture. Body-to-body contacts are increased in both force and duration. Low-intensity side-by-side postures repeatedly give way to high-intensity face-to-face contacts. These positions may be maintained for many minutes and even for several hours, during which vocal and visual signals become gradually less important and tactile signals increasingly frequent. These involve small movements and varying pressures from all parts of the body, but in particular from the fingers, hands, lips and tongue. Clothing is partially or totally removed and skin-to-skin tactile stimulation is increased over as wide an area as possible.

Mouth-to-mouth contacts reach their highest frequency and their longest duration during this phase, the pressure exerted by the lips varying from extreme gentleness to extreme violence. During the higher-intensity responses the lips are parted and the tongue is inserted into the partner's mouth. Active movements of the tongue are then used to stimulate the sensitive skin of the mouth interior. The lips and tongue are also applied to many other areas of the partner's body, especially the ear-lobes, the neck and the genitals. The male pays particular attention to the breasts and nipples of the female, and the lip and tongue contact here becomes extended into more elaborate licking and sucking. Once contacted, the partner's genitals may also become the target for repeated actions of this kind. When this occurs, the male concentrates largely on the female's clitoris, the female on the male's penis, although other areas are also involved in both cases.

In addition to kissing, licking and sucking, the mouth is also applied to various regions of the partner's body in a biting action of varying intensities. Typically this involves no more than soft nibbling of the skin, or gentle nipping, but it can sometimes develop into forceful and even painful biting.

Interspersed between bouts of oral stimulation of the partner's body, and frequently accompanying it, there is a great deal of skin manipulation. The hands and fingers explore the whole body surface, concentrating especially on the face and, at higher intensities, on the buttocks and genital region. As in oral contacts, the male pays particular attention to the female's breasts and nipples. Wherever they move, the fingers repeatedly stroke and caress. From time to time they grasp with great force and the fingernails may be dug deeply into the flesh. The female may grasp the penis of the male, or stroke it rhythmically, simulating the movements of copulation, and the male stimulates the female genitals, especially the clitoris, in a similar way, again frequently with rhythmic movements.

In addition to these mouth, hand and general body contacts, there is also a tendency at high intensities of pre-copulatory activity to rub the genitals rhythmically against the partner's body. There is also a considerable amount of twining and inter-twining of the arms and legs, with occasional powerful muscle contractions, so that the body is thrown into a state of clinging tension, followed by relaxation.

These, then, are the sexual stimuli that are given to the partner during bouts of pre-copulatory activity, and which produce sufficient physiological sexual arousal for copulation to occur. Copulation starts with the insertion of the male's penis into the female's vagina. This is most commonly performed with the couple face-to-face, the male over the female, both in a horizontal position, with the female's legs apart. There are many variations of this position, as we shall be discussing later, but this is the simplest and most typical one. The male then begins a series of rhythmic pelvic thrusts. These can vary considerably in strength and speed, but in an uninhibited situation they are usually rather rapid and deeply penetrating. As copulation progresses there is a tendency to reduce the amount of oral and manual contact, or at least to reduce its subtlety and complexity. Nevertheless these now subsidiary forms of mutual stimulation do still continue to some extent throughout most copulatory sequences.

The copulatory phase is typically much briefer than the pre-copulatory phase. The male reaches the consummatory act of sperm ejaculation within a few minutes in most cases, unless deliberate delaying tactics are employed. Other female primates do not appear to experience a climax to their sexual sequences, but the naked ape is unusual in this respect. If the male continues to copulate for a longer period of time, the female also eventually reaches a consummatory moment, an explosive orgasmic experience, as violent and tension-releasing as the male's, and physiologically identical with it in every way except for the single obvious exception of sperm ejaculation. Some females may reach this point very quickly, others not at all, but on the average it is attained between ten and twenty minutes after the start of copulation.

It is strange that there is this discrepancy between the male and female as regards the time taken to reach sexual climax and relief from tension. This is a matter that will have to be discussed in detail later when the functional significance of the various sexual patterns is being considered. Suffice it to say at this point that the male can overcome the time factor and arouse the female to orgasm either by prolonging and heightening the pre-copulatory stimulation, so that she is already strongly aroused before penis insertion takes place, or he can employ self-inhibitory tactics during copulation to delay his own climax, or he can continue to copulate immediately after ejaculation and before he loses his erection, or he can rest briefly and then copulate for a second time. In the latter case, his reduced sex drive will automatically ensure that he takes much longer to reach his next climax and this will give the female sufficient time on this occasion to reach hers.

After both partners have experienced orgasm there normally follows a considerable period of exhaustion, relaxation, rest, and frequently sleep.

From the sexual stimuli we must now turn to the sexual responses. How does the body respond to all this intensive stimulation? In both sexes there are marked increases in pulse rate, blood pressure and respiration. These changes begin during pre-copulatory activities and rise to a peak at the copulatory climax. Pulse rates which, at normal level, stand at 70 to 80 per minute, rise to 90 to 100 during the earlier phases of sexual arousal, then climb to 130 during intense arousal and attain a peak of about 150 at orgasm. Blood pressure that starts at about 120 rises to 200 or even 250 at the sexual climax. Breathing becomes deeper and more rapid as arousal develops and then, as orgasm approaches, develops into prolonged gasping often accompanied by rhythmic moaning or grunting. At climax the face may be contorted, with mouth wide open and nostrils expanded, in a manner similar to that seen in an athlete in extremis, or someone fighting for air.

Another major change that occurs during sexual arousal is a dramatic shift in the distribution of blood, from the deeper regions to the surface areas of the body. This overall forcing of additional blood into the skin leads to a number of striking results. It produces not only a body that feels generally hotter to the touch – a sexual glow, or fire – but also certain specific changes in a number of specialized areas. At high intensities of arousal a characteristic sexual flush appears. It is most commonly seen in the female, where it usually begins in the region of skin over the stomach and upper abdomen, then spreads to the upper part of the breasts, then the upper chest, then the sides and middle region of the breasts and finally the undersides of the breasts. The face and neck may also be involved. In very intensely responding females it may also spread over the lower abdomen, the shoulders, the elbows, and, with orgasm, to the thighs, buttocks and back. In certain cases it may cover almost the whole body surface. It has been described as a measles-like rash and appears to be a visual sexual signal. It also occurs, but in fewer cases, in the male where, again, it starts in the region of the upper abdomen, spreads over the chest and then the neck and face. It occasionally also covers the shoulders, forearms and thighs. Once orgasm has been reached, the sex flush rapidly disappears, vanishing in reverse order to its sequence of appearance.

In addition to the sex flush and general vaso-dilation, there is also marked vaso-congestion of various distensible organs. This blood congestion is caused by the arteries pumping blood into these organs faster than the veins can carry it away. The condition can be maintained for considerable periods of time because the engorgement of the blood vessels in the organs itself helps to close off the veins that are attempting to carry the blood away. This occurs in the lips, nose, ear-lobes, nipples and genitals of both sexes and also in the breasts of the female. The lips become swollen, redder and more protuberant than at any other time. The soft parts of the nose become swollen and the nostrils expanded. The ear-lobes also become thickened and swollen. The nipples become enlarged and erect in both sexes, but more so in the female. (This is not due to vaso-congestion alone, but also to nipple muscle contraction.) Female nipple length increases by as much as one centimetre, and nipple diameter as much as a half a centimetre. The areola region of pigmented skin around the nipples also becomes tumescent and deeper in colour in the female, but not in the male. The female breast also shows a significant increase in size. By the time orgasm has been reached the breast of the average female will have increased by anything up to 25 per cent of its normal dimensions. It becomes firmer, more rounded and more protuberant.

The genitals of both sexes undergo considerable changes as arousal proceeds. The vaginal walls of the female experience massive vaso-congestion leading to rapid lubrication of the vaginal tube. In some cases this may occur within seconds of the beginning of pre-copulatory activity. There is also a lengthening and distension of the inner two-thirds of the vaginal tube, the overall length of the vagina increasing up to ten centimetres at the phase of high sexual excitement. As orgasm approaches, there is a swelling of the outer one-third of the vaginal tube, and during orgasm itself there is a two- to four-second muscle-spasm contraction of this region, followed by rhythmic contractions at intervals of 0.8 of a second. There are from three to fifteen of these rhythmic contractions in each orgasmic experience.

During arousal the external female genitals become considerably swollen. The outer labia open and swell, and may show size increases of up to two or three times the normal proportions. The inner labia also become distended to two or three times their normal diameter and they protrude through the protective curtain of the outer labia, adding as they do so an extra centimetre to the overall vaginal length. As arousal progresses there is a second striking change in the inner labia. Having already become vaso-congested and protuberant, they now change colour, turning bright red.

The clitoris (the female counterpart of the male penis) also becomes enlarged and more protuberant as sexual arousal begins, but as higher levels of excitement are reached, the labial swelling tends to mask this change and the clitoris is retracted under the labial hood. It cannot at this later stage be stimulated directly by the male's penis, but in its swollen and sensitive condition can still be affected indirectly by the rhythmic pressures applied to that region by the thrusting movements of the male.

The penis of the male undergoes a dramatic modification with sexual arousal. From a limp, flaccid condition it expands, stiffens and erects by means of intensive vaso-congestion. Its normal, average length of nine and a half centimetres is increased by seven to eight centimetres. The diameter is also considerably increased, giving the species the largest erect penis of any living primate.

At the moment of male sexual climax there are several powerful muscle contractions of the penis that expel the seminal fluid into the vaginal tube. The first of these contractions are the strongest ones and occur at intervals of 0.8 of a second – the same rate as the orgasmic vaginal contractions of the female.

During arousal the scrotal skin of the male becomes constricted and the mobility of the testes is reduced. They are elevated by a shortening of the spermatic cords (as, indeed, they are in states of cold, fear and anger) and are held tighter against the body. Vaso-congestion of the region results in a testicular size increase of up to fifty or even a hundred per cent.

These, then, are the principal ways in which the male and female bodies become modified by sexual activity. Once the climax has been reached, all the changes noted are rapidly reversed and the resting, post-sexual individual quickly returns to the normal quiescent physiological state. There is one final, post-orgasmic response that is worth mentioning. There may be a copious sweating by both male and female immediately following sexual climax and this may occur regardless of how much or how little physical effort has been put into the preceding sexual activities. However, although it is not related to total physical expenditure, it does bear a relationship to the intensity of the orgasm itself. The film of sweat develops on the back, the thighs and the upper chest. Sweat may run from the armpits. In intense cases, the whole of the trunk, from shoulders to thighs, may be involved. The palms of the hands and soles of the feet also perspire and, where the face has become mottled with the sexual flush, there may be sweating on the forehead and upper lip.

This brief summary of the sexual stimuli of our species and the responses given to them can now serve as a basis for discussing the significance of our sexual behaviour in relation to our ancestry and our general way of life, but first it is worth pointing out that the various stimuli and responses mentioned do not all occur with equal frequency. Some occur inevitably whenever a male and female come together for sexual activity, but others appear only in a proportion of the cases. Even so, they still occur with a sufficiently high frequency to be counted as 'species characteristics'. As regards the body responses, the sex flush is seen in 75 per cent of females and about 25 per cent of males. Nipple erection is universal for females, but only occurs in 60 per cent of males. Copious sweating after orgasm is a feature of 33 per cent of both males and females. Apart from these specific cases, most of the other body responses mentioned apply in all cases, although, of course, their actual intensity and duration will vary according to the circumstances.

Another point that requires clarification is the way in which these sexual activities are distributed throughout the individual's lifetime. During the first decade of life no true sexual activity can occur in either sex. A great deal of so-called 'sex-play' can be observed in young children, but until the female has begun to ovulate and the male to ejaculate, functional sexual patterns obviously cannot occur. Menstruation begins for some females at the age of ten and by the age of fourteen 80 per cent of young females are actively menstruating. All are doing so by the age of nineteen. The development of pubic hair, the broadening of the hips, and the swelling of the breasts accompanies this change and, in fact, slightly precedes it. General body growth proceeds at a slower rate and is not completed until the twenty-second year.

The first ejaculation in boys does not usually occur until they have reached eleven years, so that they are sexually slower starters than the girls. (The earliest recorded successful ejaculation is for a boy of eight, but this is most unusual.) By the age of twelve, 25 per cent of boys have experienced their first ejaculation and by fourteen 80 per cent have done so. (At this point, therefore, they have caught up with the girls.) The mean age for the first ejaculation is thirteen years and ten months. As with the girls, there are characteristic accompanying changes. Body hair begins to grow, especially in the pubic region and on the face. The typical sequence of appearance of this hairiness is: pubic, armpit, upper lip, cheeks, chin, and then, much more gradually, the chest and other parts of the body. Instead of a broadening of the hips, there is a widening of the shoulders. The voice becomes deeper. This last change also takes place in the girls but to a much smaller extent. In both sexes there is also an acceleration of the growth of the genital organs themselves.

It is interesting that, if one measures sexual responsiveness in terms of frequency of orgasm, the male is much quicker to reach his peak of performance than the female. Although males begin their sexual maturation process a year or so behind the girls, they nevertheless attain their orgasmic peak while they are still in their teens, whereas the girls do not reach theirs until their mid-twenties or even thirties. In fact, the female of our species has to reach the age of twenty-nine before she can match the orgasm rate of the fifteen-year-old male. Only 23 per cent of fifteen-year-old females will have experienced orgasm at all, and this figure has only risen to 53 per cent by the age of twenty. By thirty-five it is 90 per cent.

The adult male achieves an average of about three orgasms a week, and over seven per cent experience daily or more than daily ejaculation. The frequency of orgasm for the average male is highest between the ages of fifteen and thirty, and then drops steadily from thirty to old age. The ability to achieve multiple ejaculation fades, and the angle at which the erect penis is carried also drops. Erection can be maintained for an average of nearly an hour in the late teens, but it has fallen to only seven minutes at the age of seventy. Nevertheless, 70 per cent of males are still sexually active at the age of seventy.

A similar picture of waning sexuality with increasing age is found in the female. The more or less abrupt cessation of ovulation at around the age of fifty does not markedly reduce the degree of sexual respons-iveness, when the population is taken as a whole. There are, however, great individual variations in its influence on sexual behaviour.

The vast majority of all the copulatory activity we have been discuss-ing occurs when the partners are in a pair-bonded state. This may take the form of an officially recognized marriage, or an informal liaison of some sort. The high frequency of non-marital copulation that is known to take place should not be taken to imply a random promiscuity. In most cases it involves typical courtship and pair-formation behaviour, even if the resulting pair-bond is not particularly long-lasting. Ap-proximately 90 per cent of the population becomes formally paired, but 50 per cent of females and 84 per cent of males will have experi-enced copulation before marriage. By the age of forty, 26 per cent of married females and 50 per cent of married males will have experi-enced extramarital copulation. Official pair-bonds also break down completely in a number of cases and are abandoned (0.9 per cent in 1956 in America, for example). The pair-bonding mechanism in our species, although very powerful, is far from perfect.

Now that we have all these facts before us we can start to ask questions. How does the way we behave sexually help us to survive? Why do we behave in the way we do, rather than in some other way? We may be helped in these questions if we ask another one: How does our sexual behaviour compare with that of other living primates?

Straight away we can see that there is much more intense sexual activity in our own species than in any other primates, including our closest relations. For them, the lengthy courtship phase is missing. Hardly any of the monkeys and apes develop a prolonged pair-bond relationship. The pre-copulatory patterns are brief and usually consist of no more than a few facial expressions and simple vocaliza-tions. Copulation itself is also very brief. (In baboons, for instance, the time taken from mounting to ejaculation is no more than seven to eight seconds, with a total of no more than fifteen pelvic thrusts, often fewer.) The female does not appear to experience any kind of climax. If there is anything that could be called an orgasm it is a trivial response when compared with that of the female of our own species.

The period of sexual receptivity of the female monkey or ape is more restricted. It usually only lasts for about a week, or a little more, of their monthly cycle. Even this is an advance on the lower mammals, where it is limited more severely to the actual time of ovulation, but in our own species the primate trend towards longer receptivity has been pushed to the very limit, so that the female is receptive at virtually all times. Once a female monkey or ape becomes pregnant, or is nursing a baby, she ceases to be sexually active. Again, our species has spread its sexual activities into these periods, so that there is only a brief time just before and just after parturition when mating is seriously limited.

Clearly, the naked ape is the sexiest primate alive. To find the reason for this we have to look back again at his origins. What happened? First, he had to hunt if he was to survive. Second, he had to have a better brain to make up for his poor hunting body. Third, he had to have a longer childhood to grow the bigger brain and to educate it. Fourth, the females had to stay put and mind the babies while the males went hunting. Fifth, the males had to co-operate with one another on the hunt. Sixth, they had to stand up straight and use weapons for the hunt to succeed. I am not implying that these changes happened in that order; on the contrary they undoubtedly all developed gradually at the same time, each modification helping the others along. I am simply enumerating the six basic, major changes that took place as the hunting ape evolved. Inherent in these changes there are, I believe, all the ingredients necessary to make up our present sexual complexity.

To begin with, the males had to be sure that their females were going to be faithful to them when they left them alone to go hunting. So the females had to develop a pairing tendency. Also, if the weaker males were going to be expected to co-operate on the hunt, they had to be given more sexual rights. The females would have to be more shared out, the sexual organization more democratic, less tyrannical. Each male, too, would need a strong pairing tendency. Furthermore, the males were now armed with deadly weapons and sexual rivalries would be much more dangerous: again, a good reason for each male being satisfied with one female. On top of that there were the much heavier parental demands being made by the slow-growing infants. Paternal behaviour would have to be developed and the parental duties shared between the mother and the father: another good reason for a strong pair-bond.

Given this situation as a starting point we can now see how other things grew from it. The naked ape had to develop the capacity for falling in love, for becoming sexually imprinted on a single partner, for

43

evolving a pair-bond. Whichever way you put it, it comes to the same thing. How did he manage to do this? What were the factors that helped him in this trend? As a primate, he will already have had a tendency to form brief mateships lasting a few hours, or perhaps even a few days, but these now had to be intensified and extended. One thing that will have come to his aid is his own prolonged childhood. During the long, growing years he will have had the chance to develop a deep personal relationship with his parents, a relationship much more powerful and lasting than anything a young monkey could experience. The loss of this parental bond with maturation and independence would create a 'relationship void' – a gap that had to be filled. He would therefore already be primed for the development of a new, equally powerful bond to replace it.

Even if this was enough to intensify his need for forming a new pair-bond, there would still have to be additional assistance to maintain it. It would have to last long enough for the lengthy process of rearing a family. Having fallen in love, he would have to stay in love. By developing a prolonged and exciting courtship phase he could ensure the former, but something more would be needed after that. The simplest and most direct method of doing this was to make the shared activities of the pair more complicated and more rewarding. In other words, to make sex sexier.

How was this done? In every possible way, seems to be the answer. If we look back now at the behaviour of the present-day naked ape we can see the pattern taking shape. The increased receptivity of the female cannot be explained only in terms of increasing the birth-rate. It is true that by being prepared to copulate while still at the maternal phase of nearing a baby, the female does increase the birth-rate. With the very long dependency period, it would be a disaster if she did not. But this cannot explain why she is ready to receive the male and become sexually aroused throughout each of her monthly cycles. She only ovulates at one point during the cycle, so that mating at all other times can have no procreative function. The vast bulk of copulation in our species is obviously concerned, not with producing offspring, but with cementing the pair-bond by providing mutual rewards for the sexual partners. The repeated attainment of sexual consummation for a mated pair is clearly, then, not some kind of sophisticated, decadent outgrowth of modern civilization, but a deep-rooted, biologically based, and evolutionarily sound tendency of our species.

Even when she has stopped going through her monthly cycles – in other words, when she is pregnant – the female remains responsive to the male. This, too, is particularly important because, with a one-male-one-female system, it would be dangerous to frustrate the male for too long a period. It might endanger the pair-bond.

In addition to increasing the amount of time when sexual activities can take place, the activities themselves have been elaborated. The hunting life that gave us naked skins and more sensitive hands has given us much greater scope for sexually stimulating body-to-body contacts. During pre-copulatory behaviour these play a major role. Stroking, rubbing, pressing and caressing occur in abundance and far exceed anything found in other primate species. Also, specialized organs such as the lips, ear-lobes, nipples, breasts and genitals are richly endowed with nerve-endings and have become highly sensitized to erotic tactile stimulation. The ear-lobes, indeed, appear to have been exclusively evolved to this end. Anatomists have often referred to them as meaningless appendages, or 'useless, fatty excrescences'. In general parlance they are explained away as 'remnants' of the time when we had big ears. But if we look at other primate species, we find that they do not possess fleshy ear-lobes. It seems that, far from being a remnant, they are something new, and when we discover that, under the influence of sexual arousal, they become engorged with blood, swollen and hyper-sensitive, there can be little doubt that their evolution has been exclusively concerned with the production of yet another erogenous zone. (Surprisingly, the humble ear-lobe has been rather overlooked in this context, but it is worth noting that there are cases on record of both males and females actually reaching orgasm as a result of ear-lobe stimulation.) It is interesting to note that the protuberant, fleshy nose of our species is another unique and mysterious feature that the anatomists cannot explain. One has referred to it as a 'mere exuberant variation of no functional significance'. It is hard to believe that something so positive and distinctive in the way of primate appendages should have evolved without a function. When one reads that the side walls of the nose contain a spongy erectile tissue that leads to nasal enlargement and nostril expansion by vaso-congestion during sexual arousal, one begins to wonder.

As well as the improved tactile repertoire, there are some rather unique visual developments. Complex facial expressions play an important part here, although their evolution is concerned with improved communication in many other contexts as well. As a primate species we have the best developed and most complex facial musculature of the entire group. Indeed, we have the most subtle and complex facial expression system of all living animals. By making tiny movements of the flesh around the mouth, nose, eyes, eyebrows, and on the forehead, and by re-combining the movements in a wide variety of ways, we can convey a whole range of complex mood-changes. During sexual

encounters, especially during the early courtship phase, these expressions are of paramount importance. (Their exact form will be discussed
in another chapter.) Pupil dilation also occurs during sexual arousal
and, although it is a small change, we may be more responsive to it
than we realize. The eye-surface also glistens.

Like the ear-lobes and the protruding nose, the lips of our species are
a unique feature, not found elsewhere in the primates. Of course, all
primates have lips, but not turned inside-out like ours. A chimpanzee
can protrude and turn back its lips in an exaggerated pout, exposing as
it does so the mucous membrane that normally lies concealed inside
the mouth. But the lips are only briefly held in this posture before the
animal reverts to its normal 'thin-lipped' face. We, on the other hand,
have permanently everted, rolled-back lips. To a chimpanzee we must
appear to be in a permanent pout. If you ever have occasion to be
embraced by a friendly chimpanzee, the kiss that it may then vigorously apply to your neck will leave you in no doubt about its ability to
deliver a tactile signal with its lips. For the chimpanzee this is a
greeting signal rather than a sexual one, but in our species it is used in
both contexts, the kissing contact becoming particularly frequent and
prolonged during the pre-copulatory phase. In connection with this
development it was presumably more convenient to have the sensitive
mucous surfaces permanently exposed, so that special muscle contractions around the mouth did not have to be maintained throughout the
prolonged kissing contacts, but this is not the whole story. The exposed, mucous lips evolved a well-defined and characteristic shape.
They did not grade off inconspicuously into the surrounding facial
skin, but developed a fixed boundary line. In this way they also became
important visual signalling devices. We have already seen that sexual
arousal produces a swelling and reddening of the lips, and the clear
demarcation of this area obviously assisted in the refinement of these
signals, making subtle changes in lip condition more easily recognizable. Also, of course, even in their un-aroused condition they are
redder than the rest of the face skin and, simply by their very existence,
without indicating changes in physiological condition, they will act as
advertising signals, drawing attention to the presence of a tactile
sexual structure.

Puzzling over the significance of our unique mucous lips, anatomists have stated that their evolution 'is not as yet clearly understood'
and have suggested that perhaps it has something to do with the
increased amount of sucking that is required of the infant at the breast.
But the young chimpanzee also does a great deal of very efficient
sucking and its more muscular and prehensile lips would seem, if
anything, to be better equipped for the job. Also, this cannot explain

the evolution of a sharp margin between the lips and the surrounding face. Nor can it explain the striking differences in the lips of light- and dark-skinned populations. If, on the other hand, the lips are regarded as visual signalling devices, these differences are easier to understand. If climatic conditions demand a darker skin, then this will work against the visual signalling capacity of the lips by reducing their colour contrast. If they really are important as visual signals, then some kind of compensating development might be expected, and this is precisely what seems to have occurred, the negroid lips maintaining their conspicuousness by becoming larger and more protuberant. What they have lost in colour contrast, they have made up for in size and shape. Also, the margins of the negroid lips are more strikingly delineated. The 'lip-seams' of the paler races become exaggerated into more prominent ridges that are lighter in colour than the rest of the skin. Anatomically, these negroid characters do not appear to be primitive, but rather represent a positive advance in the specialization of the lip region.

There are a number of other obvious visual sexual signals. At puberty, as I have already mentioned, the arrival of a fully operative breeding condition is signalled by the development of conspicuous tufts of hair, especially in the region of the genitals and armpits and, in the male, on the face. In the female there is the rapid growth of the breasts. The body shape also changes, becoming broader at the shoulders in the male and at the pelvis in the female. These changes not only differentiate the sexually mature individual from the immature, but most of them also distinguish the mature male from the mature female. They not only act as signals revealing that the sexual system is now functional, but also indicate in each case whether it is masculine or feminine.

The enlarged female breasts are usually thought of primarily as maternal rather than sexual developments, but there seems to be little evidence for this. Other species of primates provide an abundant milk supply for their offspring and yet they fail to develop clearly defined hemispherical breast swellings. The female of our species is unique amongst primates in this respect. The evolution of protruding breasts of a characteristic shape appears to be yet another example of sexual signalling. This would be made possible and encouraged by the evolution of the naked skin. Swollen breast-patches in a shaggy-coated female would be far less conspicuous as signalling devices, but once the hair has vanished they would stand out clearly. In addition to their own conspicuous shape, they also serve to concentrate visual attention on to the nipples and to make the nipple erection that accompanies sexual arousal more conspicuous. The pigmented area of skin around the nipple, that deepens in colour during sexual arousal, also helps in the same way.

The nakedness of the skin also makes possible certain colour-change signals. These occur in limited areas in other animals where there are small naked patches, but have become more extensive in our own species. Blushing occurs with particularly high frequency during the earlier courtship stages of sexual behaviour and, at the later phases of more intense arousal, there is the characteristic mottling of the sex flush. (Again, this is a form of signalling that the darker-skinned races have had to sacrifice to climatic demands. We know that they still undergo these changes, however, because, although they are invisible as colour transformations, close examination reveals significant changes in skin texture.)

Before leaving this array of visual sexual signals we must consider a rather unusual aspect of their evolution. To do so, we must take a sidelong glance at some rather strange things that have been happening to the bodies of a number of our lowlier primate cousins, the monkeys. Recent German research has revealed that certain species have started to mimic themselves. The most dramatic examples of this are the mandrill and the gelada baboon. The male mandrill has a bright red penis with blue scrotal patches on either side of it. This colour arrangement is repeated on its face, its nose being bright red and its swollen, naked cheeks an intense blue. It is as if the animal's face is mimicking its genital region by giving the same set of visual signals. When the male mandrill approaches another animal, its genital display tends to be concealed by its body posture, but it can still apparently transmit the vital messages by using its phallic face. The female gelada indulges in a similar self-copying device. Around her genitals there is a bright red skin patch, bordered with white papillae. The lips of the vulva in the centre of this area are a deeper, richer red. This visual pattern is repeated on her chest region, where again there is a patch of naked red skin surrounded by the same kind of white papillae. In the centre of this chest patch the deep red nipples have come to lie so close together that they are strongly reminiscent of the lips of the vulva. (They are indeed so close to one another that the infant sucks from both teats at the same time.) Like the true genital patch, the chest patch varies in intensity of colour during the different stages of the monthly sexual cycle.

The inescapable conclusion is that the mandrill and the gelada have brought their genital signals forward to a frontal position for some reason. We know too little about the life of mandrills in the wild to be able to speculate as to the reasons for this strange occurrence in this particular species, but we do know that wild geladas spend a great

deal more of their time in an upright sitting posture than
other similar monkey species. If this is a more typical posture for t.
then it follows that by having sexual signals on their chests they ...
more readily transmit these signals to other members of the group than
if the markings only existed on their rear ends. Many species of
primates have brightly coloured genitals, but these frontal mimics are
rare.

Our own species has made a radical change in its typical body
posture. Like geladas, we spend a great deal of time sitting up vertically. We also stand erect and face one another during social contacts.
Could it be, then, that we, too, have indulged in something similar in
the way of self-mimicry? Could our vertical posture have influenced
our sexual signals? When considered in this way the answer certainly
seems to be yes. The typical mating posture of all other primates
involves the rear approach of the male to the female. She lifts her rear
end and directs it towards the male. Her genital region is visually
presented backwards to him. He sees it, moves towards her, and
mounts her from behind. There is no frontal body contact during
copulation, the male's genital region being pressed on to the female's
rump region. In our own species the situation is very different. Not
only is there prolonged face-to-face pre-copulatory activity, but also
copulation itself is primarily a frontal performance.

There has been some argument about this last point. It is a long-
standing idea that the face-to-face mating position is the biologically
natural one for our species, and that all others should be considered
as sophisticated variations of it. Recent authorities have challenged
this and have claimed that there is no such thing as a basic posture
as far as we are concerned. Any body relationship, they feel, should
be grist to our sexual mill, and as an inventive species it should be
natural for us to experiment with any postures we like – the more the
better, in fact, because this will increase the complexity of the sexual
act, increase sexual novelty, and prevent sexual boredom between the
members of a long-mated pair. Their argument is a perfectly valid
one in the context within which they present it, but in trying to score
their point they have gone too far. Their real objection was to the
idea that any variations of the basic posture are 'sinful'. To counteract
this idea, they stressed the value of these variations, and were quite
right to do so, for the reasons given. Any improvement in sexual
rewards for the members of a mated pair will obviously be important
in strengthening the pair-bond. They are biologically sound for our
species. But in fighting this battle the authorities concerned lost
sight of the fact there is nevertheless one basic, natural mating posture
for our species – the face-to-face posture. Virtually all the sexual

signals and erogenous zones are on the front of the body – the facial expressions, the lips, the beard, the nipples, the areolar signals, the breasts of the female, the pubic hair, the genitals themselves, the major blushing areas, and the major sexual flush areas. It could be argued that many of these signals would operate perfectly well in the earlier stages, which could be face-to-face, but then, for the copulation itself, with both partners now fully aroused by frontal stimulation, the male could shift into a rear position for rear-entry copulation, or, for that matter, into any other unusual posture he cared to select. This is perfectly true, and possible as a novelty device, but it has certain disadvantages. To start with, the identity of the sexual partner is much more important to a pair-bonding species like ours. The frontal approach means that the in-coming sexual signals and rewards are kept tightly linked with the identity signals from the partner. Face-to-face sex is 'personalized sex'. In addition, the pre-copulatory tactile sensations from the frontally concentrated erogenous zones can be extended into the copulatory phase when the mating act is performed face-to-face. Many of these sensations would be lost by adopting other postures. Also, the frontal approach provides the maximum possibility for stimulation of the female's clitoris during the pelvic thrusting of the male. It is true that it will be passively stimulated by the pulling effect of the male's thrusts, regardless of his body position in relation to the female, but in a face-to-face mating there will in addition be the direct rhythmic pressure of the male's pubic region on to the clitoral area, and this will considerably heighten the stimulation. Finally, there is the basic anatomy of the female vaginal passage, the angle of which has swung forward to a marked degree, when compared with other species of primates. It has moved forward more than would be expected simply as a passive result of the process of becoming a vertical species. Undoubtedly, if it had been important for the female of our species to present her genitals to the male for rear mounting, natural selection would soon have favoured that trend and the females would by now have a more posteriorly directed vaginal tract.

So it seems plausible to consider that face-to-face copulation is basic to our species. There are, of course, a number of variations that do not eliminate the frontal element: male above, female above, side by side, squatting, standing, and so on, but the most efficient and commonly used one is with both partners horizontal, the male above the female. American investigators have estimated that in their culture 70 per cent of the population employ only this position. Even those who vary their postures still use the basic one for much of the time. Fewer than ten per cent experiment with rear-entry positions. In a massive cross-cultural survey involving nearly two hundred different societies scattered all over the world, the conclusion was that copulation with the male entering the female from the rear does not occur as the usual practice in any of the communities studied.

If we can now accept this fact, we can return from this slight digression to the original question concerning sexual self-mimicry. If the female of our species was going to successfully shift the interest of the male round to the front, evolution would have to do something to make the frontal region more stimulating. At some point, back in our ancestry, we must have been using the rear approach. Supposing we had reached the stage where the female signalled sexually to the male from behind with a pair of fleshy, hemispherical buttocks (not, incidentally, found elsewhere amongst the primates) and a pair of bright red genital lips, or labia. Supposing the male had evolved a powerful sexual responsiveness to these specific signals. Supposing that, at this point in evolution, the species became increasingly vertical and frontally orientated in its social contacts. Given this situation, one might very well expect to find some sort of frontal self-mimicry of the type seen in the gelada baboon. Can we, if we look at the frontal regions of the females of our species, see any structures that might possibly be mimics of the ancient genital display of hemispherical buttocks and red labia? The answer stands out as clearly as the female bosom itself. The protuberant, hemispherical breasts of the female must surely be copies of the fleshy buttocks, and the sharply defined red lips around the mouth must be copies of the red labia. (You may recall that, during intense sexual arousal, both the lips of the mouth and the genital labia become swollen and deeper in colour, so that they not only look alike, but also change in the same way in sexual excitement.) If the male of our species was already primed to respond sexually to these signals when they emanated posteriorly from the genital region, then he would have a built-in susceptibility to them if they could be reproduced in that form on the front of the female's body. And this, it would seem, is precisely what has happened, with the females carrying a duplicate set of buttocks and labia on their chests and mouths respectively. (The use of lipsticks and brassières immediately springs to mind, but these must be left until later, when we are dealing with the special sexual techniques of modern civilization.)

In addition to the all-important visual signals, there are certain odour stimuli that play a sexual role. Our sense of smell has been considerably reduced during evolution, but it is reasonably efficient and is more operative during sexual activities than we normally realize. We know that there are sex differences in body odours and it has been suggested that part of the process of pair-formation – falling in love – involves a kind of olfactory imprinting, a fixation on the specific individual odour of the partner's body. Connected with this is the intriguing discovery that at puberty there is a marked change in odour preferences. Before puberty there are strong preferences for sweet and fruity odours, but with the arrival of sexual maturity this response falls off and there is a dramatic shift in favour of flowery, oily and musky odours. This applies to both sexes, but the increase in musk responsiveness is stronger in males than females. It is claimed that as adults we can detect the presence of musk even when it is diluted down to one part in eight million parts of air, and it is significant that this substance plays a dominant role in the scent-signalling of many mammalian species, being produced in specialized scent-glands. Although we ourselves do not possess any large scent glands, we do have a large number of small ones – the apocrine glands. These are similar to ordinary sweat glands, but their secretions contain a higher proportion of solids. They occur on a number of parts of the body, but there are specially high concentrations of them in the regions of the armpits and the genitals. The localized hair-tufts that grow in these areas undoubtedly function as important scent-traps. It has been claimed that scent production in these areas is heightened during sexual arousal, but no detailed analysis of this phenomenon has yet been made. We do, however, know that there are 75 per cent more apocrine glands in the female of our species than in the male, and it is interesting to recall that in lower mammals during sexual encounters the male sniffs the female more than she sniffs him.

The location of our specialized odour-producing areas appears to be yet another adaptation to our frontal approach to sexual contact. There is nothing unusual about the genital centre, this we have in common with many other mammals, but the armpit concentration is a more unexpected feature. It appears to relate to the general tendency in our species to add new sexual stimulation centres to the front end of the body, in connection with the great increase in face-to-face sexual contacts. In this particular case it would result in the partner's nose being kept in close proximity with a major scent-producing area throughout much of the pre-copulatory and copulatory activity.

Up to this point we have been considering ways in which the appetitive sexual behaviour of our species has been improved and extended, so that contacts between the members of a mated pair have become increasingly rewarding, and their pair-bond therefore strengthened and maintained. But appetitive behaviour leads to a consummatory act and some improvements have been needed here too. Consider for a moment the old primate system. The adult males are sexually active all the time, except when they have just ejaculated. A consummatory orgasm is valuable for them because the relief from sexual tension that it brings damps down their sexual urges long enough for their sperm supplies to be replenished. The females, on the other hand, are sexually active only for a limited period centred around their ovulation time. During this period they are ready to receive the males at any time. The more copulations they experience, the greater the insurance that successful fertilization will be achieved. For them, there is no sexual satiation, no moment of copulatory climax that would pacify and tame their sexual urges. While they are on heat, there is no time to lose, they must keep going at all costs. If they experienced intense orgasms, they would then waste valuable potential mating time. At the end of a copulation, when the male ejaculates and dismounts, the female monkey shows little sign of emotional upheaval and usually wanders off as if nothing had happened.

With our own pair-bonding species the situation is entirely different. In the first place, as there is only a single male involved, there is no particular advantage in the female being sexually responsive at the point where he is sexually spent. So there is nothing working against the existence of a female orgasm. There are, however, two things working very much in its favour. One is the immense behavioural reward it brings to the act of sexual co-operation with the mated partner. Like all the other improvements in sexuality this will serve to strengthen the pair-bond and maintain the family unit. The other is that it considerably increases the chances of fertilization. It does this in a rather special way that applies only to our own peculiar species. Again, to understand this, we must look back at our primate relatives. When a female monkey has been inseminated by a male, she can wander away without any fear of losing the seminal fluid that now lies in the innermost part of her vaginal tract. She walks on all fours. The angle of her vaginal passage is still more or less horizontal. If a female of our own species were so unmoved by the experience of copulation that she too was likely to get up and wander off immediately afterwards, the situation would be different, for she walks bipedally and the angle of her vaginal passage during normal locomotion is almost vertical. Under the simple influence of gravity the seminal fluid would flow back down the vaginal tract and much of it would be lost. There is therefore a great advantage in any reaction that tends to keep the female horizontal when the male ejaculates and stops copulating. The violent response of female orgasm, leaving the female sexually satiated and exhausted, has precisely this effect. It is therefore doubly valuable.*

The fact that the female orgasm in our species is unique amongst

* An additional function for the female orgasm has since been suggested, namely that it causes contractions of the cervix that help to suck the sperm up into the uterus and in this way facilitates fertilization.

primates, combined with the fact that it is physiologically almost identical with the orgasmic pattern of the male, suggests that perhaps it is in an evolutionary sense a 'pseudo-male' response. In the make-up of both males and females there are latent properties belonging to the opposite sex. We know from comparative studies of other groups of animals that evolution can, if necessary, call up one of these latent qualities and bring it into the front line (in the 'wrong' sex, as it were). In this particular instance we know that the female of our species has developed a particular susceptibility to sexual stimulation of the clitoris. When we remember that this organ is the female homologue, or counterpart, of the male penis, this does seem to point to the fact that, in origin at any rate, the female orgasm is a 'borrowed' male pattern.

This may also explain why the male has the largest penis of any primate. It is not only extremely long when fully erect, but also very thick when compared with the penises of other species. (The chimpanzee's is a mere spike by comparison.) This broadening of the penis results in the female's external genitals being subjected to much more pulling and pushing during the performance of pelvic thrusts. With each inward thrust of the penis, the clitoral region is pulled downwards and then, with each withdrawal, it moves up again. Add to this the rhythmic pressure being exerted on to the clitoral region by the pubic region of the frontally copulating male, and you have a repeated massaging of the clitoris that – were she a male – would virtually be masturbatory.

So we can sum up by saying that with both appetitive and consummatory behaviour, everything possible has been done to increase the sexuality of the naked ape and to ensure the successful evolution of a pattern as basic as pair-formation, in a mammalian group where it is elsewhere virtually unknown. But the difficulties of introducing this new trend are not over yet. If we look at our naked ape couple, still successfully together and helping one another to rear the infants, all appears to be well. But the infants are growing now and soon they will have reached puberty, and then what? If the old primate patterns are left unmodified, the adult male will soon drive out the young males and mate with the young females. These will then become part of the family unit as additional breeding females along with their mother, and we shall be right back where we started. Also, if the young males are driven out into an inferior status on the edge of society, as in many primate species, then the co-operative nature of the all-male hunting group will suffer.

Clearly some additional modification to the breeding system is needed here, some kind of exogamy or out-breeding device. For the pair-bond system to survive, both the daughters and the sons will have to find mates of their own. This is not an unusual demand for pair-forming species and many examples of it can be found amongst the lower mammals, but the social nature of most primates makes it a more difficult proposition. In most pair-forming species the family splits up and spreads out when the young grow up. Because of its co-operative social behaviour the naked ape cannot afford to scatter in this way. The problem is therefore kept much more on the doorstep, but it is solved in basically the same way. As with all pair-bonded animals, the parents are possessive of one another. The mother 'owns' the father sexually and vice versa. As soon as the offspring begin to develop their sexual signals at puberty, they become sexual rivals, the sons of the father and the daughters of the mother. There will be a tendency to drive them both out. The offspring will also begin to develop a need for a home-based 'territory' of their own. The urge to do this must obviously have been present in the parents for them to have set up a breeding home in the first place, and the pattern will simply be repeated. The parental home-base, dominated and 'owned' by the mother and father, will not have the right properties. Both the place itself and the individuals living in it will be heavily loaded with both primary and associative parental signals. The adolescent will automatically reject this and set off to establish a new breeding base. This is typical of young territorial carnivores, but not of young primates, and this is one more basic behavioural change that is going to be demanded of the naked ape.

It is perhaps unfortunate that this phenomenon of exogamy is so often referred to as indicating an 'incest taboo'. This immediately implies that it is a comparatively recent, culturally controlled restriction, but it must have developed biologically at a much earlier stage, or the typical breeding system of the species could never have emerged from its primate background.

Another related feature, and one that appears to be unique to our species, is the retention of the hymen or maidenhead in the female. In lower mammals it occurs as an embryonic stage in the development of the urogenital system, but as part of the naked ape's neoteny it is retained. Its persistence means that the first copulation in the life of the female will meet with some difficulty. When evolution has gone to such lengths to render her as sexually responsive as possible, it is at first sight strange that she should also be equipped with what amounts to an anti-copulatory device. But the situation is not as contradictory as it may appear. By making the first copulation attempt difficult and even painful, the hymen ensures that it will not be indulged in lightly. Clearly, during the adolescent phase, there is going to be a period of sexual experimentation, of 'playing the field' in search of a suitable partner. Young males at this time will have no good reason for stopping short of full copulation. If a pair-bond does not form they have not committed themselves in any way and can move on until they find a suitable mate. But if young females were to go so far without pair-formation, they might very well find themselves pregnant and heading straight towards a parental situation with no partner to accompany them. By putting a partial brake on this trend in the female, the hymen demands that she shall have already developed a deep emotional involvement before taking the final step, an involvement strong enough to take the initial physical discomfort in its stride.

A word must be added here on the question of monogamy and polygamy. The development of the pair-bond, which has occurred in the species as a whole, will naturally favour monogamy, but it does not absolutely demand it. If the violent hunting life results in adult males becoming scarcer than females, there will be a tendency for some of the surviving males to form pair-bonds with more than one female. This will then make it possible to increase the breeding rate without setting up dangerous tensions by creating 'spare' females. If the pair-formation process became so totally exclusive that it prevented this, it would be inefficient. This would not be an easy development, however, because of the possessiveness of the females concerned and the dangers of provoking serious sexual rivalries between them. Also working against it would be the basic economic pressures of maintaining the larger family group with all its offspring. A small degree of polygamy could exist, but it would be severely limited. It is interesting that although it still occurs in a number of minor cultures today, all the major societies (which account for the vast majority of the world population of the species) are monogamous. Even in those that permit polygamy, it is not usually practised by more than a small minority of the males concerned. It is intriguing to speculate as to whether its omission from almost all the larger cultures has, in fact, been a major factor in the attainment of their present successful status. We can, at any rate, sum up by saying that, whatever obscure, backward tribal units are doing today, the mainstream of our species expresses its pair-bonding character in its most extreme form, namely long-term monogamous matings.

This, then, is the naked ape in all its erotic complexity: a highly sexed, pair-forming species with many unique features; a complicated blend of primate ancestry with extensive carnivore modifications. Now, to this we must add the third and final ingredient: modern civilization. The enlarged brain that accompanied the transformation of the simple forest-dweller into a co-operative hunter began to busy itself with technological improvements. The simple tribal dwelling places became great towns and cities. The axe age blossomed into the space age. But what effect did the acquisition of all this gloss and glitter have on the sexual system of the species? Very little, seems to be the answer. It has all been too quick, too sudden, for any fundamental biological advances to occur. Superficially they *seem* to have occurred, it is true, but this is largely make-believe. Behind the façade of modern city life there is the same old naked ape. Only the names have been changed: for 'hunting' read 'working', for 'hunting grounds' read 'place of business', for 'home base' read 'house', for 'pair-bond' read 'marriage', for 'mate' read 'wife', and so on. The American studies of contemporary sexual patterns, referred to earlier, have revealed that the physiological and anatomical equipment of the species is still being put to full use. The evidence of prehistoric remnants combined with comparative data from living carnivores and other living primates has given us a picture of how the naked ape must have used this sexual equipment in the distant past and how he must have organized his sex life. The contemporary evidence appears to give much the same basic picture, once one has cleaned away the dark varnish of public moralizing. As I said at the beginning of the chapter, it is the biological nature of the beast that has moulded the social structure of civilization, rather than the other way around.

Yet, although the basic sexual system has been retained in a fairly primitive form (there has been no communalization of sex to match the enlarged communities), many minor controls and restrictions have been introduced. These have become necessary because of the elaborate set of anatomical and physiological sexual signals and the heightened sexual responsiveness we have acquired during our evolution. But these were designed for use in a small, closely knit tribal unit, not in a vast metropolis. In the big city we are constantly intermixing with hundreds of stimulating (and stimulatable) strangers. This is something new, and it has to be dealt with.

In fact, the introduction of cultural restrictions must have begun much earlier, before there were strangers. Even in the simple tribal units it must have been necessary for members of a mated pair to curtail their sexual signalling in some way when they were moving about in public. If sexuality had to be heightened to keep the pair together, then steps must have been taken to damp it down when the pair were apart, to avoid the over-stimulation of third parties. In other pair-forming but communal species this is done largely by aggressive gestures, but in a co-operative species like ours, less belligerent methods would be favoured. This is where our enlarged brain can come to the rescue. Communication by speech obviously plays a vital role here ('My husband wouldn't like it'), as it does in so many facets of social contact, but more immediate measures are also needed.

The most obvious example is the hallowed and proverbial fig-leaf. Because of his vertical posture it is impossible for a naked ape to approach another member of his species without performing a genital display. Other primates, advancing on all fours, do not have this problem. If they wish to display their genitals they have to assume a special posture. We are faced with it, hour in and hour out, whatever we are doing. It follows that the covering of the genital region with some simple kind of garment must have been an early cultural development. The use of clothing as a protection against the cold no doubt developed from this as the species spread its range into less friendly climates, but that stage probably came much later.

With varying cultural conditions, the spread of the anti-sexual garments has varied, sometimes extending to other secondary sexual signals (breast coverings, lip-veils), sometimes not. In certain extreme cases the genitals of the females are not only concealed but also made completely inaccessible. The most famous example of this is the chastity belt, which covered the genital organs and anus with a metal band perforated in the appropriate places to permit the passage of body excretions. Other similar practices have included the sewing up of the genitals of young girls before marriage, or the securing of the labia with metal clasps or rings. In more recent times a case has been recorded of a male boring holes in his mate's labia and then padlocking her genitals after each copulation. Such extreme precautions as this are, of course, very rare, but the less drastic course of simply hiding the genitals behind a concealing garment is now almost universal.

Another important development was the introduction of privacy for the sexual acts themselves. The genitals not only became private parts, they also had to be privately used parts. Today this has resulted in the growth of a strong association between mating and sleeping activities. Sleeping with someone has become synonymous with copulating with them: so, the vast bulk of copulatory activity, instead of being spread out through the day, has now become limited to one particular time – the late evening.

Body-to-body contacts have, as we have seen, become such an important part of sexual behaviour that these too have to be damped down during the ordinary daily routine. A ban has to be placed on physical contact with strangers in our busy, crowded communities. Any accidental brushing against a stranger's body is immediately followed by an apology, the intensity of this apology being proportional to the degree of sexuality of the part of the body touched. Speeded-up film of a crowd moving through a street, or milling around in a large building, reveals clearly the incredible intricacy of these non-stop 'bodily-contact avoidance' manœuvres.

This contact restriction with strangers normally breaks down only under conditions of extreme crowding, or in special circumstances in relation to particular categories of individuals (hairdressers, tailors and doctors, for instance) who are socially 'licensed to touch'. Contact with close friends and relatives is less inhibited. Their social roles are already clearly established as non-sexual and there is less danger. Even so, greeting ceremonies have become highly stylized. The handshake has become a rigidly fixed pattern. The greeting kiss has developed its own ritualized form (mutual mouth-to-cheek touching) that sets it apart from mouth-to-mouth sexual kissing.

Body postures have become de-sexualized in certain ways. The female sexual invitation posture of legs apart is strongly avoided. When sitting, the legs are kept tightly together, or crossed one over the other.

If the mouth is forced to adopt a posture that is reminiscent in some way of a sexual response, it is often hidden by the hand. Giggling and certain kinds of laughing and grimacing are characteristic of the court-ship phase, and when these occur in social contexts, the hand can frequently be seen to shoot up and cover the mouth region.

Males in many cultures remove certain of their secondary sexual characters by shaving off their beards and/or moustaches. Females depilate their armpits. As an important scent-trap, the armpit hair-tuft has to be eliminated if normal dressing habits leave that region ex-posed. Pubic hair is always so carefully concealed by clothing that it does not usually warrant this treatment, but it is interesting that this area is also frequently shaved by artists' models, whose nudity is non-sexual.

In addition, a great deal of general body de-scenting is practised. The body is washed and bathed frequently – far more than is required simply by the demands of medical care and hygiene. Body odours are socially suppressed and commercial chemical deodorants are sold in large numbers.

Most of these controls are maintained by the simple, unanswerable strategy of referring to the phenomena they restrict as 'not nice', 'not done', or 'not polite'. The true anti-sexual nature of the restrictions is seldom mentioned or even considered. More overt controls are also applied, however, in the form of artificial moral codes, or sexual laws. These vary considerably from culture to culture, but in all cases the major concern is the same – to prevent sexual arousal of strangers and to curtail sexual inter-action outside the pair-bond. As an aid to this process, which is recognized to be a difficult one even by the most puritanical groups, various sublimatory techniques are employed. Schoolboy sports, for example, and other vigorous physical activities are sometimes encouraged in the vain hope that this will reduce the sexual urges. Careful examination of this concept and its application reveals that by and large it is a dismal failure. Athletes are neither more nor less sexually active than other groups. What they lose from physical exhaustion, they gain in physical fitness. The only behavioural method that seems to be of assistance is the age-old system of punishment and reward – punishment for sexual indulgence and reward for sexual restraint. But this, of course, produces suppression rather than reduction of drive.

It is quite clear that our unnaturally enlarged communities will call for some steps of this kind to prevent the intensified social exposure from leading to dangerously increased sexual activities outside the pair-bond. But the naked ape's evolution as a highly sexed primate can take only so much of this treatment. Its biological nature keeps on rebelling. As fast as artificial controls are applied in one way, counteracting improvements are made in another. This often leads to ridiculously contradictory situations.

The female covers her breasts, and then proceeds to redefine their shape with a brassière. This sexual signalling device may be padded or inflatable, so that it not only reinstates the concealed shape, but also enlarges it, imitating in this way the breast-swelling that occurs during sexual arousal. In some cases, females with sagging breasts even go to the length of cosmetic surgery, subjecting themselves to sub-cutaneous wax injections to produce similar effects on a more permanent basis.

Sexual padding has also been added to certain other parts of the body: one only has to think of male codpieces and padded shoulders, and female buttock-enlarging bustles. In certain cultures today it is possible for skinny females to purchase padded buttock-brassières, or 'bottom-falsies'. The wearing of high-heeled shoes, by distorting the normal walking posture, increases the amount of swaying in the buttock region during locomotion.

Female hip-padding has also been employed at various times and, by the use of tight belts, both the hip and breast curves can be exaggerated. Because of this, small female waists have been strongly favoured and tight corseting of this region has been widely practised. As a trend this reached its peak with the 'wasp waists' of half a century ago, at which time some females even went to the extreme of having the lower ribs removed surgically to increase the effect.

The widespread use of lipstick, rouge and perfume to heighten sexual lip signals, flushing signals, and body-scent signals respectively, provides further contradictions. The female who so assiduously washes off her own biological scent then proceeds to replace it with commercial 'sexy' perfumes which, in reality, are no more than diluted forms of the products of the scent-glands of other, totally unrelated mammalian species.

Reading through all these various sexual restrictions and the artificial counter-attractions, one cannot help feeling that it would be much easier simply to go back to square one. Why refrigerate a room and then light a fire in it? As I explained before, the reason for the restrictions is straightforward enough: it is a matter of preventing random sexual stimulation which would interfere with the pair-bonds. But why not a total restriction in public? Why not limit the sexual displays, both biological and artificial, to the moments of privacy between the members of the mated pair? Part of the answer to this is our very high level of sexuality, which demands constant expression and outlet. It was developed to keep the pair together, but now, in the stimulating atmosphere of a complex society, it is constantly being triggered off in non-pair-bond situations. But this is only part of the answer. Sex is also being used as a status device – a well-known manœuvre in other primate species. If a female monkey wants to approach an aggressive male in a non-sexual context, she may display sexually to him, not because she wants to copulate, but because by so doing she will arouse his sexual urges sufficiently to suppress his aggression. Such behaviour patterns are referred to as re-motivating activities. The female uses sexual stimulation to re-motivate the male and thereby gain a non-sexual advantage. Similar devices are used by our own species. Much of the artificial sexual signalling is being employed in this way. By making themselves attractive to members of the opposite sex, individuals can effectively reduce antagonistic feelings in other members of the social group.

There are dangers in this strategy, of course, for a pair-bonding species. The stimulation must not go too far. By conforming to the basic sexual restrictions that the culture has developed, it is possible to give clear signals that 'I am not available for copulation', and yet, at the same time, to give other signals which say that 'I am nevertheless very sexy'. The latter signals will do the job of reducing antagonism, while the former ones will prevent things from getting out of hand. In this way one can have one's cake and eat it.

This should work out very neatly, but unfortunately there are other influences at work. The pair-bonding mechanism is not perfect. It has had to be grafted on to the earlier primate system, and this still shows through. If anything goes wrong with the pair-bond situation, then the old primate urges flare up again. Add to this the fact that another of the naked ape's great evolutionary developments has been the extension of childhood curiosity into the adult phase, and the situation can obviously become dangerous.

The system was obviously designed to work in a situation where the female is producing a large family of overlapping children and the male is off hunting with other males. Although fundamentally this has persisted, two things have changed. There is a tendency to limit artificially the number of offspring. This means that the mated female will not be at full parental pressure and will be more sexually available during her mate's absence. There is also a tendency for many females to join the hunting group. Hunting, of course, has now been replaced by 'working' and the males who set off on their daily working trips are liable to find themselves in heterosexual groups instead of the old all-male parties. This means that the pair-bond has a lot to put up with on both sides. All too often it collapses under the strain. (The American figures, you will recall, indicated that 26 per cent of married females and 50 per cent of married males have experienced extra-marital copulation by the age of forty.) Frequently, though, the original pair-bond is strong enough to maintain itself during these outside activities, or to re-assert itself when they have passed. In only a small percentage of cases is there a complete and final breakdown.

To leave the matter there would overstate the case for the pair-bond, however. It may be able to survive sexual curiosity in most cases, but it is not strong enough to stamp it out. Although the powerful sexual imprinting keeps the mated pair together, it does not eliminate their interest in outside sexual activities. If outside matings conflict too strongly with the pair-bond, then some less harmful substitute for them has to be found. The solution has been voyeurism, using the term in its broadest sense, and this is employed on an enormous scale. In the strict sense voyeurism means obtaining sexual excitement from watching other individuals copulating, but it can logically be broadened out to include any non-participatory interest in any sexual activity. Almost the entire population indulges in this. They watch it, they read about it, they listen to it. The vast bulk of all television, radio, cinema, theatre and fiction-writing is concerned with satisfying this demand. Magazines, newspapers and general conversation also make a large contribution. It has become a major industry. And never once throughout all this does the sexual observer actually *do* anything. Everything is performed by proxy. So urgent is the demand that we have had to invent a special category of performers – actors and actresses – who pretend to go through sexual sequences for us, so that we can watch them at it. They court and marry, and then live again in new roles, to court and marry another day. In this way the voyeur supplies are tremendously increased.

If one looked at a wide range of animal species, one would be forced to the conclusion that this voyeurist activity of ours is biologically abnormal. But it is comparatively harmless and may actually help our species, because it satisfies to some extent the persistent demands of our sexual curiosity without involving the individuals concerned in new potential mateship relationships that could threaten the pair-bond.

Prostitution operates in much the same way. Here, of course, there is involvement, but in the typical situation it is ruthlessly restricted to the copulatory phase. The earlier courtship phase and even the pre-copulatory activities are kept to an absolute minimum. These are the stages where pair-formation begins to operate and they are duly suppressed. If a mated male indulges his urge for sexual novelty by copulating with a prostitute he is, of course, liable to damage his pair-bond, but less so than if he becomes involved in a romantic, but non-copulatory, love affair.

Another form of sexual activity that requires examination is the development of a homosexual fixation. The primary function of sexual behaviour is to reproduce the species and this is something that the formation of homosexual pairs patently fails to do. It is important to make a subtle distinction here. There is nothing biologically unusual about a homosexual act of pseudo-copulation. Many species indulge in this, under a variety of circumstances. But the formation of a homosexual pair-bond is reproductively unsound, since it cannot lead to the production of offspring and wastes potential breeding adults. To understand how this can happen it will help to look at other species.

I have already explained how a female may use sexual signals to re-motivate an aggressive male. By arousing him sexually she suppresses his antagonism and avoids being attacked. A subordinate male may use a similar device. Young male monkeys frequently adopt female sexual invitation postures and are then mounted by dominant males that would otherwise have attacked them. Dominant females may also mount subordinate females in the same way. This utilization of sexual patterns in non-sexual situations has become a common feature of the primate social scene and has proved extremely valuable in helping to maintain group harmony and organization. Because these other species of primates do not undergo a process of intense pair-bond formation, it does not lead to difficulties in the shape of long-term homosexual pairings. It simply solves immediate dominance problems, but does not have long-term sexual relationship consequences.

Homosexual behaviour is also seen in situations where the ideal sexual object (a member of the opposite sex) is unavailable. This applies in many groups of animals: a member of the same sex is used as a substitute object – 'the next best thing' for sexual activity. In total isolation animals are often driven to more extreme measures and will attempt to copulate with inanimate objects, or will masturbate. In captivity, for example, certain carnivores have been known to copulate with their food containers. Monkeys frequently develop masturbatory patterns and this has even been recorded in the case of lions. Also, animals housed with the wrong species may attempt to mate with them. But these activities typically disappear when the biologically correct stimulus – a member of the opposite sex – appears on the scene.

Similar situations occur with high frequency in our own species and the response is much the same. If either males or females cannot for some reason obtain sexual access to their opposite numbers, they will find sexual outlets in other ways. They may use other members of their own sex, or they may even use members of other species, or they may masturbate. The detailed American studies of sexual behaviour revealed that in that culture 13 per cent of females and 37 per cent of males have indulged in homosexual contacts to the point of orgasm by the age of 45. Sexual contacts with other animal species are much rarer (because, of course, they provide far fewer of the appropriate sexual stimuli) and have been recorded in only 3.6 per cent of females and 8 per cent of males. Masturbation, although it does not provide 'partner stimuli', is nevertheless so much easier to initiate that it occurs with a much higher frequency. It is estimated that 58 per cent of females and 92 per cent of males masturbate at some time in their lives.

If all these reproductively wasteful activities can take place without reducing the long-term breeding potential of the individuals concerned, then they are harmless. In fact, they can be biologically advantageous, because they can help to prevent sexual frustration which can lead to social disharmony in various ways. But the moment they give rise to sexual fixations they create a problem. In our species there is, as we have seen, a strong tendency to 'fall in love' – to develop a powerful bond with the object of our sexual attentions. This sexual imprinting process produces the all-important long-term mateship so vital to the prolonged parental demands. The imprinting is going to start operating as soon as serious sexual contacts are made, and the consequences are obvious. The earliest objects towards which we direct our sexual attentions are liable to become *the* objects. Imprinting is an associative process. Certain key stimuli that are present at the moment of sexual reward become intimately linked with the reward, and in no time at all it is impossible for sexual behaviour to occur without the presence of these vital stimuli. If we are driven by social pressures to experience our earliest sexual rewards in homosexual or masturbatory contexts, then certain elements present in these contexts are likely to assume powerful sexual significance of a lasting kind. (The more unusual forms of fetishism also originate in this way.)

One might expect these facts to lead to more trouble than actually occurs, but two things help to prevent this in most cases. Firstly, we are well equipped with a set of instinctive responses to the characteristic sexual signals of the opposite sex, so that we are unlikely to experience a powerful courtship reaction to any object lacking these signals. Secondly, our earliest sexual experiments are of a very tentative nature. We start by falling in and out of love very frequently and very easily. It is as if the process of full imprinting lags behind the other sexual developments. During this 'searching' phase we typically develop a large number of minor 'imprints', each one being counteracted by the next, until eventually we arrive at a point where we are susceptible to a major imprinting. Usually by this time we have been sufficiently exposed to a variety of sexual stimuli to have latched on to the appropriate biological ones, and mating then proceeds as a normal heterosexual process.

It will perhaps be easier to understand this if we compare it with the situation that has evolved in certain other species. Pair-forming colonial birds, for example, migrate to the breeding grounds where the nest sites will be established. Young and previously unmated birds, flying in as adults for the first time, must, like all the older birds, establish territories and form breeding pairs. This is done without much delay, soon after arrival. The young birds will select mates on the basis of their sexual signals. Their response to these signals will be inborn. Having courted a mate they will then limit their sexual advances to that particular individual. This is achieved by a process of sexual imprinting. As the pair-forming courtship proceeds, the instinctive sexual clues (which all members of each sex of each species will have in common) have to become linked with certain unique individual recognition characters. Only in this way can the imprinting process narrow down the sexual responsiveness of each bird to its mate. All this has to be done quickly, because the breeding season is limited. If, at the start of this stage, all members of one sex were experimentally removed from the colony, a large number of homosexual pair-bonds might become established, as the birds desperately tried to find the nearest thing to a correct mate that was available.

In our own species the process is much slower. We do not have to work against the deadline of a brief breeding season. This gives us time to scout around and 'play the field'. Even if we are thrown into a sexually segregated environment for considerable periods during adolescence, we do not all automatically and permanently develop homosexual pair-bonds. If we were like the colonial nesting birds, then no young male could emerge from an all-male boarding school (or other similar unisexual organization) with the slightest hope of ever making a heterosexual pair-bond. As it is, the process is not too damaging. The imprinting canvas is only lightly sketched in in most cases and can easily be erased by later, more powerful impressions.

In a minority of cases, however, the damage is more permanent. Powerful associative features will have become firmly linked with sexual expression and will always be required in later bond-forming situations. The inferiority of the basic sexual signals given by a partner of the same sex will not be sufficient to outweigh the positive imprinting associations. It is a fair question to ask why a society should expose itself to such dangers. The answer seems to be that it is caused by a need to prolong the educational phase as much as possible to cope with the enormously elaborated and complicated technological demands of the culture. If young males and females established family units as soon as they were biologically equipped to do so, a great deal of training potential would be wasted. Strong pressures are therefore put upon them to prevent this. Unfortunately, no amount of cultural restriction is going to prevent the development of the sexual system, and if it cannot take the usual route it will find some other.

There is another separate but important factor that can influence homosexual trends. If, in the parental situation, the offspring are exposed to an unduly masculine and dominant mother, or an unduly weak and effeminate father, then this will give rise to considerable confusion. Behavioural characters will point one way, anatomical ones the other. If, when they become sexually mature, the sons seek mates with the behavioural (rather than the anatomical) qualities of the mother, they are liable to take male mates rather than females. For the daughters there is a similar risk, in reverse. The trouble with sexual problems of this sort is that the prolonged period of infant dependency creates such an enormous overlap between the generations that disturbances are carried over, time after time. The effeminate father mentioned above was probably previously exposed to sexual abnormalities in the relationship between his own parents, and so on. Problems of this kind reverberate down the generations for a long time before they peter out, or before they become so acute that they solve themselves by preventing breeding altogether.

As a zoologist I cannot discuss sexual 'peculiarities' in the usual moralistic way. I can only apply anything like a biological morality in terms of population success and failure. If certain sexual patterns interfere with reproductive success, then they can genuinely be referred to as biologically unsound. Such groups as monks, nuns, long-term spinsters and bachelors and permanent homosexuals are all, in a reproductive sense, aberrant. Society has bred them, but they have failed to return the compliment. Equally, however, it should be realized that an active homosexual is no more reproductively aberrant than a monk. It must also be said that no sexual practice, no matter how disgusting and obscene it may appear to the members of a particular culture, can be criticized biologically providing it does not hinder general reproductive success. If the most bizarre elaboration of sexual performance helps to ensure either that fertilization will occur between members of a mated pair, or that the pair-bond will be strengthened, then reproductively it has done its job and is biologically just as acceptable as the most 'proper' and approved-of sexual customs.

Having said all this, I must now point out that there is an important exception to the rule. The biological morality that I have outlined above ceases to apply under conditions of population over-crowding. When this occurs the rules become reversed. We know from studies of other species in experimentally over-crowded conditions that there comes a moment when the increasing population density reaches such a pitch that it destroys the whole social structure. The animals develop diseases, they kill their young, they fight viciously and they mutilate themselves. No behaviour sequence can run through properly. Everything is fragmented. Eventually there are so many deaths that the population is cut back to a lower density and can start to breed again, but not before there has been a catastrophic upheaval. If, in such a situation, some controlled anti-reproductive device could have been introduced into the population when the first signs of over-crowding were apparent, then the chaos could have been averted. Under such conditions (serious over-crowding with no signs of any easing up in the immediate future), anti-reproductive sexual patterns must obviously be considered in a new light.

Our own species is rapidly heading towards just such a situation. We have arrived at a point where we can no longer be complacent. The solution is obvious, namely to reduce the breeding rate without interfering with the existing social structure; to prevent an increase in quantity without preventing an increase in quality. Contraceptive techniques are obviously required, but they must not be allowed to disrupt the basic family unit. Actually there should be little danger of this. Fear has been expressed that the widespread use of perfected contraceptives will lead to random promiscuity, but this is most unlikely – the powerful pair-formation tendency of the species will see to that. There may be some trouble if many mated pairs employ contraception to the point where no offspring are produced. Such couples will put heavy demands on their pair-bonds, which may break under the strain. These individuals will then constitute a greater threat to other pairs that are attempting to rear families. But extreme breeding reductions of this kind are not necessary. If every family produced two children, the parents would simply reproduce their own number and there would be no increase. Allowing for accidents and premature deaths, the average figure could be slightly higher than this without leading to further population increase and eventual species catastrophe.

The trouble is that, as a sexual phenomenon, mechanical and chemical contraception is something basically new and it will take some time before we know exactly what sort of repercussions it will have on the fundamental sexual structure of society after a large number of generations have experienced it and new traditions have gradually developed out of old ones. It may cause indirect, unforeseen distortions or disruptions of the socio-sexual system. Only time will tell. But whatever happens the alternative, if breeding limitation is not applied, is far worse.

Bearing in mind this over-crowding problem, it could be argued that the need to reduce drastically the reproduction rate now removes any biological criticism of the non-breeding categories such as the monks and nuns, the long-term spinsters and bachelors, and the permanent homosexuals. Purely on a reproductive basis this is true, but it leaves out of account the other social problems that, in certain cases, they may have to face, set aside in their special minority roles. Nevertheless, providing they are well adjusted and valuable members of society outside the reproductive sphere, they must now be considered as valuable non-contributors to the population explosion.

Looking back now on the whole sexual scene we can see that our species has remained much more loyal to its basic biological urges than we might at first imagine. Its primate sexual system with carnivore modifications has survived all the fantastic technological advances remarkably well. If one took a group of twenty suburban families and placed them in a primitive sub-tropical environment where the males had to go off hunting for food, the sexual structure of this new tribe would require little or no modification. In fact, what has happened in every large town or city is that the individuals it contains have specialized in their hunting (working) techniques, but have retained their socio-sexual system in more or less its original form. Science-fiction conceptions of baby-farms, communalized sexual activities, selective sterilization, and state-controlled division of labour in reproductive duties, have not materialized. The space ape still carries a picture of his wife and children with him in his wallet as he speeds towards the moon. Only in the field of general breeding limitation are we now coming face to face with the first major assault on our age-old sexual system by the forces of modern civilization. Thanks to medical science, surgery and hygiene, we have reached an incredible peak of breeding success. We have practised death control and now we must balance it with birth control. It looks very much as though, during the next century or so, we are going to have to change our sexual ways at last. But if we do, it will not be because they failed, but because they succeeded too well.

3

REARING

THE BURDEN OF parental care is heavier for the naked ape than for any other living species. Parental duties may be performed as *in*tensively elsewhere, but never so *ex*tensively. Before we consider the significance of this trend, we must assemble the basic facts.

Once the female has been fertilized and the embryo has started to grow in her uterus, she undergoes a number of changes. Her monthly menstrual flow ceases. She experiences early-morning nausea. Her blood pressure is lower. She may become slightly anaemic. As time passes, her breasts become swollen and tender. Her appetite increases. Typically she becomes more placid.

After a gestation period of approximately 266 days her uterus begins to contract powerfully and rhythmically. The amniotic membrane surrounding the foetus is ruptured and the fluid in which the baby has been floating escapes. Further violent contractions expel the infant from the womb, forcing it through the vaginal passage and into the outside world. Renewed contractions then dislodge and eject the placenta. The cord connecting the baby to the placenta is then severed. In other primates this breaking of the cord is achieved by the mother biting through it, and this was no doubt the method employed by our own ancestors, but today it is neatly tied and snipped through with a pair of scissors. The stump still attached to the infant's belly dries up and drops off a few days after birth.

It is a universal practice today for the female to be accompanied and aided by other adults while she is giving birth. This is probably an extremely ancient procedure. The demands of vertical locomotion have not been kind to the female of our species: the penalty for this progressive step is a sentence of several hours' hard labour. It seems likely that co-operation from other individuals was needed right back at the stages where the hunting ape was evolving from its forest-dwelling ancestors. Luckily the co-operative nature of the species was growing alongside this hunting development, so that the cause of the trouble could also provide its cure. Normally, the chimpanzee mother not only bites through the cord, she also devours all or part of the placenta, licks up the fluids, washes and cleans her newly delivered infant, and holds it protectively to her body. In our own species the exhausted mother relies on companions to perform all these activities (or modern equivalents of them).

After the birth is over it may take a day or two for the mother's milk-flow to get started, but once this has happened she then feeds the baby regularly in this way for a period of up to two years. The average suckling period is shorter than this, however, and modern practice has tended to reduce it to six to nine months. During this time the menstrual cycle of the female is normally suppressed and the menstrual flow usually reappears only when she stops breast-feeding and starts to wean the baby. If infants are weaned unusually early, or if they are bottle-fed, this delay does not, of course, occur, and the female can start breeding again more quickly. If, on the other hand, she follows the more primitive system and feeds the infant for a full two-year period, she is liable to produce offspring only about once in three years. (Suckling is sometimes deliberately prolonged in this way as a contraceptive technique.) With a reproductive life-span of approximately thirty years, this puts her natural productivity capacity at about ten offspring. With bottle-feeding or rapidly curtailed breast-feeding, the figure could theoretically rise to thirty.

The act of suckling is more of a problem for females of our species than for other primates. The infant is so helpless that the mother has to take a much more active part in the process, holding the baby to the breast and guiding its actions. Some mothers have difficulty in persuading their offspring to suck efficiently. The usual cause of this trouble is that the nipple is not protruding far enough into the baby's mouth. It is not enough for the infant's lips to close on the nipple, it must be inserted deeper into its mouth, so that the front part of the nipple is in contact with the palate and the upper surface of the tongue. Only this stimulus will release the jaw, tongue and cheek action of intense sucking. To achieve this juxtaposition, the region of breast immediately behind the nipple must be pliable and yielding. It is the length of 'hold' that the baby can manage on this yielding tissue which is critical. It is essential that suckling should be fully operative within four or five days of birth, if the breast-feeding process is to be successfully developed. If repeated failure occurs during the first week, the infant will never give the full response. It will have become fixated on the more rewarding (bottle) alternative offered.

Another suckling difficulty is the so-called 'fighting at the breast' response of certain infants. This often gives the mother the impression that the baby does not want to suck, but in reality it means that, despite desperate attempts to do so, it is failing because it is being suffocated. A slightly maladjusted posture of the baby's head at the breast will block the nose and, with the mouth full, there is no way for it to breathe. It is fighting, not to avoid sucking, but for air. There are, of course, many such problems that face the new mother, but I have selected these two because they seem to add supporting evidence for the idea of the female breast as predominantly a sexual signalling device, rather than an expanded milk machine. It is the solid, rounded shape that causes both these problems. One has only to look at the design of the teats on babies' bottles to see the kind of shape that works best. It is much longer and does not swell out into the great rounded hemisphere that causes so much difficulty for the baby's mouth and nose. It is much closer in design to the feeding apparatus of the female chimpanzee. She develops slightly swollen breasts, but even in full lactation she is flat-chested when compared with the average female of our own species. Her nipples, on the other hand, are much more elongated and protrusive and the infant has little or no difficulty in initiating the sucking activity. Because our females have rather a heavy suckling burden and because the breasts are so obviously a part of the feeding apparatus, we have automatically assumed that their protruding, rounded shape must also be part and parcel of the same parental activity. But it now looks as though this assumption has been wrong and that, for our species, breast design is primarily sexual rather than maternal in function.

Leaving the question of feeding, it is worth looking at one or two aspects of the way a mother behaves towards her baby at other times. The usual fondling, cuddling and cleaning require little comment, but the position in which she holds the baby against her body when resting is rather revealing. Careful American studies have disclosed the fact that 80 per cent of mothers cradle their infants in their left arms, holding them against the left side of their bodies. If asked to explain the significance of this preference most people reply that it is obviously the result of the predominance of right-handedness in the population. By holding the babies on their left arms, the mothers keep their dominant arm free for manipulations. But a detailed analysis shows that this is not the case. True, there is a slight difference between right-handed and left-handed females, but not enough to provide an adequate explanation. It emerges that 83 per cent of right-handed mothers hold the baby on the left side, but then so do 78 per cent of left-handed mothers. In other words, only 22 per cent of the left-handed mothers have their dominant hands free for action. Clearly there must be some other, less obvious explanation.

The only other clue comes from the fact that the heart is on the left side of the mother's body. Could it be that the sound of her heart-beat is the vital factor? And in what way? Thinking along these lines it was argued that perhaps during its existence inside the body of the mother, the growing embryo becomes fixated ('imprinted') on the sound of the heart-beat. If this is so, then the re-discovery of this familiar sound after birth might have a calming effect on the infant, especially as it has just been thrust into a strange and frighteningly new world outside. If this is so then the mother, either instinctively or by an unconscious series of trials and errors, would soon arrive at the discovery that her baby is more at peace if held on the left, against her heart, than on the right.

This may sound far-fetched, but tests have now been carried out which reveal that it is nevertheless the true explanation. Groups of newborn babies in a hospital nursery were exposed for a considerable time to the recorded sound of a heart-beat at a standard rate of 72 beats per minute. There were nine babies in each group and it was found that one or more of them was crying for 60 per cent of the time when the sound was not switched on, but that this figure fell to only 38 per cent when the heart-beat recording was thumping away. The heart-beat groups also showed a greater weight gain than the others, although the amount of food taken was the same in both cases. Clearly the beatless groups were burning up a lot more energy as a result of the vigorous actions of their crying.

Another test was done with slightly older infants at bedtime. In some groups the room was silent, in others recorded lullabies were played. In others a ticking metronome was operating at the heart-beat speed of 72 beats per minute. In still others the heart-beat recording itself was played. It was then checked to see which groups fell asleep more quickly. The heart-beat group dropped off in half the time it took for any of the other groups. This not only clinches the idea that the sound of the heart beating is a powerfully calming stimulus, but it also shows that the response is a highly specific one. The metronome imitation will not do – at least, not for young infants.

So it seems fairly certain that this is the explanation of the mother's left-side approach to baby-holding. It is interesting that when 466 Madonna-and-child paintings (dating back over several hundred years) were analysed for this feature, 373 of them showed the baby on the left breast. Here again the figure was at the 80 per cent level. This contrasts with observations of females carrying parcels, where it was found that 50 per cent carried them on the left and 50 per cent on the right.

What other possible results could this heart-beat imprinting have? It may, for example, explain why we insist on locating feelings of love in the heart rather than the head. As the song says: 'You gotta have heart!' It may also explain why mothers rock their babies to lull them to sleep. The rocking motion is carried on at about the same speed as the heart-beat, and once again it probably 'reminds' the infants of the rhythmic sensations they became so familiar with inside the womb, as the great heart of the mother pumped and thumped away above them.

Nor does it stop there. Right into adult life the phenomenon seems to stay with us. We rock with anguish. We rock back and forth on our feet when we are in a state of conflict. The next time you see a lecturer or an after-dinner speaker swaying rhythmically from side to side, check his speed for heart-beat time. His discomfort at having to face an audience leads him to perform the most comforting movements his body can offer in the somewhat limited circumstances; and so he switches on the old familiar beat of the womb.

Wherever you find insecurity, you are liable to find the comforting heart-beat rhythm in one kind of disguise or another. It is no accident that most folk music and dancing has a syncopated rhythm. Here again the sounds and movements take the performers back to the safe world of the womb. It is no accident that teenage music has been called 'rock music'. More recently it has adopted an even more revealing name – it is now called 'beat music'. And what are they singing about? 'My heart is broken', 'You gave your heart to another', or 'My heart belongs to you.'

Fascinating as this subject is, we must not stray too far from the original question of parental behaviour. Up to this point we have been looking at the mother's behaviour towards the child. We have followed her through the dramatic moments of birth, watched her feeding the child, holding it and comforting it. Now we must turn to the baby itself and study it as it grows.

The average weight of a baby at birth is just over seven pounds, which is slightly more than one-twentieth the weight of the average parent. Growth is very rapid during the first two years of life and remains reasonably fast throughout the following four years. At the age of six, however, it slows down considerably. This phase of gradual growth continues until eleven in boys and until ten in girls. Then, at puberty, it puts on another spurt. Rapid growth is seen again from eleven until seventeen in boys and from ten until fifteen in girls. Because of their slightly earlier puberty, girls tend to outstrip boys between the eleventh and fourteenth years, but then the boys pass them again and stay in front from that point on. Body growth tends to end for girls at around nineteen, and for boys much later, at about twenty-five. The first teeth usually appear around the sixth or seventh month, and the full set of milk teeth is usually complete by the end of the second year or the middle of the third. The permanent teeth erupt in the sixth year, but the final molars – the wisdom teeth – do not usually appear until about the nineteenth.

Newborn infants spend a great deal of time sleeping. It is usually claimed that they only awaken for about two hours a day during the first few weeks, but this is not the case. They are sleepy, but not that sleepy. Careful studies have revealed that the average time spent sleeping during the first three days of life is 16.6 hours out of every 24. Individuals varied a great deal, however, the sleepiest averaging 23 hours out of 24, and the most wide-awake a mere 10.5.

During childhood the sleeping-to-waking ratio gradually shrinks until, by the time the adult stage has been reached, the original sixteen-hour average has been reduced to half. Some adults vary considerably from this typical eight-hour average however. Two out of every hundred require only five hours and another two need as much as ten hours. Adult females, incidentally, have an average sleep-period that is slightly longer than that of adult males.

The sixteen-hour quota of daily sleep at birth does not occur in one long nocturnal session, it is broken up into a number of short periods scattered throughout the twenty-four hours. Even from birth, however, there is a slight tendency to sleep more at night than in the day. Gradually, as the weeks pass, one of the nocturnal sleep periods becomes longer until it dominates the scene. The infant is now taking a number of brief 'naps' during the day and a single long sleep at night. This change brings the daily sleep average down to about fourteen hours at the age of six months. In the months that follow, the short daily naps become reduced to two – one in the morning and one in the afternoon. During the second year the morning nap usually vanishes, bringing the average sleep figure down to thirteen hours a day. In the fifth year the afternoon nap disappears as well, reducing the figure still further to about twelve hours a day. From this point until puberty there is a further drop of three hours in the daily sleep requirement, so that, by the age of thirteen, children are retiring for only nine hours each night. From this point on, during adolescence, they do not show any difference from the fully adult pattern and take no more than eight hours on the average. The final sleeping rhythm, therefore, matches sexual maturity rather than final physical maturity.

It is interesting that amongst children of pre-school age, the more intelligent ones tend to sleep less than the dull ones. After the age of seven this relationship is reversed, the more intelligent schoolchildren sleeping more than the dull ones. By this stage it would seem that, instead of learning more by being more wide-awake for longer, they are being forced to learn so much that the more responsive ones are worn out by the end of the day. Amongst adults, by contrast, there appears to be no relationship between brilliance and the average amount of sleep.

The time taken to fall asleep in healthy males and females of all ages averages about twenty minutes. Waking should occur spontaneously. The need for an artificial awakening device indicates that there has been insufficient sleep, and the individual will suffer for it with reduced alertness during the waking period that follows.

During its waking periods the newborn infant moves comparatively little. Unlike other primate species its musculature is poorly developed. A young monkey can cling tightly to its mother from the moment of birth onwards. It may even clasp on to her fur with its hands while it is still in the process of being born. In our own species, by contrast, the newborn is helpless and can only make trivial movements of its arms and legs. Not until it is one month old can it, without assistance, raise its chin up off the ground when lying on its front. At two months it can raise its chest off the ground. At three months it can reach towards suspended objects. At four months it can sit up, with support from the mother. At five months it can sit up on the mother's lap and can grasp objects in the hand. At six months it can sit up in a high chair and successfully grasp dangling objects. At seven months it can sit up alone without assistance. At eight months it can stand up with support from the mother. At nine months it can stand up by holding on to furniture. At ten months it can creep along the ground on its hands and knees. At eleven months it can walk when led by the parent's hand. At twelve months it can pull itself up into a standing position with the help of solid objects. At thirteen months it can climb up a set of stairs. At fourteen months it can stand up by itself and without supporting objects to help it. At fifteen months comes the great moment when, at last, it can walk alone by itself, unaided. (These are all, of course, average figures, but they act as a good rough guide to the postural and locomotory rates of development in our species.)

At about the point where the child has started to walk unaided, it also begins to utter its first words – a few simple ones at first, but soon the vocabulary blossoms out at a startling rate. By the age of two the average child can speak nearly 300 words. By three it has tripled this figure. By four it can manage nearly 1,600 and by five it has achieved 2,100. This astonishing rate of learning in the field of vocal imitation is unique to our species and must be considered as one of our greatest achievements. It is related, as we saw in Chapter One, to the pressing need for more precise and helpful communication in connection with co-operative hunting activities. There is nothing like it, nothing even remotely approaching it, in other closely related living primates. Chimpanzees are, like us, brilliant at rapid manipulative imitation, but they cannot manage vocal imitations. One serious and painstaking attempt was made to train a young chimpanzee to speak, but with remarkably limited success. The animal was reared in a house under conditions identical with those for an infant of our own species. By combining food rewards with manipulations of its lips, prolonged attempts were made to persuade it to utter simple words. By the age of two-and-a-half the animal could say 'mama', 'papa' and 'cup'. Eventually it managed to say them in the correct contexts, whispering 'cup' when it wanted a drink of water. The arduous training continued, but by the age of six (when our own species would be well over the 2,000-word mark) its total vocabulary extended to no more than seven words.

This difference is a question of brain, not voice. The chimpanzee has a vocal apparatus that is structurally perfectly capable of making a wide variety of sounds. There is no weakness there that can explain its dumb behaviour. The weakness is centred inside its skull.

Unlike chimpanzees, certain birds have striking powers of vocal imitation. Parrots, budgerigars, mynah birds, crows, and various other species can reel off whole sentences without batting an eyelid, but unfortunately they are too bird-brained to make good use of this ability. They merely copy the complex sequences of sounds they are taught and repeat them automatically in a fixed order and without any reference to outside events. All the same, it is astonishing that chimpanzees, and monkeys for that matter, cannot achieve better things than they do. Even just a few simple, culturally determined, words would be so useful to them in their natural habitats, that it is difficult to understand why they have not evolved.

Returning to our own species again, the basic, instinctive grunts, moans and screams that we share with other primates are not thrown out by our newly won verbal brilliance. Our inborn sound signals remain, and they retain their important roles. They not only provide the vocal foundation on which we can build our verbal skyscraper, but they also exist in their own right, as species-typical communication devices. Unlike the verbal signals, they emerge without training and they mean the same in all cultures. The scream, the whimper, the laugh, the roar, the moan and the rhythmic crying convey the same messages to everyone everywhere. Like the sounds of other animals, they relate to basic emotional moods and give us an immediate impression of the motivational state of the vocalizer. In the same way we have retained our instinctive expressions, the smile, the grin, the frown, the fixed stare, the panic face and the angry face. These, too, are common to all societies and persist despite the acquisition of many cultural gestures.

It is intriguing to see how these basic species-sounds and species-faces originate during our early development. The rhythmic crying response is (as we know all too well) present from birth. Smiling arrives later, at about five weeks. Laughing and temper tantrums do not appear until the third or fourth month. It is worth taking a closer look at these patterns.

Crying is not only the earliest mood-signal we give, it is also the most basic. Smiling and laughing are unique and rather specialized signals, but crying we share with thousands of other species. Virtually all mammals (not to mention birds) give vent to high-pitched screams, squeaks, shrieks, or squeals when they are frightened or in pain. Amongst the higher mammals, where facial expressions have evolved as visual signalling devices, these messages of alarm are accompanied by characteristic 'fear-faces'. Whether performed by a young animal or an adult, these responses indicate that something is seriously wrong. The juvenile alerts its parents, the adult alerts the other members of its social group.

As infants a number of things make us cry. We cry if we are in pain, if we are hungry, if we are left alone, if we are faced with a strange and unfamiliar stimulus, if we suddenly lose our source of physical support, or if we are thwarted in attaining an urgent goal. These categories boil down to two important factors: physical pain and insecurity. In either case, when the signal is given, it produces (or should produce) protective responses in the parent. If the child is separated from the parent at the time the signal is given, it immediately has the effect of reducing the distance between them until the infant is held and either rocked, patted or stroked. If the child is already in contact with the parent, or if the crying persists after contact is made, then its body is examined for possible sources of pain. The parental response continues until the signal is switched off (and in this respect it differs fundamentally from the smiling and laughing patterns).

The action of crying consists of muscular tension accompanied by a reddening of the head, watering of the eyes, opening of the mouth, pulling back of the lips, exaggerated breathing with intense expirations and, of course, the high-pitched rasping vocalizations. With older infants it also includes running to the parent and clinging.

I have described this pattern in some detail, despite its familiarity, because it is from this that our specialized signals of laughing and smiling have evolved. When someone says 'they laughed until they cried', he is commenting on this relationship, but in evolutionary terms it is the other way round – we cried until we laughed. How did this come about? To start with, it is important to realize how similar crying and laughing are, as response patterns. Their moods are so different that we tend to overlook this. Like crying, laughing involves muscular tension, opening of the mouth, pulling back of the lips, and exaggerated breathing with intense expirations. At high intensities it also includes reddening of the face and watering of the eyes. But the vocalizations are less rasping and not so high-pitched. Above all, they are shorter and follow one another more rapidly. It is as though the long wail of the crying infant has become segmented, chopped up into little pieces, and at the same time has grown smoother and lower.

It appears that the laughing reaction evolved out of the crying one, as a secondary signal, in the following way. I said earlier that crying is present at birth, but laughing does not appear until the third or fourth month. Its arrival coincides with the development of parental recognition. It may be a wise child that knows its own father, but it is a laughing child that knows its own mother. Before it has learnt to identify its mother's face and to distinguish her from other adults, a baby may gurgle and burble, but it does not laugh. What happens when it starts to single out its own mother is that it also begins to grow afraid of other, strange adults. At two months any old face will do, all friendly adults are welcome. But now its fears of the world around it are beginning to mature and anyone unfamiliar is liable to upset it and start it crying. (Later on it will soon learn that certain other adults can also be rewarding and will lose its fear of them, but this is then done selectively on the basis of personal recognition.) As a result of this process of becoming imprinted on the mother, the infant may find itself placed in a strange conflict. If the mother does something that startles it, she gives it two sets of opposing signals. One set says, 'I am your mother – your personal protector; there is nothing to fear,' and the other set says, 'Look out, there's something frightening here.' This conflict could not arise before the mother was known as an individual, because if she had then done something startling, she would simply be the source of a frightening stimulus at that moment and nothing more. But now she can give the double signal: 'There's danger but there's no danger.' Or, to put it another way: 'There may appear to be danger, but because it is coming from me, you do not need to take it seriously.' The outcome of this is that the child gives a response that is half a crying reaction and half a parental-recognition gurgle. The magic combination produces a laugh. (Or, rather, it did, way back in evolution. It has since become fixed and fully developed as a separate, distinct response in its own right.)

So the laugh says, 'I recognize that a danger is not real,' and it conveys this message to the mother. The mother can now play with the baby quite vigorously without making it cry. The earliest causes of laughter in infants are parental games of 'peek-a-boo', hand-clapping, rhythmical knee-dropping, and lifting high. Later, tickling plays a major role, but not until after the sixth month. These are all shock stimuli, but performed by the 'safe' protector. Children soon learn to provoke them – by play-hiding, for example, so that they will experience the 'shock' of discovery, or play-fleeing so that they will be caught.

Laughter therefore becomes a play signal, a sign that the increasingly dramatic inter-actions between the child and the parent can continue and develop. If they become too frightening or painful, then, of course, the reaction can switch over into crying and immediately re-stimulate the protective response. This system enables the child to expand its exploration of its bodily capacities and the physical properties of the world around it.

Other animals also have special play signals, but compared with ours they are unimpressive. The chimpanzee, for instance, has a characteristic play-face, and a soft play-grunt which is the equivalent of our laughter. In origin these signals have the same kind of ambivalence. When greeting, a young chimpanzee protrudes its lips far forward, stretching them to the limit. When frightened, it retracts them, opening its mouth and exposing the teeth. The play-face, being motivated by both feelings of friendly greeting and fear, is a mixture of the two. The jaws open wide, as in fear, but the lips are pulled forward and keep the teeth covered. The soft grunt is halfway between the 'oo-oo-oo' greeting sound and the scream of fear. If play becomes too rough, the lips pull back and the grunt becomes a short, sharp scream. If it becomes too calm, the jaws close and the lips pull forward into the friendly chimpanzee pout. Basically the situation is the same, then, but the soft play-grunt is a puny signal when compared with our own vigorous, full-blooded laughter. As chimpanzees grow, the significance of the play signal dwindles even more, whereas ours expands and acquires still greater importance in everyday life. The naked ape, even as an adult, is a playful ape. It is all part of his exploratory nature. He is constantly pushing things to their limit, trying to startle himself, to shock himself without getting hurt, and then signalling his relief with peals of infectious laughter.

Laughing *at* someone can also, of course, become a potent social weapon among older children and adults. It is doubly insulting because it indicates that he is both frighteningly odd and at the same time not worth taking seriously. The professional comedian deliberately adopts this social role and is paid large sums of money by audiences who enjoy the reassurance of checking their group normality against his assumed abnormality.

The response of teenagers to their idols is relevant here. As an audience, they enjoy themselves, not by screaming with laughter, but screaming with screams. They not only scream, they also grip their own and one another's bodies, they writhe, they moan, they cover their faces and they pull at their hair. These are all the classic signs of intense pain or fear, but they have become deliberately stylized. Their thresholds have been artificially lowered. They are no longer cries for help, but signals to one another in the audience that they are capable of feeling an emotional response to the sexual idols which is so powerful that, like all stimuli of unbearably high intensity, they pass into the realm of pure pain. If a teenage girl found herself suddenly alone in the presence of one of her idols, it would never occur to her to scream at him. The screams were not meant for him, they were meant for the other girls in the audience. In this way young girls can reassure one another of their developing emotional responsiveness.

Before leaving the subject of tears and laughter there is one further mystery to be cleared up. Some mothers suffer agony from incessantly crying babies during the first three months of life. Nothing the parents do seems to stem the flood. They usually conclude that there is something radically, physically wrong with the infants and try to treat them accordingly. They are right, of course, there is something physically wrong; but it is probably effect rather than cause. The vital clue comes with the fact that this so-called 'colic' crying ceases, as if by magic, around the third or fourth month of life. It vanishes at just the point where the baby is beginning to be able to identify its mother as a known individual. A comparison of the parental behaviour of mothers with cry-babies and those with quieter infants gives the answer. The former are tentative, nervous and anxious in their dealings with their offspring. The latter are deliberate, calm and serene. The point is that even at this tender age, the baby is acutely aware of differences in tactile 'security' and 'safety', on the one hand, and tactile 'insecurity' and 'alarm' on the other. An agitated mother cannot avoid signalling her agitation to her newborn infant. It signals back to her in the appropriate manner, demanding protection from the cause of the agitation. This only serves to increase the mother's distress, which in turn increases the baby's crying. Eventually the wretched infant cries itself sick and its physical pains are then added to the sum total of its already considerable misery. All that is necessary to break the vicious circle is for the mother to accept the situation and become calm herself. Even if she cannot manage this (and it is almost impossible to fool a baby on this score) the problem corrects itself, as I said, in the third or fourth month of life, because at that stage the baby becomes imprinted on the mother and instinctively begins to respond to her as the 'protector'. She is no longer a disembodied series of agitating stimuli, but a familiar face. If she continues to give agitating stimuli, they are no longer so alarming because they are coming from a known source with a friendly identity. The baby's growing bond with its parent then calms the mother and automatically reduces her anxiety. The 'colic' disappears.

Up to this point I have omitted the question of smiling because it is an even more specialized response than laughing Just as laughing is a secondary form of crying, so smiling is a secondary form of laughing. At first sight it may indeed appear to be no more than a low-intensity version of laughing, but it is not as simple as that. It is true that in its mildest form a laugh is indistinguishable from a smile, and this is no doubt how smiling originated, but it is quite clear that during the course of evolution smiling has become emancipated and must now be considered as a separate entity. High-intensity smiling – the giving of a broad grin, a beaming smile – is completely different in function from high-intensity laughing. It has become specialized as a species greeting signal. If we greet someone by smiling at them, they know we are friendly, but if we greet them by laughing at them, they may have reason to doubt it.

Any social contact is at best mildly fear-provoking. The behaviour of the other individual at the moment of meeting is an unknown quantity. Both the smile and the laugh indicate the existence of this fear and its combination with feelings of attraction and acceptance. But when the laugh develops into high intensity, it signals the readiness for further 'startlement' for further exploitation of the danger-with-safety situation. If, on the other hand, the smiling expression of the low-level laugh grows instead into something else – into a broad grin – it signals that the situation is not to be extended in that way. It indicates simply that the initial mood is an end in itself, without any vigorous elaborations. Mutual smiling reassures the smilers that they are both in a slightly apprehensive, but reciprocally attracted, state of mind. Being slightly fearful means being non-aggressive and being non-aggressive means being friendly, and in this way the smile evolves as a friendly attraction device.

Why, if we have needed this signal, have other primates managed without it? They do, it is true, have friendly gestures of various kinds, but the smile for us is an additional one, and one of tremendous importance in our daily lives, both as infants and as adults. What is it about our own pattern of existence that has brought it so much to the forefront? The answer, it would seem, lies in our famous naked skins. When a young monkey is born it clings tightly to its mother's fur. There it stays, hour in and hour out, day after day. For weeks, or even months, it never leaves the snug protection of its mother's body. Later, when it is venturing away from her for the first time, it can run back to her at a moment's notice and cling on again in an instant. It has its own positive way of ensuring close physical contact. Even if the mother does not welcome this contact (as the infant grows older and heavier), she will have a hard time rejecting it. Anyone who has ever had to act as a foster-mother for a young chimpanzee can testify to this.

When *we* are born we are in a much more hazardous position. Not only are we too weak to cling, but there is nothing to cling to. Robbed of any mechanical means of ensuring close proximity with our mothers, we must rely entirely on maternally stimulating signals. We can scream our heads off to summon parental attention, but having got it we must do something more to maintain it. A young chimpanzee screams for attention just as we do. The mother rushes over and grabs it up. Instantly the baby is clinging again. This is the moment at which we need a clinging-substitute, some kind of signal that will reward the mother and make her want to stay on with us. The signal we use is the smile.

Smiling begins during the first few weeks of life, but to start with it is not directed at anything in particular. By about the fifth week it is being given as a definite reaction to certain stimuli. The baby's eyes can now fixate objects. At first it is most responsive to a pair of eyes staring at it. Even two black spots on a piece of card will do. As the weeks pass, a mouth also becomes necessary. Two black spots with a mouth-line below them are now more efficient at inciting the response. Soon a widening of the mouth becomes vital, and then the eyes begin to lose their significance as key stimuli. At this stage, around three to four months, the response starts to become more specific. It is narrowed down from any old face to the particular face of the mother. Parental imprinting is taking place.

The astonishing thing about the growth of this reaction is that, at the time when it is developing, the infant is hopeless at discriminating between such things as squares and triangles, or other sharp geometrical shapes. It seems as if there is a special advance in the maturing of the ability to recognize certain rather limited kinds of shapes – those related to human features – while other visual abilities lag behind. This ensures that the infant's vision is going to dwell on the right kind of object. It will avoid becoming imprinted on some near-by inanimate shape.

By the age of seven months the infant is completely imprinted on its mother. Whatever she does now, she will retain her mother-image for her offspring for the rest of its life. Young ducklings achieve this by the act of following the mother, young apes by clinging to her. We develop the vital bond of attachment via the smiling response.

As a visual stimulus the smile has attained its unique configuration principally by the simple act of turning up the mouth-corners. The mouth is opened to some extent and the lips pulled back, as in the face of fear, but by the addition of the curling up of the corners the character of the expression is radically changed. This development has in turn led to the possibility of another and contrasting facial posture – that of the down-turned mouth. By adopting a mouth-line that is the complete opposite of the smile shape, it is possible to signal an anti-smile. Just as laughing evolved out of crying and smiling out of laughing, so the unfriendly face has evolved, by a pendulum swing, from the friendly face.

But there is more to smiling than a mouth-line. As adults we may be able to convey our mood by a mere twist of the lips, but the infant throws much more into the battle. When smiling at full intensity, it also kicks and waves its arms about, stretches its hands out towards the stimulus and moves them about, produces babbling vocalizations, tilts back its head and protrudes its chin, leans its trunk forward or rolls it to one side, and exaggerates its respiration. Its eyes become brighter and may close slightly; wrinkles appear underneath or along-side the eyes and sometimes also on the bridge of the nose; the skin-fold between the sides of the nose and the sides of the mouth becomes more accentuated, and the tongue may be slightly protruded. Of these various elements the body movements seem to indicate a struggle on the infant's part to make contact with the mother. With its clumsy physique, the baby is probably showing us all that remains of the ancestral primate clinging response.

I have been dwelling on the baby's smile, but smiling is, of course, a two-way signal. When the infant smiles at its mother she responds with a similar signal. Each rewards the other and the bond between them tightens in both directions. You may feel that this is an obvious statement, but there can be a catch in it. Some mothers, when feeling agitated, anxious, or cross with the child, try to conceal their mood by forcing a smile. They hope that the counterfeit face will avoid upset-ting the infant, but in reality this trick may do more harm than good. I mentioned earlier that it is almost impossible to fool a baby over questions of maternal mood. In the early years of life we seem to be acutely responsive to subtle signs of parental agitation and parental calm. At the pre-verbal stages, before the massive machinery of sym-bolic, cultural communication has bogged us down, we rely much more on tiny movements, postural changes and tones of voice than we need to in later life. Other species are particularly good at this, too. The astonishing ability of 'Clever Hans', the famous counting horse, was in fact based on its acuteness in responding to minute postural changes in his trainer. When asked to do a sum, Hans would tap his foot the appropriate number of times and then stop. Even if the trainer left the room and someone else took over, it still worked, because as the vital number of taps was reached, the stranger could not help tensing his body slightly. We all have this ability ourselves, even as adults (it is used a great deal by fortune-tellers to judge when they are on the right lines), but in pre-verbal infants it appears to be especially active. If the mother is making tense and agitated movements, no matter how concealed, she will communicate these to her child. If at the same time she gives a strong smile, it does not fool the infant, it only confuses it. Two conflicting messages are being transmitted. If this is done a great deal it may be permanently damaging and cause the child serious difficulties when making social contacts and adjustments later in life.

Leaving the subject of smiling, we must now turn to a very different activity. As the months pass, a new pattern of infant behaviour begins to emerge: aggression arrives on the scene. Temper tantrums and angry crying begin to differentiate themselves from the earlier all-purpose crying response. The baby signals its aggression by a more broken, irregular form of screaming and by violent striking out with its arms and legs. It attacks small objects, shakes large ones, spits and spews, and tries to bite, scratch or strike anything in reach. At first these activities are rather random and unco-ordinated. The crying indicates that fear is still present. The aggression has not yet matured to the point of a pure attack: this will come much later when the infant is sure of itself and fully aware of its physical capacities. When it does develop, it, too, has its own special facial signals. These consist of a tight-lipped glare. The lips are pursed into a hard line, with the mouth-corners held forward rather than pulled back. The eyes stare fixedly at the opponent and the eyebrows are lowered in a frown. The fists are clenched. The child has begun to assert itself.

It has been found that this aggressiveness can be increased by raising the density of a group of children. Under crowded conditions the friendly social interactions between members of a group become reduced, and the destructive and aggressive patterns show a marked rise in frequency and intensity. This is significant when one remembers that in other animals fighting is used not only to sort out dominance disputes, but also to increase the spacing-out of the members of a species. We will return to this in Chapter Five.

Apart from protecting, feeding, cleaning and playing with the offspring, the parental duties also include the all-important process of training. As with other species, this is done by a punishment-and-reward system that gradually modifies and adjusts the trial-and-error learning of the young. But, in addition, the offspring will be learning rapidly by imitation – a process that is comparatively poorly developed in most other mammals, but superbly heightened and refined in ourselves. So much of what other animals must laboriously learn for themselves, we acquire quickly by following the example of our parents. The naked ape is a teaching ape. (We are so attuned to this method of learning that we tend to assume that other species benefit in the same way, with the result that we have grossly over-estimated the role that teaching plays in their lives.)

Much of what we do as adults is based on this imitative absorption during our childhood years. Frequently we imagine that we are behaving in a particular way because such behaviour accords with some abstract, lofty code of moral principles, when in reality all we are doing is obeying a deeply ingrained and long 'forgotten' set of purely imitative impressions. It is the unmodifiable obedience to these impressions (along with our carefully concealed instinctive urges) that makes it so hard for societies to change their customs and their 'beliefs'. Even when faced with exciting, brilliantly rational new ideas, based on the application of pure, objective intelligence, the community will still cling to its old home-based habits and prejudices. This is the cross we have to bear if we are going to sail through our vital juvenile 'blotting-paper' phase of rapidly mopping up the accumulated experiences of previous generations. We are forced to take the biased opinions along with the valuable facts.

Luckily we have evolved a powerful antidote to this weakness which is inherent in the imitative learning process. We have a sharpened curiosity, an intensified urge to explore, which work against the other tendency and produce a balance that has the potential of fantastic success. Only if a culture becomes too rigid as a result of its slavery to imitative repetition, or too daring and rashly exploratory, will it flounder. Those with a good balance between the two urges will thrive. We can see plenty of examples of the too rigid and too rash cultures around the world today. The small, backward societies, completely dominated by their heavy burden of taboos and ancient customs, are cases of the former. The same societies, when converted and 'aided' by advanced cultures, rapidly become examples of the latter. The sudden overdose of social novelty and exploratory excitement swamps the stabilizing forces of ancestral imitation and tips the scales too far the other way. The result is cultural turmoil and disintegration. Lucky is the society that enjoys the gradual acquisition of a perfect balance between imitation and curiosity, between slavish, unthinking copying and progressive, rational experimentation.

4

EXPLORATION

ALL MAMMALS HAVE a strong exploratory urge, but for some
it is more crucial than others. It depends largely on how specialized they have become during the course of evolution. If they have
put all their evolutionary effort into the perfection of one particular
survival trick, they do not need to bother so much about the general
complexities of the world around them. So long as the ant-eater has its
ants and the koala bear its gum leaves, then they are well satisfied and
the living is easy. The non-specialists, on the other hand – the opportunists of the animal world – can never afford to relax. They are never
sure where their next meal may be coming from, and they have to
know every nook and cranny, test every possibility, and keep a sharp
look-out for the lucky chance. They must explore, and keep on exploring. They must investigate, and keep on re-checking. They must have a
constantly high level of curiosity.

It is not simply a matter of feeding: self-defence can make the same
demands. Porcupines, hedgehogs and skunks can snuffle and stomp
about as noisily as they like, heedless of their enemies, but the unarmed
mammal must be forever on the alert. It must know the signs of danger
and the routes of escape. To survive it must know its home range in
every minute detail.

Looked at in this way, it might seem rather inefficient not to specialize. Why should there be any opportunist mammals at all? The
answer is that there is a serious snag in the specialist way of life.
Everything is fine as long as the special survival device works, but if the
environment undergoes a major change the specialist is left stranded.
If it has gone to sufficient extremes to outstrip its competitors, the
animal will have been forced to make major changes in its genetical
make-up, and it will not be able to reverse these quickly enough when
the crunch comes. If the gum-tree forests were swept away the koala
would perish. If an iron-mouthed killer developed the ability to munch
up porcupine quills, the porcupine would become easy prey. For the
opportunist the going may always be tough, but the creature will be
able to adapt rapidly to any quick-change act that the environment
decides to put on. Take away a mongoose's rats and mice and it will

switch to eggs and snails. Take away a monkey's fruit and nuts and it will switch to roots and shoots.

Of all the non-specialists, the monkeys and apes are perhaps the most opportunist. As a group, they have specialized in non-specialization. And among the monkeys and apes, the naked ape is the most supreme opportunist of them all. This is just another facet of his neotenous evolution. All young monkeys are inquisitive, but the intensity of their curiosity tends to fade as they become adult. With us, the infantile inquisitiveness is strengthened and stretched out into our mature years. We never stop investigating. We are never satisfied that we know enough to get by. Every question we answer leads on to another question. This has become the greatest survival trick of our species.

The tendency to be attracted by novelty has been called *neophilia* (love of the new), and has been contrasted with *neophobia* (fear of the new). Everything unfamiliar is potentially dangerous. It has to be approached with caution. Perhaps it should be avoided? But if it is avoided, then how shall we ever know anything about it? The neophilic urge must drive us on and keep us interested until the unknown has become the known, until familiarity has bred contempt and, in the process, we have gained valuable experience to be stored away and called upon when needed at a later date. The child does this all the time. So strong is his urge that parental restraint is necessary. But although parents may succeed in guiding curiosity, they can never suppress it. As children grow older their exploratory tendencies sometimes reach alarming proportions and adults can be heard referring to 'a group of youngsters behaving like wild animals'. But the reverse is actually the case. If the adults took the trouble to study the way in which adult wild animals really do behave, they would find that *they* are the wild animals. They are the ones who are trying to limit exploration and who are selling out to the cosiness of sub-human conservativism. Luckily for the species, there are always enough adults who retain their juvenile inventiveness and curiosity and who enable populations to progress and expand.

When we look at young chimpanzees at play we are immediately struck by the similarity between their behaviour and that of our own children. Both are fascinated by new 'toys'. They fall on them eagerly, lifting them, dropping them, twisting them, banging them and taking them to pieces. They both invent simple games. The intensity of their interest is as strong as ours, and during the first few years of life they do just as well – better, in fact, because their muscle system develops quicker. But after a while they begin to lose ground. Their brains are not complex enough to build on this good beginning. Their powers of concentration are weak and do not grow as their bodies grow. Above all, they lack the ability to communicate in detail with their parents about the inventive techniques they are discovering.

The best way to clarify this difference is to take a specific example. Picture-making, or graphic exploration, is an obvious choice. As a pattern of behaviour it has been vitally important to our species for thousands of years, and we have the prehistoric remnants at Altamira and Lascaux to prove it.

Given the opportunity and suitable materials, young chimpanzees are as excited as we are to explore the visual possibilities of making marks on a blank sheet of paper. The start of this interest has something to do with the investigation-reward principle of obtaining disproportionately large results from the expenditure of comparatively little energy. This can be seen operating in all kinds of play situations. A great deal of exaggerated effort may be put into the activities, but it is those actions that produce an unexpectedly increased feed-back that are the most satisfying. We can call this the play principle of 'magnified reward'. Both chimps and children like banging things and it is those objects which produce the loudest noise for the smallest effort that are preferred. Balls that bounce high when only weakly thrown, balloons that shoot across a room when only lightly touched, sand that can be moulded with the mildest pressure, toys on wheels that roll easily along at the gentlest push, these are the things that have maximum play-appeal.

When first faced with a pencil and paper the infant does not find itself in a very promising situation. The best it can do is to tap the pencil on to the surface. But this leads to a pleasant surprise. The tap does something more than simply make a noise, it produces a visual impact as well. Something comes out of the end of the pencil and leaves a mark on the paper. A line is drawn.

It is fascinating to watch this first moment of graphic discovery by a chimpanzee or a child. It stares at the line, intrigued by the unexpected visual bonus its action has brought. After viewing the result for a moment it repeats the experiment. Sure enough, it works the second time, then again, and again. Soon the sheet is covered with scribble lines. As time passes, drawing sessions become more vigorous. Single, tentative lines, placed on the paper one after the other, give way to multiple back-and-forth scribbling. If there is a choice, crayons, chalks and paints are preferred to pencils because they have an even bolder impact, produce an even bigger visual effect, as they sweep across the paper.

The first interest in this activity appears at about one-and-a-half years of age, in both chimps and children. But it is not until after the second birthday that the bold, confident, multiple scribbling really gains momentum. At the age of three the average child moves into a new graphic phase: it starts to simplify its confused scribbling. Out of the exciting chaos it begins to distil basic shapes. It experiments with crosses, then with circles, squares and triangles. Meandering lines are led round the page until they join up with themselves, enclosing a space. A line becomes an outline.

During the months that follow, these simple shapes are combined, one with another, to produce simple abstract patterns. A circle is cut through by a cross, the corners of a square are joined by diagonal lines. This is the vital stage that precedes the very first pictorial representations. In the child this great break-through comes in the second half of the third year, or the beginning of the fourth. In the chimpanzee, it never comes. The young chimp manages to make fan-patterns, crosses and circles, and it can even achieve a 'marked circle', but it can go no further. It is particularly tantalizing that the marked-circle motif is the immediate precursor of the earliest representation produced by the typical child. What happens is that a few lines or spots are placed inside the outline of the circle and then, as if by magic, a face stares back at the infant painter. There is a sudden flash of recognition. The phase of abstract experimentation, of pattern invention, is over. Now a new goal must be reached: the goal of perfected representation. New faces are made, better faces, with the eyes and mouth in the right place. Details are added – hair, ears, a nose, arms and legs. Other images are born – flowers, houses, animals, boats, cars. These are heights the young chimp can never, it seems, attain. After the peak has been reached – the circle made and its inside area marked – the animal continues to grow but its pictures do not. Perhaps one day a genius chimp will be found, but it seems unlikely.

For the child, the representational phase of graphic exploration now stretches out before it, but although it is the major area of discovery, the older abstract patterning influences still make themselves felt, especially between the ages of five and eight. During this period particularly attractive paintings are produced because they are based on the solid grounding of the abstract-shape phase. The representational images are still at a very simple stage of differentiation and they combine appealingly with the confident, well-established shape-and-pattern arrangements.

The process by which the dot-filled circle grows into an accurate full-length portrait is an intriguing one. The discovery that it represents a face does not lead to an overnight success in perfecting the process. This clearly becomes the dominant aim, but it takes time (more than a decade, in fact). To start with, the basic features have to be tidied up a little – circles for eyes, a good strong horizontal line for a mouth, two dots or a central circle for a nose. Hairs have to fringe the outer circle. And there things can pause for a while. The face, after all, is the most vital and compelling part of the mother, at least in visual terms. After a while, though, further progress is made. By the simple device of making some of the hairs longer than the rest, it is possible for this face-figure to sprout arms and legs. These in turn can grow fingers and toes. At this point the basic figure-shape is still founded on the pre-representational circle. This is an old friend and he is staying late. Having become a face he has now become a face and body combined. It does not seem to worry the child at this stage that the arms of its drawing are coming out of the side of what appears to be its head. But the circle cannot hold out for ever. Like a cell, it must divide and bud off a lower, second cell. Alternatively the two leg lines must be joined somewhere along their length, but higher than the feet. In one of these two ways, a body can be born. Whichever happens, it leaves the arms high and dry, sticking out of the side of the head. And there they stay for quite some time, before they are brought down into their more correct position, protruding from the top of the body.

It is fascinating to observe these slow steps being taken, one after the other, as the voyage of discovery tirelessly continues. Gradually more and more shapes and combinations are attempted, more diverse images, more complex colours, and more varied textures. Eventually, accurate representation is achieved and precise copies of the outside world can be trapped and preserved on paper. But at that stage the original exploratory nature of the activity becomes submerged beneath the pressing demands of pictorial communication. Earlier painting and drawing, in the young chimp and the young child, had nothing to do with the act of communicating. It was an act of discovery, of invention, of testing the possibilities of graphic variability. It was 'action-painting', not signalling. It required no reward – it was its own reward, it was play for play's sake. However, like so many aspects of childhood play, it soon becomes merged into other adult pursuits. Social communication makes a take-over bid for it and the original inventiveness is lost, the pure thrill of 'taking a line for a walk' is gone. Only in doodles do most adults allow it to re-emerge. (This does not mean that they have become uninventive, merely that the area of invention has moved on into more complex, technological spheres.)

Fortunately for the exploratory art of painting and drawing, much more efficient technical methods of reproducing images of the environment have now been developed. Photography and its offshoots have rendered representational 'information painting' obsolete. This has broken the heavy chains of responsibility that have been the crippling burden of adult art for so long. Painting can now once again explore, this time in a mature adult form. And this, one need hardly mention, is precisely what it is doing today.

I selected this particular example of exploratory behaviour because it reveals very clearly the differences between us and our nearest living relative, the chimpanzee. Similar comparisons could be made in other spheres. One or two deserve brief mention. Exploration of the world of sound can be observed in both species. Vocal invention, as we have already seen, is for some reason virtually absent in the chimpanzee, but 'percussive drumming' plays an important role in its life. Young chimpanzees repeatedly investigate the noise-potentials of acts of thumping, foot-stamping and clapping. As adults they develop this tendency into prolonged social drumming sessions. One animal after another stamps, screams and tears up vegetation, beating on tree-stumps and hollow logs. These communal displays may last for half an hour or more. Their exact function is unknown, but they have the effect of mutually arousing the members of a group. In our own species, drumming is also the most widespread form of musical expression. It begins early, as in the chimpanzee, when children begin to test out the percussive values of objects around them in much the same way. But whereas the adult chimpanzees never manage much more than a simple rhythmic tattoo, we elaborate it into complex poly-rhythms and augment it with vibrating rattles and pitch variations. We also make additional noises by blowing into hollow cavities and scraping or plucking pieces of metal. The screams and hoots of the chimpanzee become in us inventive chants. Our development of complicated musical performances appears, in simpler social groups, to have played much the same role as the drumming and hooting sessions of the chimpanzees, namely, mutual group arousal. Unlike picture-making, it was not an activity pattern that became commandeered for the transmission of detailed information on a major scale. The sending of messages by drumming sequences in certain cultures was an exception to this rule, but by and large music was developed as a communal mood-provoker and synchronizer. Its inventive and exploratory content became stronger and stronger, however, and, freed of any important 'representational' duties, it has become a major area of abstract aesthetic experimentation. (Because of its other prior information commitments, painting has only just caught up with it.)

Dancing has followed much the same course as music and singing. The chimpanzees include many swaying and jigging movements in their drumming rituals and these also accompany the mood-provoking musical performances of our own species. From there, like music, they have been elaborated and expanded into aesthetically complex performances.

Closely related to dancing has been the growth of gymnastics. Rhythmical physical performances are common in the play of both young chimps and young children. They rapidly become stylized, but retain a strong element of variability within the structured patterns they assume. But the physical games of chimpanzees do not grow and mature, they fizzle out. We, on the other hand, explore their possibilities to the full and elaborate them in our adult lives into many complex forms of exercises and sports. Again they are important as communal synchronizing devices, but essentially they are means of maintaining and expanding our exploration of our physical capacities.

Writing, a formalized offshoot of picture-making, and verbalized vocal communication have, of course, been developed as our major means of transmitting and recording information, but they have also been utilized as vehicles for aesthetic exploration on an enormous scale. The intricate elaboration of our ancestral grunts and squeaks into complex symbolic speech has enabled us to sit and 'play' with thoughts in our heads, and to manipulate our (primarily instructional) word sequences to new ends as aesthetic, experimental playthings.

So, in all these spheres – in painting, sculpture, drawing, music, singing, dancing, gymnastics, games, sports, writing and speech – we can carry on to our heart's content, all through our long lives, complex and specialized forms of exploration and experiment. Through elaborate training, both as performers and observers, we can sensitize our responsiveness to the immense exploratory potential that these pursuits have to offer. If we set aside the secondary functions of these activities (the making of money, gaining of status, and so forth), then they all emerge, biologically, either as the extension into adult life of infantile play-patterns, or as the super-imposition on to adult information-communication systems of 'play-rules'.

These rules can be stated as follows: (1) you shall investigate the unfamiliar until it has become familiar; (2) you shall impose rhythmic repetition on the familiar; (3) you shall vary this repetition in as many ways as possible; (4) you shall select the most satisfying of these variations and develop these at the expense of others; (5) you shall combine and recombine these variations one with another; and (6) you shall do all this for its own sake, as an end in itself.

These principles apply from one end of the scale to the other, whether you are considering an infant playing in the sand, or a composer working on a symphony.

The last rule is particularly important. Exploratory behaviour also plays a role in the basic survival patterns of feeding, fighting, mating and the rest. But there it is confined to the early appetitive phases of the activity sequences and is geared to their special demands. For many species of animals it is no more than this. There is no exploration for exploration's sake. But, amongst the higher mammals and to a supreme extent in ourselves, it has become emancipated as a distinct, separate drive. Its function is to provide us with as subtle and complex an awareness of the world around us, and of our own capacities in relation to it, as possible. This awareness is not heightened in the specific contexts of the basic survival goals, but in generalized terms. What we acquire in this way can then be applied anywhere, at any time, in any context.

I have omitted the growth of science and technology from this discussion because it has largely been concerned with specific improvements in the methods employed in achieving the basic survival goals, such as fighting (weapons), feeding (agriculture), nest-building (architecture) and comfort (medicine). It is interesting, though, that as time has gone by and the technical developments have become more and more interlocked with one another, the pure exploratory urge has also invaded the scientific sphere. Scientific research – the very name 're-search' gives the game away (and I mean game) – operates very much on the play-principles mentioned earlier. In 'pure' research, the scientist uses his imagination in virtually the same way as the artist. He talks of a beautiful experiment rather than of an expedient one. Like the artist, he is concerned with exploration for exploration's sake. If the results of the studies prove to be useful in the context of some specific survival goal, all to the good, but this is secondary.

In all exploratory behaviour, whether artistic or scientific, there is the ever-present battle between the neophilic and neophobic urges. The former drives us on to new experiences, makes us crave for novelty. The latter holds us back, makes us take refuge in the familiar. We are constantly in a state of shifting balance between the conflicting attractions of the exciting new stimulus and the friendly old one. If we lost our neophilia, we would stagnate. If we lost our neophobia, we would rush headlong into disaster. This state of conflict does not merely account for the more obvious fluctuations in fashions and fads, in hair-styles and clothing, in furniture and cars; it is also the very basis of our whole cultural progression. We explore and we retrench, we investigate and we stabilize. Step by step we expand our awareness and understanding both of ourselves and of the complex environment we live in.

Before leaving this topic there is one final, special aspect of exploratory behaviour that cannot go unmentioned. It concerns a critical phase of social play during the infantile period. When it is very young, the infant's social play is directed primarily at the parents, but as it grows the emphasis is shifted from them towards other children of the same age. The child becomes a member of a juvenile 'play group'. This is a critical step in its development. As an exploratory involvement it has far-reaching effects on the later life of the individual. Of course, all forms of exploration at this tender age have long-term consequences – the child that fails to explore music or painting will find these subjects difficult as an adult – but person-to-person play contacts are even more critical than the rest. An adult coming to music, say, for the first time, without childhood exploration of the subject behind him, may find it difficult, but not impossible. A child that has been severely sheltered from social contact as a member of a play group, on the other hand, will always find himself badly hampered in his adult social interactions. Experiments with monkeys have revealed that not only does isolation in infancy produce a socially withdrawn adult, but it also creates an anti-sexual and anti-parental individual. Monkeys that were reared in isolation from other youngsters failed to participate in play-group activities when exposed to them later, as older juveniles. Although the isolates were physically healthy and had grown well in their solitary states, they were quite incapable of joining in the general rough-and-tumble. Instead they crouched, immobile, in the corner of the play-room, usually clasping their bodies tightly with their arms, or covering their eyes. When they matured, again as physically healthy specimens, they showed no interest in sexual partners. If forcibly mated, female isolates produced offspring in the normal way, but then proceeded to treat them as though they were huge parasites crawling on their bodies. They attacked them, drove them away, and either killed them or ignored them.

Similar experiments with young chimpanzees showed that, in this species, with prolonged rehabilitation and special care it was possible to undo, to some extent, this behavioural damage, but, even so, its dangers cannot be over-estimated. In our own species, over-protected children will always suffer in adult social contacts. This is especially important in the case of only children, where the absence of siblings sets them at a serious initial disadvantage. If they do not experience the socializing effects of the rough-and-tumble of the juvenile play groups, they are liable to remain shy and withdrawn for the rest of their lives, find sexual pair-bonding difficult or impossible and, if they do manage to become parents, will make bad ones.

From this it is clear that the rearing process has two distinct phases – an early, inward-turning one and a later, outward-turning one. They are both vitally important and we can learn a great deal about them from monkey behaviour. During the early phase the infant is loved, rewarded and protected by the mother. It comes to understand security. In the later phase it is encouraged to be more outward-going, to participate in social contacts with other juveniles. The mother becomes less loving and restricts her protective acts to moments of serious panic or alarm, when outside dangers threaten the colony. She may now actually punish the growing offspring if it persists in clinging to her hairy apron-strings in the absence of serious panic. It now comes to understand and accept its growing independence.

The situation should be basically the same for offspring of our own species. If either of these basic phases is parentally mis-handled, the child will be in serious trouble in later life. If it has lacked the early security phase, but has been suitably active in the independence phase, it will find making new social contacts easy enough, but will be unable to maintain them or make any real depth of contact. If it has enjoyed great security in the early phase, but has been over-protected later on, it will find making new adult contacts immensely difficult and will tend to cling desperately to old ones.

If we take a closer look at the more extreme cases of social withdrawal, we can witness anti-exploratory behaviour in its most extreme and characteristic form. Severely withdrawn individuals may become socially inactive, but they are far from physically inactive. They become preoccupied with repetitive stereotypes. For hour after hour they rock or sway, nod or shake, twirl or twitch, or clasp and unclasp themselves. They may suck their thumbs, or other parts of their bodies, prod or pinch themselves, make strange and repetitive facial expressions, or tap or roll small objects rhythmically. We all exhibit 'tics' of this sort occasionally, but for them it becomes a major and prolonged form of physical expression. What happens is that they find the environment so threatening, social contacts so frightening and impossible, that they seek comfort and reassurance by super-familiarizing their behaviour. The rhythmic repetition of an act renders it increasingly familiar and 'safe'. Instead of performing a wide variety of heterogeneous activities, the withdrawn individual sticks to the few he knows best. For him the old saying: 'Nothing ventured, nothing gained' has been re-written: 'Nothing ventured, nothing lost.'

I have already mentioned the comforting regressive qualities of the heart-beat rhythm, and this applies here, too. Many of the patterns seem to operate at about heart-beat speed, but even those that do not, still act as 'comforters' by virtue of their super-familiarity achieved from constant repetition. It has been noticed that socially retarded individuals increase their stereotypes when put into a strange room. This fits in with the ideas expressed here. The increased novelty of the environment heightens the neophobic fears, and heavier demands are made on the comforting devices to counteract this.

The more a stereotype is repeated, the more it becomes like an artificially produced, maternal heart-beat. Its 'friendliness' increases and increases until it becomes virtually irreversible. Even if the extreme neophobia causing it can be removed (which is difficult enough), the stereotype may go twitching on.

As I said, socially adjusted individuals also exhibit these 'tics' from time to time. Usually they occur in stress situations and here, too, they act as comforters. We know all the signs. The executive awaiting a vital phone call taps or drums on his desk; the woman in the doctor's waiting-room clasps and unclasps her fingers around her handbag; the embarrassed child swings its body left and right, left and right; the expectant father paces back and forth; the student in the exam sucks his pencil; the anxious officer strokes his moustache. In moderation these little anti-exploratory devices are useful. They help us to tolerate the anticipated 'novelty overdose'. If used to excess, however, there is always the danger that they will become irreversible and obsessive, and will persist even when not called for.

Stereotypes also crop up in situations of extreme boredom. This can be seen very clearly in zoo animals as well as in our own species. It can sometimes reach frightening proportions. What happens here is that the captive animals would make social contacts if only they had the chance, but they are physically prevented from doing so. The situation is basically the same as in cases of social withdrawal. The restricted environment of the zoo cage blocks their social contacts and forces them into a situation of social withdrawal. The cage bars are a solid, physical equivalent of the psychological barriers facing the socially withdrawn individual. They constitute a powerful anti-exploratory device and, left with nothing to explore, the zoo animal begins to express itself in the only way possible, by developing rhythmic stereotypes. We are all familiar with the repetitive pacing to-and-fro of the caged animal, but this is only one of the many strange patterns that arise. Stylized masturbation may occur. Sometimes this no longer involves manipulation of the penis. The animal (usually a monkey) simply makes the back and forth masturbatory movements of its arm and hand, but without actually touching the penis. Some female monkeys repeatedly suck their own nipples. Young animals suck their paws. Chimpanzees may prod pieces of straw into their (previously healthy) ears. Elephants nod their heads for hours on end. Certain creatures repeatedly bite themselves, or pull their own hair out. Serious self-mutilation may occur. Some of these responses are given in stressful situations, but many of them are simply reactions to boredom. When there is no variability in the environment the exploratory urge stagnates.

Simply by looking at an isolated animal performing one of these stereotypes it is impossible to know for certain what is causing the

97

behaviour. It may be boredom, or it may be stress. If it is stress it may be the result of the immediate environmental situation, or it may be a long-term phenomenon stemming from an abnormal upbringing. A few simple experiments can give us the answer. When a strange object is placed in the cage, if the stereotypes disappear and exploration begins, then they were obviously being caused by boredom. If the stereotypes increase, however, then they were being caused by stress. If they persist after the introduction of other members of the same species, producing a normal social environment, then the individual with the stereotypes has almost certainly had an abnormally isolated infancy.

All these zoo peculiarities can be seen in our own species (perhaps because we have designed our zoos so much like our cities). They should be a lesson to us, reminding us of the enormous importance of achieving a good balance between our neophobic and neophilic tendencies. If we do not have this, we cannot function properly. Our nervous systems will do the best they can for us, but the results will always be a travesty of our true behavioural potentials.

5

FIGHTING

I F WE ARE to understand the nature of our aggressive urges, we must see them against the background of our animal origins. As a species we are so preoccupied with mass-produced and mass-destroying violence at the present time that we are apt to lose our objectivity when discussing this subject. It is a fact that the most level-headed intellectuals frequently become violently aggressive when discussing the urgent need to suppress aggression. This is not surprising. We are, to put it mildly, in a mess, and there is a strong chance that we shall have exterminated ourselves by the end of the century. Our only consolation will have to be that, as a species, we have had an exciting term of office. Not a long term, as species go, but an amazingly eventful one. But before we examine our own bizarre perfections of attack and defence, we must examine the basic nature of violence in the spearless, gunless, bombless world of animals.

Animals fight amongst themselves for one of two very good reasons: either to establish their dominance in a social hierarchy, or to establish their territorial rights over a particular piece of ground. Some species are purely hierarchical, with no fixed territories. Some are purely territorial, with no hierarchy problems. Some have hierarchies on their territories and have to contend with both forms of aggression. We belong to the last group: we have it both ways. As primates we were already loaded with the hierarchy system. This is the basic way of primate life. The group keeps moving about, rarely staying anywhere long enough to establish a fixed territory. Occasional inter-group conflict may arise, but it is weakly organized, spasmodic and of comparatively little importance in the life of the average monkey. The 'peck order' (so-called because it was first discussed in respect of chickens) is, on the other hand, of vital significance in his day-to-day – and even his minute-to-minute – living. There is a rigidly established social hierarchy in most species of monkeys and apes, with a dominant male in charge of the group, and the others ranged below him in varying degrees of subordination. When he becomes too old or weak to maintain his domination, he is overthrown by a younger, sturdier male, who then assumes the mantle of the colony boss. (In some cases the usurper literally assumes the mantle, growing one in the form of a cape of long hair.) As the troop keeps together all the time, his role as group tyrant is incessantly operative. But despite this he is invariably the sleekest, best-groomed and sexiest monkey in the community.

Not all primate species are violently dictatorial in their social organization. There is nearly always a tyrant, but he is sometimes a benign and rather tolerant tyrant, as in the case of the mighty gorilla. He shares the females amongst the lesser males, is generous at feeding times, and only asserts himself when something crops up that cannot be shared, or when there are signs of a revolt, or unruly fighting amongst the weaker members.

This basic system obviously had to be changed when the naked ape became a co-operative hunter with a fixed home base. Just as with sexual behaviour, the typical primate system had to be modified to match his adopted carnivore role. The group had to become territorial. It had to defend the region of its fixed base. Because of the co-operative nature of the hunting, this had to be done on a group basis, rather than individually. Within the group the tyrannical hierarchy system of the usual primate colony had to be modified considerably to ensure full co-operation from the weaker members when out hunting. But it could not be abolished altogether. There had to be a mild hierarchy, with stronger members and a top leader, if firm decisions were going to be taken, even if this leader was obliged to take the feelings of his inferiors more into account than his hairy, forest-dwelling equivalent would have to do.

In addition to group defence of territory and hierarchy organization, the prolonged dependency of the young, forcing us to adopt pair-bonded family units, demanded yet another form of self-assertion. Each male, as the head of a family, became involved in defending his own individual home base inside the general colony base. So for us there are three fundamental forms of aggression, instead of the usual one or two. As we know to our cost, they are all still very much in evidence today, despite the complexities of our societies.

How does the aggression work? What are the patterns of behaviour involved? How do we intimidate one another? We must look again at the other animals. When a mammal becomes aggressively aroused a number of basic physiological changes occur within its body. The whole machine has to gear itself up for action, by means of the autonomic nervous system. This system consists of two opposing and counter-balancing sub-systems – the sympathetic and the parasympathetic. The former is the one that is concerned with preparing the body for violent activity. The latter has the task of preserving and restoring bodily reserves. The former says, 'You are stripped for action, get moving'; the latter says, 'Take it easy, relax and conserve your strength.' Under normal circumstances the body listens to both these voices and maintains a happy balance between them, but when strong aggression is aroused it listens only to the sympathetic system. When this is activated, adrenalin pours into the blood and the whole circulatory system is profoundly affected. The heart beats faster and blood is transferred from the skin and viscera to the muscles and brain. There is an increase in blood pressure. The rate of production of red blood corpuscles is rapidly stepped up. There is a reduction of the time taken for blood to coagulate. In addition there is a cessation in the processes of digesting and storing food. Salivation is restrained. Movements of the stomach, the secretion of gastric juices, and the peristaltic movements of the intestines are all inhibited. Also, the rectum and bladder do not empty as easily as under normal conditions. Stored carbohydrate is rushed out of the liver and floods the blood with sugar. There is a massive increase in respiratory activity. Breathing becomes quicker and deeper. The temperature-regulating mechanisms are activated. The hair stands on end and there is profuse sweating.

All these changes assist in preparing the animal for battle. As if by magic, they instantly banish fatigue and make large amounts of energy available for the anticipated physical struggle for survival. The blood is pumped vigorously to the sites where it is most needed – to the brain, for quick thinking, and to the muscles, for violent action. The rise in blood sugars increases muscular efficiency. The speeding up of coagulation processes means that any blood spilled as a result of injury will clot more quickly and reduce wastage. The stepped-up release of red blood cells from the spleen, in combination with the increased speed of blood circulation, aids the respiratory system to boost the intake of oxygen and the removal of carbon dioxide. The full hair erection exposes the skin to the air and helps to cool the body, as does the outpouring of sweat from the sweat glands. The dangers of over-heating from excessive activity are therefore reduced.

With all the vital systems activated, the animal is ready to launch into the attack, but there is a snag. Out-and-out fighting may lead to a valuable victory, but it may also involve serious damage to the victor. The enemy invariably provokes fear as well as aggression. The aggression drives the animal on, the fear holds it back. An intense state of inner conflict arises. Typically, the animal that is aroused to fight does not go straight into an all-out attack. It begins by threatening to attack. Its inner conflict suspends it, tensed for combat, but not yet ready to begin it. If, in this state, it presents a sufficiently intimidating spectacle for its opponent, and the latter slinks away, then obviously this is preferable. The victory can be won without the shedding of blood. The species is able to settle its disputes without undue damage to its members and obviously benefits tremendously in the process.

Throughout the higher forms of animal life there has been a strong trend in this direction – the direction of ritualized combat. Threat and counter-threat has largely replaced actual physical combat. Full-blooded fighting does, of course, still take place from time to time, but only as a last resort, when aggressive signalling and counter-signalling have failed to settle a dispute. The strength of the outward signs of the physiological changes I have described indicates to the enemy just how violently the aggressive animal is preparing itself for action.

This works extremely well behaviourally, but physiologically it creates something of a problem. The machinery of the body has been geared up for a massive output of work. But the anticipated exertions do not materialize. How does the autonomic nervous system deal with this situation? It has mustered all its troops at the front line, ready for action, but their very presence has won the war. What happens now?

If physical combat followed on naturally from the massive activation of the sympathetic nervous system, all the body preparations it had made would be fully utilized. The energy would be burned up and eventually the parasympathetic system would reassert itself and gradually restore a state of physiological calm. But in the tense state of conflict between aggression and fear, everything is suspended. The result is that the parasympathetic system fights back wildly and the autonomic pendulum swings frantically back and forth. As the tense moments of threat and counter-threat tick by, we see flashes of para-sympathetic activity interspersed with the sympathetic symptoms. Dryness in the mouth may give way to excessive salivation. Tightening of the bowels may collapse and sudden defecation may occur. The urine, held back so strongly in the bladder, may be released in a flood. The removal of blood from the skin may be massively reversed, extreme pallor being replaced by intense flushing and reddening. The deep and rapid respiration may be dramatically interrupted, leading to gasps and sighs. These are desperate attempts on the part of the parasympathetic system to counteract the apparent extravagance of the sympathetic. Under normal circumstances it would be out of the question for intense reactions in one direction to occur simultaneously with intense reactions in the other, but under the extreme conditions of aggressive threat, everything gets momentarily out of phase. (This explains why, in extreme cases of shock, fainting or swooning can be observed. In such instances the blood that has been rushed to the brain is withdrawn again so violently that it leads to sudden unconsciousness.)

As far as the threat-signalling system is concerned, this physiological turbulence is a gift. It provides an even richer source of signals. During the course of evolution these mood-signs have been built on and elaborated in a number of ways. Defecation and urination have become important territorial scent-marking devices for many species of mammals. The most commonly seen example of this is the way domestic dogs cock their legs against marker-posts in their territories, and the way this activity is increased during threatening encounters between rival dogs. (The streets of our cities are excessively stimulating for this activity because they constitute overlapping territories for so many rivals, and each dog is forced to super-scent these areas in an attempt to compete.) Some species have evolved super-dunging techniques. The hippopotamus has acquired a specially flattened tail that is waggled rapidly back and forth during the act of defecating. The effect is that of shooting dung through a fan, with the result that the faeces are spread out over a wide area. Many species have developed special anal glands that add strong personal scents to the dung.

The circulatory disturbances producing extreme pallor or intense red flushes have become improved as signals by the development of bare patches of skin on the faces of many species and the rumps of others. The gasping and hissing of the respiratory disturbances have been elaborated into grunts and roars and the many other aggressive vocalizations. It has been suggested that this accounts for the origin of the whole communication system of vocal signals. Another basic trend developing out of respiratory turbulence is the evolution of inflation displays. Many species puff themselves up in threat and may inflate specialized air-sacs and pouches. (This is particularly common amongst birds, which already possess a number of air-sacs as a basic part of their respiratory systems.)

Aggressive hair-erection has led to the growth of specialized regions such as crests, capes, manes and fringes. These and other localized hair patches have become highly conspicuous. The hairs have become elongated or stiffened. Their pigmentation has often been drastically modified to produce areas of strong contrast with the surrounding fur. When aggressively aroused, with the hairs standing on end, the animal suddenly appears larger and more frightening, and the display patches become bigger and brighter.

Aggressive sweating has become another source of scent-signals. In many cases there have, once again, been specialized evolutionary trends exploiting this possibility. Certain of the sweat glands have become enormously enlarged as complex scent-glands. These can be found on the faces, feet, tails and various parts of the body of many species.

All these improvements have enriched the communication systems of animals and rendered their mood language more subtle and informative. They make the threatening behaviour of the aroused animal more 'readable' in more precise terms.

But this is only half the story. We have been considering only the autonomic signals. In addition to all these there is another whole range of signals available, which stem from the tensed-up muscular movements and postures of the threatening animal. All that the autonomic system did was to gear the body up ready for muscular action. But what did the muscles do about it? They were stiffened for the onslaught, but no onslaught came. The outcome of this situation is a series of aggressive intention movements, ambivalent actions, and conflict postures. The impulses to attack and to flee pull the body this way and that. It lunges forward, pulls back, twists sideways, crouches down, leaps up, leans in, tilts away. As soon as the urge to attack gets the upper hand, the impulse to flee immediately countermands the order. Every move to withdraw is checked by a move to attack. During the course of evolution this general agitation has become modified into specialized postures of threat and intimidation. The intention movements have become stylized, the ambivalent jerkings have become formalized into rhythmic twistings and shakings. A whole new repertoire of aggressive signals has been developed and perfected.

As a result of this we can witness, in many animal species, elaborate threat rituals and combat 'dances'. The contestants circle one another in a characteristically stilted fashion, their bodies tense and stiff. They may bow, nod, shake, shiver, swing rhythmically from side to side, or make repeated short, stylized runs. They paw the ground, arch their backs, or lower their heads. All these intention movements act as vital communication signals and combine effectively with the autonomic signals to provide a precise picture of the intensity of the aggression that has been aroused, and an exact indication of the balance between the urge to attack and the urge to flee.

But there is yet more to come. There is another important source of special signals, arising from a category of behaviour that has been named displacement activity. One of the side-effects of an intense inner conflict is that an animal sometimes exhibits strange and seemingly irrelevant pieces of behaviour. It is as if the tensed-up creature, unable to perform either of the things it is desperate to do, finds an outlet for its pent-up energy in some other, totally unrelated activity. Its urge to flee blocks its urge to attack and vice versa, so it vents its feelings in some other way. Threatening rivals can be seen suddenly to perform curiously stilted and incomplete feeding movements, and then return instantly to their full threat postures. Or they may scratch or clean themselves in some way, interspersing these movements with the typical threat manœuvring. Some species perform displacement nest-building actions, picking up pieces of nest material that happen to lie near by and dropping them on to imaginary nests. Others indulge in 'instant sleep', momentarily tucking their heads into a snoozing position, yawning or stretching.

There has been a great deal of controversy about these displacement activities. It has been argued that there is no objective justification for referring to them as irrelevancies. If an animal feeds, it is hungry, and if it scratches it must itch. It is stressed that it is impossible to prove that a threatening animal is not hungry when it performs so-called displacement feeding actions, or that it is not itching when it scratches. But this is armchair criticism, and to anyone who has actually observed and studied aggressive encounters in a wide variety of species, it is patently absurd. The tension and drama of these moments is such that it is ridiculous to suggest that the contestants would break off, even momentarily, to feed for the sake of feeding, or scratch for the sake of scratching, or sleep for the sake of sleeping.

Despite the academic arguments about the causal mechanisms involved in the production of displacement activities, one thing is clear, namely that in functional terms they provide yet one more source for the evolution of valuable threat signals. Many animals have exaggerated these actions in such a way that they have become increasingly conspicuous and showy.

All these activities, then, the autonomic signals, the intention movements, the ambivalent postures and the displacement activities, become ritualized and together provide the animals with a comprehensive repertoire of threat signals. In most encounters they will be sufficient to resolve the dispute without the contestants coming to blows. But if this system fails, as it often does under conditions of extreme crowding, for example, then real fighting follows and the signals give way to the brutal mechanics of physical attack. Then, the teeth are used to bite, slash and stab, the head and horns to butt and spear, the body to ram, bump and push, the legs to claw, kick and swipe, the hands to grasp and squeeze, and sometimes the tail to thrash and whip. Even so, it is extremely rare for one contestant to kill the other. Species that have evolved special killing techniques for dealing with their prey seldom employ these when fighting their own kind. (Serious errors have sometimes been made in this connection, with false assumptions about the presumed relationship between prey-attacking behaviour and rival-attacking activities. The two are quite distinct in both motivation and performance.) As soon as the enemy has been sufficiently subdued, it ceases to be a threat and is ignored. There is no point in wasting additional energy on it, and it is allowed to slink away without further damage or persecution.

Before relating all these belligerent activities to our own species, there is one more aspect of animal aggression that must be examined. It concerns the behaviour of the loser. When his position has become untenable, the obvious thing for him to do is to remove himself as fast as he can. But this is not always possible. His escape route may be physically obstructed, or, if he is a member of a tightly knit social group, he may be obliged to stay within range of the victor. In either of these cases, he must somehow signal to the stronger animal that he is no longer a threat and that he does not intend to continue the fight. If he leaves it until he is badly damaged or physically exhausted, this will become obvious enough, and the dominant animal will wander off and leave him in peace. But if he can signal his acceptance of defeat before his position has deteriorated to this unfortunate extreme, he will be able to avoid further serious punishment. This is achieved by the performance of certain characteristic submissive displays. These appease the attacker and rapidly reduce his aggression, speeding up the settlement of the dispute.

They operate in several ways. Basically, they either switch off the signals that have been arousing the aggression, or they switch on other, positively non-aggressive signals. The first category simply serve to calm the dominant animal down, the latter help by actively changing his mood into something else. The crudest form of submission is gross inactivity. Because aggression involves violent movement, a static pose automatically signals non-aggression. Frequently this is combined with crouching and cowering. Aggression involves expanding the body to its maximum size, and crouching reverses this and therefore acts as an appeasement. Facing away from the attacker also helps, being the opposite of the posture of frontal attack. Other threat-opposites are also used. If a particular species threatens by lowering its head, then raising the head can become a valuable appeasement gesture. If an attacker erects its hair, then compressing it will serve as a submission device. In certain rare cases a loser will admit defeat by offering a vulnerable area to the attacker. A chimpanzee, for example, will hold out its hand as a gesture of submission, rendering it extremely vulnerable for serious biting. Because an aggressive chimpanzee would never do such a thing, this begging gesture serves to appease the dominant individual.

The second category of appeasement signals operate as remotivating devices. The subordinate animal sends out signals that stimulate a non-aggressive response and, as this wells up inside the attacker, his urge to fight is suppressed and subdued by it. This is done in one of three main ways. A particularly widespread re-motivator is the adoption of juvenile food-begging postures. The weaker individual crouches and begs from the dominant one in the infantile posture characteristic of the particular species – a device especially favoured by females when they are being attacked by males. It is often so effective that the male responds by regurgitating some food to the female, who then completes the food-begging ritual by swallowing it. Now in a thoroughly paternal, protective mood, the male loses his aggression and the pair calm down together. This is the basis of courtship feeding in many species, especially with birds, where the early stages of pair-formation involve a great deal of aggression on the part of the male. Another re-motivating activity is the adoption of a female sexual posture by the weaker animal. Regardless of its sex, or its sexual condition, it may suddenly assume the female rump-presentation posture. When it displays towards the attacker in this way, it stimulates a sexual response which damps down the mood of aggression. In such situations, a dominant male *or* female will mount and pseudo-copulate with either a submissive male or a submissive female.

A third form of re-motivation involves the arousal of the mood to groom or be groomed. A great deal of social or mutual grooming goes on in the animal world and it is strongly associated with the calmer, more peaceful moments of community life. The weaker animal may either invite the winner to groom it, or may make signals requesting permission to perform the grooming itself. Monkeys make great use of this device and have a special facial gesture to go with it, consisting of rapidly smacking the lips together – a modified, ritualized version of part of the normal grooming ceremony. When one monkey grooms another it repeatedly pops fragments of skin and other detritus into its mouth, smacking its lips as it does so. By exaggerating the smacking movements and speeding them up, it signals its readiness to perform this duty and frequently manages in this way to suppress the aggression of the attacker and persuade it to relax and allow itself to be groomed. After a while the dominant individual is so lulled by this procedure that the weakling can slip away unharmed.

These, then, are the ceremonies and devices by which animals order their aggressive involvements. The phrase 'nature red in tooth and claw' was originally intended to refer to the brutal prey-killing activities of the carnivores, but it has been applied incorrectly in general terms to the whole subject of animal fighting. Nothing could be further from the truth. If a species is to survive, it simply cannot afford to go around slaughtering its own kind. Intra-specific aggression has to be inhibited and controlled, and the more powerful and savage the prey-killing weapons of a particular species are, the stronger must be the inhibitions about using them to settle disputes with rivals. This is the 'law of the jungle' where territorial and hierarchy disagreements are concerned. Those species that failed to obey this law have long since become extinct.

How do we, as a species, measure up to this situation? What is our own special repertoire of threatening and appeasing signals? What are our fighting methods, and how do we control them?

Aggressive arousal produces in us all the same physiological upheavals and muscular tensions and agitations that were described in the general animal context. Like other species, we also show a variety of displacement activities. In some respects we are not as well equipped as other species to develop these basic responses into powerful signals. We cannot intimidate our opponents, for example, by erecting our body hair. We still do it in moments of great shock ('My hair stood on end'), but as a signal it is of little use. In other respects we can do much better. Our very nakedness, which prevents us from bristling effectively, gives us the chance to send powerful flushing and paling signals. We can go 'white with rage', 'red with anger', or 'pale with fear'. It is the white colour we have to watch for here: this spells activity. If it is combined with other actions that signal attack, then it is a vital danger signal. If it is combined with other actions that signal fear, then it is a panic signal. It is caused, you will recall, by the activation of the sympathetic nervous system, the 'go' system, and it is not to be treated lightly. The reddening, on the other hand, is less worrying: it is caused by the frantic counter-balancing attempts of the parasympathetic system, and indicates that the 'go' system is already being undermined. The angry, red-faced opponent who faces you is far less likely to attack than the white-faced, tight-lipped one. Red-face's conflict is such that he is all bottled up and inhibited, but white-face is still ready for action. Neither can be trifled with, but white-face is much more likely to spring in to the attack unless he is immediately appeased or counter-threatened even more strongly.

In a similar vein, rapid deep breathing is a danger signal, but it has already become less of a threat when it develops into irregular snorts and gurgles. The same relationship exists between the dry mouth of incipient attack and the slobbering mouth of the more intensely inhibited assault. Urination, defecation and fainting usually arrive a little later on the scene, following in the wake of the massive shock-wave that accompanies moments of immense tension.

When the urge to attack and escape are both strongly activated simultaneously, we exhibit a number of characteristic intention movements and ambivalent posturings. The most familiar of these is the raising of a clenched fist – a gesture that has become ritualized in two ways. It is performed at some distance from the opponent, at a point where it is too far away to be carried through into a blow. Thus its function is no longer mechanical; instead it has become a visual signal. (With the arm bent and held sideways it has now become the defiant formalized gesture of communist regimes.) It has become further ritualized by the addition of back-and-forth striking movements of the forearm. Fist-shaking of this kind is again visual rather than mechanical in its impact. We perform rhythmically repeated 'blows' with the fist, but still at a safe distance.

While doing this, the whole body may make short approach-intention movements, actions which repeatedly check themselves from going too far. The feet may be stamped forcibly and loudly and the fist brought down and thumped on any near-by object. This last action is an example of something seen frequently in other animals, where it is referred to as a redirection activity. What happens is that, because the object (the opponent) stimulating the attack is too frightening to be directly assaulted, the aggressive movements are released, but have to be re-directed towards some other, less intimidating object, such as a harmless bystander (we have all suffered from this at one time or another), or even an inanimate object. If the latter is used it may be viciously pulverized or destroyed. When a wife smashes a vase to the floor, it is, of course, really her husband's head that lies there broken into small pieces. It is interesting that chimpanzees and gorillas frequently perform their own versions of this display, when they tear up, smash, and throw around branches and vegetation. Again, it has a powerful visual impact.

A specialized and important accompaniment to all these aggressive displays is the making of threatening facial expressions. These, along with our verbalized vocal signals, provide our most precise method of communicating our exact aggressive mood. Although our smiling face, discussed in an earlier chapter, is unique to our species, our aggressive faces, expressive though they may be, are much the same as those of all the other higher primates. (We can tell a fierce monkey or a scared monkey at a glance, but we have to learn the friendly monkey face.) The rules are quite simple: the more the urge to attack dominates the urge to flee, the more the face pulls itself forwards. When the reverse is the case and fear gets the upper hand, then all the facial details are pulled back. In the attack face, the eyebrows are brought forward in a frown, the forehead is smooth, the mouth-corners are held forward, and the lips make a tight, pursed line. As fear comes to dominate the mood, a scared-threat face appears. The eyebrows are raised, the forehead wrinkles, the mouth-corners are pulled back and the lips part, exposing the teeth. This face often accompanies other gestures that appear to be very aggressive, and such things as forehead-wrinkling and teeth-baring are sometimes thought of as 'fierce' signals because of this. But in fact they are fear signs, the face providing an early-warning signal that fear is very much present, despite the persistence of intimidating gestures by the rest of the body. It is still, of course, a threatening face and cannot be treated smugly. If full fear were being expressed, the face-pulling would be abandoned and the opponent would be retreating.

All this face-making we share with the monkeys, a fact that is worth remembering if ever you come face to face with a large baboon, but there are other faces that we have invented culturally, such as sticking out the tongue, puffing out the cheeks, thumbing the nose, and exaggeratedly screwing up the features, that add considerably to our threat repertoire. Most cultures have also added a variety of threatening or insulting gestures employing the rest of the body. Aggressive intention movements ('hopping mad') have been elaborated into violent war-dances of many different and highly stylized kinds. The function here has become communal arousal and synchronization of strong aggressive feelings, rather than direct visual display to the enemy.

Because, with the cultural development of lethal artificial weapons, we have become such a potentially dangerous species, it is not surprising to find that we have an extraordinarily wide range of appeasement signals. We share with the other primates the basic submissive response of crouching and screaming. In addition we have formalized a whole variety of subordinating displays. Crouching itself has become extended into grovelling and prostrating. Minor intensities of it are expressed in the form of kneeling, bowing and curtsying. The key signal here is the lowering of the body in relation to the dominant individual. When threatening, we puff ourselves up to our greatest height, making our bodies as tall and as large as possible. Submissive behaviour must therefore take the opposite course and bring the body down as far as possible. Instead of doing this in a random way, we have stylized it at a number of characteristic, fixed stages, each with its own special signal meaning. The act of saluting is interesting in this context, because it shows how far from the original gesture formalization can carry our cultural signs. At first sight a military salute looks like an aggressive movement. It is similar to the signal version of raising-an-arm-to-strike-a-blow. The vital difference is that the hand is not clenched and it points towards the cap or hat. It is, of course, a stylized modification of the act of removing the hat, which itself was originally part of the procedure of lowering the height of the body.

The distillation of the bowing movement from the original, crude, primate crouch is also interesting. The key feature here is the lowering of the eyes. A direct stare is typical of the most out-and-out aggression. It is part of the fiercest facial expressions and accompanies all the most belligerent gestures. (This is why the children's game of 'stare you out' is so difficult to perform and why the simple curiosity staring of a young child – 'It's rude to stare' – is so condemned.) No matter how reduced in extent the bow becomes by social custom, it always retains the face-lowering element. Male members of a royal court, for example, who, through constant repetition, have modified their bowing reactions, still lower the face, but instead of bending from the waist they now bow stiffly from the neck, lowering only the head region.

On less formal occasions the anti-stare response is given by simple looking-away movements, or a 'shifty-eyed' expression. Only a truly aggressive individual can fix you in the eye for any length of time. During ordinary face-to-face conversations we typically look away from our companions when we are talking, then glance back at them at the end of each sentence, or 'paragraph', to check their response to what we have said. A professional lecturer takes some time to train himself to look directly at the members of his audience, instead of over their heads, down at his rostrum, or out towards the side or back of the hall. Even though he is in such a dominant position, there are so many of them, all staring (from the safety of their seats) at him, that he experiences a basic and initially uncontrollable fear of them. Only after a great deal of practice can he overcome this. The simple, aggressive, physical act of being stared at by a large group of people is also the cause of the fluttering 'butterflies' in the actor's stomach before he makes his entrance on to the stage. He has all his intellectual worries about the qualities of his performance and its reception, of course, but the massed threat-stare is an additional and more fundamental hazard for him. (This is again a case of the curiosity stare being confused at an unconscious level with the threat-stare.) The wearing of spectacles and sunglasses makes the face appear more aggressive because it artificially and accidentally enlarges the pattern of the stare. If we are looked at by someone wearing glasses, we are being given a super-stare. Mild-mannered individuals tend to select thin-rimmed or rimless spectacles (probably without realizing why they do so), because this enables them to see better with the minimum of stare exaggeration. In this way they avoid arousing counter-aggression.

A more intense form of anti-stare is covering the eyes with the hands, or burying the face in the crook of the elbow. The simple act of closing the eyes also cuts off the stare, and it is intriguing that certain individuals compulsively and repeatedly shut their eyes briefly whilst facing and talking to strangers. It is as though their normal blinking responses have become lengthened into extended eye-masking moments. The response vanishes when they are conversing with close friends in a situation where they feel at ease. Whether they are trying to shut off the 'threatening' presence of the stranger, or whether they are attempting to reduce their staring rate, or both, is not always clear.

Because of their powerful intimidating affect, many species have evolved staring eye-spots as self-defence mechanisms. Many moths have a pair of startling eye-markings on their wings. These lie concealed until the creatures are attacked by predators. The wings then open and flash the bright eye-spots in the face of the enemy. It has been proved experimentally that this exerts a valuable intimidating influence on the would-be killers, who frequently flee and leave the insects unmolested. Many fish and some species of birds and even mammals have adopted this technique. In our own species, commercial products have sometimes used the same device (perhaps knowingly, perhaps not). Motor-car designers employ headlamps in this way and frequently add to the overall aggressive impression by sculpturing the line of the front of the bonnet into the shape of a frown. In addition they add 'bared teeth' in the form of a metal grille between the 'eye-spots'. As the roads have become increasingly crowded and driving an increasingly belligerent activity, the threat-faces of cars have become progressively improved and refined, imparting to their drivers a more and more aggressive image. On a smaller scale certain products have been given threat-face brand names, such as OXO, OMO, OZO, and OVO. Fortunately for the manufacturers, these do not repel customers: on the contrary, they catch the eye and, having caught it, reveal themselves to be no more than harmless cardboard boxes. But the impact has already worked, the attention has already been drawn to *that* product rather than to its rivals.

I mentioned earlier that chimpanzees appease by holding out a limp hand towards the dominant individual. We share this gesture with them, in the form of the typical begging or imploring posture. We have also adapted it as a widespread greeting gesture in the shape of the friendly handshake. Friendly gestures often grow out of submissive ones. We saw earlier how this happened with the smiling and laughing responses (both of which, incidentally, still appear in appeasing situations as the timid smile and the nervous titter). Handshaking occurs as a mutual ceremony between individuals of more or less equal rank, but is transformed into bowing to kiss the held hand when there is strong inequality between the ranks. (With increasing 'equality' between the sexes and the various classes, this latter refinement is now becoming rarer, but still persists in certain specialized spheres where formal dominance hierarchies are rigidly adhered to, as in the case of the Church.) In certain instances handshaking has become modified into self-shaking or hand-wringing. In some cultures this is the standard greeting appeasement, in others it is performed only in more extreme 'imploring' contexts.

There are many other cultural specialities in the realm of submissive behaviour, such as throwing in the towel or showing the white flag, but these need not concern us here. One or two of the simpler re-motivating devices do, however, deserve a mention, if only because they bear an interesting relationship to similar patterns in other species. You will recall that certain juvenile, sexual or grooming patterns were performed towards aggressive or potentially aggressive individuals as a method of arousing non-aggressive feelings that competed with the more violent ones and suppressed them. In our own species, juvenile behaviour on the part of submissive adults is particularly common during courtship. The courting pair often adopt 'baby-talk', not because they are heading towards parentalism themselves, but because it arouses tender, protective maternal or paternal feelings in the partner and thereby suppresses more aggressive feelings (or, for that matter, more fearful ones). It is amusing, when thinking back to the development of this pattern into courtship-feeding in birds, to notice the extraordinary increase in mutual feeding that goes on in our own courtship phase. At no other time in our lives do we devote so much effort to popping tasty morsels into one another's mouths, or offering one another boxes of chocolates.

As regards re-motivation in a sexual direction, this occurs wherever a subordinate (male or female) adopts a generalized attitude of 'femininity' towards a dominant individual (male or female) in an aggressive rather than a truly sexual context. This is widespread, but the more specific case of the adoption of the female sexual rump-presentation posture as an appeasement gesture has virtually vanished, along with the disappearance of the original sexual posture itself. It is largely confined now to a form of schoolboy punishment, with rhythmic whipping replacing the rhythmic pelvic thrusts of the dominant male. It is doubtful whether schoolmasters would persist in this practice if they fully appreciated the fact that, in reality, they were performing an ancient primate form of ritual copulation with their pupils. They could just as well inflict pain on their victims without forcing them to adopt the bent-over submissive female posture. (It is significant that schoolgirls are rarely, if ever, beaten in this way – the sexual origins of the act would then become too obvious.) It has been imaginatively suggested by one authority that the reason for sometimes forcing schoolboys to lower their trousers for the administration of the punishment is not related to increasing the pain, but rather to enabling the dominant male to witness the reddening of the buttocks as the beating proceeds, which so vividly recalls the flushing of the primate female hindquarters when in full sexual condition. Whether this is so or not, one thing is certain about this extraordinary ritual, namely that as a re-motivating appeasement device it is a dismal failure. The more the unfortunate schoolboy stimulates the dominant male crypto-sexually, the more likely he is to perpetuate the ritual and, because the rhythmic pelvic thrusts have become symbolically modified into rhythmic blows of the cane, the victim is right back where he started. He has managed to switch a direct attack into a sexual one, but has then been double-crossed by the symbolic conversion of this sexual one back into another aggressive pattern.

The third re-motivating device, that of grooming, plays a minor but useful role in our species. We frequently employ stroking and patting movements to soothe an agitated individual, and many of the more dominant members of society spend long hours having themselves groomed and fussed over by subordinates. But we shall return to this subject in another chapter.

Displacement activities also play a part in our aggressive encounters, appearing in almost any situation of stress or tension. We differ from other animals, however, in that we do not restrict ourselves to a few species-typical displacement patterns. We make use of virtually any trivial actions as outlets for our pent-up feelings. In an agitated state of conflict we may rearrange ornaments, light a cigarette, clean our spectacles, glance at a wrist-watch, pour a drink, or nibble a piece of food. Any of these actions may, of course, be performed for normal functional reasons, but in their displacement activity roles they no longer serve these functions. The ornaments that are rearranged were already adequately displayed. They were not in a muddle and may, indeed, be in a worse state after their agitated rearrangement. The cigarette that is lit in a tense moment may be started when a perfectly good and unfinished one has just been nervously stubbed out. Also, the rate of smoking during tension bears no relation to the physiological addictive nicotine demands of the system. The spectacles that are so laboriously polished are already clean. The watch that is wound up so vigorously does not need winding, and when we glance at it our eyes do not even register what time it tells. When we sip a displacement drink it is not because we are thirsty. When we nibble displacement food it is not because we are hungry. All these actions are performed, not for the normal rewards they bring, but simply for the sake of doing something in an attempt to relieve the tension. They occur with particularly high frequency during the initial stages of social encounters, where hidden fears and aggressions are lurking just below the surface. At a dinner party, or any small social gathering, as soon as the mutual appeasement ceremonies of handshaking and smiling are over, displacement cigarettes, displacement drinks and displacement food-snacks are immediately offered. Even at large-scale entertainments such as the theatre and cinema the flow of events is deliberately broken up by short intervals when the audience can indulge in brief bouts of their favourite displacement activities.

When we are in more intense moments of aggressive tension, we tend to revert to displacement activities of a kind that we share with other primate species, and our outlets become more primitive. A chimpanzee in such a situation can be seen to perform repeated and agitated scratching movements, which are of a rather special kind and different from the normal response to an itch. It is confined largely to the head region, or sometimes the arms. The movements themselves are rather stylized. We behave in much the same way, performing stilted displacement grooming actions. We scratch our heads, bite our nails, 'wash' our faces with our hands, tug at our beards or moustaches if we have them, or adjust our coiffure, rub, pick, sniff or blow our noses, stroke our ear-lobes, clean our ear-passages, rub our chins, lick our lips, or rub our hands together in a rinsing action. If moments of great conflict are studied carefully, it can be observed that these activities are all carried out in a ritual fashion without the careful localized adjustments of the true cleaning actions. The displacement head-scratch of one individual may differ markedly from its equivalent in another, but each scratcher develops his own rather fixed and characteristic way of doing it. As real cleaning is not involved, it is of little importance that one region gets all the attention while others are ignored. In any social interaction between a small group of individuals the subordinate members of the group can easily be identified by the higher frequency of their displacement self-grooming activities. The truly dominant individual can be recognized by the almost complete absence of such actions. If the ostensibly dominant member of the group does, in fact, perform a larger number of small displacement activities, then this means that his official dominance is being threatened in some way by the other individuals present.

In discussing all these aggressive and submissive behaviour patterns, it has been assumed that the individuals concerned have been 'telling the truth' and have not been consciously and deliberately modifying their actions to achieve special ends. We 'lie' more with our words than our other communication signals, but even so the phenomenon cannot be overlooked entirely. It is extremely difficult to 'utter' untruths with the kind of behaviour patterns we have been discussing, but not impossible. As I have already mentioned, when parents adopt this procedure towards their young children, it usually fails much more drastically than they realize. Between adults, however, who are much more preoccupied with the verbalized information content of the social interactions, it can be more successful. Unfortunately for the behaviour-liar, he typically lies only with certain selected elements of his total signalling system. Others, which he is not aware of, give the game away. The most successful behaviour-liars are those who, instead of consciously concentrating on modifying specific signals, think themselves into the basic mood they wish to convey and then let the small details take care of themselves. This method is frequently used with great success by professional liars, such as actors and actresses. Their entire working lives are spent performing behavioural lies, a process which can sometimes be extremely damaging to their private lives. Politicians and diplomats are also required to perform an undue amount of behavioural lying, but unlike the actors they are not socially 'licensed to lie', and the resultant guilt feelings tend to interfere with their performances. Also, unlike the actors, they do not undergo prolonged training courses.

Even without professional training, it is possible, with a little effort, and a careful study of the facts presented in this book, to achieve the desired effect. I have deliberately tested this out on one or two occasions, with some degree of success, when dealing with the police. I have reasoned that if there is a strong biological tendency to be appeased by submissive gestures, then this predisposition should be open to manipulation if the proper signals are used. Most drivers, when caught by the police for some minor motoring offence, immediately respond by arguing their innocence, or making excuses of some sort for their behaviour. In doing this they are defending their (mobile) territory and are setting themselves up as territorial rivals. This is the worst possible course of action. It forces the police to counter-attack. If, instead, an attitude of abject submission is adopted, it will become increasingly difficult for the police officer to avoid a sensation of appeasement. A total admission of guilt based on sheer stupidity and inferiority puts the policeman into a position of immediate dominance from which it is difficult for him to attack. Gratitude and admiration must be expressed for the efficiency of his action in stopping you. But words are not enough. The appropriate postures and gestures must be added. Fear and submission in both body posture and facial expression must be clearly demonstrated. Above all, it is essential to get quickly out of the car and move away from it towards the policeman. He must not be allowed to approach you, or you have forced him to go out of his way and thereby threatened him. Furthermore, by staying in the car you are remaining in your own territory. By moving away from it you are automatically weakening your territorial status. In addition to this, the sitting posture inside the car is an inherently dominant one. The power of the seated position is an unusual element in our behaviour. No one may sit if the 'king' is standing. When the 'king' rises, everyone rises. This is a special exception to the general rule about aggressive verticality, which states that increasing submissiveness goes with decreasing posture-height. By leaving the car you therefore shed both your territorial rights and your dominant seated position, and put yourself into a suitably weakened state for the submissive actions that follow. Having stood up, however, it is important not to brace the body erect, but to crouch, lower the head slightly and generally sag. The tone of voice is as important as the words used. Anxious facial expressions and looking-away movements are also valuable and a few displacement self-grooming activities can be added for good measure.

Unfortunately, as a driver of a car, one is in a basically aggressive mood of territorial defence, and it is extremely difficult to lie about this mood. It requires either considerable practice, or a good working knowledge of non-verbal behaviour signals. If you are a little short on personal dominance in your ordinary life, the experience, even when consciously and deliberately designed, may be too unpleasant, and it will be preferable to pay the fine.

Although this is a chapter about fighting behaviour, we have so far only dealt with methods of avoiding actual combat. When the situation does finally deteriorate into physical contact, the naked ape – unarmed – behaves in a way that contrasts interestingly with that seen in other primates. For them the teeth are the most important weapons, but for us it is the hands. Where they grab and bite, we grab and squeeze, or strike out with clenched fists. Only in infants or very young children does biting play a significant role in unarmed combat. They, of course, have not yet been able to develop their arm and hand muscles sufficiently to make a great impact with them.

We can witness adult unarmed combat today in a number of highly stylized versions, such as wrestling, judo and boxing, but in its original, unmodified form it is now rare. The moment that serious combat begins, artificial weapons of one sort or another are brought into play. In their crudest form, these are thrown or used as extensions of the fist for delivering heavy blows. Under special circumstances chimpanzees have been able to extend their attacks this far. In conditions of semi-captivity they have been observed to pick up a branch and slam it down hard on to the body of a stuffed leopard, or to tear up clods of earth and hurl them across a water ditch at passers-by. But there is little evidence that they use these methods to any extent in the wild state, and none at all that they use them on one another during disputes between rivals. Nevertheless, they give us a glimpse of the way we probably began, with artificial weapons being developed primarily as a means of defence against other species and for the killing of prey. Their use in intra-specific fighting was almost certainly a secondary trend, but once the weapons were there, they became available for dealing with any emergency, regardless of the context.

The simplest form of artificial weapon is a hard, solid, but unmodified, natural object of wood or stone. By simple improvements in the shapes of these objects, the crude actions of throwing and hitting became augmented with the addition of spearing, slashing, cutting and stabbing movements.

The next great behavioural trend in attacking methods was the extension of the distance between the attacker and his enemy, and it is this step that has nearly been our undoing. Spears can work at a distance, but their range is too limited. Arrows are better, but they lack accuracy. Guns widen the gap dramatically, but bombs dropped from the sky can be delivered at an even greater range, and ground-to-ground rockets can carry the attacker's 'blow' further still. The outcome of this is that the rivals, instead of being defeated, are indiscriminately destroyed. As I explained earlier, the proper business of intra-specific aggression at a biological level is the subduing and not the killing of the enemy. The final stages of destruction of life are avoided because the enemy either flees or submits. In both cases the aggressive encounter is then over: the dispute is settled. But the moment that attacking is done from such a distance that the appeasement signals of the losers cannot be read by the winners, then violent aggression is going to go raging on. It can only be consummated by a direct confrontation with abject submission, or the enemy's headlong flight. Neither of these can be witnessed in the remoteness of modern aggression, and the result is wholesale slaughter on a scale unheard of in any other species.

Aiding and abetting this mayhem is our specially evolved co-operativeness. When we improved this important trait in connection with hunting prey, it served us well, but it has now recoiled upon us. The strong urge towards mutual assistance to which it gave rise has become susceptible to powerful arousal in intra-specific aggressive contexts. Loyalty on the hunt has become loyalty in fighting, and war is born. Ironically, it is the evolution of a deep-seated urge to help our fellows that has been the main cause of all the major horrors of war. It is this that has driven us on and given us our lethal gangs, mobs, hordes and armies. Without it they would lack cohesion and aggression would once again become 'personalized'.

It has been suggested that because we evolved as specialized prey-killers, we automatically became rival-killers, and that there is an inborn urge within us to murder our opponents. The evidence, as I have already explained, is against this. Defeat is what an animal wants, not murder; domination is the goal of aggression, not destruction, and basically we do not seem to differ from other species in this respect. There is no good reason why we should. What has happened, however, is that because of the vicious combination of attack remoteness and group co-operativeness, the original goal has become blurred for the individuals involved in the fighting. They attack now more to support their comrades than to dominate their enemies, and their inherent susceptibility to direct appeasement is given little or no chance to express itself. This unfortunate development may yet prove to be our undoing and lead to the rapid extinction of the species.

Not unnaturally, this dilemma has given rise to a great deal of displacement head-scratching. A favourite solution is massive mutual disarmament; but to be effective this would have to be carried to an almost impossible extreme, one that would ensure that all future fighting was carried out as close-contact combat where the automatic, direct appeasement signals could come into operation again. Another solution is to de-patriotize the members of the different social groups; but this would be working against a fundamental biological feature of our species. As fast as alliances could be forged in one direction, they would be broken in another. The natural tendency to form social in-groups could never be eradicated without a major genetical change in our make-up, and one which would automatically cause our complex social structure to disintegrate.

A third solution is to provide and promote harmless, symbolic substitutes for war; but if these really are harmless they will inevitably only go a very small way towards resolving the real problem. It is worth remembering here that this problem, at a biological level, is one of group territorial defence and, in view of the gross overcrowding of our species, also one of group territorial expansion. No amount of boisterous international football is going to solve this.

A fourth solution is the improvement of intellectual control over aggression. It is argued that, since our intelligence has got us into this mess, it is our intelligence that must get us out. Unhappily, where matters as basic as territorial defence are concerned, our higher brain centres are all too susceptible to the urgings of our lower ones. Intellectual control can help us just so far, but no further. In the last resort it is unreliable, and a single, unreasoned, emotional act can undo all the good it has achieved.

The only sound biological solution to the dilemma is massive depopulation, or a rapid spread of the species on to other planets, combined if possible with assistance from all four of the courses of action already mentioned. We already know that if our populations go on increasing at their present terrifying rate, uncontrollable aggressiveness will become dramatically increased. This has been proved conclusively with laboratory experiments. Gross over-crowding will produce social stresses and tensions that will shatter our community organizations long before it starves us to death. It will work directly against improvements in intellectual control and will savagely heighten the likelihood of emotional explosion. Such a development can be prevented only by a marked drop in the breeding rate. Unfortunately there are two serious snags here. As already explained, the family unit – which is still the basic unit of all our societies – is a rearing device. It has evolved into its present, advanced and complex state as a system for producing, protecting and maturing offspring. If this function is seriously curtailed or temporarily eliminated, the pair-bonding pattern will suffer, and this will bring its own brand of social chaos. If, on the other hand, a selective attempt is made to stem the breeding flood, with certain pairs permitted to breed freely and others prevented from doing so, then this will work against the essential co-operativeness of society.

What it amounts to, in simple numerical terms, is that if all adult members of the population form pairs and breed, they can only afford to produce two offspring per pair if the community is to be maintained at a steady level. Each individual will then, in effect, be replacing him- or herself. Allowing for the fact that a small percentage of the population already fails to mate and breed, and that there will always be a number of premature deaths from accidental injury or other causes, the average family size can, in fact, be slightly larger. Even so, this will put a heavier burden on the pair-bond mechanism. The lighter offspring-load will mean that greater efforts will have to be made in other directions to keep the pair-bonds tightly tied. But this is a much smaller hazard, in the long term, than the alternative of suffocating overcrowding.

To sum up then, the best solution for ensuring world peace is the widespread promotion of contraception or abortion. Abortion is a drastic measure and can involve serious emotional disturbance. Furthermore, once a zygote has been formed by the act of fertilization it constitutes a new individual member of society, and its destruction is, in effect, an act of aggression, which is the very pattern of behaviour that we are attempting to control. Contraception is obviously preferable, and any religious or other 'moralizing' factions that oppose it must face the fact that they are engaged in dangerous war-mongering.

Having brought up the question of religion, it is perhaps worthwhile taking a closer look at this strange pattern of animal behaviour, before going on to deal with other aspects of the aggressive activities of our species. It is not an easy subject to deal with, but as zoologists we must do our best to observe what actually happens rather than listen to what is supposed to be happening. If we do this, we are forced to the conclusion that, in a behavioural sense, religious activities consist of the coming together of large groups of people to perform repeated and prolonged submissive displays to appease a dominant individual. The dominant individual concerned takes many forms in different cultures, but always has the common factor of immense power. Sometimes it takes the shape of an animal from another species, or an idealized version of it. Sometimes it is pictured more as a wise and elderly member of our own species. Sometimes it becomes more abstract and is referred to as simply as 'the state', or in other such terms. The submissive responses to it may consist of closing the eyes, lowering the head, clasping the hands together in a begging gesture, kneeling, kissing the ground, or even extreme prostration, with the frequent accompaniment of wailing or chanting vocalizations. If these submissive actions are successful, the dominant individual is appeased. Because its powers are so great, the appeasement ceremonies have to be performed at regular and frequent intervals, to prevent its anger from rising again. The dominant individual is usually, but not always, referred to as a god.

Since none of these gods exist in a tangible form, why have they been invented? To find the answer to this we have to go right back to our ancestral origins. Before we evolved into co-operative hunters, we must have lived in social groups of the type seen today in other species of apes and monkeys. There, in typical cases, each group is dominated by a single male. He is the boss, the overlord, and every member of the group has to appease him or suffer the consequences. He is also most active in protecting the group from outside hazards and in settling squabbles between lesser members. The whole life of a member of such a group revolves around the dominant animal. His all-powerful role gives him a god-like status. Turning now to our immediate ancestors, it is clear that, with the growth of the co-operative spirit so vital for successful group hunting, the application of the dominant individual's authority had to be severely limited if he was to retain the active, as opposed to passive, loyalty of the other group members. They had to want to help him instead of simply fear him. He had to become more 'one of them'. The old-style monkey tyrant had to go, and in his place there arose a more tolerant, more co-operative naked ape leader. This step was essential for the new type of 'mutual-aid' organization that was evolving, but it gave rise to a problem. The total dominance of the Number I member of the group having been replaced by a qualified dominance, he could no longer command unquestioning allegiance. This change in the order of things, vital as it was to the new social system, nevertheless left a gap. From our ancient background there remained a need for an all-powerful figure who could keep the group under control, and the vacancy was filled by the invention of a god. The influence of the invented god-figure could then operate as a force additional to the now more restricted influence of the group leader.

At first sight, it is surprising that religion has been so successful, but its extreme potency is simply a measure of the strength of our fundamental biological tendency, inherited directly from our monkey and ape ancestors, to submit ourselves to an all-powerful, dominant member of the group. Because of this, religion has proved immensely valuable as a device for aiding social cohesion, and it is doubtful whether our species could have progressed far without it, given the unique combination of circumstances of our evolutionary origins. It has led to a number of bizarre by-products, such as a belief in 'another life' where we will at last meet up with the god figures. They were, for reasons already explained, unavoidably detained from joining us in the present life, but this omission can be corrected in an after-life. In order to facilitate this, all kinds of strange practices have been developed in connection with the disposal of our bodies when we die. If we are going to join our dominant overlords, we must be well prepared for the occasion and elaborate burial ceremonies must be performed.

Religion has also given rise to a great deal of unnecessary suffering and misery, wherever it has become over-formalized in its application, and whenever the professional 'assistants' of the god figures have been unable to resist the temptation to borrow a little of his power and use it themselves. But despite its chequered history it is a feature of our social life that we cannot do without. Whenever it becomes unacceptable, it is quietly, or sometimes violently, rejected, but in no time at all it is back again in a new form, carefully disguised perhaps, but containing all the same old basic elements. We simply have to 'believe in something'. Only a common belief will cement us together and keep us under control. It could be argued that, on this basis, any belief will do, so long as it is powerful enough; but this is not strictly true. It must be impressive and it must be seen to be impressive. Our communal nature demands the performance of and participation in elaborate group ritual. Elimination of the 'pomp and circumstance' will leave a terrible cultural gap and the indoctrination will fail to operate properly at the deep, emotional level so vital to it. Also, certain types of belief are more wasteful and stultifying than others and can side-track a community into rigidifying patterns of behaviour that hamper its qualitative development. As a species we are a predominantly intelligent and exploratory animal, and beliefs harnessed to this fact will be the most beneficial for us. A belief in the validity of the acquisition of knowledge and a scientific understanding of the world we live in, the creation and appreciation of aesthetic phenomena in all their many forms, and the broadening and deepening of our range of experiences in day-to-day living, is rapidly becoming the 'religion' of our time. Experience and understanding are our rather abstract god-figures, and ignorance and stupidity will make them angry. Our schools and universities are our religious training centres, our libraries, museums, art galleries, theatres, concert halls and sports arenas are our places of communal worship. At home we worship with our books, newspapers, magazines, radios and television sets. In a sense, we still believe in an after-life, because part of the reward obtained from our creative works is the feeling that, through them, we will 'live on' after we are dead. Like all religions, this one has its dangers, but if we have to have one, and it seems that we do, then it certainly appears to be the one most suitable for the unique biological qualities of our species. Its adoption by an ever-growing majority of the world population can serve as a compensating and reassuring source of optimism to set against the pessimism expressed earlier concerning our immediate future as a surviving species.

Before we embarked on this religious discourse, we had been examining the nature of only one aspect of the organization of aggressiveness in our species, namely the group defence of a territory. But as I explained at the beginning of this chapter, the naked ape is an animal with three distinct social forms of aggression, and we must now consider the other two. They are the territorial defence of the family-unit within the larger group-unit, and the personal, individual maintenance of hierarchy positions.

The spatial defence of the home site of the family unit has remained with us through all our massive architectural advances. Even our largest buildings, when designed as living-quarters, are assiduously divided into repetitive units, one per family. There has been little or no architectural 'division of labour'. Even the introduction of communal eating or drinking buildings, such as restaurants and bars, has not eliminated the inclusion of dining-rooms in the family-unit quarters. Despite all the other advances, the design of our cities and towns is still dominated by our ancient, naked-ape need to divide our groups up into small, discrete, family territories. Where houses have not yet been squashed up into blocks of flats, the defended area is carefully fenced, walled, or hedged off from its neighbours, and the demarcation lines are rigidly respected and adhered to, as in other territorial species.

One of the important features of the family territory is that it must be easily distinguished in some way from all the others. Its separate location gives it a uniqueness, of course, but this is not enough. Its shape and general appearance must make it stand out as an easily identifiable entity, so that it can become the 'personalized' property of the family that lives there. This is something which seems obvious enough, but which has frequently been overlooked or ignored, either as a result of economic pressures, or the lack of biological awareness of architects. Endless rows of uniformly repeated, identical houses have been erected in cities and towns all over the world. In the case of blocks of flats the situation is even more acute. The psychological damage done to the territorialism of the families forced by architects, planners and builders to live under these conditions is incalculable. Fortunately, the families concerned can impose territorial uniqueness on their dwellings in other ways. The buildings themselves can be painted different colours. The gardens, where there are any, can be planted and landscaped in individual styles. The insides of the houses or flats can be decorated and filled with ornaments, bric-à-brac and personal belongings in profusion. This is usually explained as being done to make the place 'look nice'. In fact, it is the exact equivalent to another territorial species depositing its personal scent on a landmark near its den. When you put a name on a door, or hang a painting on a wall, you are, in dog or wolf terms, for example, simply cocking your leg on them and leaving your personal mark there. Obsessive 'collecting' of specialized categories of objects occurs in certain individuals who, for some reason, experience an abnormally strong need to define their home territories in this way.

Bearing this in mind, it is amusing to note the large number of cars that contain small mascots and other personal identification symbols, or to watch the business executive moving into a new office and immediately setting out on his desk his favourite personal pen-tray, paper-weight and perhaps a photograph of his wife. The motor-car and the business office are sub-territories, offshoots of his home base, and it is a great relief to be able to cock his leg on these as well, making them into more familiar, 'owned' spaces.

This leaves us with the question of aggression in relation to the social dominance hierarchy. The individual, as opposed to the places he frequents, must also be defended. His social status must be maintained and if possible improved, but it must be done cautiously, or he will jeopardize his co-operative contacts. This is where all the subtle aggressive and submissive signalling described earlier comes into play. Group co-operativeness demands and gets a high degree of conformity in both dress and behaviour, but within the bounds of this conformity there is still great scope for hierarchy competitiveness. Because of these conflicting demands it reaches almost incredible degrees of subtlety. The exact form of the knotting of a tie, the precise arrangement of the exposed section of a breast-pocket handkerchief, minute distinctions in vocal accent, and other such seemingly trivial characteristics, take on a vital social significance in determining the social standing of the individual. An experienced member of society can read them off at a glance. He would be totally at a loss to do so if suddenly jettisoned into the social hierarchy of New Guinea tribesmen, but in his own culture he is rapidly forced to become an expert. In themselves these tiny differences of dress and habit are utterly meaningless, but in relation to the game of juggling for position and holding it in the dominance hierarchy they are all-important.

We did not evolve, of course, to live in huge conglomerations of thousands of individuals. Our behaviour is designed to operate in small tribal groups probably numbering well under a hundred individuals. In such situations every member of the tribe will be known personally to every other member, as is the case with other species of apes and monkeys today. In this type of social organization it is easy enough for the dominance hierarchy to work itself out and become stabilized, with only gradual changes as members become older and die. In a massive city community the situation is much more stressful. Every day exposes the urbanite to sudden contacts with countless strangers, a situation unheard-of in any other primate species. It is impossible to enter into personal hierarchy relationships with all of them, although this would be the natural tendency. Instead they are allowed to go scurrying by, undominated and undominating. In order to facilitate this lack of social contact, anti-touching behaviour patterns develop. This has already been mentioned when dealing with sexual behaviour, where one sex accidentally touches another, but it applies to more than simply the avoidance of sexual behaviour. It covers the whole range of social-relationship initiation. By carefully avoiding staring at one another, gesturing in one another's direction, signalling in any way, or making physical bodily contact, we manage to survive in an otherwise impossibly over-stimulating social situation. If the no-touching rule is broken, we immediately apologize to make it clear that it was purely accidental.

Anti-contact behaviour enables us to keep our number of acquaintances down to the correct level for our species. We do this with remarkable consistency and uniformity. If you require confirmation, take the address or phone books of a hundred widely different types of city-dwellers and count up the number of personal acquaintances listed there. You will find that nearly all of them know well about the same number of individuals, and that this number approximates to what we would think of as a small tribal group. In other words, even in our social encounters we are obeying the basic biological rules of our ancient ancestors.

There will of course be exceptions to this rule – individuals who are professionally concerned with making large numbers of personal contacts, people with behaviour defects that make them abnormally shy or lonely, or people whose special psychological problems render them unable to obtain the expected social rewards from their friends and who try to compensate for this by frantic 'socializing' in all directions. But these types account for only a small proportion of the town and city populations. All the rest happily go about their business in what seems to be a great seething mass of bodies, but which is in reality an incredibly complicated series of interlocking and overlapping tribal groups. How little, how very little, the naked ape has changed since his early, primitive days.

6

FEEDING

THE FEEDING BEHAVIOUR of the naked ape appears at first sight to be one of his most variable, opportunistic, and culturally susceptible activities, but even here there are a number of basic biological principles at work. We have already taken a close look at the way his ancestral fruit-picking patterns had to become modified into co-operative prey-killing. We have seen how this led to a number of fundamental changes in his feeding routine. Food-seeking had to become more elaborate and carefully organized. The urge to kill prey had to become partially independent of the urge to eat. Food was taken to a fixed home base for consumption. Greater food preparation had to be carried out. Meals became larger and more spaced out in time. The meat component of the diet became dramatically increased. Food storage and food sharing was practised. Food had to be provided by the males for their family units. Defecation activities had to be controlled and modified.

All these changes were taking place over a very long period of time, and it is significant that, despite the great technological advances of recent years, we are still faithful to them. It would seem that they are rather more than mere cultural devices, to be buffeted this way and that by the whims of fashion. Judging by our present-day behaviour, they must, to some extent at any rate, have become deep-seated biological characteristics of our species.

As we have already noted, the improved food-collecting techniques of modern agriculture have left the majority of the adult males in our societies without a hunting role. They compensate for this by going out to 'work'. Working has replaced hunting, but has retained many of its basic characteristics. It involves a regular trip from the home base to the 'hunting' grounds. It is a predominantly masculine pursuit, and provides opportunities for male-to-male interaction and group activity. It involves taking risks and planning strategies. The pseudo-hunter speaks of 'making a killing in the City'. He becomes ruthless in his dealings. He is said to be 'bringing home the bacon'.

When the pseudo-hunter is relaxing he goes to all-male 'clubs', from which the females are completely excluded. Younger males tend to form into all-male gangs, often 'predatory' in nature. Throughout the whole range of these organizations, from learned societies, social clubs, fraternities, trade unions, sports clubs, masonic groups, secret societies, right down to teenage gangs, there is a strong emotional feeling of male 'togetherness'. Powerful group loyalties are involved. Badges, uniforms and other identification labels are worn. Initiation ceremonies are invariably carried out with new members. The unisexuality of these groupings must not be confused with homosexuality: They have basically nothing to do with sex. They are all primarily concerned with the male-to-male bond of the ancient co-operative hunting group. The important role they play in the lives of the adult males reveals the persistence of the basic, ancestral urges. If this were not so, the activities they promote could just as well be carried on without the elaborate segregation and ritual, and much of it could be done within the sphere of the family units. Females frequently resent the departure of their males to 'join the boys', reacting to it as though it signified some kind of family disloyalty. But they are wrong to do so. All they are witnessing is the modern expression of the age-old male-grouping hunting tendency of the species. It is just as basic as the male-female bonding of the naked ape and, indeed, evolved in close conjunction with it. It will always be with us, at least until there has been some new and major genetic change in our make-up.

Although working has largely replaced hunting today, it has not completely eliminated the more primitive forms of expression of this basic urge. Even where there is no economic excuse for participating in the pursuit of animal prey, this activity still persists in a variety of forms. Big-game hunting, stag-hunting, fox-hunting, coursing, falconry, wild-fowling, angling and the hunting-play of children are all contemporary manifestations of the ancient hunting urge.

It has been argued that the true motivation behind these present-day activities has more to do with the defeating of rivals than the hunting down of prey; that the desperate creature at bay represents the most hated member of our own species, the one we would so like to see in the same situation. There is undoubtedly an element of truth in this, at least for some individuals, but when these patterns of activity are viewed as a whole it is clear that it can provide only a partial explanation. The essence of 'sport-hunting' is that the prey should be given a fair chance of escaping. (If the prey is merely a substitute for a hated rival, then why give him any chance at all?) The whole procedure of sport-hunting involves a deliberately contrived inefficiency, a self-imposed handicap, on the part of the hunters. They could easily use machine-guns, or more deadly weapons, but that would not be 'playing the game' – the hunting game. It is the challenge that counts, the complexities of the chase and the subtle manœuvres that provide the rewards.

One of the essential features of the hunt is that it is a tremendous gamble and so it is not surprising that gambling, in the many stylized forms it takes today, should have such a strong appeal for us. Like primitive hunting and sport-hunting, it is predominantly a male pursuit and, like them, it is surrounded by seriously observed social rules and rituals.

An examination of our class structure reveals that both sport-hunting and gambling are more the concern of the lower and upper social classes than of the middle classes, and there is a very good reason for this if we accept them as expressions of a basic hunting drive. I pointed out earlier that work has become the major substitute for primitive hunting, but as such it has most benefited the middle classes. For the average lower-class male, the nature of the work he is required to do is poorly suited to the demands of the hunting drive. It is too repetitive, too predictable. It lacks the elements of challenge, luck and risk so essential to the hunting male. For this reason, lower-class males share with the (non-working) upper-class males a greater need to express their hunting urges than do the middle classes, the nature of whose work is much more suited to its role as a hunting substitute.

Leaving hunting and turning now to the next act in the general feeding pattern, we come to the moment of the kill. This element can find a certain degree of expression in the substitute activities of work, sport-hunting and gambling. In sport-hunting the action of killing still occurs in its original form, but in working and gambling contexts it is transformed into moments of symbolic triumph that lack the violence of the physical act. The urge to kill prey is therefore considerably modified in our present-day way of life. It keeps reappearing with startling regularity in the playful (and not so playful) activities of young boys, but in the adult world it is subjected to powerful cultural suppression.

Two exceptions to this suppression are (to some extent) condoned. One is the sport-hunting already mentioned, and the other is the spectacle of bullfighting. Although vast numbers of domesticated animals are slaughtered daily, their killing is normally concealed from the public gaze. With bullfighting the reverse is the case, huge crowds gathering to watch and experience by proxy the acts of violent prey-killing.

Within the formal limits of blood-sports these activities are permitted to continue, but not without protest. Outside these spheres, all forms of cruelty to animals are forbidden and punished. This has not always been the case. A few hundred years ago the torture and killing of 'prey' was regularly staged as a public entertainment in Britain and many other countries. It has since been recognized that participation in violence of this kind is liable to blunt the sensitivities of the individuals concerned towards all forms of blood-letting. It therefore constitutes a potential source of danger in our complex and crowded societies, where territorial and dominance restrictions can build up to an almost unbearable degree, sometimes finding release in a flood of pent-up aggression of abnormal savagery.

We have so far been dealing with the earlier stages of the feeding sequence and their ramifications. After hunting and killing, we come to the meal itself. As typical primates we ought to find ourselves munching away on small, non-stop snacks. But we are not typical primates. Our carnivorous evolution has modified the whole system. The typical carnivore gorges itself on large meals, well spaced out in time, and we clearly fall in with this pattern. The tendency persists even long after the disappearance of the original hunting pressures that demanded it. Today it would be quite easy for us to revert to our old primate ways if we had the inclination to do so. Despite this, we stick to well-defined feeding times, just as though we were still engaged in active prey-hunting. Few, if any, of the millions of naked apes alive today indulge in the typical, scattered feeding routine of the other primates. Even in conditions of plenty, we rarely eat more than three, or at the very most, four times a day. For many people, the pattern involves only one or two large daily meals. It could be argued that this is merely a case of cultural convenience, but there is little evidence to support this. It would be perfectly possible, given the complex organization of food supplies that we now enjoy, to devise an efficient system whereby all food was taken in small snacks, scattered throughout the day. Spreading feeding out in this way could be achieved without any undue loss of efficiency once the cultural pattern became adjusted to it, and it would eliminate the need for the major disruptions in other activities caused by the present 'main meal' system. But, because of our ancient predatory past, it would fail to satisfy our basic biological needs.

It is also relevant to consider the question of why we heat our food and eat it while it is still hot. There are three alternative explanations. One is that it helps to simulate 'prey temperature'. Although we no longer consume freshly killed meat, we nevertheless devour it at much the same temperature as other carnivore species. Their food is hot because it has not yet cooled down: ours is hot because we have re-heated it. Another interpretation is that we have such weak teeth that we are forced to 'tenderize' the meat by cooking it. But this does not explain why we should want to eat it while it is still hot, or why we should heat up many kinds of food that do not require 'tenderizing'. The third explanation is that, by increasing the temperature of the food, we improve its flavour. By adding a complicated range of tasty subsidiaries to the main food objects, we can take this process still further. This relates back, not to our adopted carnivory, but to our more ancient primate past. The foods of typical primates have a much wider variety of flavours than those of carnivores. When a carnivore has gone through its complex sequence of hunting, killing and preparing its food, it behaves much more simply and crudely at the actual crunch. It gobbles; it bolts its food down. Monkeys and apes, on the other hand, are extremely sensitive to the subtleties of varying tastiness in their food morsels. They relish them and keep on moving from one flavour to another. Perhaps, when we heat and spice our meals, we are harking back to this earlier primate fastidiousness. Perhaps this is one way in which we resisted the move towards full-blooded carnivory.

Having raised the question of flavour, there is a misunderstanding that should be cleared up concerning the way we receive these signals. How do we taste what we taste? The surface of the tongue is not smooth, but covered with small projections, called papillae, which carry the taste buds. We each possess approximately 10,000 of these taste buds, but in old age they deteriorate and decrease in number, hence the jaded palate of the elderly gastronome. Surprisingly, we can only respond to four basic tastes. They are: sour, salt, bitter and sweet. When a piece of food is placed on the tongue, we register the proportions of these four properties contained in it, and this blending gives the food its basic flavour. Different areas of the tongue react more strongly to one or other of the four tastes. The tip of the tongue is particularly responsive to salt and sweet, the sides of the tongue to sour and the back of the tongue to bitter. The tongue as a whole can also judge the texture and the temperature of the food, but beyond that it cannot go. All the more subtle and varied 'flavours' that we respond to so sensitively are not, in fact, tasted, but smelt. The odour of the food diffuses up into the nasal cavity, where the olfactory membrane is located. When we remark that a particular dish 'tastes' delicious, we are really saying that it tastes and smells delicious. Ironically, when we are suffering from a heavy head cold and our sense of smell is severely reduced, we say that our food is tasteless. In reality, we are tasting it as clearly as we ever did. It is its lack of odour that is worrying us.

Having made this point, there is one aspect of our true tasting that requires special comment, and that is our undeniably prevalent 'sweet-tooth'. This is something alien to the true carnivore, but typically primate-like. As the natural food of primates becomes riper and more suitable for consumption, it usually becomes sweeter, and monkeys and apes have a strong reaction to anything that is strongly endowed with this taste. Like other primates, we find it hard to resist 'sweets'.

Our ape ancestry expresses itself, despite our strong meat-eating tendency, in the seeking out of specially sweetened substances. We favour
this basic taste more than the others. We have 'sweet shops', but no
'sour shops'. Typically, when eating a full-scale meal, we *end* the often
complex sequence of flavours with some sweet substance, so that this
is the taste that lingers on afterwards. More significantly, when we
occasionally take small, inter-meal snacks (and thereby revert, to a
slight extent, to an ancient, primate scatter-feeding pattern), we nearly
always choose primate-sweet food objects, such as candy, chocolate,
ice-cream, or sugared drinks.

So powerful is this tendency that it can lead us into difficulties. The
point is that there are two elements in a food object that make it
attractive to us: its nutritive value and its palatability. In nature, these
two factors go hand in hand, but in artificially produced foodstuffs
they can be separated, and this can be dangerous. Food objects that are
nutritionally almost worthless can be made powerfully attractive simply by adding a large amount of artificial sweetener. If they appeal to
our old primate weakness by tasting 'super-sweet', we will lap them up
and so stuff ourselves with them that we have little room left for
anything else: thus the balance of our diet can be upset. This applies
especially in the case of growing children. In an earlier chapter I
mentioned recent research which has shown that the preference for
sweet and fruity odours falls off dramatically at puberty, when there is
a shift in favour of flowery, oily and musky odours. The juvenile
weakness for sweetness can be easily exploited, and frequently is.

Adults face another danger. Because their food is in general made so
tasty – so much more tasty than it would be in nature – its palatability
value rises sharply, and eating responses are over-stimulated. The result
is in many cases an unhealthily overweight condition. To counteract
this, all kinds of bizarre 'dieting' regimes are invented. The 'patients'
are told to eat this or that, cut down on this or on that, or to exercise
in various ways. Unfortunately there is only one true answer to the
problem: to eat less. It works like a charm, but since the subject
remains surrounded by super-palatability signals, it is difficult for him,
or her, to maintain this course of action for any length of time. The
overweight individual is also bedevilled by a further complication. I
mentioned earlier the phenomenon of 'displacement activities' – trivial, irrelevant actions performed as tension-relievers in moments of
stress. As we saw, a very frequent and common form of displacement
activity is 'displacement feeding'. In tense moments we nibble small
morsels of food or sip unneeded drinks. This may help to relax the
tension in us, but it also helps us to put on weight, especially as the
'trivial' nature of the displacement feeding action usually means that

we select for the purpose something sweet. If practised repeatedly over a long period, this leads to the well-known condition of 'fat anxiety', and we can witness the gradual emergence of the familiar, rounded contours of guilt-edged insecurity. For such a person, slimming routines will work only if accompanied by other behavioural changes that reduce the initial state of tension. The role of chewing-gum deserves a mention in this context. This substance appears to have developed exclusively as a displacement feeding device. It provides the necessary tension-relieving 'occupational' element, without contributing damagingly to the overall food intake.

Turning now to the variety of foodstuffs eaten by a present-day group of naked apes, we find that the range is extensive. By and large, primates tend to have a wider range of food objects in their diets than carnivores. The latter have become food specialists, whereas the former are opportunists. Careful field studies of a wild population of Japanese macaque monkeys, for example, have revealed that they consume as many as 119 species of plants, in the shape of buds, shoots, leaves, fruits, roots and barks, not to mention a wide variety of spiders, beetles, butterflies, ants and eggs. A typical carnivore's diet is more nutritious, but also much more monotonous.

When we became killers, we had the best of both worlds. We added meat with a high nutritive value to our diet, but we did not abandon our old primate omnivory. During recent times – that is, during the last few thousand years – food-obtaining techniques have improved considerably, but the basic position remains the same. As far as we can tell, the earliest agricultural systems were of a kind that can loosely be described as 'mixed farming'. The domestication of animals and plants advanced side by side. Even today, with our now immensely powerful dominance over our zoological and botanical environments, we still keep both strings to our bow. What has stopped us from swinging further in one direction or the other? The answer seems to be that, with vastly increasing population densities, an all-out reliance on meat would give rise to difficulties in terms of quantity, whereas an exclusive dependence on crops would be dangerous in terms of quality.

It could be argued that, since our primate ancestors had to make do without a major meat component in their diets, we should be able to do the same. We were driven to become flesh-eaters only by environmental circumstances, and now that we have the environment under control, with elaborately cultivated crops at our disposal, we might be expected to return to our ancient primate feeding patterns. In essence, this is the vegetarian (or, as one cult calls itself, fruitarian) creed, but it has had remarkably little success. The urge to eat meat appears to have become too deep-seated. Given the opportunity to devour flesh, we are loth to relinquish the pattern. In this connection, it is significant that vegetarians seldom explain their chosen diet simply by stating that they prefer it to any other. On the contrary, they construct an elaborate justification for it, involving all kinds of medical inaccuracies and philosophical inconsistencies.

Those individuals who are vegetarian by choice ensure a balanced diet by utilizing a wide variety of plant substances, like the typical primates. But for some communities a predominantly meatless diet has become a grim practical necessity rather than an ethical minority-preference. With advancing crop-cultivation techniques and the concentration on a very few staple cereals, a kind of low-grade efficiency has proliferated in certain cultures. The large-scale agricultural operations have permitted the growth of big populations, but their dependency on a few basic cereals has led to serious malnutrition. Such people may breed in large numbers, but they produce poor physical specimens. They survive, but only just. In the same way that abuse of culturally developed weapons can lead to aggressive disaster, abuse of culturally developed feeding techniques can lead to nutritional disaster. Societies that have lost the essential food balance in this way may be able to survive, but they will have to overcome the widespread ill-effects of deficiencies in proteins, minerals and vitamins if they are to progress and develop qualitatively. In all the healthiest and most go-ahead societies today, the meat-and-plant diet balance is well maintained and, despite the dramatic changes that have occurred in the methods of obtaining the nutritional supplies, the progressive naked ape of today is still feeding on much the same basic diet as his ancient hunting ancestors. Once again, the transformation is more apparent than real.

7

COMFORT

THE PLACE WHERE the environment comes into direct contact with an animal – its body surface – receives a great deal of rough treatment during the course of its life. It is astonishing that it survives the wear and tear and lasts so well. It manages to do so because of its wonderful system of tissue replacement and also because animals have evolved a variety of special comfort movements that help to keep it clean. We tend to think of these cleaning actions as comparatively trivial when considered alongside such patterns as feeding, fighting, fleeing and mating, but without them the body could not function efficiently. For some creatures, such as small birds, plumage maintenance is a matter of life and death. If the feathers are allowed to become bedraggled, the bird will be unable to take off fast enough to avoid its predators and will be unable to keep up its high body temperature if conditions become cold. Birds spend many hours bathing, preening, oiling and scratching themselves and carry out this performance in a long and complicated sequence. Mammals are slightly less complex in their comfort patterns, but nevertheless indulge in a great deal of grooming, licking, nibbling, scratching and rubbing. Like feathers, the hair has to be maintained in good order if it is to keep its owner warm. If it becomes clogged and dirty, it will also increase the risk of disease. Skin parasites have to be attacked and reduced in numbers as far as possible. Primates are no exception to this rule.

In the wild state, monkeys and apes can frequently be seen to groom themselves, systematically working through the fur, picking out small pieces of dried skin or foreign bodies. These are usually popped into the mouth and eaten, or at least tasted. These grooming actions may go on for many minutes, the animal giving an impression of great concentration. The grooming bouts may be interspersed with sudden scratchings or nibblings, directed at specific irritations. Most mammals only scratch with the back foot, but a monkey or ape can use either back or front. Its front limbs are ideally suited to the cleaning tasks. The nimble fingers can run through the fur and locate specific trouble spots with great accuracy. Compared with claws and hooves, the primate's hands are precision 'cleaners'. Even so, two hands are better than one, and this creates something of a problem. The monkey or ape can manage to bring both its hands into play when dealing with its legs, flanks, or front, but cannot really get to grips efficiently in this way with its back, or the arms themselves. Also, lacking a mirror, it cannot see what it is doing when it is concentrating on the head region. Here, it can use both hands, but it must work blind. Obviously, the head, back and arms are going to be less beautifully groomed than the front, sides and legs, unless something special can be done for them.

The solution is social grooming, the development of a friendly mutual-aid system. This can be seen in a wide range of both bird and mammal species, but it reaches a peak of expression amongst the higher primates. Special grooming invitation signals have been evolved here and social 'cosmetic' activities are prolonged and intense. When a groomer monkey approaches a groomee monkey, the former signals its intentions to the latter with a characteristic facial expression. It performs a rapid lip-smacking movement, often sticking its tongue out between each smack. The groomee can signal its acceptance of the groomer's approach by adopting a relaxed posture, perhaps offering a particular region of its body to be groomed. As I explained in an earlier chapter, the lip-smacking action has evolved as a special ritual out of the repeated particle-tasting movements that take place during a bout of fur-cleaning. By speeding them up and making them more exaggerated and rhythmic, it has been possible to convert them into a conspicuous and unmistakable visual signal.

Because social grooming is a co-operative, non-aggressive activity, the lip-smacking pattern has become a friendly signal. If two animals wish to tighten their bond of friendship, they can do so by repeatedly grooming one another, even if the condition of their fur hardly warrants it. Indeed, there seems to be little relationship today between the amount of dirt on the coat, and the amount of mutual grooming that takes place. Social grooming activities appear to have become almost independent of their original stimuli. Although they still have the vital task of keeping the fur clean, their motivation now appears to be more social than cosmetic. By enabling two animals to stay close together in a non-aggressive, co-operative mood they help to tie tighter the interpersonal bonds between the individuals in the troop or colony.

Out of this friendly signalling system have grown two remotivating devices, one concerned with appeasement and the other with reassurance. If a weak animal is frightened of a stronger one, it can pacify the latter by performing the lip-smacking invitation signal and then proceed to groom its fur. This reduces the aggression of the dominant animal and helps the subordinate one to become accepted. It is permitted to remain 'in the presence' because of services rendered. Conversely, if a dominant animal wishes to calm the fears of a weaker one, it can do so in the same way. By lip-smacking at it, it can underline the fact that it is not aggressive. Despite its dominant aura, it can show that it means no harm. This particular pattern – a reassurance display – is less often seen than the appeasement variety, simply because primate social life requires it less. There is seldom anything that a weak animal has which a dominant might want and could not take by a direct use of aggression. One exception to this can be seen when a dominant but childless female wants to approach and cuddle an infant belonging to another member of the troop. The young monkey is naturally rather frightened by the approach of the stranger and retreats. On such occasions it is possible to observe the large female attempting to reassure the tiny infant by making the lip-smacking face at it. If this calms the youngster's fears, the female can then fondle it and continues to calm it by gently grooming it.

Clearly, if we turn now to our own species, we might expect to see some manifestation of this basic primate grooming tendency, not only as a simple cleaning pattern, but also in a social context. The big difference, of course, is that we no longer have a luxuriant coat of fur to keep clean. When two naked apes meet and wish to reinforce their friendly relationship they must therefore find some kind of substitute for social grooming. If one studies those situations where, in another primate species, one would expect to see mutual grooming, it is intriguing to observe what happens. To start with it is obvious that smiling has replaced lip-smacking. Its origin as a special infantile signal has already been discussed and we have seen how, in the absence of the clinging response, it became necessary for the baby to have some way of attracting and pacifying the mother. Extended into adult life, the smile is clearly an excellent 'grooming-invitation' substitute. But, having invited friendly contact, what next? Somehow it has to be maintained. Lip-smacking is reinforced by grooming, but what reinforces smiling? True, the smiling response can be repeated and extended in time long after the initial contact, but something else is needed, something more 'occupational'. Some kind of activity, like grooming, has to be borrowed and converted. Simple observations reveal that the plundered source is verbalized vocalization.

The behaviour pattern of talking evolved originally out of the increased need for the co-operative exchange of information. It grew out of the common and widespread animal phenomenon of non-verbal mood vocalization. From the typical, inborn mammalian repertoire of grunts and squeals there developed a more complex series of learnt sound signals. These vocal units and their combinations and re-combinations became the basis of what we can call *information talking*. Unlike the more primitive non-verbal mood signals, this new method of communication enabled our ancestors to refer to objects in the environment and also to the past and the future as well as to the present. To this day, information talking has remained the most important form of vocal communication for our species. But, having evolved, it did not stop there. It acquired additional functions. One of these took the form of *mood talking*. Strictly speaking, this was unnecessary, because the non-verbal mood signals were not lost. We still can and do convey our emotional states by giving vent to ancient primate screams and grunts, but we augment these messages with verbal confirmation of our feelings. A yelp of pain is closely followed by a verbal signal that 'I am hurt'. A roar of anger is accompanied by the message 'I am furious'. Sometimes the non-verbal signal is not performed in its pure state but instead finds expression as a tone of voice. The words 'I am hurt' are whined or screamed. The words 'I am furious' are roared or bellowed. The tone of voice in such cases is so unmodified by learning and so close to the ancient non-verbal mammalian signalling system that even a dog can understand the message, let alone a foreigner from another race of our own species. The actual words used in such instances are almost superfluous. (Try snarling 'good dog', or cooing 'bad dog' at your pet, and you will see what I mean.) At its crudest and most intense level, mood talking is little more than a 'spilling over' of verbalized sound signalling into an area of communication that is already taken care of. Its value lies in the increased possibilities it provides for more subtle and sensitive mood signalling.

A third form of verbalization is *exploratory talking*. This is talking for talking's sake, aesthetic talking, or, if you like, play talking. Just as that other form of information-transmission, picture-making, became used as a medium for aesthetic exploration, so did talking. The poet paralleled the painter. But it is the fourth type of verbalization that we are concerned with in this chapter, the kind that has aptly been described recently as *grooming talking*. This is the meaningless, polite chatter of social occasions, the 'nice weather we are having' or 'have you read any good books lately' form of talking. It is not concerned with the exchange of important ideas or information, nor does it reveal the true mood of the speaker, nor is it aesthetically pleasing. Its function is to reinforce the greeting smile and to maintain the social togetherness. It is our substitute for social grooming. By providing us with a non-aggressive social preoccupation, it enables us to expose ourselves communally to one another over comparatively long periods, in this way enabling valuable group bonds and friendships to grow and become strengthened.

Viewed in this way, it is an amusing game to plot the course of grooming talk during a social encounter. It plays its most dominant role immediately after the initial greeting ritual. It then slowly loses ground, but has another peak of expression as the group breaks up. If the group has come together for purely social reasons, grooming talk may, of course, persist throughout to the complete exclusion of any kind of information, mood or exploratory talk. The cocktail party is a good example of this, and on such occasions 'serious' talking may even be actively suppressed by the host or hostess, who repeatedly intervene to break up long conversations and rotate the mutual-groomers to ensure maximum social contact. In this way, each member of the party is repeatedly thrown back into a state of 'initial contact', where the stimulus for grooming talk will be strongest. If these non-stop social-grooming sessions are to be successful, a sufficiently large number of guests must be invited in order to prevent new contacts from running out before the party is over. This explains the mysterious minimum size that is always automatically recognized as essential for gatherings of this kind. Small, informal dinner parties provide a slightly different situation. Here the grooming talk can be observed to wane as the evening progresses and the verbal exchange of serious information and ideas can be seen to gain in dominance as time passes. Just before the party breaks up, however, there is a brief resurgence of grooming talk prior to the final parting ritual. Smiling also reappears at this point, and the social bonding is in this way given a final farewell boost to help carry it over to the next encounter.

If we switch our observations now to the more formal business encounter, where the prime function of the contact is information talking, we can witness a further decline in the dominance of grooming talk, but not necessarily a total eclipse of it. Here its expression is almost entirely confined to the opening and closing moments. Instead of waning slowly, as at the dinner party, it is suppressed rapidly, after a few polite initial exchanges. It reappears again, as before, in the closing moments of the meeting, once the anticipated moment of parting has been signalled in some way. Because of the strong urge to perform grooming talk, business groups are usually forced to heighten the formalization of their meetings in some way, in order to suppress it. This explains the origin of committee procedure, where formality reaches a pitch rarely encountered on other private social occasions.

Although grooming talk is the most important substitute we have for social grooming, it is not our only outlet for this activity. Our naked skin may not send out very exciting grooming signals, but other more stimulating surfaces are frequently available and are used as sub-stitutes. Fluffy or furry clothing, rugs, or furniture often release a strong grooming response. Pet animals are even more inviting, and few naked apes can resist the temptation to stroke a cat's fur, or scratch a dog behind the ear. The fact that the animal appreciates this social-grooming activity provides only part of the reward for the groomer. More important is the outlet the pet animal's body surface gives us for our ancient primate grooming urges.

As far as our own bodies are concerned, we may be naked over most of our surfaces, but in the head region there is still a long and luxuriant growth of hair available for grooming. This receives a great deal of attention – far more than can be explained on a simple hygienic basis – at the hands of the specialist groomers, the barbers and hairdressers. It is not immediately obvious why mutual hairdressing has not become part of our ordinary domestic social gatherings. Why, for instance, have we developed grooming talk as our special substitute for the more typical primate friendship grooming, when we could so easily have concentrated our original grooming efforts in the head region? The explanation appears to lie in the sexual significance of the hair. In its present form the arrangement of the head hair differs strikingly between the two sexes and therefore provides a secondary sexual characteristic. Its sexual associations have inevitably led to its involvement in sexual behaviour patterns, so that stroking or manipulating the hair is now an action too heavily loaded with erotic significance to be permissible as a simple social friendship gesture. If, as a result of this, it is banned from communal gatherings of social acquaintances, it is necessary to find some other outlet for it. Grooming a cat or a sofa may provide an outlet for the urge to groom, but the need to *be* groomed requires a special context. The hairdressing salon is the perfect answer. Here the customer can indulge in the groomee role to his or her heart's content, without any fear of a sexual element creeping into the proceedings. By making the professional groomers into a separate category, completely dissociated from the 'tribal' acquaintanceship group, the dangers are eliminated. The use of male groomers for males and female groomers for females reduces the dangers still further. Where this is not done, the sexuality of the groomer is reduced in some way. If a female is attended by a male hairdresser, he usually behaves in an effeminate manner, regardless of his true sexual personality. Males are nearly always groomed by male barbers, but if a female masseuse is employed, she is typically rather masculine.

As a pattern of behaviour, hairdressing has three functions. It not only cleans the hair and provides an outlet for social grooming, but it also decorates the groomee. Decoration of the body for sexual, aggressive, or other social purposes is a widespread phenomenon in the case of the naked ape, and it has been discussed under other headings in other chapters. It has no real place in a chapter on comfort behaviour, except that it so often appears to grow out of some kind of grooming activity. Tattooing, shaving and plucking of hair, manicuring, ear-piercing and the more primitive forms of scarification all seem to have their origin in simple grooming actions. But, whereas grooming talk has been borrowed from elsewhere and utilized as a grooming sub-stitute, here the reverse process has taken place and grooming actions have been borrowed and elaborated for other uses. In acquiring a display function, the original comfort actions concerned with skin care have been transformed into what amounts to skin mutilation.

This trend can also be observed in certain captive animals in a zoo. They groom and lick with abnormal intensity until they have plucked bare patches or inflicted small wounds, either on their own bodies or those of companions. Excessive grooming of this kind is caused by conditions of stress or boredom. Similar conditions may well have provoked members of our own species to mutilate their body surfaces, with the already exposed and hairless skin aiding and abetting the process. In our case, however, our inherent opportunism enabled us to exploit this otherwise dangerous and damaging tendency and press it into service as a decorative display device.

Another and more important trend has also developed out of simple skin care, and that is medical care. Other species have made little progress in this direction, but for the naked ape the growth of medical practice out of social grooming behaviour has had an enormous in-fluence on the successful development of the species, especially in more recent times. In our closest relatives, the chimpanzees, we can already witness the beginning of this trend. In addition to the general skin care of mutual grooming, one chimpanzee has been seen to attend to the minor physical disabilities of another. Small sores or wounds are carefully examined and licked clean. Splinters are carefully removed by pinching the companion's skin between two forefingers. In one in-stance a female chimpanzee with a small cinder in her left eye was seen to approach a male, whimpering and obviously in distress. The male sat down and examined her intently and then proceeded to remove the cinder with great care and precision, gently using the tips of one finger from each hand. This is more than simple grooming. It is the first sign of true co-operative medical care. But for chimpanzees, the incident described is already the peak of its expression. For our own species, with greatly increased intelligence and co-operativeness, specialized grooming of this kind was to be the starting point of a vast technology of mutual physical aid. The medical world today has reached a condi-tion of such complexity that it has become, in social terms, the major expression of our animal comfort behaviour. From coping with minor discomforts it has expanded to deal with major diseases and gross bodily damage. As a biological phenomenon its achievements are unique, but in becoming rational, its irrational elements have been somewhat overlooked. In order to understand this, it is essential to distinguish between serious and trivial cases of 'indisposition'. As with any other species, a naked ape can break a leg or become infected with a vicious parasite on a purely accidental or chance basis. But in the case of trivial ailments, all is not what it seems. Minor infections and sicknesses are usually treated rationally, as if they are simply mild versions of serious illnesses, but there is strong evidence to suggest that they are in reality much more related to primitive 'grooming demands'. The medical symptoms reflect a behavioural problem that has taken a physical form, rather than a true physical problem.

Common examples of 'grooming invitation ailments', as we can call them, include coughs, colds, influenza, backache, headache, stomach upsets, skin rashes, sore throats, biliousness, tonsillitis and laryngitis. The condition of the sufferer is not serious, but sufficiently unhealthy to justify increased attention from social companions. The symptoms act in the same way as grooming invitation signals, releasing comfort behaviour from doctors, nurses, chemists, relations and friends. The groomee provokes friendly sympathy and care and this alone is usually enough to cure the illness. The administering of pills and medicines replaces the ancient grooming actions and provides an occupational ritual that sustains the groomee-groomer relationship through this special phase of social interaction. The exact nature of the chemicals prescribed is almost irrelevant and there is little difference at this level between the practices of modern medicine and those of ancient witch-doctoring.

The objection to this interpretation of minor ailments is likely to be based on the observation that real viruses or bacteria can be proved to be present. If they are there and can be shown to be the medical cause of the cold or stomach ache, then why should we seek for a behavioural explanation? The answer is that in any large city, for example, we are all exposed to these common viruses and bacteria all the time, but we only occasionally fall prey to them. Also, certain individuals are much more susceptible than others. Those members of a community who are either very successful or socially well adjusted rarely suffer from 'grooming invitation ailments'. Those that have temporary or long-standing social problems are, by contrast, highly susceptible. The most intriguing aspect of these ailments is the way they are tailored to the special demands of the individual. Supposing an actress, for example, is suffering from social tensions and strains, what happens? She loses her voice, develops laryngitis, so that she is forced to stop work and take a rest. She is comforted and looked after. The tension is resolved (for the time being, at least). If instead she had developed a skin rash on her body, her costume would have covered it and she could have gone on working. The tension would have continued. Compare her situation with that of an all-in wrestler. For him a loss of voice would be useless as a 'grooming invitation ailment', but a skin rash would be ideal, and it is precisely this ailment that wrestlers' doctors find is the muscle-men's most common complaint. In this connection it is amusing that one famous actress, whose reputation relies on her nude appearances in her films, suffers under stress not from laryngitis, but from skin rash. Because, like the wrestlers, her skin exposure is vital, she falls into their ailment category rather than into that of other actresses.

If the need for comfort is intense, then the ailment becomes more intense. The time in our lives when we receive the most elaborate care and protection is when we are infants in our cots. An ailment that is severe enough to put us helplessly to bed, therefore, has the great advantage of recreating for us all the comforting attention of our secure infancy. We may think we are taking a strong dose of medicine, but in reality it is a strong dose of security that we need and that cures us. (This does not imply malingering. There is no need to malinger. The symptoms are real enough. It is the cause that is behavioural, not the effects.)

We are all to some extent frustrated groomers, as well as groomees, and the satisfaction that can be obtained from caring for the sick is as basic as the cause of the sickness. Some individuals have such a great need to care for others that they may actively promote and prolong sickness in a companion in order to be able to express their grooming urges more fully. This can produce a vicious circle, with the groomer-groomee situation becoming exaggerated out of all proportion, to the extent where a chronic invalid demanding (and getting) constant attention is created. If a 'mutual grooming pair' of this type were faced with the behavioural truth concerning their reciprocal conduct, they would hotly deny it. Nevertheless, it is astonishing what miraculous cures can sometimes be worked in such instances when a major social upheaval occurs in the groomer-groomee (nurse-patient) environment that has been created. Faith-healers have occasionally exploited this situation with startling results, but unfortunately for them many of the cases they encounter have physical causes as well as physical effects. Also working against them is the fact that the physical effects of behaviourally produced 'grooming invitation ailments' can easily create irreversible body damage if sufficiently prolonged or intense. Once this has happened, serious, rational medical treatment is required.

Up to this point I have been concentrating on the social aspects of comfort behaviour in our species. As we have seen there have been great developments in that direction, but this has not excluded or replaced the simpler kinds of self-cleaning and self-comfort. Like other primates we still scratch ourselves, rub our eyes, pick our sores, and lick our wounds. We also share with them a strong tendency to sunbathe. In addition we have added a number of specialized cultural patterns, the most common and widespread of which is washing with water. This is rare in other primates, although certain species bathe occasionally, but for us it now plays the major role in body-cleaning in most communities.

Despite its obvious advantages, frequent cleansing with water nevertheless puts a severe strain on the production of antiseptic and protective oils and salts by the skin glands, and to some extent it is bound to make the body surface more susceptible to diseases. It survives this disadvantage only because, at the same time that it eliminates the natural oils and salts, it removes the dirt that is the source of these diseases.

In addition to problems of keeping clean, the general category of comfort behaviour also includes those patterns of activity concerned with the task of maintaining a suitable body temperature. Like all mammals and birds, we have evolved a constant, high body temperature, giving us greatly increased physiological efficiency. If we are healthy, our deep body temperature varies no more than $3°$ Fahrenheit, regardless of the outside temperature. This internal temperature fluctuates with a daily rhythm, the highest level occurring in the late afternoon and the lowest at around 4 a.m. If the external environment becomes too hot or too cold we quickly experience acute discomfort. The unpleasant sensations we receive act as an early-warning system, alerting us to the urgent need to take action to prevent the internal body organs from becoming disastrously chilled or overheated. In addition to encouraging intelligent, voluntary responses, the body also takes certain automatic steps to stabilize its heat level. If the environment becomes too hot, vasodilation occurs. This gives a hotter body surface and encourages heat loss from the skin. Profuse sweating also takes place. We each possess approximately two million sweat glands. Under conditions of intense heat these are capable of secreting a maximum of one litre of sweat per hour. The evaporation of this liquid from the body surface provides another valuable form of heat loss. During the process of acclimatization to a generally hotter environment, we undergo a marked increase in sweating efficiency. This is vitally important because, even in the hottest climates, our internal body temperature can only stand an upward shift of $0.4°$ Fahrenheit, regardless of our racial origin.

If the environment becomes too cold, we respond with vasoconstriction and with shivering. The vasoconstriction helps to conserve the body heat and the shivering can provide up to three times the resting heat production. If the skin is exposed to the intense cold for any length of time, there is a danger that the prolonged vasoconstriction will lead to frostbite. In the hand region there is an important, built-in anti-frostbite system. The hands at first respond to intense cold by drastic vasoconstriction; then, after about five minutes, this is reversed and there is strong vasodilation, the hands becoming hot and flushed. (Anyone who has been snowballing in winter will have experienced this.) The constriction and dilation of the hand region then continues to alternate, the constriction phases curtailing heat loss and the dilation phases preventing frostbite. Individuals living permanently in a cold climate undergo various forms of bodily acclimatization, including a slightly increased basal metabolic rate.

As our species has spread over the globe, important cultural additions have been made to these biological temperature control mechanisms. The development of fire, clothing, and insulated dwelling-houses have combated heat loss, and ventilation and refrigeration have been used against heat gain. Impressive and dramatic as these advances have been, they have in no way altered our internal body temperature. They have merely served to control the external temperature, so that we can continue to enjoy our primitive primate temperature level in a more diverse range of external conditions. Despite recent claims, suspended animation experiments involving special freezing techniques are still confined to the realms of science fiction.

Before leaving the subject of temperature responses, there is one particular aspect of sweating that should be mentioned. Detailed studies of sweating responses in our species have revealed that they are not as simple as they may first appear. Most areas of the body surface begin to perspire freely under conditions of increased heat, and this is undoubtedly the original, basic response of the sweat-gland system. But certain regions have become reactive to other types of stimulation and sweating can occur there regardless of the external temperature. The eating of highly spiced foods, for example, produces its own special pattern of facial sweating. Emotional stress quickly leads to sweating on the palms of the hands, the soles of the feet, the armpits and sometimes also the forehead, but not on other parts of the body. There is a further distinction in the areas of emotional sweating, the palms and the soles differing from the armpits and the forehead. The first two regions respond well *only* to emotional situations, whereas the last two react to both emotional and to temperature stimuli. It is clear from this that the hands and feet have 'borrowed' sweating from the temperature control system and are now using it in a new functional context. The moistening of the palms and soles during stress appears to have become a special feature of the 'ready for anything' response that the body gives when danger threatens. Spitting on the hands before wielding an axe is, in a sense, the non-physiological equivalent of this process.

So sensitive is the palmar sweating response that whole communities or nations may show sudden increases in this reaction if their group security is threatened in some way. During a recent political crisis, when there was a temporary increase in the likelihood of nuclear war, all experiments into palmar sweating at a research institute had to be abandoned because the base level of the response had become so abnormal that the tests would have been meaningless. Having our palms read by a fortune-teller may not tell us much about the future, but having them read by a physiologist can certainly tell us something about our fears for the future.

8

ANIMALS

U P TO THIS point we have been considering the naked ape's behaviour towards himself and towards members of his own species – his intra-specific behaviour. It now remains to examine his activities in relation to other animals – his inter-specific behaviour.

All the higher forms of animal life are aware of at least some of the other species with which they share their environment. They regard them in one of five ways: as prey, symbionts, competitors, parasites, or predators. In the case of our own species, these five categories may be lumped together as the 'economic' approach to animals, to which may be added the scientific, aesthetic and symbolic approaches. This wide range of interests has given us an inter-specific involvement unique in the animal world. In order to unravel it and understand it objectively we must tackle it step by step, attitude by attitude.

Because of his exploratory and opportunist nature, the naked ape's list of prey species is immense. At some place, at some time, he has killed and eaten almost any animal you care to mention. From a study of prehistoric remains we know that about half a million years ago, at one site alone, he was hunting and eating species of bison, horse, rhino, deer, bear, sheep, mammoth, camel, ostrich, antelope, buffalo, boar and hyaena. It would be pointless to compile a 'species menu' for more recent times, but one feature of our predatory behaviour does deserve mention, namely our tendency to domesticate certain selected prey species. For, although we are likely to eat almost anything palatable on occasion, we have nevertheless limited the bulk of our feeding to a few major animal forms.

Domestication of livestock, involving the organized control and selective breeding of prey, is known to have been practised for at least ten thousand years and, in certain cases, probably much longer. Goats, sheep and reindeer appear to have been the earliest prey species dealt with in this way. Then, with the development of settled agricultural communities, pigs and cattle, including Asiatic buffalo and yak, were added to the list. We have evidence that, in the case of cattle, several distinct breeds had already been developed four thousand years ago. Whereas the goats, sheep and reindeer were transformed directly from hunted prey to herded prey, it is thought that the pigs and cattle began their close association with our species as crop-robbers. As soon as cultivated crops were available, they moved in to take advantage of this rich new food supply, only to be taken over by the early farmers and brought under domestic control themselves.

The only small mammalian prey species to undergo prolonged domestication was the rabbit, but this was apparently a much later development. Amongst the birds, important prey species domesticated thousands of years ago were the chicken, the goose and the duck, with later minor additions of the pheasant, guinea fowl, quail and turkey. The only prey fish with a long history of domestication are the Roman eel, the carp and the goldfish. The latter, however, soon became ornamental rather than gastronomic. The domestication of these fish is limited to the last two thousand years and has played only a small role in the general story of our organized predation.

The second category in our list of inter-specific relationships is that of the symbiont. Symbiosis is defined as the association of two different species to their mutual benefit. Many examples of this are known from the animal world, the most famous being the partnership between the tick birds and certain large ungulates such as the rhinoceros, giraffe and buffalo. The birds eat the skin parasites of the ungulates, helping to keep the bigger animals healthy and clean, while the latter provide the birds with a valuable source of food.

Where we ourselves are one of the members of a symbiotic pair, the mutual benefit tends to become biased rather heavily in our favour, but it is nevertheless a separate category, distinct from the more severe prey-predator relationship, since it does not involve the death of the other species concerned. They are exploited, but in exchange for the exploitation we feed and care for them. It is a biased symbiosis because we are in control of the situation and our animal partners usually have little or no choice in the matter.

The most ancient symbiont in our history is undoubtedly the dog. We cannot be certain exactly when our ancestors first began to domesticate this valuable animal, but it appears to be at least ten thousand years ago. The story is a fascinating one. The wild, wolf-like ancestors of the domestic dog must have been serious competitors with our hunting forebears. Both were co-operative pack-hunters of large prey and, at first, little love can have been lost between them. But the wild dogs possessed certain special refinements that our own hunters lacked. They were particularly adept at herding and driving prey during hunting manœuvres and could carry this out at high speed. They also had more delicate senses of smell and hearing. If these attributes could be exploited in exchange for a share in the kill, then the bargain was a good one. Somehow – we do not know exactly how

– this came about and an inter-specific bond was forged. It is probable that it began as a result of young puppies being brought in to the tribal home base to be fattened as food. The value of these creatures as alert nocturnal watch-dogs would have scored a mark in their favour at an early stage. Those that were allowed to live in a now tamed condition and permitted to accompany the males on their hunting trips would soon show their paces in assisting to track down the prey. Having been hand-reared, the dogs would consider themselves to be members of the naked-ape pack and would co-operate instinctively with their adopted leaders. Selective breeding over a number of generations would soon weed out the trouble-makers and a new, improved stock of increasingly restrained and controllable domestic hunting dogs would arise.

It has been suggested that it was this progression in the dog relationship that made possible the earliest forms of ungulate prey domestication. The goats, sheep and reindeer were under some degree of control before the advent of the true agricultural phase, and the improved dog is envisaged as the vital agent that made this feasible by assisting in the large-scale and long-term herding of these animals. Studies of the driving behaviour of present-day sheepdogs and of wild wolves reveal many similarities in technique and provide strong support for this view.

During more recent times, intensified selective breeding has produced a whole range of symbiotic dog specializations. The primitive all-purpose hunting dog assisted in all stages of the operation, but his later descendants were perfected for one or other of the different components of the overall behaviour sequence. Individual dogs with unusually well-developed abilities in a particular direction were inbred to intensify their special advantages. As we have already seen, those with good qualities in manœuvring became herding dogs, their contribution being confined largely to the rounding up of domesticated prey (sheepdogs). Others, with a superior sense of smell, were inbred as scent-trackers (hounds). Others, with an athletic turn of speed, became coursing dogs and were employed to chase after prey by sight (greyhounds). Another group were bred as prey-spotters, their tendency to 'freeze' when locating the prey being exploited and intensified (setters and pointers). Yet another line was improved as prey-finders and carriers (retrievers). Small breeds were developed as vermin-killers (terriers). The primitive watch-dogs were genetically improved as guard-dogs (mastiffs).

In addition to these widespread forms of exploitation, other dog lines have been selectively bred for more unusual functions. The most extraordinary example is the hairless dog of the ancient New World Indians, a genetically naked breed with an abnormally high skin temperature that was used as a primitive form of hot-water bottle in their sleeping quarters.

In more recent times, the symbiotic dog has earned his keep as beast of burden, pulling sledges or carts, as a messenger or a mine-detector during times of war, as a rescue operator, locating climbers buried under snow, as a police dog, tracking or attacking criminals, as a guide dog, leading the blind, and even as a substitute space-traveller. No other symbiotic species has served us in such a complex and varied way. Even today, with all our technological advances, the dog is still actively employed in most of his functional roles. Many of the hundreds of breeds that can now be distinguished are purely ornamental, but the day of the dog with a serious task to perform is far from over.

So successful has the dog been as a hunting companion that few attempts have been made to domesticate other species in this particular form of symbiosis. The only important exceptions are the cheetah and certain birds of prey, especially the falcon, but in neither case has any progress been made with regard to controlled breeding, let alone selective breeding. Individual training has always been required. In Asia the cormorant, a diving bird, has been used as an active companion in the hunt for fish. Cormorant eggs are taken and hatched out under domestic chickens. The young sea-birds are then hand-reared and trained to catch fish on the end of a line. The fish are brought back to the boats and disgorged, the cormorants having been fitted with a collar to prevent them swallowing their prey. But here again no attempt has been made to improve the stock by selective breeding.

Another ancient form of exploitation involves the use of small carnivores as pest-destroyers. This trend did not gain momentum until the agricultural phase of our history. With the development of large-scale grain storage, rodents became a serious problem and rodent-killers were encouraged. The cat, the ferret and the mongoose were the species that came to our aid and in the first two cases full domestication with selective breeding followed.

Perhaps the most important kind of symbiosis has been the utilization of certain larger species as beasts of burden. Horses, onagers (Asiatic wild asses), donkeys (African wild asses), cattle, including the water buffalo and the yak, reindeer, camels, llamas and elephants have all been subjected to massive exploitation in this way. In most of these cases the original wild types have been 'improved' by careful selective breeding, the exceptions to this rule being the onager and the elephant. The onager was being used as a beast of burden by the ancient Sumerians over four thousand years ago, but was rendered obsolete by the introduction of a more easily controlled species, the horse. The elephant, although still employed as a working animal, has always offered too big a challenge to the stock-breeder and has never been submitted to the pressures of selective breeding.

A further category concerns the domestication of a variety of species as sources of produce. The animals are not killed, so that in this role they cannot be considered as prey. Only certain parts are taken from them: milk from cattle and goats, wool from sheep and alpaca, eggs from chickens and ducks, honey from bees and silk from silk-moths.

In addition to these major categories of hunting companions, pest-destroyers, beasts of burden, and sources of produce, certain animals have entered a symbiotic relationship with our species on a more unusual and specialized basis. The pigeon has been domesticated as a message-carrier. The astonishing homing abilities of this bird have been exploited for thousands of years. This relationship became so valuable in times of war that, during recent epochs, a counter-symbiosis was developed in the form of falcons trained to intercept the message-carriers. In a very different context, Siamese fighting fish and fighting cocks have been selectively bred over a long period as gambling devices. In the realm of medicine, the guinea-pig and the white rat have been widely employed as 'living test-beds' for laboratory experiments.

These, then, are the major symbionts, animals that have been forced into some form of partnership with our ingenious species. The advantage to them is that they cease to be our enemies. Their numbers are dramatically increased. In terms of world populations they are tremendously successful. But it is a qualified success. The price they have paid is their evolutionary freedom. They have lost their genetic independence and, although well fed and cared for, are now subject to our breeding whims and fancies.

The third major category of animal relationships, after prey and symbionts, is that of competitors. Any species which competes with us for food or space, or interferes with the efficient running of our lives, is ruthlessly eliminated. There is no point in listing such species. Virtually any animal that is either inedible or symbiotically useless is attacked and exterminated. This process is continuing today in all parts of the world. In the case of minor competitors, the persecution is haphazard, but serious rivals stand little chance. In the past our closest primate relatives have been our most threatening rivals and it is no accident that today we are the only species surviving in our entire family. Large carnivores have been our other serious competitors and these too have been eliminated wherever the population density of our species has risen above a certain level. Europe, for example, is now virtually denuded of all large forms of animal life, save for a great seething mass of naked apes.

For the next major category, that of parasites, the future looks even more bleak. Here the fight is intensified and although we may mourn the passing of an attractive food rival, no one will shed a tear over the increasing rarity of the flea. As medical science progresses, the grip of the parasites dwindles. In its wake this brings an added threat to all the other species, for as the parasites go and our health increases, our populations can swell at an even more startling rate, thus accentuating the need to eliminate all the milder competitors.

The fifth major category, the predators, are also on the way out. We have never really constituted a main diet component for any species, and our numbers have never been seriously reduced by predation at any stage in our history, as far as we can tell. But the larger carnivores, such as the big cats and the wild dogs, the bigger members of the crocodile family, the sharks and the more massive birds of prey have nibbled away at us from time to time and their days are clearly numbered. Ironically, the killer that has accounted for more naked-ape deaths than any other (parasites excepted) is one that cannot devour the nutritious corpses it produces. This deadly enemy is the venomous snake and, as we shall see later, this has become the most hated of all higher forms of animal life.

These five categories of inter-specific relationships – prey, symbiont, competitor, parasite and predator – are the ones that can be found to exist between other pairs of species. Basically, we are not unique in these respects. We carry the relationships much further than other species, but they are the same types of relationships. As I said earlier, they can be lumped together as the economic approach to animals. In addition we have our own special approaches, the scientific, the aesthetic and the symbolic.

The scientific and aesthetic attitudes are manifestations of our powerful exploratory drive. Our curiosity, our inquisitiveness, urges us on to investigate all natural phenomena and the animal world has naturally been the focus of much attention in this respect. To the zoologist, all animals are, or should be, equally interesting. To him there are no bad species or good species. He studies them all, exploring them for their own sake. The aesthetic approach involves the same basic exploration, but with different terms of reference. Here, the enormous variety of animal shapes, colours, patterns and movements are studied as objects of beauty rather than as systems for analysis.

The symbolic approach is entirely different. In this case, neither economics nor exploration are involved. The animals are employed instead as personifications of concepts. If a species looks fierce, it becomes a war-symbol. If it looks clumsy and cuddly, it becomes a child-symbol. Whether it is genuinely fierce or genuinely cuddly matters little. Its true nature is not investigated in this context, for this is not a scientific approach. The cuddly animal may be bristling with razor-sharp teeth and be endowed with a vicious aggressiveness, but providing these attributes are not obvious and its cuddliness is, it is perfectly acceptable as the ideal child-symbol. For the symbolic animal, justice does not have to be done, it has only to appear to be done.

The symbolic attitude to animals was originally christened the 'anthropoidomorphic' approach. Mercifully, this ugly term was later contracted to 'anthropomorphic' which, although still clumsy, is the expression in general use today. It is invariably used in a derogatory sense by scientists who, from their point of view, are fully justified in scorning it. They must retain their objectivity at all costs if they are to make meaningful explorations into the animal world. But this is not as easy as it may sound.

Quite apart from deliberate decisions to use animal forms as idols, images and emblems, there are also subtle, hidden pressures working on us all the time that force us to see other species as caricatures of ourselves. Even the most sophisticated scientist is liable to say, 'Hallo, old boy' when greeting his dog. Although he knows perfectly well that the animal cannot understand his words, he cannot resist the temptation. What is the nature of these anthropomorphic pressures and why are they so difficult to overcome? Why do some creatures make us say 'Aah' and others make us say 'Ugh!'? This is no trivial consideration. A vast amount of our present culture's inter-specific energies is involved here. We are passionate animal lovers and animal haters, and these involvements cannot be explained on the basis of economic and exploratory considerations alone. Clearly some kind of unsuspected, basic response is being triggered off inside us by the specific signals we are receiving. We delude ourselves that we are responding to the animal as an animal. We declare that it is charming, irresistible, or horrible, but what makes it so?

In order to find the answer to this question we must first assemble some facts. What exactly are the animal loves and animal hates of our culture and how do they vary with age and sex? Quantitative evidence is required on a large scale if reliable statements are to be made on this topic. To obtain such evidence an investigation was carried out involving 80,000 British children between the ages of four and fourteen. During a zoo television programme they were asked the simple questions: 'Which animal do you like most?' and 'Which animal do you dislike most?' From the massive response to this inquiry a sample of 12,000 replies to each question was selected at random and analysed.

Dealing first with the inter-specific 'loves', how did the various groups of animals fare? The figures are as follows: 97.15 per cent of all the children quoted a mammal of some kind as their top favourite. Birds accounted for only 1.6 per cent, reptiles 1.0 per cent, fish 0.1 per cent, invertebrates 0.1 per cent, and amphibians 0.05 per cent. Obviously there is something special about mammals in this context.

(It should perhaps be pointed out that the replies to the questions were written, not spoken, and it was sometimes difficult to identify the animals from the names given, especially in the case of very young children. It was easy enough to decipher loins, hores, bores, penny kings, panders, tapers and leapolds, but almost impossible to be certain of the species referred to as bettle twigs, the skipping worm, the otamus, or the coco-cola beast. Entries supporting these appealing creatures were reluctantly rejected.)

If we now narrow our sights to the 'top ten animal loves' the figures emerge as follows: 1. Chimpanzee (13.5 per cent). 2. Monkey (13 per cent). 3. Horse (9 per cent). 4. Bushbaby (8 per cent). 5. Panda (7.5 per cent). 6. Bear (7 per cent). 7. Elephant (6 per cent). 8. Lion (5 per cent). 9. Dog (4 per cent). 10. Giraffe (2.5 per cent).

It is immediately clear that these preferences do not reflect powerful economic or aesthetic influences. A list of the ten most important economic species would read very differently. Nor are these animal favourites the most elegant and brightly coloured of species. They include instead a high proportion of rather clumsy, heavy-set and dully coloured forms. They are, however, well endowed with anthropomorphic features and it is to these that the children are responding when making their choices. This is not a conscious process. Each of the species listed provides certain key stimuli strongly reminiscent of special properties of our own species, and to these we react automatically without any realization of what it is exactly that appeals to us. The most significant of these anthropomorphic features in the top ten animals are as follows:

1. They all have hair, rather than feathers or scales. 2. They have rounded outlines (chimpanzee, monkey, bushbaby, panda, bear, elephant). 3. They have flat faces (chimpanzee, monkey, bushbaby, bear, panda, lion). 4. They have facial expressions (chimpanzee, monkey, horse, lion, dog). 5. They can 'manipulate' small objects (chimpanzee, monkey, bushbaby, panda, elephant). 6. Their postures are in some ways, or at some times, rather vertical (chimpanzee, monkey, bushbaby, panda, bear, giraffe).

The more of these points a species can score, the higher up the top ten list it comes. Non-mammalian species fare badly because they are weak in these respects. Amongst the birds, the top favourites are the penguin (0.8 per cent) and the parrot (0.2 per cent). The penguin achieves the number one avian position because it is the most vertical of all the birds. The parrot also sits more vertically on its perch than most birds and it has several other special advantages. Its beak shape gives it an unusually flattened face for a bird. It also feeds in a strange way, bringing its foot up to its mouth rather than lowering its head, and it can mimic our vocalizations. Unfortunately for its popularity, it lowers itself into a more horizontal posture when walking and in this way loses points heavily to the vertically waddling penguin.

Amongst the top mammals there are several special points worth noting. Why, for instance, is the lion the only one of the big cats to be included? The answer appears to be that it alone, in the male, has a heavy mane of hair surrounding the head region. This has the effect of flattening the face (as is clear from the way lions are portrayed in children's drawings) and helps to score extra points for this species.

Facial expressions are particularly important, as we have already seen in earlier chapters, as basic forms of visual communication in our species. They have evolved in a complex form in only a few groups of mammals – the higher primates, the horses, the dogs and the cats. It is no accident that five of the top ten favourites belong to these groups. Changes in facial expression indicate changes in mood and this provides a valuable link between the animal and ourselves, even though the correct significance of the expressions is not always precisely understood.

As regards manipulative ability, the panda and the elephant are unique cases. The former has evolved an elongated wrist bone with which it can grasp the thin bamboo sticks on which it feeds. A structure of this kind is found nowhere else in the animal kingdom. It gives the flat-footed panda the ability to hold small objects and bring them up to its mouth while sitting in a vertical posture. Anthropomorphically this scores heavily in its favour. The elephant is also capable of 'manipulating' small objects with its trunk, another unique structure, and taking them up to its mouth.

The vertical posture so characteristic of our species gives any other animal that can adopt this position an immediate anthropomorphic advantage. The primates in the top ten list, the bears and the panda all sit up vertically on frequent occasions. Sometimes they may even stand vertically or go so far as to take a few faltering steps in this position, all of which helps them to score valuable points. The giraffe, by virtue of its unique body proportions, is, in a sense, permanently vertical. The dog, which achieves such a high anthropomorphic score for its social behaviour, has always been something of a postural disappointment. It is uncompromisingly horizontal. Refusing to accept defeat on this point, our ingenuity went to work and soon solved the problem – we taught the dog to sit up and beg. In our urge to anthropomorphize the poor creature, we went further still. Being tailless ourselves, we started docking its tail. Being flat-faced ourselves, we employed selective breeding to reduce the bone structure in the snout region. As a result, many dog breeds are now abnormally flat-faced. Our anthropomorphic desires are so demanding that they have to be satisfied, even at the expense of the animals' dental efficiency. But then we must recall that this approach to animals is a purely selfish one. We are not seeing animals as animals, but merely as reflections of ourselves, and if the mirror distorts too badly we either bend it into shape or discard it.

So far we have been considering the animal loves of children of all ages between four and fourteen. If we now split up the responses to these favourite animals, separating them into age groups, some remarkably consistent trends emerge. For certain of the animals there is a steady decrease in preference with the increasing age of the children. For others there is a steady rise.

The unexpected discovery here is that these trends show a marked relationship with one particular feature of the preferred animals, namely their body size. The younger children prefer the bigger animals and the older children prefer the smaller ones. To illustrate this we can take the figures for the two largest of the top ten forms, the elephant and the giraffe, and two of the smallest, the bushbaby and the dog. The elephant, with an overall average rating of 6 per cent, starts out at 15 per cent with the four-year-olds and then falls smoothly to 3 per cent with the fourteen-year-olds. The giraffe shows a similar drop in popularity from 10 per cent to 1 per cent. The bushbaby, on the other hand, starts at only 4.5 per cent with the four-year-olds and then rises gradually to 11 per cent with the fourteen-year-olds. The dog rises from 0.5 to 6.5 per cent. The medium-sized animals amongst the top ten favourites do not show these marked trends.

We can sum up the findings so far by formulating two principles. The first law of animal appeal states that 'The popularity of an animal is directly correlated with the number of anthropomorphic features it possesses.' The second law of animal appeal states that 'The age of a child is inversely correlated with the size of the animal it most prefers.'

How can we explain the second law? Remembering that the preference is based on a symbolic equation, the simplest explanation is that the smaller children are viewing the animals as parent-substitutes and the older children are looking upon them as child-substitutes. It is not enough that the animal must remind us of our own species, it must remind us of a special category within it. When the child is very young, its parents are all-important protective figures. They dominate the child's awareness. They are large, friendly animals, and large friendly animals are therefore easily identified with parental figures. As the child grows it starts to assert itself, to compete with its parents. It sees itself in control of the situation, but it is difficult to control an elephant or a giraffe. The preferred animal has to shrink down to a manageable size. The child, in a strangely precocious way, becomes the parent itself. The animal has become the symbol of *its* child. The real child is too young to be a real parent, so instead it becomes a symbolic parent. Ownership of the animal becomes important and pet-keeping develops as a form of 'infantile parentalism'. It is no accident that, since becoming available as an exotic pet, the animal previously known as the galago has now acquired the popular name of bush*baby*. (Parents should be warned from this that the pet-keeping urge does not arrive until late in childhood. It is a grave error to provide pets for very young children, who respond to them as objects for destructive exploration, or as pests.)

There is one striking exception to the second law of animal appeal and that concerns the horse. The response to this animal is unusual in two ways. When analysed against increase in age of children, it shows a smooth rise in popularity followed by an equally smooth fall. The peak coincides with the onset of puberty. When analysed against the different sexes, it emerges that it is three times as popular with girls as with boys. No other animal love shows anything approaching this sex difference. Clearly there is something unusual about the response to the horse and it requires separate consideration.

The unique feature of the horse in the present context is that it is something to be mounted and ridden. This applies to none of the other top ten animals. If we couple this observation with the facts that its popularity peak coincides with puberty and that there is a strong sexual difference in its appeal, we are forced to the conclusion that the response to the horse must involve a strong sexual element. If a symbolic equation is being made between mounting a horse and sexual mounting, then it is perhaps surprising that the animal has a greater appeal for girls. But the horse is a powerful, muscular and dominant animal and is therefore more suited to the male role. Viewed objectively, the act of horse-riding consists of a long series of rhythmic movements with the legs wide apart and in close contact with the body of the animal. Its appeal for girls appears to result from the combination of its masculinity and the nature of the posture and actions performed on its back. (It must be stressed here that we are dealing with the child population as a whole. One child in every eleven preferred the horse to all other animals. Only a small fraction of this percentage would ever actually own a pony or a horse. Those that do, quickly learn the many more varied rewards that go with this activity. If, as a result, they become addicted to horse-riding, this is not, of course, necessarily significant in the context we have been discussing.)

It remains to explain the fall in popularity of the horse following puberty. With increasing sexual development, it might be expected to show further increases in popularity, rather than a decrease. The answer can be found by comparing the graph for horse love with the curve for sex play in children. They match one another remarkably well. It would seem that, with the growth of sexual awareness, and the characteristic sense of privacy that comes to surround teenage sexual feelings, the response to the horse declines along with the decline in overt sex-play 'romping'. It is significant here that the appeal of monkeys also suffers a decline at this point. Many monkeys have particularly obtrusive sexual organs, including large pink sexual swellings. For the younger child these have no significance and the monkeys' other powerful anthropomorphic features can operate unhindered. But for older children the conspicuous genitals become a source of embarrassment and the popularity of these animals suffers as a consequence.

This, then, is the situation with regard to animal 'loves' in children. For adults, the responses become more varied and sophisticated, but the basic anthropomorphism persists. Serious naturalists and zoologists bewail this fact, but providing it is fully realized that symbolic responses of this kind tell us nothing about the true nature of the different animals concerned, they do little harm and provide a valuable subsidiary outlet for emotional feelings.

Before considering the other side of the coin – the animal 'hates' – there is one criticism that must be answered. It could be argued that the results discussed above are of purely cultural significance, and have no meaning for our species as a whole. As regards the exact identity of the animals involved this is true. To respond to a panda, it is obviously necessary to learn of its existence. There is no inborn panda response. But this is not the point. The choice of the panda may be culturally determined, but the *reasons* for choosing it do reflect a deeper, more biological process at work. If the investigation were repeated in another culture, the favourite species might be different, but they would still be selected according to our fundamental symbolic needs. The first and second law of animal appeal would still operate.

Turning now to animal 'hates', we can subject the figures to a similar analysis. The top ten most disliked animals are as follows: 1. Snake (27 per cent). 2. Spider (9.5 per cent). 3. Crocodile (4.5 per cent). 4. Lion (4.5 per cent). 5. Rat (4 per cent). 6. Skunk (3 per cent). 7. Gorilla (3 per cent). 8. Rhinoceros (3 per cent). 9. Hippopotamus (2.5 per cent). 10. Tiger (2.5 per cent).

These animals share one important feature: they are dangerous. The crocodile, the lion and the tiger are carnivorous killers. The gorilla, the rhinoceros and the hippopotamus can easily kill if provoked. The skunk indulges in a violent form of chemical warfare. The rat is a pest that spreads disease. There are venomous snakes and poisonous spiders.

Most of these creatures are also markedly lacking in the anthropomorphic features that typify the top ten favourites. The lion and the gorilla are exceptions. The lion is the only form to appear in both the top ten lists. The ambivalence of the response to this species is due to this animal's unique combination of attractive anthropomorphic characters and violent predatory behaviour. The gorilla is strongly endowed with anthropomorphic characters, but unfortunately for him his facial structure is such that he appears to be in a constantly aggressive and fearsome mood. This is merely an accidental outcome of his bone structure and bears no relationship to his true (and rather gentle) personality, but combined with his great physical strength it immediately converts him into a perfect symbol of savage brute force.

The most striking feature of the list of top ten hates is the massive response to the snake and the spider. This cannot be explained solely on the basis of the existence of dangerous species. Other forces are at work. An analysis of the reasons given for hating these forms reveals that snakes are disliked because they are 'slimy and dirty' and spiders are repulsive because they are 'hairy and creepy'. This must mean either that they have a strong symbolic significance of some kind, or alternatively that we have a powerful inborn response to avoid these animals.

The snake has long been thought of as a phallic symbol. Being a poisonous phallus, it has represented unwelcome sex, which may be a partial explanation for its unpopularity; but there is more to it than this. If we examine the different levels of snake hatred in children between the ages of four and fourteen, it emerges that the peak of unpopularity comes early, long before puberty is reached. Even at four, the hate level is high – around 30 per cent – and it then climbs slightly, reaching its peak at age six. From then on it shows a smooth decline, sinking to well below 20 per cent by the age of fourteen. There is little difference between the sexes, although at each age level the response from girls is slightly stronger than the response from boys. The arrival of puberty appears to have no impact on the response in either sex.

From this evidence it is difficult to accept the snake simply as a strong sexual symbol. It seems more likely that we are dealing here with an inborn aversion response of our species towards snake-like forms. This would explain not only the early maturation of the reaction, but also the enormously high level of the response when compared with all other animal hates and loves. It would also fit with what we know of our closest living relatives, the chimpanzees, gorillas and orang-utans. These animals also exhibit a great fear of snakes and here again it matures early. It is not seen in the very young apes, but is fully developed by the time they are a few years old and have reached the stage where they are beginning to make brief sorties away from the security of their mothers' bodies. For them an aversion response clearly has an important survival value and would also have been a great benefit to our early ancestors. Despite this, it has been argued that the snake reaction is not inborn, but merely a cultural phenomenon resulting from individual learning. Young chimpanzees reared under abnormally isolated conditions have reputedly failed to show the fear response when first exposed to snakes. But these experiments are not very convincing. In some instances, the chimpanzees have been too young when first tested. Had they been retested a few years later, the reaction may well have been present. Alternatively, the effects of isolation may have been so severe that the young animals in question were virtually mental defectives. Such experiments are based on a fundamental misconception about the nature of inborn responses, which do not mature in an encapsulated form, irrespective of the outside environment. They should be thought of more as inborn susceptibilities. In the case of the snake response, it may be necessary for the young chimpanzee, or child, to encounter a number of different frightening objects in its early life and to learn to respond negatively to these. The inborn element in the snake case would then manifest itself in the form of a much more massive response to this stimulus than to others. The snake fear would be out of all proportion to the other fears, and this disproportionateness would be the inborn factor. The terror produced in normal young chimpanzees by exposure to a snake and the intense hatred of snakes exhibited by our own species are difficult to explain in any other way.

The reaction of children to spiders takes a rather different course. Here there is a marked sex difference. In boys there is an increase in spider hatred from age four to fourteen, but it is slight. The level of the reaction is the same for girls up to the age of puberty, but it then shows a dramatic rise, so that by the age of fourteen it is double that of the boys. Here we do seem to be dealing with an important symbolic factor. In evolutionary terms, poisonous spiders are just as dangerous to males as to females. There may or may not be an inborn response to these creatures in both sexes, but it cannot explain the spectacular leap in spider hatred that accompanies female puberty. The only clue here is the repeated female reference to spiders being nasty, hairy things.

Puberty is, of course, the stage when tufts of body hair are beginning to sprout on both boys and girls. To children, body hairiness must appear as an essentially masculine character. The growth of hair on the body of a young girl would therefore have a more disturbing (unconscious) significance for her than it would in the case of a boy. The long legs of a spider are more hair-like and more obvious than those of other small creatures such as flies, and it would as a result be the ideal symbol in this role.

These, then, are the loves and the hatreds we experience when encountering or contemplating other species. Combined with our economic, scientific and aesthetic interests, they add up to a uniquely complex inter-specific involvement, and one which changes as we grow older. We can sum this up by saying that there are 'seven ages' of inter-specific reactivity. The first age is the *infantile phase*, when we are completely dependent on our parents and react strongly to very big animals, employing them as parent symbols. The second is the *infantile-parental phase*, when we are beginning to compete with our parents and react strongly to small animals that we can use as child-substitutes. This is the age of pet-keeping. The third age is the *objective pre-adult phase*, the stage where the exploratory interests, both scientific and aesthetic, come to dominate the symbolic. It is the time for bug-hunting, microscopes, butterfly-collecting and aquaria. The fourth is the *young adult phase*. At this point the most important animals are members of the opposite sex of our own species. Other species lose ground here, except in a purely commercial or economic context. The fifth is the *adult parental phase*. Here symbolic animals enter our lives again, but this time as pets for our children. The sixth age is the *post-parental phase*, when we lose our children and may turn once more to animals as child-substitutes to replace them. (In the case of childless adults, the use of animals as child-substitutes may, of course, begin earlier.) Finally, we come to the seventh age, the *senile phase*, which is characterized by a heightened interest in animal preservation and conservation. At this point the interest is focused on those species which are in danger of extermination. It makes little difference whether, from other points of view, they are attractive or repulsive, useful or useless, providing their numbers are few and becoming fewer. The increasingly rare rhinoceros and gorilla, for example, that are so disliked by children, become the centre of attention at this stage. They have to be 'saved'. The symbolic equation involved here is obvious enough: the senile individual is about to become personally extinct and so employs rare animals as symbols of his own impending doom. His emotional concern to save them from extinction reflects his desire to extend his own survival.

During recent years, interest in animal conservation has spread to some extent into the lower age groups, apparently as a result of the development of immensely powerful nuclear weapons. Their huge destructive potential threatens all of us, regardless of age, with the possibility of immediate extermination, so that now we all have an emotional need for animals that can serve as rarity symbols.

This observation should not be interpreted as implying that this is the only reason for the conservation of wild life. There are, in addition, perfectly valid scientific and aesthetic reasons why we should wish to give aid to unsuccessful species. If we are to continue to enjoy the rich complexities of the animal world and to use wild animals as objects of scientific and aesthetic exploration, we must give them a helping hand. If we allow them to vanish, we shall have simplified our environment in a most unfortunate way. Being an intensely investigatory species, we can ill afford to lose such a valuable source of material.

Economic factors are also sometimes mentioned when conservation problems are under discussion. It is pointed out that intelligent protection and controlled cropping of wild species can assist the protein-starved populations in certain parts of the world. While this is perfectly true on a short-term basis, the long-term picture is more gloomy. If our numbers continue to increase at the present frightening rate, it will eventually become a matter of choosing between us and them. No matter how valuable they are to us symbolically, scientifically or aesthetically, the economics of the situation will shift against them. The blunt fact is that when our own species density reaches a certain pitch, there will be no space left for other animals. The argument that they constitute an essential source of food does not, unhappily, stand up to close scrutiny. It is more efficient to eat plant food direct, than to convert it into animal flesh and then eat the animals. As the demand for living space increases still further, even more drastic steps will ultimately have to be taken and we shall be driven to synthesizing our foodstuffs. Unless we can colonize other planets on a massive scale and spread the load, or seriously check our population increase in some way, we shall, in the not-too-far-distant future, have to remove all other forms of life from the earth.

If this sounds rather melodramatic, consider the figures involved. At the end of the seventeenth century the world population of naked apes was only 500 million. It has now risen to 3,000 million.* Every twenty-four hours it increases by another 150,000. (The inter-planetary emigration authorities would find this figure a daunting challenge.) In 260 years' time, if the rate of increase stays steady – which is unlikely –

---

* In the quarter of a century since this book was written, the world population has nearly doubled and is now well over 5,000 million.

there will be a seething mass of 400,000 million naked apes crowding the face of the earth. This gives a figure of 11,000 individuals to every square mile of the entire land surface. To put it another way, the densities we now experience in our major cities would exist in every corner of the globe. The consequence of this for all forms of wild life is obvious. The effect it would have on our own species is equally depressing.

We need not dwell on this nightmare: the possibility of its becoming a reality is remote. As I have stressed throughout this book, we are, despite all our great technological advances, still very much a simple biological phenomenon. Despite our grandiose ideas and our lofty self-conceits, we are still humble animals, subject to all the basic laws of animal behaviour. Long before our populations reach the levels envisaged above we shall have broken so many of the rules that govern our biological nature that we shall have collapsed as a dominant species. We tend to suffer from a strange complacency that this can never happen, that there is something special about us, that we are somehow above biological control. But we are not. Many exciting species have become extinct in the past and we are no exception. Sooner or later we shall go, and make way for something else. If it is to be later rather than sooner, then we must take a long, hard look at ourselves as biological specimens and gain some understanding of our limitations. This is why I have written this book, and why I have deliberately insulted us by referring to us as naked apes, rather than by the more usual name we use for ourselves. It helps to keep a sense of proportion and to force us to consider what is going on just below the surface of our lives. In my enthusiasm I may, perhaps, have overstated my case. There are many praises I could have sung, many magnificent achievements I could have described. By omitting them I have inevitably given a one-sided picture. We are an extraordinary species and I do not wish to deny it, or to belittle us. But these things have been said so often. When the coin is tossed it always seems to come up heads, and I have felt that it was high time we turned it over and looked at the other side. Unfortunately, because we are so powerful and so successful when compared with other animals, we find the contemplation of our humble origins somehow offensive, so that I do not expect to be thanked for what I have done. Our climb to the top has been a get-rich-quick story, and, like all *nouveaux riches*, we are very sensitive about our background. We are also in constant danger of betraying it.

Optimism is expressed by some who feel that since we have evolved a high level of intelligence and a strong inventive urge, we shall be able to twist any situation to our advantage; that we are so flexible that we can re-mould our way of life to fit any of the new demands made by our rapidly rising species-status; that when the time comes, we shall manage to cope with the over-crowding, the stress, the loss of our privacy and independence of action; that we shall re-model our behaviour patterns and live like giant ants; that we shall control our aggressive and territorial feelings, our sexual impulses and our parental tendencies; that if we have to become battery chicken-apes, we can do it; that our intelligence can dominate all our basic biological urges. I submit that this is rubbish. Our raw animal nature will never permit it. Of course, we are flexible. Of course, we are behavioural opportunists, but there are severe limits to the form our opportunism can take. By stressing our biological features in this book, I have tried to show the nature of these restrictions. By recognizing them clearly and submitting to them, we shall stand a much better chance of survival. This does not imply a naive 'return to nature'. It simply means that we should tailor our intelligent opportunist advances to our basic behavioural requirements. We must somehow improve in quality rather than in sheer quantity. If we do this, we can continue to progress technologically in a dramatic and exciting way without denying our evolutionary inheritance. If we do not, then our suppressed biological urges will build up and up until the dam bursts and the whole of our elaborate existence is swept away in the flood.

CHAPTER REFERENCES

It is impossible to list all the many works that have been of assistance in writing *The Naked Ape*, but some of the more important ones are arranged below on a chapter-by-chapter and topic-by-topic basis. Detailed references for these publications are given in the bibliography.

1 ORIGINS

Classification of primates: Morris, 1965. Napier and Napier, 1967.
Evolution of primates: Dart and Craig, 1959. Eimerl and DeVore, 1965. Hooton, 1947. Le Gros Clark, 1959. Morris and Morris, 1966. Napier and Napier, 1967. Oakley, 1961. Read, 1925. Washburn, 1962 and 1964. Tax, 1960.
Carnivore behaviour: Guggisberg, 1961. Kleiman, 1966. Kruuk, 1966. Leyhausen, 1956. Lorenz, 1954. Moulton, Ashton and Eayrs, 1960. Neuhaus, 1953. Young and Goldman, 1944.
Primate behaviour: Morris, 1967. Morris and Morris, 1966. Schaller, 1963. Southwick, 1963. Yerkes and Yerkes, 1929. Zuckerman, 1932.

2 SEX

Animal courtship: Morris, 1956.
Sexual responses: Masters and Johnson, 1966.
Sexual pattern frequencies: Kinsey *et al.*, 1948 and 1953.
Self-mimicry: Wickler, 1963 and 1967.
Mating postures: Ford and Beach, 1952.
Odour preferences: Monicreff, 1965.
Chastity devices: Gould and Pyle, 1896.
Homosexuality: Morris, 1955.

3 REARING

Suckling: Gunther, 1955. Lipsitt, 1966.
Heart-beat response: Salk, 1966.

Growth rates: Harrison, Weiner, Tanner and Barnicott, 1964.
Sleep: Kleitman, 1963.
Stages of development: Shirley, 1933.
Development of vocabulary: Smith, 1926.
Chimpanzee vocal imitations: Hayes, 1952.
Crying, smiling and laughing: Ambrose, 1960.
Facial expressions in primates: van Hooff, 1962.
Group density in children: Hutt and Vaizey, 1966.

4 EXPLORATION

Neophilia and neophobia: Morris, 1964.
Ape picture-making: Morris, 1962.
Infant picture-making: Kellogg, 1955.
Chimpanzee exploratory behaviour: Morris and Morris, 1966.
Isolation during infancy: Harlow, 1958.
Stereotyped behaviour: Morris, 1964 and 1966.

5 AGGRESSION

Primate aggression: Morris and Morris, 1966.
Autonomic changes: Cannon, 1929.
Origin of signals: Morris, 1956 and 1957.
Displacement activities: Tinbergen, 1951.
Facial expressions: van Hooff, 1962.
Eye-spot signals: Coss, 1965.
Reddening of buttocks: Comfort, 1966.
Redirection of aggression: Bastock, Morris and Moynihan, 1953.
Over-crowding in animals: Calhoun, 1962.

6 FEEDING

Male association patterns: Tiger, 1967.
Organs of taste and smell: Wyburn, Pickford and Hirst, 1964.
Cereal diets: Harrison, Weiner, Tanner and Barnicott, 1964.

7 COMFORT

Social grooming: van Hooff, 1962. Sparks, 1963. (I am particularly indebted to Jan van Hooff for inventing the term 'Grooming talk'.)
Skin glands: Montagna, 1956.

Temperature responses: Harrison, Weiner, Tanner and Barnicott, 1964.
'Medical' aid in chimpanzees: Miles, 1963.

8 ANIMALS

Domestication: Zeuner, 1963.
Animal likes: Morris and Morris, 1966.
Animal dislikes: Morris and Morris, 1965.
Animal phobias; Marks, 1966.
Population explosion: Fremlin, 1965.

THE HUMAN ZOO

CONTENTS

INTRODUCTION

When the pressures of modern living become heavy, the harassed city-dweller often refers to his teeming world as a concrete jungle. This is a colourful way of describing the pattern of life in a dense urban community, but it is also grossly inaccurate, as anyone who has studied a real jungle will confirm.

Under normal conditions, in their natural habitants, wild animals do not mutilate themselves, masturbate, attack their offspring, develop stomach ulcers, become fetishists, suffer from obesity, form homosexual pair-bonds, or commit murder. Among human city-dwellers, needless to say, all of these things occur. Does this, then, reveal a basic difference between the human species and other animals? At first glance it seems to do so. But this is deceptive. Other animals do behave in these ways under certain circumstances, namely when they are confined in the unnatural conditions of captivity. The zoo animal in a cage exhibits all these abnormalities that we know so well from our human companions. Clearly, then, the city is not a concrete jungle, it is a human zoo.

The comparison we must make is not between the city-dweller and the wild animal, but between the city-dweller and the captive animal. The modern human animal is no longer living in conditions natural for his species. Trapped, not by a zoo collector, but by his own brainy brilliance, he has set himself up in a huge, restless menagerie where he is in constant danger of cracking under the strain.

Despite the pressures, however, the benefits are great. The zoo world, like a gigantic parent, protects its inmates: food, drink, shelter, hygiene and medical care are provided; the basic problems of survival are reduced to a minimum. There is time to spare. How this time is used in a non-human zoo varies, of course, from species to species. Some animals quietly relax and doze in the sun; others find prolonged inactivity increasingly difficult to accept. If you are an inmate of a human zoo, you inevitably belong to this second category. Having an essentially exploratory, inventive brain, you will not be able to relax for very long. You will be driven on and on to more and more elaborate activities. You will investigate, organize and create and, in the end, you will have plunged yourself deeper still into an even more captive zoo world. With each new complexity, you will find yourself one step farther away from your natural tribal state, the state in which your ancestors existed for a million years.

The story of modern man is the story of his struggle to deal with the consequences of this difficult advance. The picture is confused and confusing, partly because we are involved in it in a dual role, being, at the same time, both spectators and participants. Perhaps it will become clearer if we view it from the zoologist's standpoint, and this is what I shall attempt to do in the pages that follow. In most cases I have deliberately selected examples which will be familiar to Western readers. This does not mean, however, that I intend my conclusions to relate only to Western cultures. On the contrary, there is every indication that the underlying principles apply equally to city-dwellers throughout the world.

If I seem to be saying 'Go back, you are heading for disaster,' let me assure you that I am not. We have, in our relentless social progress, gloriously unleashed our powerful inventive, exploratory urges. They are a basic part of our biological inheritance. There is nothing artificial or unnatural about them. They provide us with our great strength as well as our great weaknesses. What I am trying to show is the increasing price we have to pay for indulging them and the ingenious ways in which we contrive to meet that price, no matter how steep it becomes. The stakes are rising higher all the time, the game becoming more risky, the casualties more startling, the pace more breathless. But despite the hazards it is the most exciting game the world has ever seen. It is foolish to suggest that anyone should blow a whistle and try to stop it. Nevertheless, there are different ways of playing it, and if we can understand better the true nature of the players it should be possible to make the game even more rewarding, without at the same time becoming more dangerous and, ultimately, disastrous for the whole species.

TRIBES AND SUPER-TRIBES

IMAGINE a piece of land twenty miles long and twenty miles wide. Picture it wild, inhabited by animals small and large. Now visualize a compact group of sixty human beings camping in the middle of this territory. Try to see yourself sitting there, as a member of this tiny tribe, with the landscape, your landscape, spreading out around you farther than you can see. No one apart from your tribe uses this vast space. It is your exclusive home-range, your tribal hunting ground. Every so often the men in your group set off in pursuit of prey. The women gather fruits and berries. The children play noisily around the camp site, imitating the hunting techniques of their fathers. If the tribe is successful and swells in size, a splinter group will set off to colonize a new territory. Little by little the species will spread.

Imagine a piece of land twenty miles long and twenty miles wide. Picture it civilized, inhabited by machines and buildings. Now visualize a compact group of six million human beings camping in the middle of this territory. See yourself sitting there, with the complexity of the huge city spreading out all around you, farther than you can see.

Now compare these two pictures. In the second scene there are a hundred thousand individuals for every one in the first scene. The space has remained the same. Speaking in evolutionary terms, this dramatic change has been almost instantaneous; it has taken a mere few thousand years to convert scene one into scene two. The human animal appears to have adapted brilliantly to his extraordinary new condition, but he has not had time to change biologically, to evolve into a new, genetically civilized species. This civilizing process has been accomplished entirely by learning and conditioning. Biologically he is still the simple tribal animal depicted in scene one. He lived like that, not for a few centuries, but for a million hard years. During that period he did change biologically. He evolved spectacularly. The pressures of survival were great and they moulded him.

So much has happened in the past few thousand years, the urban years, the crowded years of civilized man, that we find it hard to grasp the idea that this is no more than a minute part of the human story. It is so familiar to us that we vaguely imagine we grew into it gradually and that, as a result, we are biologically fully equipped to deal with all the new social hazards. If we force ourselves to be coolly objective about it, we are bound to admit that this is not so. It is only our incredible plasticity, our ingenious adaptability, that makes it seem so. The simple tribal hunter is doing his best to wear his new trappings lightly and proudly; but they are complex, cumbersome garments and he keeps tripping over them. However, before we examine the way he trips and so frequently loses his balance, we must first see how he contrived to stitch together his fabulous cloak of civilization.

We must begin by lowering the temperature until we are back in the grip of the Ice Age, say, twenty thousand years ago. Our early hunting ancestors had already succeeded in spreading throughout much of the Old World and were soon to trek across from eastern Asia to the New World. To have achieved such a shattering expansion must have meant that their simple hunting way of life was already more than a match for their carnivorous rivals. But this is not so surprising when one stops to think that our Ice Age ancestors' brains were already as big and highly developed as ours are today. Skeletally, there is little to choose between us. The modern man, physically speaking, had already arrived on the scene. In fact, if it were possible, with the aid of a time machine, to take the newborn child of an Ice Age hunter into your home and rear it as your own, it is doubtful if anyone would detect the deception.

In Europe the climate was hostile, but our ancestors fought it well. With the simplest of technologies they were able to slay huge game animals. Happily they have left us a testimony of their hunting skills, not only in the accidental remnants that we can scratch up from the floors of their caves, but also in the staggering murals painted on their walls. The shaggy mammoths, woolly rhinos, bison and reindeer portrayed there leave no doubt as to the nature of the climate. As one emerges from the darkness of the caves today and steps out into the baked countryside, it is difficult to imagine it inhabited by these heavy-furred creatures. The contrast between the temperature then and now comes vividly to mind.

As the last glaciation came towards its end, the ice began to retreat northwards at a rate of fifty yards a year and the cold-country animals moved north with it. Rich forests took the place of the cold tundra landscape. The great Ice Age ended about ten thousand years ago and heralded a new epoch in human development.

The breakthrough was to come at the point where Africa, Asia and Europe meet. There, at the eastern end of the Mediterranean, there was a small change in human feeding behaviour that was to alter the whole course of man's progress. It was trivial enough and simple enough in itself, but its impact was to be enormous. Today we take it for granted: we call it farming.

Previously all human tribes had filled their bellies in one of two ways: the men had hunted for animal foods and the women had gathered plant foods. The diet was balanced by the sharing of the spoils. Virtually all the active adult members of the tribe were food-getters. There was comparatively little food-storing. They merely went out and collected what they wanted when they wanted it. This was less hazardous than it sounds because, of course, the whole world population of our species was then minute, compared with the massive numbers of today. However, although these early hunter/gatherers were highly successful and spread to cover a large part of the globe, their tribal units remained small and simple. During the hundreds of thousands of years of human evolution, men had become increasingly adapted, both physically and mentally, both structurally and behaviourally, to this hunting way of life. The new step they took, the step to farming and food-production, swept them over an unexpected threshold and threw them so rapidly into an unfamiliar form of social existence that there was no time for them to evolve new, genetically-controlled qualities to go with it. From now on, their adaptability and behavioural plasticity, their ability to learn and adjust to novel and more complex ways, were going to be tested to the full. Urbanization and the intricacies of town-living were only one more step away.

Luckily the long hunting apprenticeship had developed ingenuity and a mutual-aid system. The human hunters, it is true, were still innately competitive and self-assertive, like their monkey ancestors, but their competitiveness had become forcibly tempered by an increasingly basic urge to co-operate. It had been their only hope of succeeding in their rivalry with the long-established, sharp-clawed, professional killers of the carnivore world, such as the big cats. The human hunters had evolved their co-operativeness alongside their intelligence and their exploratory nature, and the combination had proved effective and deadly. They learned quickly, remembered well, and were adept at bringing together separate elements of their past learning to solve new problems. If this quality had been helpful to them in the early days, when on their arduous hunting trips, it was even more essential to them now, nearer to home, as they stood on the threshold of a new and vastly more complex form of social life.

The lands around the eastern end of the Mediterranean were the natural homes of two vital plants: wild wheat and wild barley. Also found in this region were wild goats, wild sheep, wild cattle and wild pigs. The human hunter/gatherers who settled in this area had already domesticated the dog, but it was used primarily as a hunting companion and watchdog, rather than as a direct source of food. True farming began with the cultivation of the two plants, wheat and barley. It was followed soon after by the domestication, first, of goats and sheep, and then, a little later, of cattle and pigs. In all probability, the animals were first attracted by the cultivated crops, came to eat, and stayed to be bred and eaten themselves.

It is no accident that the other two regions of the earth which, later on, saw the birth of independent ancient civilizations (southern Asia and central America) were also places where the hunter/gatherers found wild plants suitable for cultivation: rice in Asia and maize in America.

So successful were these Late Stone Age cultivations that, from that day to this, the plants and animals that were domesticated then have remained the major food sources of all large-scale agricultural operations. The great new advance in farming has been mechanical rather than biological. But it was what started out as the mere left-overs of early farming that were to have the truly shattering impact on our species.

Retrospectively it is easy to explain. Before the arrival of farming, everyone who was going to eat had to do their share of food-finding. Virtually the whole tribe was involved. But when the forward-thinking brains that had planned and schemed hunting manœuvres turned their attention to the problems of organizing the cultivation of crops, the irrigation of land and the breeding of captive animals, they achieved two things. Such was their success that they created, for the first time, not only a constant food supply, but also a regular and reliable food *surplus*. The creation of this surplus was the key that was to unlock the gateway to civilization. At last, the human tribe could support more members than were required to find food. The tribe could not only become bigger, but it could free some of its people for other tasks: not part-time tasks, fitted in around the priority demands of food-finding, but full-time activities that could flourish in their own right. An age of specialization was born.

From these small beginnings grew the first towns.

I have said that it is easy to explain, but all this means is that it is not difficult for us, in looking back, to pick out the vital factor that led on to the next great step in the human story. It does not, of course, mean that it was an easy step to take at the time. It is true that the human hunter/gatherer was a magnificent animal, full of untapped potentialities and capabilities. The fact that we are here today is proof enough. But he had evolved as a tribal hunter, not as a patient, sedentary farmer. It is also true that he had a far-seeing brain, capable of planning a hunt and understanding the seasonal changes in his environment. But to be a successful farmer, he had to stretch his far-seeingness beyond anything he had previously experienced. The tactics of hunting had to become the strategy of farming. This accomplished, he had to push his brain even farther to contend with the added social complexities that were to follow his new-found affluence, as villages grew into towns.

It is important to realize this when talking of an 'urban revolution'. The use of this phrase gives the impression that towns and cities began to spring up all over the place in an overnight rush towards an impressive new social life. But it was not like that. The old ways died hard and slowly. Indeed, in many parts of the world they are still with us today. Numerous contemporary cultures are still operating at virtually Neolithic farming levels, and in certain regions, such as the Kalahari Desert, Northern Australia, and the Arctic, we can still observe Palaeolithic-style communities of hunter/gatherers.

The first urban developments, the first towns and cities, grew, not as a sudden rash on the skin of prehistoric society, but as a few tiny, isolated spots. They appeared at sites in south-west Asia as dramatic exceptions to the general rule. By present-day standards they were very small and the pattern spread slowly, very slowly. Each was based on a highly localized organization, intimately connected with and bound up in the surrounding farmlands.

At first there was little trade or inter-action between one urban centre and the others. This was to be the next great advance, and it took time. The psychological barrier to such a step was obviously the loss of local identity. It was not so much a case of 'the tribe that lost its head', as the human head refusing to lose its tribe. The species had evolved as a tribal animal and the basic characteristic of a tribe is that it operates on a localized, inter-personal basis. To abandon this fundamental social pattern, so typical of the ancient human condition, was going to go against the grain. But it was the grain, in another sense, so efficiently harvested and transported, that was forcing the pace. As agriculture advanced and the urban elite, liberated from the labours of production, began to concentrate their brain-power on other, newer problems, it was inevitable that there would eventually emerge an urban network, a hierarchically organized interconnection between neighbouring towns and cities.

The oldest known town arose at Jericho more than 8,000 years ago, but the first fully urban civilization developed farther east, across the Syrian desert in Sumer. There, between 5,000 and 6,000 years ago, the first empire was born, and the 'pre' was taken out of prehistory with the invention of writing. Inter-city co-ordination developed, leaders became administrators, professions became established, metalwork and transport advanced, beasts of burden (as distinct from food animals) were domesticated and monumental architecture arose.

By our standards the Sumerian cities were small, with populations ranging from 7,000 to no more than 20,000. Nevertheless, our simple tribesman had already come a long way. He had become a citizen, a super-tribesman, and the key difference was that *in a super-tribe he no longer knew personally each member of his community*. It was this change, the shift from the personal to the impersonal society, that was going to cause the human animal its greatest agonies in the millennia ahead. As a species we were not biologically equipped to cope with a mass of strangers masquerading as members of our tribe. It was something we had to learn to do, but it was not easy. As we shall see later, we are still fighting against it today in all kinds of hidden ways – and some that are not so hidden.

As a result of the artificiality of the inflation of human social life to the super-tribal level, it became necessary to introduce more elaborate forms of controls to hold the bulging communities together. The enormous material benefits of super-tribal life had to be paid for in discipline. In the ancient civilizations which began to develop around the Mediterranean, in Egypt, Greece and Rome and elsewhere, administration and law grew heavier and more complex alongside the increasingly flourishing technologies and arts.

It was a slow process. The magnificence of the remnants of these civilizations that we marvel at today tends to make us think of them as comprising vast populations, but this was not so. In heads per super-tribe, the growth was gradual. As late as 600 B.C., the largest city, Babylon, contained no more than 80,000 people. Classical Athens had a citizen population of only 20,000, and only a quarter of these were members of the true urban elite. The total population of the whole city state, including foreign merchants, slaves, and rural as well as urban residents, has been estimated at no more than 70,000 to 100,000. To put this into perspective, it is slightly smaller than present-day university towns such as Oxford or Cambridge. The great modern metropolises, of course, bear no comparison: there are over a hundred today boasting populations exceeding one million, with the biggest exceeding ten million. Modern Athens itself contains no fewer than 1,850,000 people.

If they were to continue to grow in splendour, the ancient urban states could no longer rely on local produce. They had to augment their supplies in one of two ways: by trade or by conquest. Rome did both, but put the emphasis on conquest and carried it out with such devastating administrative and military efficiency that it was able to create the greatest city the world had seen, containing a population approaching half a million, and setting a pattern that was to echo widely down the centuries that followed. These echoes exist today, not only in the brain-straining toil of the organizers, manipulators and creative talents, but also in the increasingly idle, sensation-seeking urban elite, who have grown so numerous that they can easily turn sour with shattering effect and must be kept amused at all costs. In the sophisticated city-dweller of Imperial Rome, we can already see a prototype of the present-day super-tribesman.

In unfolding our urban tale we have, with ancient Rome, come to a stage where the human community has grown so big and is so densely packed that, zoologically speaking, we have already arrived at the modern condition. It is true that, during the centuries that followed, the plot thickened, but it was essentially the same plot. The crowds became denser, the elite became eliter, the technologies became more technical. The frustrations and stresses of city life became greater. Super-tribal clashes became bloodier. There were too many people and that meant there were people to spare, people to waste. As human relationships, lost in the crowd, became ever more impersonal, so man's inhumanity to man increased to horrible proportions. However, as I have said before, an impersonal relationship is not a biologically human one, so this is not surprising. What is surprising is that the bloated super-tribes have survived at all and, what is more, survived so well. This is not something we should accept simply because we are sitting here in the twentieth century, it is something we should marvel at. It is an astonishing testimony to our incredible ingenuity, tenacity and plasticity as a species. How on earth did we manage it? All we had to go on, as animals, was a set of biological characteristics evolved during our long hunting apprenticeship. The answer must lie in the nature of these characteristics and the way we have been able to exploit and manipulate them without distorting them as badly as we (superficially) seem to have done. We must take a closer look at them.

Bearing in mind our monkey ancestry, the social organization of surviving monkey species can provide us with some revealing clues. The existence of powerful, dominant individuals, lording it over the rest of the group, is a widespread phenomenon amongst higher primates. The weaker members of the group accept their subordinate roles. They do not rush off into the undergrowth and set up on their own. There is strength and security in numbers. When these numbers become too great, then, of course, a splinter group will form and depart, but isolated individual monkeys are abnormalities. The groups move about together and keep together at all times. This allegiance is not merely the result of an enforced tyranny on the part of the leaders, the dominant males. Despots they may be, but they also play another role, that of guardians and protectors. If there is a threat to the group from without, such as an attack from a hungry predator, it is they who are most active in defence. In the face of an external challenge, the top males must get together to meet it, their internal squabbles forgotten. But on other occasions active co-operation within the group is at a minimum.

Returning to the human animals, we can see that this basic system – social co-operation when facing outward, social competition when facing inward – also applies to us, although our early human ancestors were forced to shift the balance somewhat. Their gargantuan struggle to convert from fruit-eaters to hunters required much greater, more active, internal co-operation. The external world, in addition to providing occasional panics, now threw an almost non-stop challenge in the face of the emerging hunter. The result was a basic shift towards mutual aid, towards sharing and combining resources. This does not mean that early man began to move as one, like a shoal of fish; life was too complex for that. Competition and leadership remained, helping to provide impetus and reduce indecisiveness, but despotic authority was severely curtailed. A delicate balance was achieved and, as we have already seen, one that was to prove immensely successful, enabling the early human hunters to spread over most of the earth's surface, with only the minimum of technology to help them on their way.

What happened to this delicate balance as the tiny tribes blossomed into giant super-tribes? With the loss of the person-to-person tribal pattern, the competitive/co-operative pendulum began to swing dangerously back and forth, and it has been oscillating damagingly ever since. Because the subordinate members of the super-tribes became impersonal crowds, the most violent swings of the pendulum have been towards the domineering, competitive side. The over-grown urban groups rapidly and repeatedly fell prey to exaggerated forms of tyranny, despotism and dictatorship. The super-tribes gave rise to super-leaders, exercising powers that make the old monkey tyrants look positively benign. They also gave rise to super-subordinates in the form of slaves, who suffered subservience of a kind more extreme than anything even the most lowly of monkeys would have known.

It took more than a single despot to dominate a super-tribe in this way. Even with deadly new technologies – weapons, dungeons, tortures – to aid him in forcibly maintaining conditions of widespread subjugation, he also required a massive following if he was to succeed in holding the biological pendulum so far to one side. This was possible because the followers, like the leaders, were infected by the impersonality of the super-tribal condition. They calmed their co-operative consciences to some extent by the device of setting up sub-groups, or pseudo-tribes, within the main body of the super-tribe. Each individual established personal relationships of the old, biological type with a small, tribe-sized group of social or professional companions. Within that group he could satisfy his basic urges towards mutual aid and sharing. Other sub-groups – the slave class, for example – could then be looked upon more comfortably as outsiders beyond his protection. The social 'double standard' was born. The insidious strength of these new sub-divisions lay in the fact that they even made it possible for personal relationships to be carried on in an impersonal way. Although a subordinate – a slave, a servant, or a serf – might be personally known to a master, the fact that he had been neatly placed into another social category meant that he could be treated as badly as one of the impersonal mob.

It is only a partial truth to say that power corrupts. Extreme subjugation can corrupt equally effectively. When the bio-social pendulum swings away from active co-operation towards tyranny, the whole society becomes corrupt. It may make great material strides. It may shift 4,883,000 tons of stone to build a pyramid; but with its deformed social structure its days are numbered. You can dominate just so much, just so long and just so many, but even within the hot-house atmosphere of a super-tribe, there is a limit. If, when that limit is reached, the bio-social pendulum tilts gently back to its balanced mid-point, the society can count itself lucky. If, as is more likely, it swings wildly back and forth, the blood will flow on a scale our primitive hunting ancestor would never have dreamt of.

It is the miracle of civilized survival that the human co-operative urge reasserts itself so strongly and so repeatedly. There is so much working against it and yet it keeps on coming back. We like to think of this as the conquest of bestial weaknesses by the powers of intellectual altruism, as if ethics and morality were some kind of modern invention. If this were really true, it is doubtful if we would be here today to proclaim it. If we did not carry in us the basic biological urge to co-operate with our fellow men, we would never have survived as a species. If our hunting ancestors had really been ruthless, greedy tyrants, loaded with 'original sin', the human success story would have petered out long ago. The only reason why we are always having the doctrine of original sin instilled into us, in one form or another, is that the artificial conditions of the super-tribe keep on working against our biological altruism, and it needs all the help it can get.

I am aware that there are some authorities who will disagree violently with what I have just said. They see men as naturally inclined to be weak, greedy and wicked, requiring stern codes of imposed control to make them strong, kind and good. But when they deride the concept of the 'noble savage' they confuse the issue. They point out that there was nothing noble about ignorance or superstition, and in that respect they are right. But it is only part of the story. The other part concerns the early hunter's conduct towards his companions. Here the situation must have been different. Compassion, kindness, mutual assistance, a fundamental urge to co-operate within the tribe *must* have been the pattern for the early groups of men to survive in their precarious environment. It was only when the tribes expanded into impersonal super-tribes that the ancient pattern of conduct came under pressure and began to break down. Only then did artificial laws and codes of discipline have to be imposed to correct the balance. If these had been imposed to a degree to match the new pressures, all would have been well; but in the early civilizations men were novices at achieving this delicate balance. They failed repeatedly, with lethal results. We are more expert now, but the system has never been perfected because, as the super-tribes have continued to swell, the problem has re-set itself.

Let me put it in another way. It has often been said that 'the law forbids men to do only what their instincts incline them to do'. It follows from this that if there are laws against theft, murder and rape, then the human animal must, by nature, be a thieving, murderous rapist. Is this really a fair description of man as a social biological species? Somehow, it does not fit the zoological picture of the emerging tribal species. Sadly, however, it does fit the super-tribal picture.

Theft, perhaps the most common of crimes, is a good example. A member of a super-tribe is under pressure, suffering from all the stresses and strains of his artificial social condition. Most people in his super-tribe are strangers to him; he has no personal, tribal bond with them. The typical thief is not stealing from one of his known companions. He is not breaking the old, biological tribal code. In his mind, he is simply setting his victim outside his tribe altogether. To counteract this, a super-tribal law has to be imposed. It is relevant here that we sometimes talk of 'honour among thieves' and the 'code of the underworld'. This underlines the fact that we look upon criminals as belonging to a separate and distinct pseudo-tribe within the super-tribe. In passing, it is interesting to note how we deal with the criminal: we shut him away in a confined, all-criminal community. As a short-term solution it works well enough, but the long-term effect is that it strengthens his pseudo-tribal identity instead of weakening it and furthermore helps him to widen his pseudo-tribal social contacts.

Reconsidering the idea that 'the law forbids men to do only what their instincts incline them to do', we might re-word it to the effect that 'the law forbids men to do only what the artificial conditions of civilization drive them to do'. In this way we can see the law as a balancing device, tending to counteract the distortions of super-tribal existence and helping to maintain, in unnatural conditions, the forms of social conduct natural to the human species.

However, this is an over-simplification. It implies perfection in the leaders, the law-makers. Tyrants and despots can, of course, impose harsh and unreasonable laws restraining the population to a greater extent than is justified by the prevailing super-tribal conditions. A weak leadership may impose a system of law that lacks the strength to hold together a teeming populace. Either way lies cultural disaster or decline.

There is also another kind of law that has little to do with the argument I have been putting forward, except that it serves to hold society together. It is an 'isolating law', one which helps to make one culture different from another. It gives cohesion to a society by providing it with a unique identity. These laws play only a minor role in the law courts. They are more the concern of religion and social custom. Their function is to increase the illusion that one belongs to a unified tribe rather than a sprawling, seething super-tribe. If they are criticized because they seem arbitrary or meaningless, the answer comes back that they are traditional and must be obeyed without question. It is as well not to question them because, in themselves, they *are* arbitrary and frequently meaningless. Their value lies in the fact that they are shared by all the members of the community. When they fade, the unity of the community fades a little, too. They take many forms: the elaborate procedures of social ceremonies – marriages, burials, celebrations, parades, festivals and the rest; the intricacies of social etiquette, manners and protocol; the complexities of social costume, uniform decoration, adornment and display.

These subjects have been studied in detail by ethnologists and cultural anthropologists, who have been fascinated by their great diversity. Diversity, the differentiation of one culture from another, has, of course, been the very function of these patterns of behaviour. But in marvelling at their variety, one must not overlook their fundamental similarities. The customs and costumes may be strikingly different *in detail* from culture to culture, but they have the same basic function and the same basic forms. If you start by making a list of all the social customs of one particular culture, you will find equivalents to nearly all of them in nearly all other cultures. Only the details will differ, and they will differ so wildly that they will sometimes obscure the fact that you are dealing with the same basic social patterns.

To give an example: in some cultures ceremonies of mourning involve the wearing of black costumes; in others, by complete contrast, the mourning dress is white. Furthermore, if you cast the net wide enough, you can find still other cultures that employ dark blue, or grey, or yellow, or natural brown sackcloth. Having grown up yourself in a culture where, from early childhood, one of these colours, say black, has been heavily associated with death and mourning, it will be startling to think of wearing such colours as yellow or blue in this context. Therefore, your immediate reaction on discovering that these colours *are* worn as mourning dress in other places is to remark on how different they are from your own familiar custom. Herein lies the trap, so neatly set by the demands of cultural isolation. The superficial observation that the colours vary so dramatically obscures the more basic fact that all these cultures share the performance of a mourning 'display', and that in all of them it involves the wearing of a costume that is strikingly different from the non-mourning costume.

In the same way, when an Englishman first visits Spain, he is surprised to find the public spaces of the towns and villages thronged with people in the early evening, all wandering up and down in an apparently aimless way. His immediate reaction is not that this is their cultural equivalent of his more familiar cocktail parties, but rather that it is some sort of strange local custom. Again, the basic social pattern is the same, but the details differ.

Similar examples could be given to cover almost all forms of communal activity, the principle being that the more social the occasion, the more variable the details and the stranger the other culture's behaviour appears to be at first sight. The greatest social occasions of all, such as coronations, state funerals, balls, banquets, independence celebrations, investitures, great sporting events, military parades, festivals and garden parties (or their equivalents), are the ones where the isolating laws play the strongest part. They vary from case to case in a thousand tiny details, each of which is scrupulously attended to, as though the very lives of the participants depended upon it. In a sense, of course, their social lives do depend on it, for it is only by their conduct in public places that they can strengthen and support their feelings of social identity, of belonging to a cultural group, and the grander the occasion, the stronger the boost.

This is a fact that successful revolutionaries sometimes overlook or underestimate. In ridding themselves of the old power structure that they have come to detest, they are forced to sweep away with it most of the old ceremonials. Even though these ritual procedures may have nothing directly to do with the overthrown power system, they are too strongly reminiscent of it and must go. A few hurriedly improvised performances may be put in their place, but it is difficult to invent rituals overnight. (It is an interesting sidelight on the Christian movement that its early success depended to some extent on its making a take-over bid for many of the old pagan ceremonies and incorporating them, suitably disguised, into their own festive occasions.) When all the excitement and upheaval of the revolution is over, the eventual unhappiness of many a disgruntled post-revolutionary is due, in a concealed way, to his sense of loss of social occasion and pageantry. Revolutionary leaders would do well to anticipate this problem. It is not the chains of social identity that their followers will want to break, it is the chains of a *particular* social identity. As soon as these are smashed, they will need new ones and will soon become dissatisfied merely with an abstract sense of 'freedom'. Such are the demands of the isolating laws.

Other aspects of social behaviour are also brought into play as cohesive forces. Language is one of these. We tend to think of language exclusively as a communication device, but it is more than that. If it were not, we should all be speaking with the same tongue. Looking back through super-tribal history, it is easy to see how the anti-communication function of language has been almost as important as its communication function. More than any social custom, it has set up enormous inter-group barriers. More than anything else, it has identified an individual as a member of a particular super-tribe, and put obstacles in the way of his defecting to another group.

As the super-tribes have grown and merged, so local languages have been merged, or submerged, and the total number in the world is being reduced. But as this happens, a counter-trend develops: accents and dialects become more socially significant; slang, cant and jargon are invented. As the members of a massive super-tribe attempt to strengthen their tribal identities by setting up sub-groups, so a whole spectrum of 'tongues' develops within the official, major language. Just as English and German act as identity badges and isolating mechanisms between an Englishman and a German, so an upper-class English accent isolates its owner from a lower-class one, and the jargon of chemistry and psychiatry isolates chemists from psychiatrists. (It is a sad fact that the academic world which, in its educational role, should be devoted to communication above all else, exhibits pseudo-tribal isolating languages as extreme as criminal slang. The excuse is that precision of expression demands it. This is true up to a point, but the point is frequently and blatantly exceeded.)

The jargon of slang words can become so specialized that it is almost as if a new language is being born. It is typical of slang expressions that once they spread and become common property, they are replaced with new terms by the originating sub-group. If they are adopted by the whole super-tribe and slide into the official language, then they have lost their original function. (It is doubtful whether you are using the same slang expression for, say, an attractive girl, a policeman, or a sexual act, that your parents employed when they were your age. But you still use the same official words.) In extreme cases, one sub-group will adopt an entirely foreign language. The Russian court at one point, for example, spoke in French. In Britain one still sees remnants of this kind of behaviour at the more expensive restaurants, where the menus are usually in French.

Religion has operated in much the same way as language, strengthening bonds within a group and weakening them between groups. It acts on the single, simple premise that there are powerful forces at work above and beyond the ordinary human members of the group and that these forces, these super-leaders or gods, must be pleased, appeased and obeyed without question. The fact that they are never around to be questioned helps them to maintain their position.

At first the powers of the gods were limited and their spheres of influence divided, but as the super-tribes swelled to increasingly unmanageable proportions, greater cohesive forces were needed. A government of minor gods was not strong enough. A massive super-tribe required a single, all-powerful, all-wise, all-seeing god, and from the ancient candidates it was this type that won through and survived the passage of the centuries. In the smaller and more backward cultures today the minor gods still rule, but members of all the major cultures have turned to the single super-god.

It is a common observation that the power of religion as a social force has been weakening during recent years. There are two reasons for this. In the first place, it is failing to serve its double function as a cohesive influence. As the populations continued to grow and swell, the ancient empires became unmanageable and split up into national groups. The new super-tribes fought to establish their identities, using all the usual devices. But many of them now shared a common religion. This meant that, for them, religion, although still a powerful force for bonding together the members of a nation, failed in its other cohesive function, namely to weaken the bonds between nations. A compromise was reached by the formation of sects within a major religion. Although sectarianism put back some of the isolating qualities and helped to tribalize, or localize, religious ceremonies again, it was only a partial solution.

The second reason for its loss of power was the growing standard of widespread scientific education, with its increasing demands that the individual *should* ask questions, rather than blindly accept dogmas. The Christian religion, in particular, has suffered serious setbacks. The increasingly logical mind of the Western super-tribesman cannot help but notice certain glaring illogicalities. Perhaps the most important of these is the great discrepancy between, on the one hand, the teaching of humility and gentleness and, on the other, the elaborate finery, pomp and power of the Church leaders.

In addition to law, custom, language and religion there is another, more violent form of cohesive force that helps to bind the members of a super-tribe together, and that is war. To put it cynically, one could say that nothing helps a leader like a good war. It gives him his only chance of being a tyrant and being loved for it at the same time. He can introduce the most ruthless forms of control and send thousands of his followers to their deaths and still be hailed as a great protector. Nothing ties tighter the in-group bonds than an out-group threat.

The fact that internal squabbles are suppressed by the existence of a common enemy has not escaped the attention of rulers past and present. If an overgrown super-tribe is beginning to split at the seams, the splits can rapidly be stitched up by the appearance of a powerful hostile THEM that converts us into a unified US. Just how often leaders have deliberately manipulated an inter-group clash with this in mind, it is hard to say, but whether the move is consciously deliberate or not, the cohesive reaction nearly always occurs. It takes a remarkably inefficient leader to bungle it. Of course, he has to have an enemy who is capable of being painted in sufficiently villainous colours, or he is likely to be in trouble. The disgusting horrors of war become converted into glorious battles only when the threat from outside is really serious, or can be made to seem so.

Despite its attractions to a ruthless leader, war has an obvious disadvantage: one side is liable to be utterly defeated and it might be his. The super-tribesman can be grateful for this unfortunate drawback.

These, then, are the cohesive forces that are brought to bear on the great urban societies. Each has developed its own specialized kind of leader: the administrator, the judge, the politician, the social leader, the high priest, the general. In simpler times they were all rolled into one, an omnipotent emperor or king who was able to cope with the whole range of leadership. But as time has passed and the groups have expanded, the true leadership has shifted from one sphere to another, moving to whichever category happens to contain the most exceptional individual.

In more recent times it has often become the practice to allow the populace to have a say in the election of a new leader. This political device has, in itself, been a valuable cohesive force, giving the super-tribesmen a greater feeling of 'belonging' to their group and having some influence over it. Once the new leader has been elected it soon becomes clear that the influence is slighter than imagined, but nevertheless at the time of the election itself· a valuable ripple of social identity runs through the community.

As an aid to this process, local pseudo-tribal sub-leaders are sent to participate in the government of the land. In some countries this has become little more than a ritual act, since the 'local' representatives are no more than imported professionals. However, this type of distortion is inevitable in a community as complex as a modern super-tribe.

The aim of government by elected representatives is fine and clear, even if it is difficult to match in practice. It is based on a partial return to the 'politics' of the original human tribal system, where each member of the tribe (or at least, the adult males) had a say in the running of the society. They were, in a sense, communists, with the emphasis on sharing and with little regard for the rigid protection of personal property. Property was as much for giving as for keeping. But as I have said before, the tribes were small and everyone knew everyone else. They may have prized individual possessions, but doors and locks were things of the future. As soon as the tribe had become an impersonal super-tribe, with strangers in its midst, the rigorous protection of property became necessary and began to play a much larger role in social life. Any political attempt to ignore this fact would meet with considerable difficulties. Modern communism is beginning to find this out and has already started to adjust its system accordingly.

Another adjustment was also necessary in all cases where the aim was to reinstate the ancient tribal-hunter pattern of 'government of the people, by the people'. The super-tribes were simply too big and the problems of government too complex, too technical. The situation demanded a system of representation and it demanded a professional class of experts. Just how far this can get away from 'government by the people' was clearly illustrated in England recently when it was suggested that parliamentary debates should be televised so that, thanks to modern science, the populace could at last play a more intimate role in the affairs of state. But this would have interfered with the specialized, professional atmosphere and was hotly opposed and rejected. So much for government by the people. However, this is not surprising. Running a super-tribe is like trying to balance an elephant on a tightrope. It seems that the best that any modern political system can hope for is to use right-wing methods to implement left-wing policies. (This is, in effect, what is being done both in the East and the West at the present moment.) It is a difficult trick to pull off and requires great professional finesse, not to mention double-talk. If modern politicians are frequently the subject of scorn and satire, it is because too many people see through the trick too often. But given the size of the present super-tribes there appears to be no alternative.

Because the modern super-tribes are in so many ways socially unmanageable, there has been a great tendency for them to fragment. I have already mentioned the way in which specialized pseudo-tribes crystallize inside the main body, as social groups, class groups, professional groups, academic groups, sports groups and so on, reinstating for the urban individual various forms of tribal identity. These groups remain happily enough within the main community, but more drastic splits than this frequently occur. Empires split into independent countries; countries split into self-ruling sectors. Despite improved communications, despite increasingly shared aims and common policies, the splits go on. Alliances can be quickly forged under the cohesive pressure of war, but during peace, separations and divisions are the order of the day. When splinter-groups desperately struggle to forge some kind of local identity, it simply means that the cohesive forces of the super-tribe to which they belonged were not strong enough or exciting enough to hold them together.

The dream of a peaceful, global super-tribe is repeatedly being shattered. It seems as if only an alien threat from another planet would provide the necessary cohesive force, and then only temporarily. It remains to be seen whether, in the future, man's ingenuity will introduce some new factor into his social existence that will solve the problem. At the moment it appears unlikely.

There has been a great deal of debate recently concerning the way in which modern mass-communication devices, such as television, are 'shrinking' the social surface of the world, creating a global televillage. It has been suggested that this trend will aid the move towards a genuinely international community. Unhappily this is a myth, for the single reason that television, unlike personal social intercourse, is a *one-way* system. I can listen to and get to know a tele-speaker, but he cannot listen to or get to know me. True, I can learn what he is thinking and doing and this is admittedly a great advantage, broadening my range of social information, but it is no substitute for the two-way relationships of real social contacts.

Even if startlingly new and at present unimagined advances in mass-communication techniques are made in the years to come, they will continue to be hampered by the bio-social limitations of our species. We are not equipped, like termites, to become willing members of a vast community. We are and probably always will be, at base, simple tribal animals.

Yet despite this, and despite the spasmodic fragmentations that are constantly occurring around the globe, we are bound to face the fact that the major trend is still to maintain the massive super-tribal levels. While splits are occurring in one part of the world, mergers are developing in another. If the situation remains as unstable today as it has been for centuries, then why do we persist with it? If it is so dangerous, why do we keep it up?

It is far more than just an international power game. There is an intrinsic, biological property of the human animal that obtains deep satisfaction from being thrown into the urban chaos of a super-tribe. That quality is man's insatiable curiosity, his inventiveness, his intellectual athleticism. The urban turmoil seems to energize this quality. Just as colony-nesting sea-birds are reproductively aroused by massing in dense breeding communities, so the human animal is intellectually aroused by massing in dense urban communities. They are breeding colonies of human ideas. This is the credit side of the story. It keeps the system going despite its many disadvantages.

We have looked at some of these disadvantages on the social level, but they exist on the personal level as well. Individuals living in a large urban complex suffer from a variety of stresses and strains: noise, polluted air, lack of exercise, cramping of space, overcrowding, over-stimulation and, paradoxically, for some, isolation and boredom.

You may think that the price the super-tribesman is paying is too high; that a quiet, peaceful, contemplative life would be far preferable. He thinks so, too, of course, but like that physical exercise he is always going to take, he seldom does anything about it. Moving to the suburbs is about as far as he goes. There he can create a pseudo-tribal atmosphere away from the strains of the big city, but come Monday morning and he is dashing back into the fray again. He could move away, but he would miss the excitement, the excitement of the neo-hunter, setting off to capture the biggest game in the biggest and best hunting grounds his environment has to offer.

On this basis one might expect every great city to be a raging inferno of novelty and inventiveness. Compared with a village it may seem to be so, but it is very far from reaching its exploratory limits. This is because there is a fundamental clash between the cohesive and inventive forces of society. The one tends to keep things steady and therefore repetitive and static. The other strains on to new developments and the inevitable rejection of old patterns. Just as there is a conflict between competition and co-operation, so there is a fight between conformity and innovation. Only in the city does sustained innovation stand a real chance. Only the city is strong enough and secure enough in its amassed conformity to tolerate the disruptive forces of rebellious originality and creativity. The sharp swords of iconoclasm are mere pin-pricks in the giant's flesh, giving it a pleasant tingling sensation, rousing it from sleep and urging it into action.

This exploratory excitement, then, with the help of the cohesive forces I have described, is what keeps so many modern city-dwellers voluntarily locked inside their human zoo cages. The exhilarations and challenges of super-tribal living are so great that with a little assistance they can outweigh the enormous dangers and disadvantages. But how do the drawbacks measure up to those of the animal zoo?

The animal zoo inmate finds itself in solitary confinement, or in an abnormally distorted social group. Alongside, in other cages, it may be able to see or hear other animals but it cannot make any real contact with them. Ironically, the super-social conditions of human urban life can work in much the same way. The loneliness of the city is a well-known hazard. It is easy to become lost in the great impersonal crowd. It is easy for natural family groupings and personal tribal relationships to become distorted, crushed or fragmented. In a village all the neighbours are personal friends or, at worst, personal enemies; none are strangers. In a large city many people do not even know the names of their neighbours.

This de-personalizing does help to support the rebels and innovators who, in a smaller, tribal community, would be subjected to much greater cohesive forces. They would be flattened by the demands of conformity. But at the same time the paradox of the social isolation of the teeming city can cause a great deal of stress and misery for many of the human zoo inmates.

Apart from the personal isolation there is also the direct pressure of physical crowding. Each kind of animal has evolved to exist in a certain amount of living space. In both the animal zoo and the human zoo this space is severely curtailed and the consequences can be serious. We think of claustrophobia as an abnormal response. In its extreme form it is, but in a milder, less clearly recognized form it is a condition from which all city-dwellers suffer. Half-hearted attempts are made to correct this. Special sections of the city are set aside as a token gesture towards providing open spaces – small bits of 'natural environment' called parks. Originally parks were hunting-grounds containing deer and other prey species, where rich super-tribesmen could re-live their ancestral patterns of hunting behaviour; but in modern city parks only the plant life remains.

In terms of quantity of space, the city park is a joke. It would have to cover thousands of square miles to provide a truly natural amount of wandering space for the huge city population it serves. The best that can be said for it is that it is decidedly better than nothing.

The alternative for the urban space-seekers is to make brief sorties into the countryside, and this they do with great vigour. Bumper to bumper the cars set off each weekend, and bumper to bumper they return. But no matter, they have wandered – they have patrolled a broader home range – and in so doing have kept up the fight against the unnatural spatial cramping of the city. If the crowded roads of the modern super-tribe have turned this into something of a ritual, it is still preferable to giving up. The position is even worse for the inmates of the animal zoo. Their version of bumper-to-bumper patrolling is the even more stultified pacing to and fro across their cage floor. But they do not give up either. We should be thankful that we can do more than pace back and forth across our living-room floors.

Having now traced the course of events that has led us to our present social condition, we can start to examine in more detail the various ways in which our behaviour patterns have succeeded in adjusting to life in the human zoo, or, in some instances, how they have disastrously failed to do so.

STATUS AND SUPER-STATUS

I N ANY ORGANIZED group of mammals, no matter how co-operative, there is always a struggle for social dominance. As he pursues
this struggle, each adult individual acquires a particular social rank,
giving him his position, or status, in the group hierarchy. The situation never remains stable for very long, largely because all the status
strugglers are growing older. When the overlords, or 'top dogs', become
senile, their seniority is challenged and they are overthrown by their
immediate subordinates. There is then renewed dominance squabbling
as everyone moves a little farther up the social ladder. At the other end
of the scale, the younger members of the group are maturing rapidly,
keeping up the pressure from below. In addition, certain members of the
group may suddenly be struck down by disease or accidental death,
leaving gaps in the hierarchy that have to be quickly filled.

The general result is a constant condition of *status tension*. Under
natural conditions this tension remains tolerable because of the limited
size of the social groupings. If, however, in the artificial environment
of captivity, the group size becomes too big, or the space available too
small, then the status 'rat race' soon gets out of hand, dominance
battles rage uncontrollably, and the leaders of the packs, prides, colonies or tribes come under severe strain. When this happens, the
weakest members of the group are frequently hounded to their deaths,
as the restrained rituals of display and counter-display degenerate into
bloody violence.

There are further repercussions. So much time has to be spent sorting
out the unnaturally complex status relationships that other aspects of
social life, such as parental care, become seriously and damagingly
neglected.

If the settling of dominance disputes creates difficulties for the
moderately crowded inmates of the animal zoo, then it is obviously
going to provide an even greater dilemma for the vastly overgrown
super-tribes of the human zoo. The essential feature of the status
struggle in nature is that it is based on the *personal* relationships of the
individuals inside the social group. For the primitive human tribesman
the problem was therefore a comparatively simple one, but when the
tribes grew into super-tribes and relationships became increasingly

impersonal, the problem of status rapidly expanded into the nightmare of super-status.

Before we probe this tender area of urban life, it will be helpful to take a brief look at the basic laws which govern the dominance struggle. The best way to do this is to survey the battlefield from the viewpoint of the dominant animal.

If you are to rule your group and to be successful in holding your position of power, there are ten golden rules you must obey. They apply to all leaders, from baboons to modern presidents and prime ministers. The ten commandments of dominance are these:

1. *You must clearly display the trappings, postures and gestures of dominance.*

For the baboon this means a sleek, beautifully groomed, luxuriant coat of hair; a calm, relaxed posture when not engaged in disputes; a deliberate and purposeful gait when active. There must be no outward signs of anxiety, indecision or hesitancy.

With a few superficial modifications, the same holds true for the human leader. The luxuriant coat of fur becomes the rich and elaborate costume of the ruler, dramatically excelling those of his subordinates. He assumes postures unique to his dominant role. When he is relaxing, he may recline or sit, while others must stand until given permission to follow suit. This is also typical of the dominant baboon, who may sprawl out lazily while his anxious subordinates hold themselves in more alert postures near by. The situation changes once the leader stirs into aggressive action and begins to assert himself. Then, be he baboon or prince, he must rise into a more impressive position than that of his followers. He must literally rise above them, matching his psychological status with his physical posture. For the baboon boss this is easy: a dominant monkey is nearly always much larger than his underlings. He has only to hold himself erect and his greater body size does the rest. The situation is enhanced by cringing and crouching on the part of his more fearful subordinates. For the human leader, artificial aids may be necessary. He can magnify his size by wearing large cloaks or tall headgear. His height can be increased by mounting a throne, a platform, an animal, or a vehicle of some kind, or by being carried aloft by his followers. The crouching of the weaker baboons becomes stylized in various ways: subordinate humans lower their height by bowing, curtsying, kneeling, kowtowing, salaaming or prostrating.

The ingenuity of our species permits the human leader to have it both ways. By sitting on a throne on a raised platform, he can enjoy both the relaxed position of the passive dominant *and* the heightened position of the active dominant at one and the same time, thus providing himself with a doubly powerful display posture.

The dignified displays of leadership that the human animal shares with the baboon are still with us in many forms today. They can be seen in their most primitive and obvious conditions in generals, judges, high priests and surviving royalty. They tend to be more limited to special occasions than they once were, but when they do occur they are as ostentatious as ever. Not even the most learned academics are immune to the demands of pomp and finery on their more ceremonial occasions.

Where emperors have given way to elected presidents and prime ministers, personal dominance displays have, however, become less overt. There has been a shift of emphasis in the role of leadership. The new-style leader is a servant of the people who happens to be dominant, rather than a dominator of the people who also serves them. He underlines his acceptance of this situation by wearing a comparatively drab costume, but this is only a trick. It is a minor dishonesty that he can afford, to make him seem more 'one of the crowd', but he dare not carry it too far or, before he knows it, he really will have become one of the crowd again. So, in other, less blatantly personal ways, he must continue to perform the outward display of his dominance. With all the complexities of the modern urban environment at his disposal, this is not difficult. The loss of grandeur in his dress can be compensated for by the elaborate and exclusive nature of the rooms in which he rules and the buildings in which he lives and works. He can retain ostentation in the way he travels, with motorcades, outriders and personal planes. He can continue to surround himself with a large group of 'professional subordinates' – aides, secretaries, servants, personal assistants, bodyguards, attendants, and the rest – part of whose job is merely to be seen to be servile towards him, thereby adding to his image of social superiority. His postures, movements and gestures of dominance can be retained unmodified. Because the power signals they transmit are so basic to the human species, they are accepted unconsciously and can therefore escape restriction. His movements and gestures are calm and relaxed, or firm and deliberate. (When did you last see a president or a prime minister running, except when taking voluntary exercise?) In conversation he uses his eyes like weapons, delivering a fixed stare at moments when subordinates would be politely averting their gaze, and turning his head away at moments when subordinates would be watching intently. He does not scrabble, twitch, fidget or falter. These are essentially the reactions of subordinates. If the leader performs them there is something seriously wrong with him in his role as the dominant member of the group.

2. *In moments of active rivalry you must threaten your subordinates*
   aggressively.

At the slightest sign of any challenge from a subordinate baboon, the
group leader immediately responds with an impressive display of
threatening behaviour. There is a whole range of threat displays available, varying from those motivated by a lot of aggression tinged with
a little fear to those motivated by a lot of fear and only a little
aggression. The latter – the 'scared threats' of weak-but-hostile individuals – are never shown by a dominant animal unless his leadership
is tottering. When his position is secure he shows only the most
aggressive threat displays. He can be so secure that all he needs to do
is to indicate that he is about to threaten, without actually bothering
to carry it through. A mere jerk of his massive head in the direction of
the unruly subordinate may be sufficient to subdue the inferior individual. These actions are called 'intention movements', and they operate
in precisely the same way in the human species. A powerful human
leader, irritated by the actions of a subordinate, need only jerk his head
in the latter's direction and fix him with a hard stare to assert his
dominance successfully. If he has to raise his voice or repeat an order,
his dominance is slightly less secure, and he will, on eventually regain-
ing control, have to re-establish his status by administering a rebuke or
a symbolic punishment of some kind.

The act of raising his voice, or raging, is only a weak sign in a
leader when it occurs as a reaction to an immediate threat. It may also
be used spontaneously or deliberately by a strong ruler as a general
device for reaffirming his position. A dominant baboon may behave in
the same way, suddenly charging at his subordinates and terrorizing
them, reminding them of his powers. It enables him to chalk up a few
points, and after that he can more easily get his own way with the
merest nod of his head. Human leaders perform in this manner from
time to time, issuing stern edicts, making lightning inspections, or
haranguing the group with vigorous speeches. If you are a leader, it is
dangerous to remain silent, unseen or unfelt for too long. If natural
circumstances do not prompt a show of power, the circumstances must
be invented that do. It is not enough to have power, one must be
observed to have power. Therein lies the value of spontaneous threat
displays.

3. *In moments of physical challenge you (or your delegates) must be*
   able forcibly to overpower your subordinates.

If a threat display fails, then a physical attack must follow. If you are a baboon boss this is a dangerous step to take, for two reasons. Firstly, in a physical fight even the winner may be damaged, and injury is more serious for a dominant animal than for a subordinate. It makes him less daunting for a subsequent attacker. Secondly, he is always outnumbered by his subordinates, and if they are driven too far they may gang up on him and overpower him in a combined effort. It is these two facts that make threat rather than actual attack the preferred method for dominant individuals.

The human leader overcomes this to some extent by employing a special class of 'suppressors'. They, the military or police, are so specialized and professional at their task that only a general uprising of the whole populace would be strong enough to beat them. In extreme cases, a despot will employ a further, even more specialized class of suppressors (such as secret police), whose job it is to suppress the ordinary suppressors if they happen to get out of line. By clever manipulation and administration it is possible to run an aggressive system of this kind in such a way that only the leader knows enough of what is happening to be able to control it. Everyone else is in a state of confusion unless they have orders from above and, in this way, the modern despot can hold the reins and dominate effectively.

4. *If a challenge involves brain rather than brawn you must be able to outwit your subordinates.*

The baboon boss must be cunning, quick and intelligent as well as strong and aggressive. This is obviously even more important for a human leader. In cases where there is a system of inherited leadership, the stupid individual is quickly deposed or becomes the mere figurehead and pawn of the true leaders.

Today the problems are so complex that the modern leader is forced to surround himself with intellectual specialists, but despite this he cannot escape the need for quick-wittedness. It is he who must make the final decisions, and make them sharply and clearly, without faltering. This is such a vital quality in leadership that it is more important to make a firm, unhesitating decision than it is to make the 'right' one. Many a powerful leader has survived occasional wrong decisions, made with style and forcefulness, but few have survived hesitant indecisiveness. The golden rule of leadership here, which in a rational age is an unpleasant one to accept, is that it is the manner in which you do something that really counts, rather than what you do. It is a sad truth that a leader who does the wrong things in the right way will, up to a certain point, gain greater allegiance and enjoy more success than one who does the right things in the wrong way. The progress of civilization has repeatedly suffered as a result of this. Lucky indeed is the society whose leader does the right things and at the same time obeys the ten golden rules of dominance; lucky – and rare, too. There appears to be a sinister, more-than-chance relationship between great leadership and aberrant policies.

It seems as if one of the curses of the immense complexity of the super-tribal condition is that it is almost impossible to make sharp, clear-cut decisions, concerning major issues, on a rational basis. The evidence available is so complicated, so diverse and frequently so contradictory that any reasonable, rational decision is bound to involve undue hesitancy. The great super-tribal leader cannot enjoy the luxury of ponderous restraint and 'further examination of the facts' so typical of the great academic. The biological nature of his role as a dominant animal forces him to make a snap decision or lose face.

The danger is obvious: the situation inevitably favours, as great leaders, rather abnormal individuals, fired by some kind of obsessive fanaticism, who will be prepared to cut through the mass of conflicting evidence that the super-tribal condition throws up. This is one of the prices that the biological tribesman must pay for becoming an artificial super-tribesman. The only solution is to find a brilliant, rational, balanced, deep-thinking brain housed in a glamorous, flamboyant, self-assertive, colourful personality. Contradictory? Yes. Impossible? Perhaps; but there is a glimmer of hope in the fact that the very size of the super-tribe, which causes the problem in the first place, also offers literally millions of potential candidates.

5. *You must suppress squabbles that break out between your subordinates.*

If a baboon leader sees an unruly squabble taking place he is likely to interfere and suppress it, even though it does not in any way constitute a direct threat to himself. It gives him another opportunity of displaying his dominance and at the same time helps to maintain order inside the group. Interference of this kind from the dominant animal is directed particularly at squabbling juveniles, and helps to instil in them, at an early age, the idea of a powerful leader in their midst.

The equivalent of this behaviour for the human leader is the control and administration of the laws of his group. The rulers of the earlier and smaller super-tribes were powerfully active in this respect, but there has been increasing delegation of these duties in modern times, due to the increasing weight of other burdens that relate more directly to the status of the leader. Nevertheless, a squabbling community is an inefficient one and some degree of control and influence has to be retained.

6. *You must reward your immediate subordinates by permitting them to enjoy the benefits of their high ranks.*

The sub-dominant baboons, although they are the leader's worst rivals, are also of great help to him in times of threat from outside the group. Further, if they are too strongly suppressed they may gang up on him and depose him. They therefore enjoy privileges which the weaker members of the group cannot share. They have more freedom of action and are permitted to stay closer to the dominant animal than are the junior males.

Any human leader who has failed to obey this rule has soon found himself in difficulties. He needs more help from his sub-dominants, and is in greater danger of a 'palace revolt', than his baboon equivalent. So much more can go on behind his back. The system of rewarding the sub-dominants requires brilliant expertise. The wrong sort of reward gives too much power to a serious rival. The trouble is that a true leader cannot enjoy true friendship. True friendship can only be fully expressed between members of roughly the same status level. A partial friendship can, of course, occur between a dominant and a subordinate, at any level, but it is always marred by the difference in rank. No matter how well meaning the partners in such a friendship may be, condescension and flattery inevitably creep in to cloud the relationship. The leader, at the very peak of the social pyramid, is, in the full sense of the word, permanently friendless; and his partial friends are perhaps more partial than he likes to think. As I said, the giving of favours requires an expert hand.

7. *You must protect the weaker members of the group from undue persecution.*

Females with young tend to cluster around the dominant male baboon. He meets any attack on these females or on unprotected infants with a savage onslaught. As a defender of the weak he is ensuring the survival of the future adults of the group. Human leaders have increasingly extended their protection of the weak to include also the old, the sick and the disabled. This is because efficient rulers not only need to defend the growing children, who will one day swell the ranks of their followers, but also need to reduce the anxieties of the active adults, all of whom are threatened with eventual senility, sudden sickness, or possible disability. With most people the urge to give aid in such cases is a natural development of their biologically co-operative nature. But for the leaders it is also a question of making people work more efficiently by taking a serious weight off their minds.

8. *You must make decisions concerning the social activities of your group.*

When the baboon leader decides to move, the whole group moves. When he rests, the group rests. When he feeds, the group feeds. Direct control of this kind is, of course, lost to the leader of a human super-tribe, but he can nevertheless play a vital role in encouraging the more abstract directions his group takes. He may foster the sciences or push towards a greater military emphasis. As with the other golden rules of leadership, it is important for him to exercise this one even when it does not appear to be strictly necessary. Even if a society is cruising happily along on a set and satisfactory course, it is vital for him to change that course in certain ways in order to make his impact felt. It is not enough simply to alter it as a reaction to something that is going wrong. He must spontaneously, of his own volition, insist on new lines of development, or he will be considered weak and colourless. If he has no ready-made preferences and enthusiasms, he must invent them. If he is seen to have what appear to be strong convictions on certain matters, he will be taken more seriously on *all* matters. Many modern leaders seem to overlook this and their political 'platforms' are desperately lacking in originality. If they win the battle for leadership it is not because they are more inspiring than their rivals but simply because they are less uninspiring.

9. *You must reassure your extreme subordinates from time to time.*

If a dominant baboon wishes to approach a subordinate peacefully, it may have difficulty doing so, because its close proximity is inevitably threatening. It can overcome this by performing a reassurance display. This consists of a very gentle approach, with no sudden or harsh movements, accompanied by facial expressions (called lip-smacking) which are typical of friendly subordinates. This helps to calm the fears of the weaker animal and the dominant one can then come near.

Human leaders, who may be characteristically tough and unsmiling with their immediate subordinates, frequently adopt an attitude of friendly submissiveness when coming into personal contact with their extreme subordinates. Towards them they offer a front of exaggerated courtesy, smiling, waving, shaking hands interminably and even fond-

ling babies. But the smiles soon fade as they turn away and disappear
back inside their ruthless world of power.

10. *You must take the initiative in repelling threats or attacks arising
from outside your group.*

It is always the dominant baboon that is in the forefront of the defence
against an attack from an external enemy. He plays the major role as
the protector of the group. For the baboon, the enemy is usually a
dangerous member of another species, but for the human leader it
takes the form of a rival group of the same species. At such moments,
his leadership is put to a severe test, but, in a sense, it is less severe than
during times of peace. The external threat, as I pointed out in the last
chapter, has such a powerful cohesive effect on the members of the
threatened group that the leader's task is in many ways made easier.
The more daring and reckless he is, the more fervently he seems to be
protecting the group who, caught up in the emotional fray, never dare
question his actions (as they would in peace-time), no matter how
irrational these actions may be. Carried along on the grotesque tidal-
wave of enthusiasm that war churns up, the strong leader comes into
his own. With the greatest of ease he can persuade the members of his
group, deeply conditioned as they are to consider the killing of another
human being as the most hideous crime known, to commit this same
action as an act of honour and heroism. He can hardly put a foot
wrong, but if he does, the news of his blunder can always be suppressed as bad for national morale. Should it become public, it can still
be put down to bad luck rather than bad judgment. Bearing all this in
mind, it is little wonder that, in times of peace, leaders are prone to
invent, or at least to magnify, threats from foreign powers that they
can then cast in the role of potential enemies. A little added cohesion
goes a long way.

These, then, are the patterns of power. I should make it clear that I
am not implying that the dominant baboon/human ruler comparison
should be taken as meaning that we evolved from baboons, or that our
dominance behaviour evolved from theirs. It is true that we shared a
common ancestor with baboons, way back in our evolutionary history,
but that is not the point. The point is that baboons, like our early
human forebears, have moved out of the lush forest environment into
the tougher world of the open country, where tighter group control is
necessary. Forest-living monkeys and apes have a much looser social
system; their leaders are under less pressure. The dominant baboon has
a more significant role to play and I selected him as an example for this
reason. The value of the baboon/human comparison lies in the way it

reveals the very basic nature of human dominance patterns. The striking parallels that exist enable us to view the human power game with a fresh eye and see it for what it is: a fundamental piece of animal behaviour. But we must leave the baboons to their simpler tasks and take a closer look at the complications of the human situation.

For the modern human leader there are clearly difficulties in performing his dominant role efficiently. The grotesquely inflated power which he wields means that there is the ever-present danger that only an individual with an equally grotesquely inflated ego will successfully be able to hold the super-tribal reins. Also, the immense pressures will easily push him into initiating acts of violence, an all-too-natural response to the strains of super-status. Furthermore, the absurd complexity of his task is bound to absorb him to such an extent that it inevitably makes him remote from the ordinary problems of his followers. A good tribal leader knows exactly what is happening in every corner of his group. A super-tribal leader, hopelessly isolated by his lofty position of super-status, and totally preoccupied by the machinery of power, rapidly becomes cut off.

It has been said that to be a successful leader in the modern world a man has to be prepared to make major decisions with the minimum of information. This is a frightening way to run a super-tribe, and yet it happens all the time. There is too much information available for any one individual to assimilate and, in addition, there is a great deal more, hidden in the super-tribal labyrinth, that can never be made available. A rational solution is to do away with the powerful leader-figure, to relegate him to the ancient, tribal past where he belonged, and to replace him with a computer-fed organization of interdependent, specialized experts.

Something approaching such an organization already exists, of course, and in England any civil servant will tell you without hesitation that it is the civil service that really runs the country. To emphasize his point he will inform you that when parliament is in session his work is seriously hampered; only during parliamentary recesses can serious progress be made. All this is very logical, but unfortunately it is not bio-logical, and the country he claims to be running happens to be made up of biological specimens – the super-tribesmen. True, a super-tribe needs super-control, and if it is too much for one man it might seem reasonable to solve the problem by converting a power-figure into a power-organization. This does not, however, satisfy the biological demands of the followers. They may be able to reason super-tribally, but their feelings are still tribal, and they will continue to demand a real leader in the form of an identifiable, solitary individual. It is a fundamental pattern of their species, and there is no avoiding it.

Institutions and computers may be valuable servants to the masters, but they can never themselves become masters (science-fiction stories notwithstanding). A diffuse organization, a faceless machine, lacks the essential properties: it cannot inspire and it cannot be deposed. The single dominant human is therefore doomed to struggle on, behaving publicly like a tribal leader, with panache and assurance, while in private he grapples laboriously with the almost impossible tasks of super-tribal control.

Despite the great burdens of present-day leadership and despite the daunting fact that an ambitious male member of a modern super-tribe has a chance smaller than one in a million of becoming the dominant individual of his group, there has been no observable lessening of desire to achieve high status. The urge to climb the social ladder is too ancient, too deeply ingrained to be weakened by a rational assessment of the new situation.

Throughout the length and breadth of our massive communities there are, then, hundreds of thousands of frustrated would-be leaders with no real hope of leading. What happens to their thwarted ladder-climbing? Where does all the energy go? They can, of course, give up and drop out, but this is a depressing condition. The flaw in the social dropout's solution is that he does not really drop out at all: he stays put and pours scorn on the rat race that surrounds him. This unhappy state is avoided by the great majority of the super-tribesmen by the simple device of competing for leadership in specialized sub-groups of the super-tribe. For some this is easier than others. A competitive profession or craft automatically provides it own social hierarchy. But even there the odds against achieving true leadership may be too great. This gives rise to the almost arbitrary invention of new sub-groups where competition may prove more rewarding. All kinds of extraordinary cults are set up – everything from canary-breeding and train-spotting to UFO-watching and body-building. In each case the overt nature of the activity is comparatively unimportant. What is really important is that the pursuit provides a new social hierarchy where one did not exist before. Inside it a whole range of rules and procedures is rapidly developed, committees are formed and – most important of all – leaders emerge. A champion canary-breeder or body-builder would, in all probability, have no chance whatsoever of enjoying the heady fruits of dominance, were it not for his involvement in his specialized sub-group.

In this way the would-be leader can fight back against the depressingly heavy social blanket that falls over him as he struggles to rise in his massive super-tribe. The vast majority of all sports, pastimes, hobbies and 'good works' have as their principal function not their specifically avowed aims, but the much more basic aim of follow-the-leader-and-beat-him-if-you-can. However, this is a description and not a criticism. In fact, the situation would be much more grave if this multitude of harmless sub-groups, or pseudo-tribes, did not exist. They funnel off a great deal of the frustrated ladder-climbing that might otherwise cause considerable havoc.

I have said that the nature of these activities is of little significance, but nevertheless it is intriguing to notice how many sports and hobbies involve an element of ritualized aggression, over and above simple competitiveness. To take a single example: the act of 'taking aim' is, in origin, a typically aggressive pattern of co-ordination. It reappears, suitably transformed, in a whole range of pastimes, including bowling, billiards, darts, table tennis, croquet, archery, squash, netball, cricket, tennis, football, hockey, polo, shooting and spear-fishing. In children's toys and fairgrounds it abounds. In a slightly heavier disguise it accounts for a great deal of the appeal of amateur photography: we 'shoot' film, 'capture' on celluloid, take snap 'shots', and our cameras = pistols, rolls of film = bullets, cameras with long telescopic lenses = rifles, and cine-cameras = machine-guns. However, although these symbolic equations may be helpful, they are by no means essential in the search for 'pastime dominance'. Matchbox-top collecting will do almost as well, providing, of course, you can make contact with suitable rivals, similarly preoccupied, whose matchbox-top collections you can then seek to dominate.

The setting up of specialist sub-groups is not the only solution to the super-status dilemma. Localized geographical pseudo-tribes also exist. Each village, town, city and country within a super-tribe develops its own regional hierarchy, providing further substitutes for thwarted super-tribal leadership.

On a smaller scale still, each individual has his own closely knit 'social circle' of personal acquaintances. The list of non-commercial names in his private phone-book or address-book gives a good indication of the extent of this kind of pseudo-tribe. It is particularly important because, as in a true tribe, all its members are personally known to him. Unlike a true tribe, however, all the members are not necessarily known to one another. The social groups overlap and interlock with one another in a complex network. Nevertheless, for each individual his social pseudo-tribe provides one more sphere in which he can assert himself and express his leadership.

Another major super-tribal pattern that has helped to split the group up without destroying it has been the system of social classes. These have existed in much the same basic form from the times of the earliest civilizations: an upper or ruling class, a middle class comprising merchants and specialists, and a lower class of peasants and labourers. Subdivisions have appeared as the groups have swollen, and the details have varied, but the principle has remained the same.

The recognition of distinct classes has made it possible for members of classes below the top one to strive for a more realistic dominance status at their particular class level. Belonging to a class is much more than a mere question of money. A man at the top of his social class may earn more than a man at the bottom of the class above. The rewards of being dominant at his own level may be such that he has no wish to abandon his class-tribe. Overlaps of this sort indicate just how strongly tribal the classes can become.

The class-tribe system of splitting up the super-tribe has, however, suffered serious setbacks in recent years. As the super-tribes grew to even bigger proportions and technologies became more and more complex, so the standard of mass education had to be raised to keep pace with the situation. Education, combined with improvements in mass communication and especially the pressures of mass advertising, led to a major breakdown in class barriers. The comforts of 'knowing your own station' in life were replaced by the exciting and increasingly real possibilities of exceeding that station. Despite this, the old class-tribe system kept on fighting back and is still doing so. We can see the outward signs of this running battle very clearly today in the ever-increasing speed of fashion cycles. New styles of clothing, furniture, decoration, music and art replace one another more and more quickly. It is often suggested that this is the result of commercial interests and pressures, but it would be just as easy – easier, in fact – to go on selling new variations of old themes rather than introducing new themes. Yet new themes are continually demanded, because the old ones permeate so rapidly through the social system. The quicker they reach the lower strata, the quicker they must be replaced by something new and exclusive at the top. History has never before witnessed such an incredible turnover in styles and tastes. The result, of course, is a major loss of the pseudo-tribal identity provided by the old, rigid social class system.

Replacing this loss, to some extent, is a new super-tribal splitting system that has recently developed. Age classes are emerging. A widening gap has appeared between what we must now call a young-adult pseudo-tribe and an old-adult pseudo-tribe. The former possesses its own customs and its own dominance system which are increasingly distinct from those of the latter. The entirely new phenomenon of powerful teenage idols and student leaders has ushered in a major new pseudo-tribal division. Desultory attempts on the part of the old-adult pseudo-tribe to encompass the new group have met with very limited success. The piling of old-adult honours on the heads of young-adult leaders, or the tolerant acceptance of the extremes of young-adult fashions and styles, has only led to further rebellious excesses. (If cannabis smoking is ever legalized and widely adopted, for example, an immediate replacement will be required, just as alcohol had to be replaced by cannabis itself.) When these excesses reach a point that the old adults cannot engulf, or refuse to copy, then the young adults can rest easy for a while. Safely flying their new pseudo-tribal flags, they can enjoy the satisfactions of their new pseudo-tribal independence and their more manageable, self-contained dominance system.

The sobering lesson to be learnt from all this is that the ancient biological need of the human species for a distinct tribal identity is a powerful force that cannot be subdued. As fast as one super-tribal split is invisibly mended, another one appears. Well-meaning authorities talk airily about 'hopes for a global society'. They see clearly the technical possibility of such a development, given the marvels of modern communication, but they stubbornly overlook the biological difficulties.

A pessimistic view? Certainly not. The prospects will remain gloomy only as long as there is a failure to come to terms with the biological demands of the species. Theoretically there is no good reason why small groupings, satisfying the requirements of tribal identity, should not be constructively interrelated inside thriving super-tribes which, in turn, constructively interact to form a massive, global mega-tribe. Failures to date have largely been due to attempts to *suppress* the existing differences between the various groups, rather than to improve the nature of these differences by converting them into more rewarding and peaceful forms of competitive social interaction. Attempts to iron out the whole world into one great expanse of uniform monotony are doomed to disaster. This applies at all levels, from breakaway nations to tearaway gangs. When the sense of social identity is threatened, it fights back. The fact that it has to fight for its existence means, at the least, social upheaval and, at the worst, bloodshed. We shall be taking a closer look at this in a later chapter, but for the moment we must return to the question of social status and examine it at the level of the individual.

Where exactly does he stand, this modern status-seeker? First, he has his personal friends and acquaintances. Together they form his social pseudo-tribe. Second, he has his local community – his regional pseudo-tribe. Third, he has his specializations: his profession, craft or employment, and his pastimes, hobbies or sports. They make up his specialist pseudo-tribes. Fourth, he has the remnants of a class-tribe and a new age-tribe.

Put together, these sub-groupings provide him with a much greater chance of achieving some sort of dominance and of satisfying his basic status urge than if he were simply a tiny unit in a homogeneous mass, a human ant crawling about in a gigantic, super-tribal ant-hill. So far, so good; but there are snags.

To begin with, the dominance achieved in a limited sub-group is itself limited. It may be real, but it is only a partial solution. It is impossible to ignore the fact that there are bigger things going on all around. Being a big fish in a little pond cannot blot out dreams of a bigger pond. In the past this was not such a problem, because the rigid class system, ruthlessly applied, kept everyone in his 'place'. This may have been very neat, but it could all too easily lead to super-tribal stagnation. Individuals with minor talents were well served, but many of those with greater talents were held back, frittering their energies away on strictly limited goals. It was possible for a potential genius from the lower class to stand less chance of success than a raging idiot from the upper class.

The rigid class structure had its value as a splitting device, but it was a grotesquely wasteful system, and it is not surprising that it eventually succumbed. Its ghost goes marching on, but it has largely been replaced today by a much more efficient meritocracy, in which each individual is theoretically able to find his optimum level. Once there he can consolidate his social identity by means of the various pseudo-tribal groupings.

This meritocratic system provides an exciting format, but there is another side to it. With excitement goes strain. An essential feature of a meritocracy is that, although it avoids waste of talent, it also opens up a clear channel from the very bottom to the very top of the enormous super-tribal community. If any small boy can, on his personal merits, eventually become the greatest of leaders, then for every one who succeeds there will be vast numbers of failures. These failures can no longer put the blame on the external forces of the wicked class system. They must place it firmly where it belongs, on their own personal shortcomings.

It seems, therefore, that any large-scale, lively, progressive super-tribe must inevitably contain a high proportion of intensely frustrated status-seekers. The dumb contentment of a rigid, stagnant society is replaced by the feverish longings and anxieties of a mobile, developing one. How do the struggling status-seekers react to this situation? The answer is that, if they cannot get to the top, they do their best to create the illusion of being less subordinate than they really are. To understand this, it will help at this point to take a sidelong glance at the world of insects.

Many kinds of insects are poisonous, and larger animals learn to avoid eating them. It is in the interests of these insects to show a warning flag of some kind. The typical wasp, for example, carries a conspicuous colour pattern of black and yellow bands on its body. This is so distinctive that it is easy for a predatory animal to remember it. After a few unfortunate experiences it quickly learns to avoid insects bearing this pattern. Other, unrelated, poisonous insect species may also carry a similar pattern. They become members of what has been called a 'warning club'.

The important point for us, in the present context, is that some *harmless* species of insects have taken advantage of this system by developing colour patterns similar to those of the poisonous members of the 'warning club'. Certain innocuous flies, for instance, display black and yellow bands on their bodies that mimic the colour patterns of the wasps. By becoming fake members of the 'warning club' they reap the benefits without having to possess any real poison. The killers dare not attack them, even though they would, in reality, make a pleasant meal.

We can use this insect example as a crude analogy to help us to understand what has happened to the human status-seeker. All we have to do is to substitute the possession of dominance for the possession of poison. Truly dominant individuals will display their high status in many visible ways. They will wave their dominance flags in the form of the clothes they wear, the houses they live in, the way they travel, talk, entertain and eat. By wearing the social badges of the 'dominance club' they make their senior status immediately obvious, both to subordinates and to one another, so that they do not have constantly to reassert their dominance in a more direct way. Like the poisonous insects, they do not have to keep on 'stinging' their enemies, they only have to wave the flag that says they could if they wanted to.

It follows, naturally enough, that harmless subordinates can join the 'dominance club' and enjoy its benefits if only they can display the same flags. If, like the black-and-yellow flies, they can mimic the black-and-yellow wasps, they can at least create the illusion of dominance.

Dominance mimicry has, in fact, become a major preoccupation of the super-tribal status-seekers, and it is important to examine it more closely. First, it is essential to make a clear distinction between a status symbol and a dominance mimic. A status symbol is an outward sign of the true level of social dominance you have attained. A dominance mimic is an outward sign of the level of dominance you would like to attain, but have not yet reached. In terms of material objects, a status symbol is something you can afford; a dominance mimic is something you cannot quite afford, but buy all the same. Dominance mimics therefore frequently involve making major sacrifices in other directions, whereas true status symbols do not.

Earlier societies, with their more rigid class structures, clearly did not give so much scope for dominance mimicry. As I have already pointed out, people were much more content to 'know their station'. But the upgrading urge is a powerful force, and there were always exceptions, no matter how rigid the class structure. The dominant individuals, seeing their position weakened by imitation, reacted harshly. They introduced strict regulations and even laws to curb the mimicry.

The various rules of costume give a good example. In England, the law of the Westminster parliament of 1363 was concerned chiefly with regulating the fashion of dress in the different social classes, so important had this subject become. In Renaissance Germany, a woman who dressed above her station was liable to have a heavy wooden collar locked around her neck. In India, strict rules were introduced relating the way you folded your turban to your particular caste. In the England of Henry VIII no woman whose husband could not afford to maintain a light horse for the king's service was allowed to wear velvet bonnets or golden chains. In America, in early New England, a woman was forbidden to wear a silk scarf unless her husband was worth a thousand dollars. The examples are endless.

Today, with the breakdown of the class structure, these laws have become severely curtailed. They are limited now to a few special categories such as medals, titles and regalia, which it is still illegal, or at least socially unacceptable, to adopt without the appropriate status. In general, however, the dominant individual is far less protected against the practices of dominance mimicry than he once was.

He has retaliated with ingenuity. Accepting the fact that lower-status individuals are determined to copy him, he has responded by making available cheap, mass-produced imitations of high-status goods. The bait is tempting and has been eagerly swallowed. An example will explain how the trap works.

High-status wife wears a diamond necklace. Low-status wife wears a bead necklace. Both necklaces are well made; the beads are inexpensive, but they are gay and attractive and make no pretence to be anything other than what they are. Unfortunately, they have low status value, and the low-status wife wants something more. There is no law or social edict preventing her from wearing a diamond necklace. By working hard, saving every penny, and eventually spending more than she can afford, she may be able to acquire a necklace of small but real diamonds. If she takes this step, adorning her neck with a dominance mimic, she starts to become a threat to the high-status wife. The difference in their status displays becomes blurred. High-status husband therefore puts on the market necklaces of large, fake diamonds. They are inexpensive and superficially so attractive that the low-status wife abandons her struggle for real diamonds and settles for the fake ones instead. The trap is sprung. True dominance mimicry has been averted.

On the surface this is not apparent. The low-status wife, sporting her flashy fake necklace, *seems* to be mimicking her dominant rival, but this is an illusion. The point is that the fake necklace is too good to be true, when judged against her general way of life. It fools no one, and therefore fails to act as an aid in raising her status.

It is surprising that the trick works so well and so often, but it does. It has infiltrated many spheres of life and has not been without its repercussions. It has destroyed a great deal of genuine but overtly low-status art and craft. Native folk art has been replaced by cheap reproductions of the great masters; folk music has been replaced by the gramophone record; peasant craftsmanship has been replaced by mass-produced plastic imitations of more expensive goods.

Folklore societies have been rapidly formed to bewail and reverse this trend, but the damage has already been done. At best, all they can achieve is to act as folk-culture taxidermists. Once the status race was opened up from the bottom to the top of society, there was no turning back. If, as I suggested earlier, society is repeatedly going to rebel against the dreary uniformity of this 'new monotony', then it will do so by giving birth to new cultural patterns rather than by propping up old, dead ones.

For the really serious status-climber, however, there is no rebellion. Nor, for him, do the cheap fakes provide a satisfactory answer. He sees them for what they are, a clever side-track, a mere fantasy version of true dominance mimicry. For him, the dominance mimics must be genuine articles, and he must always go one step further than he can afford, when purchasing them, in order to give the impression that he is slightly more socially dominant than he in fact is. Only then does he stand a chance of getting away with it.

For safety's sake, he tends to concentrate on areas where cheap fakes are out of the question. If he can afford a small motor car, he buys a medium-sized one; if he can afford a medium-sized one, he buys a large one; if he can afford a single large one, he buys a second car as a runabout; if large cars become too common, he buys a small but wildly expensive foreign sports car; if large rear-lights become the fashion, he buys the latest model with even bigger ones, 'to let the people behind know he is in front', as the advertisers so succinctly express it. The one thing he does not do is to buy a row of life-sized, cardboard models of Rolls-Royces and display them outside his garage. There are no fake diamonds in the world of the status-climbing fanatic.

Motor cars are a single example, and an important one because they are so public, but the ardent status-struggler cannot stop there. He must extend himself and his bank balance in all directions if he is to paint a convincing picture for his higher-status rivals. The whole hire-purchase, mortgage and overdraft system depends for its survival on this expression of the powerful upgrading urge in terms of dominance mimics.

Unhappily, the extravagant trappings of the unrelenting status-seeker acquire such an importance that they appear to be more than they are. They are, after all, only mimics of dominance, not dominance itself. True dominance, true social status, is related to the possession of power and influence over super-tribal subordinates, not to the possession of a second colour television set. Of course, if you can easily afford a second colour television set, then it is a natural reflection of your status and acts as a true status symbol. A second colour television set, when you can only just afford the first one, is a different matter. It may help to impress on the members of the social level above you that you are ready to join them, but it in no way *ensures* that you will do so. All your rivals, at your own level, will be busily installing their own second colour television sets with the same idea in mind, but it is the fundamental law of the hierarchy that only a few from your level will make the grade to the one above. They, the lucky ones, can justifiably hang wreaths around their second colour television sets. Their dominance mimics worked the trick. All the rest, the power failures, must sit there, surrounded by the expensive clutter of the dominance mimics that have suddenly revealed themselves for what they are: illusions of grandeur. The realization that although they are valuable aids to successful dominance ladder-climbing, they do not actually guarantee it, is a bitter pill to swallow.

The damage caused by the exaggerated pursuit of dominance mimicry can be enormous. It not only leads to a condition of depressing disillusionment for the less successful status-seekers, it also demands such great efforts of the super-tribesman that he may have little time or energy for anything else.

The male status-seeker who indulges in an excess of dominance mimicry is often driven to neglect his family. This forces his mate to take over the masculine parental role in the home. Taking such a step provides a psychologically damaging atmosphere for the children, which can easily warp their own sexual identities when they mature. All that the young child will see is that its father has lost his leading role inside the family. The fact that he has sacrificed it to a struggle for dominance outside, in the larger sphere of the super-tribe, will mean little or nothing in the child's brain. If it matures with a well-balanced state of mental health it will be surprising. Even the older child, who comes to understand the super-tribal status race and boasts about his father's status achievements, will find them small compensation for the absence of an active paternal influence. Despite his mounting status in the outside world, the father can easily become a family joke.

It is very bewildering for our struggling super-tribesman. He has obeyed all the rules, but something has gone wrong. The super-status demands of the human zoo are cruel indeed. Either he fails and becomes disillusioned, or he succeeds and loses control of his family. Worse still, he can work so hard that he loses control of his family and *still* fails.

This brings us to another and more violent way in which certain members of the super-tribe can react to the frustrations of the dominance struggle. Students of animal behaviour refer to it as the re-direction of aggression. At the best of times it is an unpleasant phenomenon; at worst, it is literally lethal. One can see it very clearly when two rival animals meet. Each wants to attack the other and each is afraid to do so. If the aroused aggression cannot find an outlet against the frightening opponent who caused it, then it will find expression elsewhere. A scapegoat is sought, a milder, less intimidating individual, and the pent-up anger is vented in his direction. He has done nothing to warrant it. His only crime was to be weaker and less frightening than the original opponent.

In the status race it frequently occurs that a subordinate dare not express his anger openly towards a dominant. Too much is at stake. He has to re-direct it elsewhere. It may land on his unfortunate children, his wife, or his dog. In former times, the flanks of his horse also suffered; today, it is the gearbox of his car. He may have the luxury of staff subordinates of his own that he can lash with his tongue. If he has inhibitions in all these directions there is always one person left: himself. He can give himself ulcers.

In extreme cases, when everything seems utterly hopeless, he can increase his self-inflicted aggression to the maximum: he can kill himself. (Zoo animals have been known to inflict serious mutilations on themselves, biting their flesh to the bone when unable to get at their enemies through the bars, but suicide seems to be a uniquely human activity.) Views concerning the main causes of suicide have differed widely, but hardly anyone denies that re-directed aggression is a major factor. One authority went so far as to claim that: 'Nobody kills himself unless he also wants to kill others or at least wishes some other person to die.' This is perhaps slightly over-stating the case. A man who kills himself because of the pain of an incurable disease scarcely falls into this category. It would be fanciful to suggest that he wants to kill the doctor who has failed to cure him. What he wants is release from pain. But re-direction of aggression does seem to account for a large number of cases. Here are some of the facts that support this idea.

There is a higher suicide rate in big towns and cities than in rural areas. In other words, where the status race is hottest, the suicide rate is highest. There are more male suicides than female suicides, but the females are catching up fast. In other words, the sex that is most involved in the status race has the highest suicide rate, and now that females are becoming increasingly emancipated and joining in the race more, they are sharing its hazards. There is a higher rate of suicide during periods of economic crisis. In other words, when the status race gets into difficulties at the top, there is an increase of re-directed aggression down the hierarchy, with disastrous results.

There is a lower rate of suicide during times of war. The suicide curves for the present century show two huge dips during the periods of the two world wars. In other words, why kill yourself if you can kill someone else? It is the inhibitions about killing the people who are dominating and frustrating the potential suicide that force him to re-direct his violence. He has the choice of killing a less daunting scapegoat, or himself. In peace-time, inhibitions about killing make him turn most often towards himself, but during war-time he is or-dered to kill, and the suicide rate goes down.

The relationship between suicide and murder is a close one. To a certain extent they are two sides of the same coin. Countries with a high murder rate tend to have a low suicide rate, and vice versa. It is as if there is just so much intense aggression to be let loose, and if it does not take the one form it will take the other. Which way it goes will depend on how inhibited a particular community is about committing murder. If the inhibitions are weak, then the suicide rate goes down. It is similar to the war-time situation, where inhibitions against killing were actively and purposely reduced.

By and large, however, our modern super-tribes are remarkably heavily inhibited where acts of murder are concerned. It is difficult for the majority of us, who have never had to toss the murder/suicide coin, to appreciate the conflict, although in theory it seems biologically more unnatural to kill oneself rather than someone else. Despite this the figures go the other way. In Britain during recent times the yearly suicide figures have hovered around the 5,000 mark, while the yearly (detected) murders have kept below the 200 level. What is more, if we look at these murders, we find something unexpected. Most of us gain our ideas about murder from newspaper reports and detective novels, but newspapers and thriller-writers tend to concentrate on murders that will sell copies of papers and books. In reality the most common form of homicide is an unglamorous and squalid little family affair in which the victim is a close relative. There were 172 murders in Britain in 1967, and 81 of these were of this type. Furthermore, in 51 cases the murderer followed his act of homicide by committing suicide. Many of these latter cases are of the kind where a man, driven to turn his frustrated aggression on to himself, first kills his loved ones and then himself. Often, it appears that he cannot bear to leave them behind to suffer from the mess he has made, and so dispatches them first. Students of murder have discovered that an interesting change may then come over the killer. If he does not finish the job off and add his own corpse quickly to the rest, he is likely to experience such an enormous relief from tension that he suddenly finds he no longer wants to kill himself. Society dominated and frustrated him to the point where he was ready to take his own life, but now the slaying of his family consummates his revenge on society so effectively that his depression lifts and he feels released. This leaves him in a difficult situation. There are bodies lying about and all the signs that he has committed a multiple murder, when in fact it was only part of a desperate suicide. Such are the nightmare extremes of re-directed aggression.

Most of us, happily, do not reach such extremes. Our families may experience nothing more than our arrival home occasionally in a disgruntled mood. Many super-tribesmen can find an outlet by watching other people kill villains on television or at the cinema. It is significant that in strongly subordinated or suppressed communities, the local cinemas show a remarkably high proportion of films of violence. In fact, it can be argued that the thrills of fictional violence have an appeal that is directly proportional to the degree of dominance frustration that is being experienced in real life.

Since all the large super-tribes, by their very size, involve extensive dominance frustration, the prevalence of fictional violence is widespread. To prove the point it is only necessary to compare the international sales of books by authors of violent fiction with those of other writers. In a recent survey of the all-time best-sellers in the world of fiction, the name of one author who specializes in extremes of violence appeared seven times in the top twenty, with a total score of over 34 million copies sold. In the world of television the picture is much the same. A detailed analysis of transmissions in the New York area in 1954 revealed that there were no less than 6,800 aggressive incidents in a single week.

Clearly there is a powerful urge to watch other people being subjected to the most extreme forms of domination. Whether this acts as
a valuable and harmless outlet for suppressed aggression is a hotly
debated point. As with dominance mimicry, the cause of violence-
watching is obvious, but the value is dubious. Reading about or watching an act of persecution does not alter the reader's or watcher's
real-life situation. He may enjoy the experience of the fiction while he
is involved in it, but when it is over and he re-emerges into the cold
light of reality, he is still as dominated as he was before. The relief from
tension is therefore only a temporary one, like scratching an insect
bite. What is more, scratching a bite is likely to increase the inflamma-
tion. Repeated involvement in fictional mayhem tends to intensify the
preoccupation with the whole phenomenon of violence. The best that
can be said for it is that, while it is going on, the audience itself is not
performing acts of violence.

The action of re-directing aggression has often been referred to as the
' . . . and the office-boy kicked the cat' phenomenon. This implies that
only the lowest members of a hierarchy will turn their blocked anger
on to an animal. Unhappily for animals this is not the case, and animal
protection societies have the figures to prove it. Cruelty to animals has
provided a major outlet for re-directed aggression from the times of the
earliest civilizations, right up to the present day, and it has certainly
not been confined to the lowest levels in the social hierarchy. From the
slaughters of the Roman amphitheatres, to the bear-baiting of the
Middle Ages and the bull-fighting of modern times, the infliction of
pain and death on animals has undeniably had a mass appeal for
members of super-tribal communities. It is true that ever since our
early ancestors turned to hunting as a method of survival, man has
inflicted pain and death on other animal species, but the motives were
different in prehistoric times. In the strict sense, there was no cruelty
then, the definition of cruelty being 'taking delight in another's pain'.

In super-tribal times we have killed animals for four reasons: to
obtain food, clothing and other materials; to exterminate pests and
vermin; to further scientific knowledge; and to experience the pleasure
of killing. The first and second of these two reasons we share with our
early hunting ancestors, the third and fourth are novelties of the
super-tribal condition. It is the fourth that concerns us here. The others
may, of course, involve elements of cruelty, but it is not their primary
characteristic.

The history of deliberate cruelty to other species has taken a strange
course. The early hunter had a kinship with animals. He respected
them. So, rather naturally, did the early farming peoples. But the
moment that urban populations began to develop, large groups of

human beings became cut off from direct contact with animals, and the respect was lost. As civilizations grew, so did man's arrogance. He shut his eyes to the fact that he was just as much an animal as any other species. A great gulf appeared: now only *he* had a soul and other animals did not. They were no more than brute beasts put on earth for his pleasure. With the spreading influence of the Christian religion, animals were in for a rough passage. We need not go into the details, but it is worth noting that as late as the middle of the nineteenth century, Pope Pius IX refused permission for the opening of an animal protection office in Rome on the grounds that man owed duties to his fellow men, but none to the lower animals. Later in the same century a Jesuit lecturer wrote: 'Brute beasts, not having understanding and therefore not being persons, cannot have any rights . . . We have, then, no duties of charity nor duties of any kind to the lower animals, as neither to stocks and stones.'

Many Christians were beginning to have doubts about this attitude, but it was not until Darwin's theory of evolution began to have a major impact on human thought that man and the animals came closer together again. The reacceptance of man's affinity with animals, which had been so natural to the early hunters, led to a second era of respect. As a result, our attitude towards deliberate cruelty to animals has been changing rapidly during the past hundred years; but despite increasingly powerful disapproval, the phenomenon is still very much with us. Public displays are rare, but private savageries persist. We may respect animals today, but they are still our subordinates, and as such are highly vulnerable objects for the unloading of re-directed aggression.

Next to animals, children are the most vulnerable subordinates and, despite greater inhibitions here, they too are subjected to a great deal of re-directed violence. The viciousness with which animals, children and other helpless subordinates are subjected to persecution is a measure of the weight of the dominance pressures imposed on the persecutors.

Even in war, where killing is glorified, this mechanism can be seen in operation. Sergeants and other N.C.O.s frequently dominate their men with extreme ruthlessness, not merely to produce discipline, but also to arouse hatred, with the deliberate intention of seeing this hatred re-directed at the enemy in battle.

Looking back, we can see now how the unnaturally heavy weight of dominance from above, which is an inevitable characteristic of the super-tribal condition, has taken its toll. The abnormality of the situation for the human animal, who only a few thousand years ago was a simple tribal hunter, has produced patterns of behaviour which, by animal standards, are also abnormal: the exaggerated preoccupation with dominance mimicry, the excitement of watching acts of violence, the deliberate cruelty towards animals, children and other extreme subordinates, the acts of murder and, if all else fails, the acts of self-cruelty and self-destruction. Our super-tribesman, neglecting his family to drag himself one more rung up the social ladder, gloating over the brutalities in his books and films, kicking his dogs, beating his children, persecuting his underlings, torturing his victims, killing his enemies, giving himself stress diseases and blowing his brains out, is not a pretty sight. He has often boasted about being unique in the animal world, and on this score he certainly is.

It is true that other species also indulge in intense status struggles and that the attaining of dominance is often a time-consuming element in their social lives. In their natural habitats, however, wild animals never carry such behaviour to the extreme limits observable in the modern human condition. As I said at the outset, only in the cramped quarters of zoo cages do we find anything approaching the human state. If, in captivity, a group of animals is assembled which is too numerous for the species concerned, and they are packed too tightly together, then, with an inadequate cage environment, serious trouble will certainly develop. Persecutions, mutilations and killings will occur. Neuroses will appear. But even the least experienced zoo director would never contemplate crowding and cramping a group of animals to the extent that man has crowded and cramped himself in his modern cities and towns. *That* level of abnormal grouping, the director would predict with confidence, would cause a complete fragmentation and collapse of the normal social pattern of the animal species concerned. He would be astonished at the folly of suggesting that he should attempt such an arrangement with, say, his monkeys, his carnivores, or his rodents. Yet mankind does this willingly to himself; he struggles under just these conditions and somehow manages to survive. By all the rules, the human zoo should be a screaming mad-house by now, disintegrating into complete social confusion. Cynics might argue that this is indeed the case, but plainly it is not. The trend towards denser living, far from abating, is ever gaining momentum. The various kinds of behaviour disorders I have outlined in this chapter are startling, not so much for their existence as for their rarity in relation to the population sizes involved. Remarkably few of the struggling super-tribesmen succumb to the extreme forms of action I have discussed. For every desperate status-seeker, home-wrecker, murderer, suicide, persecutor, or ulcer-nurser, there are hundreds of men and women who not only survive, but thrive under the extraordinary conditions of the super-tribal assemblages. This, more than anything else, is a truly astonishing testimony to the enormous tenacity, resilience and ingenuity of our species.

3

SEX AND SUPER-SEX

WHEN YOU PUT a piece of food into your mouth it does not necessarily mean that you are hungry. When you take a drink it does not inevitably indicate that you are thirsty. In the human zoo, eating and drinking have come to serve many functions. You may be nibbling peanuts to kill time, or you may be sucking sweets to soothe your nerves. Like a wine-taster, you may merely savour the flavour and then spit the liquid out, or you may down ten pints of beer to win a wager. Under certain circumstances, you may be prepared to swallow a sheep's eyeball in order to maintain your social status.

In none of these cases is the nourishment of the body the true value of the activity. This multi-functional utilization of basic behaviour patterns is not unknown in the world of animals, but, in the human zoo, man's ingenious opportunism extends and intensifies the process. In theory, this should fall on the credit side of our super-tribal existence. There can, however, be drawbacks if we handle the process clumsily. If we eat too much to soothe our nerves, we become overweight and unhealthy; if we drink too much of certain liquids we damage our livers or develop addictions; if we experiment too wildly with new tastes we get indigestion. These difficulties arise because we fail to separate non-nutritional feeding and drinking from their primary nutritional roles. We baulk at the ancient Roman habit of tickling the throat with a feather to make the stomach disgorge unwanted food, and the avoidance of swallowing practised by the wine-taster is no more than an isolated exception to the general rule. Nevertheless, with appropriate caution, we can indulge in multi-functional feeding and drinking to a considerable extent without coming to any serious harm.

Where sexual behaviour is concerned the situation is similar, but it is much more complicated and deserves our special attention. Here there has been an even greater failure to separate non-reproductive sexual activities from their primary reproductive functions. This has not, however, prevented the human zoo from converting sex into multi-functional super-sex, despite the fact that the results are sometimes disastrous for the human animals concerned. Man's opportunism knows no bounds and it is inconceivable that an activity so basic and so deeply rewarding should have escaped diversification. In fact, of all our activities, it has, regardless of the dangers, become functionally the most elaborate, with no fewer than ten major categories.

In order to clarify the picture, it will help if we examine the different functions of sexual behaviour one by one. It is important to realize at the outset that, although these functions are separate and distinct, and sometimes clash with one another, they are not all mutually exclusive. Any particular act of courtship or copulation may serve several functions simultaneously.

The ten functional categories are these:

1. Procreation Sex

There can be no argument that this is the most basic function of sexual behaviour. It has sometimes been mistakenly argued that it is the only natural and therefore proper role. Paradoxically, some of the religious groups that claim this do not practise what they preach, monks, nuns and many priests denying themselves the very activity which they hold to be so uniquely natural.

An important point to be added here is that when a population becomes seriously overcrowded, the value of the procreative function of sex becomes greatly reduced. Eventually it becomes a nuisance. Instead of being a fundamental mechanism of survival, it changes into a potential mechanism of destruction. This happens occasionally with such species as lemmings and voles which, when conditions are exceptionally lush, breed themselves up to such a density that their populations explode in chaos, with an enormous loss in lives. It is also happening to the human species at this very moment and the human animal may soon have to face the imposition of obtaining a breeding licence before being permitted to indulge in procreation.

This is not a matter that can be treated lightly and in recent years it has led to a great deal of agitated debate. It is worth looking at both sides of the argument, an exercise that has become increasingly rare as the protagonists have pushed one another into more and more extreme positions.

The basic question is: dare we tamper with the procreative process? Or, as the other side would put it: dare we *not* tamper with it? The arguments usually rage at a philosophical, ethical, or religious level, but how do they appear when we view them biologically?

If a human group opposes efficient techniques for limiting procreativity, it gains two advantages. Firstly, it will breed more rapidly than the groups that do employ modern contraceptive devices. By gaining in numbers, it can hope eventually to swamp the others out of existence –

a fact that cannot fail to appeal to its leaders, whether military or religious. Secondly, it will ensure that its basic social units – the family groups – are strong. A mated pair is not only a sexual unit, it is also a parental unit and the more parentally occupied it is, the more stable it becomes.

These are strong arguments, but so also are the opposing ones. Proponents of efficient contraception can point out that it is no longer a question of one group gaining on another. Over-population has become a world-wide problem and must be viewed as such. In this respect we are one vast, global lemming colony, and if the explosion comes it will affect us all. Indeed, it is already doing so.

As regards the family unit, it can be argued that contraception is not creating an unnatural situation, but merely recreating a natural one. Before medical care, hygiene, and other security devices of modern living existed, the family unit may have produced large numbers of offspring, but it also lost a high proportion of them. All that contraception does, when applied in moderation, is to advance these losses to a point in time before the human egg has been fertilized.

If we do not pursue a world-wide policy of contraception, then some other unavoidable population limiting factor will step in. As a species we are rapidly reaching saturation point and if we fail to reduce our fecundity by voluntary means, the existing populations will suffer for it. If prevention is better than cure, then contraception is the obvious choice. It is difficult to see how anyone could argue that preventing someone from living is worse than curing someone of being alive. The individual human being is not a simple organism, to be squandered carelessly. It is a high-quality product requiring years of growth and development, and it needs all the protection it can get. Yet the opponents of contraception persist in their views. If they win, the swarms of non-contraceived offspring they are encouraging into the world may live to see the total collapse of the whole of human society.

2. *Pair-formation Sex*

The human animal is basically and biologically a pair-forming species. As the emotional relationship develops between a pair of potential mates it is aided and abetted by the sexual activities they share. The pair-formation function of sexual behaviour is so important for our species that nowhere outside the pairing phase do sexual activities regularly reach such a high intensity.

It is this function that causes so much trouble when it clashes with the various non-reproductive forms of sex. Even if Procreation Sex is successfully avoided and no fertilization takes place, a pair-bond may still automatically start to form where none is intended. It is because of this that casual copulations frequently create so many problems.

If a copulator has had his or her pair-forming mechanism damaged in some way during childhood, so that he or she is incapable of 'falling in love', or if there is a temporary and deliberate suppression of the pair-forming urge, then a casual copulation may succeed and be enjoyed without any later repercussions. But it takes two to copulate, and the partner in such an encounter may not be so lucky. If his or her pair-forming mechanism is more active, a one-sided pair-bond may start to form as a result of the emotional intensity of the sexual actions. The inevitable outcome of this is that society becomes littered with 'broken hearts', 'hang-ups' and 'abandoned lovers' who subsequently find it extremely difficult to form a new pair-bond with a fresh partner.

Only when the pair-bonding mechanism has been equally damaged or is equally suppressed in both partners can a casual human copulation be performed without undue risk. Even then, there is always the danger that the strength of the sexual response of one partner may be such that, for him or her, it will start to repair the bonding damage or disinhibit the bonding urge.

3. Pair-maintenance Sex

Once a pair-bond has been successfully formed, sexual activities still function to maintain and reinforce the bond. Although these activities may become more elaborate and *ex*tensive, they usually become less *in*tensive than those of the pair-forming stage, because the pair-forming function is no longer operating.

This distinction between the pair-forming and pair-maintaining functions of sexual activity is clearly illustrated whenever the members of a long-established mated pair are separated from one another for a period of time by war, business, or some other external demand. When they are reunited there is typically a resurgence of high sexual intensity on the first nights they are together again, as they go through a minor re-bonding process.

There is one apparent contradiction that must be disposed of here. In some cultures, where the natural biological process of 'falling in love' is interfered with by arranged marriages or by anti-sexual propaganda, a young couple may find themselves newly married without even the beginnings of pair-bonding, or with a strongly inhibited approach to copulatory activity. In such cases they may report that (if they are lucky) their sexual behaviour becomes more intense at a later stage. For them, the pair-maintenance phase seems, at first sight, to be more sexually intense than the pair-formation stage, apparently reversing the correlation I have described. But this is not a real contradiction, it is simply that the true pair-forming stage has been artificially delayed.

Such couples are not always this fortunate. What frequently happens in such cases is that the family unit has to rely on external social pressures to hold it together, rather than the more basic, and more reliable, internal bonding process. If a marriage partner remains biologically 'unbonded' in this way, there is a considerable danger that a powerful extra-marital pair-bond will suddenly form. The true pair-forming capacity will be lying idle, so to speak, and will be all too ready to leap into action, causing havoc to the officially recognized 'pseudo-bond'.

There is a different kind of hazard for the young couple that *does* manage to base its marriage on the formation of a true pair-bond. This hazard is not caused by anti-sexual propaganda, but rather by a surfeit of pro-sexual propaganda, which can lead them to suppose that the very high intensity of the pair-formation stage should persist even after the pair has been fully formed. When it inevitably fails to do so, they imagine that something has gone wrong, whereas in reality they have merely reached the normal pair-maintenance sexual phase. The case for reproductive sex can be overstated as well as understated, and either way may lead to trouble.

These first three categories – Procreation, Pair-formation and Pair-maintenance Sex – together make up the primary reproductive functions of human sexual behaviour. Before moving on to examine the non-reproductive patterns, there is one final, general comment that is relevant here. Individuals whose pair-bonding mechanism has run into some sort of trouble have occasionally found it convenient to argue that there is no such thing as a biological pairing urge in the human species. 'Romantic love', as they prefer to call it, is looked upon as a recent and highly artificial invention of modern living. Man, they argue, is basically promiscuous, like so many of his monkey relatives. The facts, however, are against this. It is true that, in many cultures, economic considerations have led to a gross distortion of the pair-forming pattern, but even where this pattern's interference with officially planned 'pseudo-bonds' has been most rigorously suppressed, with savage penalties and punishments, it has always shown signs of re-asserting itself. From ancient times, young lovers who have known that the law may demand no less than their lives if they are caught, have nevertheless found themselves driven to take the risk. Such is the power of this fundamental biological mechanism.

4. *Physiological Sex*

In the healthy adult human male and female there is a basic physiological requirement for repeated sexual consummation. Without such consummation, a physiological tension builds up and eventually the body demands relief. Any sexual act that involves an orgasm provides the orgasmic individual with this relief. Even if a copulation fails to fulfil any of the other nine functions of sexual behaviour, it can at least satisfy this basic physiological need. For an unmated and otherwise sexually unsuccessful male, a visit to a prostitute can serve this function. A more widespread solution, and one that is indulged in by both sexes, is masturbation.

A recent American study revealed that as many as 58 per cent of females and 92 per cent of males in that culture masturbate to orgasm at some time in their lives. Because this sexual act does not involve a partner and cannot therefore lead to fertilization, puritanical attempts have been made to suppress it at various times in the past, and all kinds of strange superstitions have grown up around it. The list of disasters that were supposed to threaten the masturbator included: desiccation, sterility, emaciation, frigidity, paroxysm, pallid complexion, hysteria, dizziness, jaundice, deformed figure, insanity, insomnia, exhaustion, pimples, pain, death, cancer, stomach ulcers, genital cancer, digestive upsets, headaches, appendicitis, weak hearts, kidney troubles, lack of hormones and blindness. This incredible collection of catastrophes would be amusing were it not for the untold miseries and fears the dire warnings must have caused, year after year and century after century. Happily these totally false superstitions are at last beginning to lose ground and a great deal of unnecessary anxiety is fading with them.

If no active sexual outlet is obtained, the body may take charge of the situation itself. Both male and female celibates are likely to undergo spontaneous orgasms while sleeping. Both sexes experience erotic dreams which may be accompanied by full orgasmic muscle responses and genital secretions in the female, and by 'nocturnal emissions' in the male.

Spontaneous orgasms appear to occur in even the most strictly abstemious and devoutly religious individuals, when they are described in rather different terms and referred to as religious frenzies, ecstasies, or trances. St Theresa, for instance, described how a vision of an angel came to her: 'In his hands I saw a long golden spear and at the end of the iron tip I seemed to see a point of fire. With this he seemed to pierce my heart several times so that it penetrated to my entrails. When he drew it out I thought he was drawing them out with it and he left me completely afire with a great love of God. The pain was so sharp that it made me utter several sharp moans; and so excessive was the sweetness caused me by this intense pain that one can never wish to lose it.'

Unfortunately we know far too little about the spontaneous sexual outlets of extreme celibates to be able to make any firm statements concerning how widespread or how frequent these orgasms are. We do, however, know that individuals who have carried on an active sex life and are then confined in prison frequently show a marked increase in orgasmic dreaming. In one study, involving 208 female prisoners, this was found to be true for over 60 per cent of the group.

It would be wrong to give the impression, however, that orgasmic dreaming acts *solely* as a compensating device helping to keep up the sexual output when other more active outlets are missing. There is more to it than that, as there is, of course, with prostitution and masturbation, which serve other sexual functions as well. Some individuals, for example, show an increase in the frequency of orgasmic dreaming during periods when they are experiencing an unusually *high* frequency of active copulation, on the hypersensitizing principle of 'the more you get, the more you want'. However, this does not invalidate the evidence that spontaneous orgasm can and does occur as a response to sexual deprivation. It merely means that the phenomenon is more complex. But here we are only concerned with the simple, 'relief from physiological tension' function of sexual behaviour, and it is clear that this must be included as one of the ten basic functional categories of human sexual behaviour.

Physiological Sex can also be observed in other animal species and it is worth taking a look at a few examples. As one might expect, they are most readily encountered in the animal zoo, rather than in the wild state. Many zoo animals have been seen to masturbate when kept in isolation. This is most commonly observed in captive monkeys and apes. In males, the penis is stimulated sometimes by the hand or foot, sometimes by the mouth, and sometimes by the tip of the prehensile tail. Male elephants stimulate their penises with their trunks and female elephants kept in a group without a male stimulate one another's genitals with their trunks. Even a male lion, kept in isolation in a zoo cage, has been seen to heave itself up into an inverted position against a wall and masturbate with its paws. Male porcupines have been observed to walk around on three legs, holding one forepaw on their genitals. One male dolphin developed the pattern of holding its erect penis in the powerful jet of the water intake of its pool. Sex-dreaming also seems to occur in animals and in domestic cats the erection of the penis while asleep has been observed to lead to full ejaculation.

5. *Exploratory Sex*

One of man's greatest qualities is his inventiveness. In all probability our monkey ancestors were already endowed with a reasonably high level of curiosity; it is a characteristic of the whole primate group. However, when our early human ancestors took to hunting, they undoubtedly had to develop and strengthen this quality and magnify their basic urge to explore all the details of their environment. It is clear that exploration became an end in itself, leading man on to fresh pastures and fresh achievements, always investigating, always asking new questions, never satisfied with old answers. So powerful did this urge become that it soon began to spread into all other areas of behaviour. With the arrival of the super-tribal condition, even simple patterns like locomotion were explored for possible variations. Instead of being satisfied with walking and running, we tried out hopping, skipping, leaping, marching, dancing, handstanding, vaulting, diving and swimming. Half the reward was in the experimentation itself, the actual discovering of a new variation. (Repeatedly indulging in it, following the discovery, was the second half of the reward, but we are not concerned with that for the moment.)

In the sexual sphere, this trend led to a wide range of variations on the sexual theme. Sexual partners began to experiment with new forms of mutual stimulation. Ancient sexual writings record in detail the great diversity of novel sexual movements, pressures, sounds, contacts, scents and copulatory positions that were the subject of erotic experimentation.

Although this was an inevitable development, paralleled by similar sensory explorations in other patterns, such as feeding behaviour, there were repeated attempts to suppress it in various cultures. The official reason given was often the one we have heard already: namely, that it represented an elaboration of sexual behaviour beyond what was necessary for the act of procreation. The significance of the development of exploratory sexual behaviour as an aid to the cementing of the pair-bond and the subsequent strengthening of the vital family unit was ignored. This was unfortunate, for one particularly important reason. As I have already mentioned, the intensity of love-making during the pair-forming stage wanes slightly after the pair-bond is fully formed. Theoretically, if the family unit is successful and remains unharassed by external forces, all should be well. It is an adaptive system because, if the exhausting intensiveness of the love-making of the young couple during pair-formation were prolonged indefinitely, it could well impair their efficiency in other activities. But the stresses and strains of the super-tribal condition *do* tend to harass the family

unit. The external pressures are strong. The replacement of pair-forming intensiveness with exploratory extensiveness in later sexual activities is the ideal solution, and despite its repeated suppression it is still
very much with us today.

There is only one drawback. The excitement of exploring novel
forms of sexual stimulation, when practised between the members of a
mated pair, serves the family unit well. But it can take another form.
The urge for novelty can be satisfied not only by exploring new
patterns with a familiar partner, but also by exploring a new partner
with familiar patterns, or, even more so, by exploring a new partner with new patterns.

The development of Exploratory Sex emerges, therefore, as a double-
edged sword. Because our super-tribal cultures have laid increasing
stress on the benefits of exploratory behaviour – our educational
system, our great learning, our arts, sciences and technologies all
depend on this – the exploratory urges in all our other patterns of
behaviour have been similarly strengthened. In the sexual sphere this
has frequently led to difficulties. The idea of a mated female attending
practical classes in copulatory technique, or a mated male limbering up
in a sexual gymnasium, is deeply offensive to their long-term sexual
partners, since it interferes with the inherent exclusivity of the pair-
bonding mechanism. Sexual experiments away from the mate therefore
have to be made privately and in secret, and the new hazard of
pair-bond betrayal enters the scene. The ancient and fundamental
social nucleus of our species – the family unit – has suffered as a result,
but somehow it has managed to survive.

These difficulties would not arise if we were a different kind of
species, if we laid eggs in the sand like a turtle and left them to hatch
out by themselves. But for us, with our heavy parental duties, sexual
experiments outside the pair-bond have two dangers. They not only
provoke powerful sexual jealousies, but they also encourage the accidental formation of new pair-bonds, to the long-lasting detriment
of the offspring of the family units involved. Complex sexual combinations and communes may have worked from time to time, but un-
qualified successes seem to have been isolated rarities, limited to
exceptional and unusual personalities. Only the most ruthless intellectual control by all parties concerned will permit sexual experiments of
this kind to operate smoothly.

Even the rather widespread harem system, when viewed against the
broader background of super-tribal success, has not fared well, and
some scholars have pointed an accusing finger at it as an important
factor in the social decline of the cultures concerned.

As with the other nine categories of sexual behaviour, the exploratory function is basic enough to be observable in other animal species. Since it requires a high level of inventiveness, it is not surprising that it is limited largely to the higher primates. The great apes, in particular, show a considerable range of sexual experiments when living under conditions of captivity, including a number of copulatory postures not seen in their wild counterparts.

6. Self-rewarding Sex

It is impossible to draw up a complete list of the functions of sex without including a category based on the idea that there is such a thing as 'sex for sex's sake'; sexual behaviour, the performance of which brings its own reward, regardless of any other considerations. This function is closely related to the last one, but they are nevertheless distinct.

The relationship between Exploratory Sex and Self-rewarding Sex is rather like the relationship between exploring and playing a game, or between random play and structured play in children. When children burst out into a new play environment, they usually start off with a great deal of erratic rushing about and investigating. As time goes by, this almost random behaviour settles down into a patterned sequence. A play-structure emerges and a 'game' is born. A particular environment may lend itself to a climbing game, or a hiding game, or a hunting game, and once such a game has been developed it may be repeated eagerly on later occasions without undue variation. If it proves to be a rewarding pattern it will be returned to over and over again, even though it is no longer a novelty. The initial, erratic behaviour was exciting because it was Exploratory Play; the later, repeated pattern is exciting as Self-rewarding Play.

The parallel with Exploratory Sex and Self-rewarding Sex is obvious enough. Many highly satisfying copulatory incidents occur between the members of a mated pair that are deliberately not aimed at procreation, that are far in excess of the demands of pair-maintenance, and that do not involve the introduction of novel experiments. They therefore fall into the present functional category. They represent Self-rewarding Sex, or, if you prefer, pure eroticism. They are to the copulator what gastronomy is to the feeder, or what aesthetics is to the artist. It is inconsistent to sing the praises of exquisite gastronomic experiences, or of sublime aesthetic experiences, while at the same time condemning beautiful erotic experiences. Yet this has often been done. It is true that undue excess can sometimes create problems, but then so can undue excesses with gastronomy or aesthetics. Extreme cases of sexual athleticism can prove so exhausting that there is little energy left for anything else, and the pattern of living becomes unbalanced, just as extreme indulgence in feeding can cause serious obesity and loss of physical health, and extreme obsession with aesthetic problems can lead to a damaging disregard for other aspects of social life. The same basic rules apply in each case.

Preoccupation with action for action's sake implies the existence of some degree of spare time and spare energy. This in turn implies that the basic survival needs are being taken care of. In humans this means an urban society. In animals it means life in a zoo, with food supplied and enemies eliminated, and it is there, not surprisingly, that we find the examples of animal hyper-sexuality.

7. Occupational Sex

This is sex operating as occupational therapy, or, if you prefer, as an anti-boredom device. It is closely related to the last category, but again can be clearly distinguished from it. There is a difference between having spare time and being bored. Self-rewarding Sex can occur as just one of many ways of constructively utilizing the spare time available, with not the slightest sign of any boredom syndrome on the horizon. The function is the positive pursuit of sensory rewards. Occupational Sex, by contrast, functions as a therapeutic remedy for the negative condition produced by a sterile and monotonous environment. Mild boredom produces listlessness and a lack of direction or motivation. Intense boredom, in a really bleak, empty environment, has a different impact. It creates anxiety and agitation, irritability and eventually anger.

Experiments with students who were placed alone in featureless cubicles, wearing opaque goggles and heavy gloves that made small hand actions impossible, produced startling results. As the hours went by, they became increasingly unable to relax. They went to extreme lengths to invent any kind of trivial action they could perform in the limited circumstances. They began to whistle, talk to themselves, tap out rhythms, anything at all to break the monotony, no matter how absurd the activity. After several days they suffered from signs of severe stress and found the conditions so unbearable that they could not continue.

Intense boredom is not, therefore, a matter of lying around doing nothing, it is precisely the opposite. A point is reached where *any* activity will do, just so long as some kind of behaviour output can be achieved. The situation is too threatening to enjoy the sensitive pleasures typical of self-rewarding activities; it is more a question of stopping the pain of gross inactivity. Being under-active is damaging to the nervous system and the brain does its utmost to protect itself.

Under normal conditions of boredom – that is to say an empty environment, but not as deliberately empty as the one in the student experiments – the object most readily available for breaking the monotony is the subject's own body. If there is nothing else, there is always that. Nails can be bitten, noses can be picked, hair can be scratched; and the body can always be provoked to produce a sexual response. Since the goal is to produce the maximum amount of stimulation, sexual activities in this situation often become brutal and painful and sometimes even lead to severe mutilation, or physical injury of the genitals. The pain they cause is, in a sense, a bizarre part of the therapy, rather than an accidental outcome of it. Savage and prolonged masturbation is typical of this phenomenon, perhaps involving tearing of the skin, or the insertion of sharp objects into the genital tracts.

Extreme forms of Occupational Sex can be observed in human prisoners who have been forcibly cut off from their normal, stimulating environments. This is not a matter of Physiological Sex – a much smaller amount of indulgence would satisfy the specific physiological demands.

The phenomenon can also be seen in the case of pathological introverts. Here it may occur in environments that appear, superficially, to be adequately stimulating. A closer examination soon reveals, however, that although the individuals concerned seem to be surrounded by exciting stimuli, they are cut off from them by their abnormal psychological condition. They are psychologically starving in the midst of plenty. If for some reason they have become intensely antisocial and mentally isolated, unable to make contact with the ordinary world around them, they may be suffering from under-stimulation just as intense as that experienced by the physical prisoners in their cells. For the extreme isolates, whether physical or mental, the painful excesses of Occupational Sex become a lesser evil than total, moribund inactivity.

Zoo animals kept in sterile cages exhibit similar responses. When isolated from their mates they may exhibit Physiological Sex. Free from the pressures of finding food and avoiding enemies, and with spare time on their hands, they may indulge in Self-rewarding Sex. But driven to extremes of boredom, they may resort to Occupational Sex of a drastic kind. Some male monkeys become obsessional masturbators. Male ungulates kept with females, but with nothing else to do, may literally worry their mates to death, harrying them and chasing them beyond all natural limits. Apes have been known to behave in the same way. One male orang-utan living in an empty cage, when provided with a female, mated with her and embraced her so persistently that she temporarily lost the use of her arms and had to be removed. Monkeys or apes that have been reared away from their own kind frequently find it impossible to adjust to social life when introduced as adults to a group of their own species. Like the mentally disturbed human being who 'lives in a world of his own', they may huddle in a corner and continue to indulge in solitary Occupational Sex while only a few feet away from a receptive mate. This is very common in zoo chimpanzees, which are all too often reared in isolation as pets and then thrown together as adults. One pair with abnormal childhoods, that were kept as a 'married couple' in a cage with no other companions, repeatedly engaged in a great deal of sexual behaviour, but it was never directed at one another. Although they shared the enclosure, they were both mentally isolated. Sitting apart from each other, they would both masturbate regularly in a variety of ways. The female used small branches, or pieces of wood that she tore off the walls with her teeth and inserted into her vagina, performing these actions while the male stimulated his penis in another corner.

8. *Tranquillizing Sex*

Just as the nervous system cannot tolerate gross inactivity, so it rebels against the strains of excessive over-activity. Tranquillizing Sex is the other side of the coin from Occupational Sex. Instead of being anti-boredom, it is anti-turmoil. When faced with an overdose of strange, conflicting, unfamiliar or frightening stimuli, the individual seeks escape in the performance of friendly old familiar patterns that serve to calm his shattered nerves. When the pressures of living are heavy, the stressed victim can tranquillize himself by resorting to actions that he knows will bring him the satisfaction of a consummatory reward. In his stressed, over-active state he is unable to push anything through to a conclusion. He is tugged this way and that, never able to resolve specific problems because of constant interferences and the confusion of blocked pathways. His frustrations mount until any simple familiar act, no matter how irrelevant to the major preoccupations, will provide a welcome release, if only it can be performed without obstruction.

Trivial actions such as smoking a cigarette, chewing gum, or taking a drink, help to pacify the anxious. Tranquillizing Sex operates in the same way. The soldier at war, waiting for battle, or the business executive in the middle of a crisis, may seek momentary peace in the arms of a responsive female. The personal, emotional involvement can be at a minimum, the actions stereotyped. In a way, the more automatic it is the better, because his brain is already over-involved and seeks only simplicity.

This is similar to the animal activity known as 'displacement activity'. When two rival animals meet and come into conflict with one another, each wants to attack the other, but each is afraid to do so. Their behaviour is blocked and in their thwarted, frustrated condition they may turn aside to perform simple, irrelevant actions, such as grooming themselves, nibbling at food or fiddling with nest material. These displacement actions do not, of course, resolve the original conflict, but provide a momentary respite from the stressed condition. If a female happens to be near by she may be briefly mounted, and as in the human cases the action is usually stereotyped and simple.

9. *Commercial Sex*

Prostitution has already been mentioned, but only from the point of view of the customer. For the prostitute herself the function of copulation is different. Subsidiary factors may be operating, but primarily and overwhelmingly it is a straightforward commercial transaction. Commercial Sex of a kind also figures as an important function in many marriage situations, where a one-sided pair-bond exists: one partner simply provides a copulatory service for the other in exchange for money and shelter. The provider who has developed a true pair-bond has to accept a mock one in return. The woman (or man) who marries for money is, of course, functioning as a prostitute. The only difference is that whereas she, or he, receives indirect payment, the ordinary prostitute has to operate on a pay-as-you-lay basis. But whether the system is organized on long-term or short-term contracts, the function of the sexual behaviour involved is fundamentally the same.

A milder form of sex-for-material-gain is executed by strip-teasers, dance hostesses, beauty queens, club girls, dancers, models and many actresses. For payment, they provide ritualized performances of the earlier stages of the sexual sequence, but (in their official capacities) stop short of copulation itself. Compensating for the incompleteness of their sexual patterns, they frequently exaggerate and elaborate the preliminaries that they offer. Their sexual postures and movements, their sexual personality and anatomy, all tend to become magnified in an attempt to make up for the strict limitations of the sexual services they provide.

Commercial Sex appears to be rare in other species, even in zoos, but a form of 'prostitution' has been observed in certain primates. Female monkeys in captivity have been seen to offer themselves sexually to a male as a means of obtaining food morsels scattered on the ground, the sexual actions distracting the male from the business of competing for food.

10. *Status Sex*

With this, the final functional category of sexual behaviour, we enter a strange world, full of unexpected developments and ramifications. Status Sex infiltrates and pervades our lives in many hidden and unrecognized ways. Because of its complexity, I omitted it from the last chapter so that I could deal with it more fully here. It will help if we start by examining the form it assumes in other species, before we take a look at it in the human animal.

Status Sex is concerned with dominance, not with reproduction, and to understand how this link is forged we must consider the differing roles of the sexual female and the sexual male. Although a full expression of sexuality involves the active participation of both sexes, it is nevertheless true to say that, for the mammalian female, the sexual role is essentially a submissive one, and for the male it is essentially an aggressive one. (It is no accident of legal jargon that when a man sexually 'fondles' an unwilling female, his action is referred to as an indecent 'assault'.) This is not merely due to the fact that the male is physically stronger than the female. The relationship is an integral part of the nature of the copulatory act. It is the male mammal who has to mount the female. It is he who has to penetrate and invade his partner's body. An over-submissive female and an over-aggressive male are simply exaggerating their natural roles, but an aggressive female and a submissive male are completely reversing their roles.

The sexual action of a female monkey is to 'present' herself to the male by turning her rump towards him, raising it up conspicuously and lowering the head end of her body. The sexual action of the male monkey is to mount the female's back, insert his penis and make pelvic thrusts. Because, in a sexual encounter, the female submits herself and the male imposes himself, these actions have been 'borrowed' for use in primarily non-sexual situations requiring more general signals of submissiveness and aggressiveness. If female sexual 'presenting' signifies submissiveness, then it can be used in this way in a purely hostile encounter. A non-sexual female monkey can present her rump to a male as a sign that she is simply not aggressive. It acts as an appeasement gesture and functions as *an indication of her subordinate status*. In response, the male can mount her and make a few cursory pelvic thrusts, using these actions *purely to display his dominant status*.

234

Status Sex, used in this way, is a valuable device in the social lives of monkeys and apes. As a ritual of subordination and dominance, it avoids bloodshed. A male approaches a female aggressively, spoiling for a fight. Instead of screaming or attempting to flee, which would only feed the fire of his aggression, the female 'presents' herself to him sexually, the male responds, and they part, their relative dominance positions reaffirmed.

This is only the beginning. The value of Status Sex is such that it has spread to cover virtually all forms of aggressive encounter within the group. If a weak male is threatened by a strong one, the underling can protect himself by behaving as a pseudo-female. He signals his subordination by adopting the female sexual posture, offering his rump to the dominant male. The latter asserts his dominant status by mounting the weaker male, just as if he were dealing with a submissive female.

Precisely the same interaction can be observed between two females. An inferior female, threatened by a superior one, will 'present' to her and be mounted by her. Even juvenile monkeys will go through the same ritual, although they have not yet reached the adult sexual condition. This underlines the extent to which Status Sex has become divorced from its original sexual condition. The actions performed are still sexual actions, but they are no longer sexually motivated. Dominance has made a take-over bid for them.

The fact that sexual activities are being used repeatedly and frequently in this additional context explains the apparently orgiastic condition of some monkey colonies. Visitors to zoos often come away with the idea that monkeys are insatiable sexual athletes, ready, at the flick of a rump, to mate with anything, be it male or female, adult or juvenile. In one sense, of course, this is true. The observation is accurate enough. It is the interpretation that is wrong. Only when one understands the non-sexual motivation of Status Sex does the picture become more balanced.

It may help to give an example from nearer home. Nearly everyone is familiar with the friendly, submissive greeting of a domestic cat, as it rubs the side of its body against a human leg, with its tail held stiffly upwards and the rear end of its body raised high. Both male and female cats do this and if, in response, we stroke their backs, we can feel them pushing the rear ends of their bodies up against the pressure of our hands. Most people accept this simply as a feline greeting gesture and do not question its origin or significance. In reality, it is another example of Status Sex. It is derived from the sexual presentation display of the female cat towards the male, its original function being the pre-copulatory exposure of the vulva. But like the Status Sex presentation action of the monkeys and apes, it has now become emancipated from its purely sexual role and is performed by either sex when in a friendly, submissive condition. Because of the human cat-owner's size and strength, he is inevitably and permanently dominant as far as his pet is concerned. If contact is made after a temporary absence, the cat feels the need to re-establish its subordinate role, hence the greeting ceremony utilizing a submissive Status Sex display.

The feline pattern is a fairly simple one, but returning to the monkeys again, there are some striking anatomical extensions of Status Sex that we should examine before investigating the human condition. Many female monkeys possess bright red patches of swollen, naked skin in the rump region. These are conspicuously displayed to the male during the sexual rump-presentation action. They are also, of course, displayed when a female offers her rump submissively in Status Sex encounters. It has recently been pointed out that the males of some species have evolved mimics of these red patches on *their* rumps, as an enhancement of their submissive Status Sex displays towards more dominant individuals. For the females, the red rump patches serve a dual purpose, but for the males their function is *exclusively* concerned with Status Sex.

Switching from the submissive to the dominant Status Sex display, we can see a similar development. The dominant action involves erection of the penis, and this too has been elaborated by the addition of conspicuous colours. In a number of species the males possess bright red penises, often surrounded by a vivid blue patch of skin over the scrotal region. This makes the male genitals as conspicuous as possible and males can frequently be seen sitting with their legs spread apart, displaying these bright colours to the maximum. In this way they can signal their high status without even moving. In some monkey species, males displaying in this manner sit at the edge of their group and, if another group comes near by, the red penis becomes fully erect and may be repeatedly raised to strike its owner's stomach. In ancient Egypt, the sacred baboon was seen as the embodiment of masculine sexuality. It was not only depicted in its Status Sex display posture in Egyptian paintings and carvings, but was even embalmed and buried in that posture, seventy days being spent on the embalming procedure and two days on the funeral ceremony. Obviously, the dominant Status Sex display of this species came through loud and clear, not only to other baboons, but also to ancient Egyptians. This was no accident, as we shall see in a moment.

Just as in some species the males have mimicked the submissive female displays, evolving their own red rump patches, so the females have, in some cases, mimicked the dominant displays of the males. Some female South American monkeys have evolved an elongated clitoris, which has virtually become a pseudo-penis. In certain species it is so similar in appearance to the true penis of the male that it is hard to tell the sexes apart. This has given rise to a number of native legends in the areas where these animals live wild. Because they all appear to be male, the local populations believe them to be exclusively homosexual. (Strangely enough the female hyena has also evolved a similar pseudo-penis, but the myth that has emerged in Africa is that this species is hermaphrodite, each individual enjoying both male and female sexual activities.)

In a few species of monkeys the females have evolved a pseudo-scrotum as well as a pseudo-penis. As yet, we have little information about the way these fake male genitals are employed in the wild. We do know that certain male South American monkeys use penis erection as a direct threat to a subordinate. In the case of the little squirrel monkey, it has become the most important dominance signal in the animal's repertoire. It is more than merely sitting around with the legs apart. When in a threatening mood, a superior male of this species approaches close to an inferior and obtrusively erects his penis in the inferior's face. The pseudo-penis of the female monkeys does not, however, appear to be erectile; perhaps it is enough simply to display it as it is, towards an inferior monkey.

This, then, is the Status Sex situation in our closest relatives, the monkeys and apes. I have gone into it in some detail because it provides a useful evolutionary background against which to examine Status Sex developments in the human species. It makes some of the extraordinary lengths to which the human animal has gone, in this direction, a little easier to comprehend. Already, while reading the details of the monkey behaviour, you will, like the ancient Egyptians, have noted certain similarities to the human condition. With men, as with monkeys, the submissive female sexual patterns and the dominant male patterns have come to stand for submissiveness and dominance in non-sexual contexts.

The ancient female pattern of presenting the rump to the male still survives as a gesture of subordination. Children are often forced to bend over for punishment in this posture. Also the buttocks are generally regarded as the most 'ridiculous' part of the human body, to be joked about and laughed at, or have pins stuck in them. Helpless victims in sado-masochistic pornography – not to mention popular comedy cartoon films and drawings – are frequently trapped with their buttocks in the air.

It is in the realm of dominant male patterns, however, that the human imagination has really run riot. The art and literature of civilization, since its earliest days, have been strewn with phallic symbols of all kinds. In recent times these have often become highly cryptic and far removed from their original source, the erect human penis, but it is still possible to observe more direct and overt phallic displays in the most primitive surviving cultures. Amongst New Guinea tribesmen, for example, the males make war wearing long tubes fitted on to their penises. These extensions, often well over a foot in length, are held in a near-vertical position by cords attached to the wearer's body. In other cultures, too, the penis is ornamented and artificially enlarged in a variety of ways.

Clearly, if the erection of the penis is used as a threatening display of male dominance, then it follows that the greater the erection, the greater the threat. The visual signals transmitting the intensity of the threat are of four kinds: as the penis becomes erect, it alters its angle, it changes from soft to hard, it increases in width, and it grows in length. If all four of these qualities can be artificially exaggerated, then the impact of the display will be maximally enhanced. There is a limit to what can be done on the human body itself (which is more or less reached by the New Guinea tribesmen), but there is no such limit where human effigies are concerned. In drawings, paintings and sculptures of the human form the phallic display can be magnified at will. The average length of the erect penis in real life is $6\,^{1}/_{4}$ inches, which is less than one-tenth of the height of an adult male. In phallic statues, the length of the penis often exceeds the height of the figure.

Exaggerating the phallus still further, the depiction of a body is omitted altogether and the drawing or sculpture simply shows a huge, vertical, disembodied penis. Ancient sculptures of this kind, often rising many feet into the air, have been found in numerous parts of the world. Giant phallic statues nearly two hundred feet high guarded the temple of Venus at Hierapolis, but even these were exceeded in size by another ancient phallus that reputedly towered to the height of three hundred and sixty feet, or roughly seven hundred times the length of the physical organ it represented. It is said to have been covered in pure gold.

From obvious representations of this kind, it is only one more step to the world of phallic symbolism, where almost any long, stiff, erect object can take on a phallic role. We know from the dream studies of the psychoanalysts just how varied these symbols can be. But they are not confined to dreams. They are deliberately used by advertisers, artists and writers. They appear in films, plays and almost all forms of entertainment. Even when they are not consciously understood they can still make their impact, because of the very basic signal they transmit. They include everything from candles, bananas, neckties, broom-handles, eels, walking-sticks, snakes, carrots, arrows, water-hoses and fireworks, to obelisks, trees, whales, lamp-posts, sky-scrapers, flag-poles, cannons, factory chimneys, space rockets, light-houses and towers. These all have symbolic value because of their general shape, but in some cases a more specific property is involved. Fish have become phallic symbols because of their texture as well as their shape, and because they thrust themselves through the water; elephants because of their erectile trunks; rhinos because of their horns; birds because they rise up against gravity; magic wands because they give special powers to the magician; swords, spears and lances because they penetrate into the body; champagne bottles because they ejaculate when opened; keys because they are inserted into keyholes; and cigars because they are tumescent cigarettes. The list is almost endless, the scope for imaginative symbolic equations enormous.

All these images have been used, and in most cases used frequently, as objects representing masculinity. The tough, dominant male (or would-be tough, dominant male) who chews on his fat cigar and thrusts it into his companion's face, is fundamentally performing the same Status Sex display as the little squirrel monkey that spreads its legs and thrusts its erect penis into the face of a subordinate. Social taboos have forced us to employ cryptic substitutes for our aggressive sexual displays, but the imagination of man being what it is, this has not reduced the phenom-enon, it has only diversified and elaborated it. As I explained in the chapter dealing with status and super-status, we have good reason in our super-tribal condition to make great play with our status devices, and this is precisely what we do in the case of Status Sex.

We can find examples of various kinds of improvements in phallic symbols taking place almost as we watch. The design of sports cars illustrates this well. They have always radiated bold, aggressive masculinity and have been considerably aided in this by their phallic qualities. Like a baboon's penis, they stick out in front, they are long, smooth and shiny, they thrust forward with great vigour and they are frequently bright red in colour. A man sitting in his open sports car is like a piece of highly stylized phallic sculpture. His body has dis-appeared and all that can be seen are a tiny head and hands surmount-ing a long, glistening penis. (The argument may be used that the shape of sports cars is controlled purely by the technical demands of stream-lining, but the crowded traffic conditions of modern driving and the increasingly strict speed limits render this explanation nonsensical.) Even ordinary motor cars have their phallic qualities, and this may explain to some extent why male drivers become so aggressive and so eager to overtake one another, despite considerable risks and despite the fact that they all meet up again at the next set of traffic-lights, or, at best, only cut a few seconds off their journey.

Another illustration comes from the world of popular music, where the guitar has recently undergone a change of sex. The old-fashioned guitar, with a curvaceous, waisted body, was symbolically essentially female. It was held close to the chest, its strings lovingly caressed. But times have changed and its femininity has vanished. Ever since groups of masculine 'sex idols' have taken to playing electric guitars, the designers of these instruments have been at pains to improve their masculine, phallic qualities. The body of the guitar (now its symbolic testicles) has become smaller, less waisted, and more brightly coloured, making it possible for the neck (its new symbolic penis) to become longer. The players themselves have helped by wearing the guitars lower and lower until they are now centred on the genital region. The angle at which they are played has also been altered, the neck being held in an increasingly erect position. With the combination of these modifications, the modern pop-group can stand on stage and go through the movements of masturbating their giant electric phalluses while they dominate their devoted 'slaves' in the audience. (The singer has to make do with fondling a phallic microphone.)

Contrasting with these phallic 'improvements' are a number of cases where phallic symbols have gone into decline or eclipse. As early civilizations (which, as I have said, were much more open in their use of phallic symbolism) were superseded, their obvious images were often cloaked and distorted. Perhaps the most startling example of this is the Christian cross. In earlier days this was a straightforward phallic symbol, the vertical piece representing the penis and the side pieces the testicles. It sometimes appears in a more explicit form in ancient, pre-Christian images, with a man's head at the top of the upright, his body being completely replaced by the stylized representation of his sexual organs in the form of the cross. It has been pointed out by one author that the acceptance of this symbol in a new role by Christians was probably facilitated by its earlier significance as a symbol of the 'life force'.

Another cross that has long since lost its original meaning is the famous Maltese cross. The ancient prehistoric ruins of Malta were full of phalluses, most of which have been lost, stolen, or destroyed. Amongst them was a cross consisting of four huge stone phalluses which, as one writer put it, 'were subsequently metamorphosed by the virtuous Knights of St John' and served as their arms.

Easter festivities have also shown a reduction in phallic overtness. In ancient cultures this was often a time for baking phallic cakes. They were made in the shape of both male and female genitals, but today they have been transformed, appearing in certain countries as sweet-meats made in the shape of a fish (the male cake) and a doll (the female cake). The phallic nature of the fish symbol was also originally involved in the ritual of eating fish on Fridays, but this has long ago lost its sexual significance.

There are many other examples one could give. The bonfire, for instance, although still retaining an almost magical, ritual quality on certain occasions, has lost its sexual properties. It was originally lit in a special way by rubbing a 'male' stick against a 'female' stick in an act of symbolic copulation, until a spark was generated and the bonfire burst into sexual flames.

Many buildings used to display carved phalluses on their outer walls to protect them against the 'evil eye' and other imagined dangers. These symbols, being aggressive, dominant Status Sex threats to the outside world, guarded the buildings and their occupants. In certain Mediterranean countries today one can still see symbols of this kind, but they have become less overtly sexual. They now usually consist of a pair of bull's horns, firmly fixed high up on an outer surface or the corner of a roof. However, despite these expurgations and censorships, which have turned the tree of carnal knowledge into the simple tree of knowledge and have replaced the obvious codpiece with the less obvious necktie, there are still areas where aggressive phallic symbols retain their original overt properties. In the realm of *insults* we find them still very much in evidence.

Verbal insults frequently take a phallic form. Almost all the really vicious swearwords we can use to hurl abuse at someone are sexual words. Their literal meanings relate to copulation or to various parts of the genital anatomy, but they are used predominantly in moments of extreme aggression. This again is typical of Status Sex and demonstrates very clearly the way in which sex is borrowed for use in a dominance context.

Visual insults follow the same trend, several kinds of phallic actions being employed as hostile gestures. Sticking out the tongue originated in this way, with the protruded tongue symbolizing the erect penis. Hostile gestures known as giving the 'phallic hand' have existed in various forms for at least two thousand years. One of the most ancient consists of aiming the middle (i.e. second) finger, stiffly and fully extended, at the person being subjected to contempt. The rest of the fist is clenched. Symbolically, the middle finger represents the penis, the clenched thumb and first finger together represent one testicle, and the clenched third and fourth fingers together represent the other testicle. This gesture was popular in Roman times, when the middle finger was known as the *digitus impudicus*, or *digitus infamis*. It has become modified over the centuries, but can still be seen in many parts of the world. Sometimes the first finger is used instead of the middle finger, probably because this is a slightly easier posture to hold. Sometimes the first and second finger are extended together, emphasizing the size of the symbolic penis. It is usual today for this type of phallic hand to be jerked upwards into the air one or more times, in the direction of the insulted person, symbolizing the action of pelvic thrusting. The two extended fingers may be held together or separated into a V shape.

An interesting corruption of this last form appeared in recent times as the victory-V sign, which did far more than merely copy the first letter of the word victory. Its phallic properties also helped it. It differed from the insult-V by the position of the hand. In the insult-V the palm of the hand is held towards the insulter's face; in the victory-V it is held towards the admiring crowd of onlookers. This means, in effect, that the dominant individual making the victory-V sign is really making the insult-V, but on *their* behalf; for them, not against them. What they see as they watch their signalling leader is the same hand position as they themselves would see if *they* were making the insult-V. By the simple device of rotating the hand, the phallic insult becomes a phallic protection. As we have already observed, threat and protection are two of the most vital aspects of dominance. If a dominant individual performs a threat towards a member of his group, it insults the latter, but if the dominant one performs the same threat *away* from the group towards an enemy, or an imagined enemy, then his subordinates will cheer him for the protective role he is playing. It is astonishing to think that a leader can change his whole image simply by twisting his hand through 180 degrees, but such are the refinements of modern Status Sex signalling.

Another ancient form of 'phallic hand', also dating back at least two thousand years, is the so-called 'fig'. In this, the whole fist is clenched, but as it is aimed at the insulted person the thumb is pushed through between the base of the curled first and second fingers. The tip of the thumb then protrudes slightly, like the head of a penis, pointing at the subordinate or enemy. This gesture has spread throughout much of the world and almost everywhere is known as 'making the fig'. In English, the phrase 'I don't give a fig for him' means that he is not even worthy of an insult.

Many examples of these phallic hands have been found on ancient amulets and other ornaments. They were worn as protections against the 'evil eye'. Some people today might look upon such emblems as improper or obscene, but this was not the role they played when they were worn. They were used then, quite properly, as protective Status Sex symbols. In specific contexts the symbolic phallus was seen as something to be acclaimed and even worshipped as a magical guardian ready to ravage, not the members of the group, but threats coming from outside it. At the Roman festival of Liberalia an enormous phallus was carried in procession on a magnificent chariot into the middle of the public place of the town where, with great ceremony, the females, including even the most respectable matrons, hung garlands around it to 'remove enchantments from the land'. In the Middle Ages many churches had phalluses on their outer walls to protect them from evil influences, but in almost all cases these were later destroyed as 'depraved'.

Even plants were called into phallic service. The mandrake, a plant with phallus-shaped roots, was widely used as a protective amulet. It was improved for its symbolic role by embedding grains of millet or barley in it in the appropriate area, re-burying it for twenty days so that the grains sprouted, then digging it up again and trimming the sprouts to make them look like pubic hair. Kept in this form it was said to be so effective in dominating outside forces that it would double its owner's money each year.

It would be possible to go on and fill a whole book with examples of phallic symbolism, but the few I have selected are sufficient, I think, to show how widespread and varied this phenomenon is. We arrived at this topic by singling out just one of the elements of the aggressive male Status Sex display, namely penis erection. Other important developments have also occurred, however, which we should not overlook. The original, straightforward pattern of copulation is, for the male, as I have already stressed, a fundamentally assertive and aggressive act of penetration. Under certain conditions it can therefore function as a Status Sex device. A male can copulate with a female primarily to boost his masculine ego, rather than to achieve any of the other nine sexual goals I have listed in this chapter. In such cases he may speak of making a 'conquest', as if he has been fighting a battle rather than making love. And when I say he *speaks* of it I mean this literally, for boasting to other males is an important part of the Status Sex victory. If he keeps quiet about it, it can always feed his ego privately, but it provides him with a much stronger status boost if he tells his friends. Any female who finds out about this can be reasonably sure what kind of copulation she has been involved with. The details of pair-forming copulations are, by contrast, strictly private.

The male who uses females for Status Sex purposes is more concerned, in fact, with showing them off than with anything else. He may even be content to display his dependent females to his group without bothering to copulate with them. Providing they are clearly seen to be subordinate to him, this will often suffice.

The huge harems that were assembled by the rulers of certain cultures

acted largely as a Status Sex device. They did not indicate the existence
of multiple pair-bonding. Frequently a favourite wife emerged from the
group of females, with whom some sort of pair-bond developed, but
the business of Status Sex really came to dominate the whole scene.
There was a simple equation: power = number of females in harem.
Sometimes there were so many that the ruler had neither the time
nor the energy to copulate with all of them, but as a symbol of virility
he did attempt to sire as many offspring as possible. The would-be
present-day harem overlord usually has to make do with a long series
of females, dominating them one at a time, instead of gathering them
around him all at once. He has to rely on his verbal reputation rather
than on a massive visual display of sexual power.

It is relevant here to mention the special attitude of heterosexual
Status Sex devotees towards homosexual males. It is an attitude of
increased hostility and contempt, caused by the unconscious realization that 'if they won't join the game, they can't be beaten'. In other
words, the homosexual male's lack of sexual interest in females gives
him an unfair advantage in the Status Sex battle, for no matter how
many females the heterosexual expert subdues, the homosexual will
fail to be impressed. It then becomes necessary to defeat him by
ridicule. Inside the homosexual world there will, of course, be Status
Sex competition as vigorous as that found in the heterosexual sphere,
but this in no way improves the understanding between the two
groups, since the objects competed for are so different in the two cases.

If the modern Status Sex practitioner is unable to achieve real conquests, there are still a number of alternatives available to him. A
mildly insecure male can express himself by telling dirty jokes. These
carry the implication that he is aggressively sexual, but an obsessive,
persistent dirty-joke-teller begins to arouse suspicions in his companions. They detect a compensation mechanism.

Males with a greater inferiority problem can frequent prostitutes. I
have already mentioned other functions of this sexual activity, but
status-boosting is perhaps the most important. The essential property
of this form of Status Sex is that the female is being degraded. The
male, providing he has a small amount of cash, can *demand* sexual
submission. The fact that he knows the girl does not welcome his
advances, but submits to them anyway, can even help to increase his
feeling of power over her. Another alternative is the strip-tease display.
The female, again for a small sum of money, has to strip herself naked
in front of him, debasing herself and thereby raising the relative status
of the watching males.

There is a savage satirical drawing on the subject of strip-tease,
captioned simply 'tripes-tease'. It shows a naked girl who, having

removed all her clothes and still being faced with shouts for 'more', makes an incision in her belly, and, with a seductive smile, starts to pull out her entrails to the beat of the music. This brutal comment reveals that with the subject of strip-tease we are moving into the realm of that extreme form of Status Sex expression, the realm of sadism.

It is an unpalatable but obvious fact that the more drastic the need for male ego-boosting, the more desperate the measures; the more degrading and violent the act, the greater the boost. For the vast majority of males, these extreme measures are unnecessary. The level of self-assertion achieved in ordinary social life is sufficiently rewarding. But under the heavy status pressures of super-tribal living, where there can be so few dominants and must be so many suppressed subordinates, sadistic thoughts nevertheless tend to proliferate. For most men they remain nothing more than thoughts; sadistic fantasies that never see the light of day. Some individuals go further, avidly studying the details of whippings, beatings and tortures in sadistic books, pictures and films. A few attend pseudo-sadistic exhibitions, and a very, very few become actual practising sadists. It is true that many men may be mildly brutal in their love-play and that some perform mock-sadist rituals with their mates, but the full-blooded sadist is fortunately a rare beast.

One of the most common forms of sadism is rape. Perhaps the reason for this is that it is so exclusively an act of the male that it expresses aggressive masculinity better than other types of sadistic activity. (Males can torture females and females can torture males. Males can rape females, but females cannot rape males.) In addition to the total domination and degradation of the female, one of the bizarre satisfactions of rape for the sadist is that the writhings and facial expressions of pain he produces in the female are somewhat similar to the writhings and facial expressions of a female experiencing an intense orgasm. Furthermore, if he then kills his victim, her immediately limp and passive condition presents a gruesome mimic of post-orgasmic collapse and relaxation.

An alternative pattern for meeker males is what might be described as 'visual rape'. Usually referred to as exhibitionism, this consists of suddenly exposing the genitals to a strange female, or females. No attempt is made to effect physical contact. The aim is to produce shame and confusion on the part of the unwilling female spectators, by presenting them with the most basic form of Status Sex threat display. Here we are right back to the penis threat of the little squirrel monkey.

Perhaps the most extreme kind of sadism is the torture, rape and murder of a small child by an adult male. Sadists of this type must suffer from feelings of the most intense status inferiority known to man. In order to obtain their ego boost, they are forced to select the weakest and most helpless individuals in their society and impose upon them the most violent form of domination they can perform. Happily, these extreme measures are rarely taken. They appear to be more common than they in fact are, because of the enormous publicity that such cases receive, but in reality they comprise only a minute fraction of the total picture of 'crimes of violence'. All the same, any super-tribe that contains even only a few individuals that are driven to dominance excesses of this kind must constitute a society operating under immense status pressures.

One final point on Status Sex: it is intriguing to discover that certain individuals with a demonstrably vast lust for power suffered from physical sexual abnormalities. The autopsy on Hitler, for instance, revealed that he had only one testicle. The autopsy on Napoleon noted the 'atrophied proportions' of his genitals. Both had unusual sex lives, and one is left to wonder just how much the course of European history would have been changed had they been sexually normal. Being structurally sexually inferior, they were perhaps driven back on to more direct forms of aggressive expression. But no matter how extreme their domination became, their urge for super-status could never be satiated, because no matter how much they achieved it could never give them the perfect genitals of the typical dominant male. This is Status Sex going full circle. First, the dominant male sexual condition is borrowed as an expression of dominant aggression. Then it becomes so important in this role that if there is something wrong with the sexual equipment, it becomes necessary to compensate by putting a stronger emphasis on pure aggression.

Perhaps there is something to be said for Status Sex (in its very mildest forms) after all. In its more ritualized and symbolic varieties it does at least provide comparatively harmless outlets for otherwise potentially damaging aggressions. When a dominant monkey mounts a subordinate, it manages to assert itself without recourse to sinking its teeth into the weaker animal's body. Swapping sex jokes in a bar causes less injury than having a punch-up or a brawl. Making an obscene gesture at someone does not give him a black eye. Status Sex has, in fact, evolved as a bloodless substitute for the bloody violence of direct domination and aggression. It is only in our overgrown super-tribes, where the status ladder stretches right up into the clouds and the pressures of maintaining or improving a position in the social hierarchy have become so immense, that Status Sex has got out of hand and gone to lengths that are as bloody as pure aggression itself. This is yet another of the prices that the super-tribesman has to pay for the great achievements of his super-tribal world and the excitements of living in it.

In surveying these ten basic functions of sexual behaviour we have seen very clearly the way in which, for the modern urban human animal, sex has become super-sex. Although he shares these ten functions with other animals, he has pushed most of them much farther than the other species have ever done. Even in the most puritanical cultures, sex has played a major role, if only because there it was constantly on people's minds as something that needed suppressing. It is probably true to say that no one is quite so obsessed by sex as a fanatical puritan.

The influences at work in the trend towards super-sex have been interwoven with one another. The main factor was the evolution of a giant brain. On the one hand this led to a prolonged childhood and this in turn meant a long-term family unit. A pair-bond had to be forged and maintained. Pair-formation Sex and Pair-maintenance Sex were added to primary Procreation Sex. If active sexual outlets were not available, the ingenuity of the giant brain made it possible for various techniques to be employed to obtain relief from physiological sexual tension. Man's strengthened urge for novelty, his heightened curiosity and inventiveness, gave rise to a massive increase in Exploratory Sex. Because of its efficiency, the giant brain organized his life in such a way that man had more and more time to spare and a greater sensitivity available to him while filling it. Self-rewarding Sex, sex for sex's sake, was able to blossom. If there was too much time to spare, then Occupational Sex could step in. If, by contrast, the increased strain of super-tribal pressures and stresses became too heavy, then there was always Tranquillizing Sex. The added complexities of super-tribal life brought increased division of labour and trading, and sexual activity became involved here too, in the form of Commercial Sex. Finally, with the vastly magnified dominance and status problems of the huge super-tribal structure, sex was increasingly borrowed for use in a non-sexual context, as all-pervading Status Sex.

The greatest sexual complication to arise has been the clash between the primarily reproductive categories (Procreation, Pair-formation and Pair-maintenance Sex), on the one hand, and the primarily non-reproductive categories, on the other. In the pre-pill days, when contraception was forbidden, rare, or inefficient, Procreation Sex provided a major hazard for Exploratory Sex, Self-rewarding Sex and the rest. Even in the so-called 'post-pill paradise', which some have seen as heralding an epoch of wild promiscuity, the problem is far from solved, because of the persistence of the fundamental pair-bonding properties of human sexual encounters. Widespread, trouble-free promiscuity is a myth and always will be. It is a myth born of the wishful thinking of Status Sex, but it will for ever remain a wishful thought. Man's strong pair-forming urge stemming, in evolutionary terms, from his greatly increased parental duties, will persist no matter what technical advances are achieved with perfected contraception in the years to come. This does not mean that such advances will have no impact on our sexual activities. On the contrary, they will profoundly alter our behaviour. The triple pressure of improved contraception, dwindling venereal disease and ever swelling human population will together work towards a dramatic increase in non-reproductive forms of copulatory indulgence. There can be no doubt of this. Equally there can be no doubt that this will intensify the clash between these forms of sex and the demands of the pair-bond. Unhappily, as a result, the children will suffer along with their sexually confused parents.

It would be much easier if, like our monkey relatives, we had a lighter parental burden and were more truly biologically promiscuous. Then we could extend and intensify our sexual activities with the same facility that we magnify our body-cleaning behaviour. Just as we harmlessly spend hours in the bathroom, visit masseurs, beauty parlours, hairdressers, Turkish baths, swimming pools, sauna baths, or Oriental bath-houses, so we could indulge in lengthy erotic escapades with anyone, at any time, without the slightest repercussions. As it is, it seems as if our basic animal nature will always stand in the way of this development, or will at least discourage it until such time as we have undergone some radical genetical change.

The only hope is that, as the clashing demands of super-sex grow more intense, we shall learn to play the game more deftly. It is, after all, possible to indulge oneself gastronomically without growing either fat or sick. With sex the trick is more difficult to accomplish, and society is littered with the bitter jealousies, forlorn heartbreaks, miserable, shattered families, and unwanted offspring to prove it.

No wonder super-sex has become such a problem for the Urban Super-ape. No wonder it has so often been abused. It is capable of providing man with his most intense physical and emotional rewards. When it goes wrong, it is also capable of causing him his greatest miseries. As he has expanded it, elaborated it and manipulated it, he has magnified its potentials both as a reward and a punishment. But sadly there is nothing unusual in this. In many departments of human behaviour we find the same development. Even in medical care, for example, where the rewards are so obvious, the punishments are still there: it can so easily contribute to overcrowding which, in turn, leads to the proliferation of new stress diseases. It can also lead to hypersensitivity to pain. A New Guinea tribesman can have a spear removed from his thigh with more aplomb than a super-tribesman having a small splinter removed from his finger. But this is no reason for wanting to turn back. If our increased sensitivities can work both ways, we must make sure that they work the right way. The big change is that matters are now in our hands, or rather our brains. The tightrope of survival which has been set up, and on which our species performs its daring tricks, has been raised higher and higher. The dangers have become greater, but then so have the thrills. The only snag is that when the tribes became super-tribes, someone took away our biological safety-net. It is up to us now to make sure that we do not crash to our deaths. We have taken over evolution and have no one to blame but ourselves. The strength of our animal properties is still carried securely within us, but so are our animal weaknesses. The better we understand them and the enormous challenges they are facing in the unnatural world of the human zoo, the better our chances of success.

4

IN-GROUPS AND OUT-GROUPS

QUESTION: What is the difference between black natives slicing up a white missionary, and a white mob lynching a helpless Negro? Answer: very little – and, for the victims, none at all. Whatever the reasons, whatever the excuses, whatever the motives, the basic behaviour mechanism is the same. They are both cases of members of the in-group attacking members of the out-group.

In plunging into this subject we are entering an area where it is difficult for us to maintain our objectivity. The reason is obvious enough: we are, each one of us, a member of some particular in-group and it is difficult for us to view the problems of inter-group conflict without, however unconsciously, taking sides. Somehow, until I have finished writing and you have finished reading this chapter, we must try to step outside our groups and gaze down on the battlefields of the human animal with the unbiased eyes of a hovering Martian. It is not going to be easy, and I must make it clear at the outset that nothing I say should be construed as implying that I am favouring one group as against another, or suggesting that one group is inevitably superior to another.

Using a harsh evolutionary argument, it might be suggested that if two human groups clash and one exterminates the other, the winner is biologically more successful than the loser. But if we view the species as a whole this argument no longer applies. It is a small view. The bigger view is that if they had contrived to live competitively but peacefully alongside one another, the species as a whole would be that much more successful.

It is this large view that we must try to take. If it seems an obvious one, then we have some rather difficult explaining to do. We are not a mass-spawning species like certain kinds of fish that produce thousands of young in one go, most of which are doomed to be wasted and only a few survive. We are not quantity breeders, we are quality breeders, producing few offspring, lavishing more care and attention on them and looking after them for a longer period than any other animal. After devoting nearly two decades of parental energy to them it is, apart from anything else, grotesquely *inefficient* to send them off to be knifed, shot, burned and bombed by the offspring of other men.

Yet, in little more than a single century (from 1820 to 1945), no less than 59 million human animals were killed in inter-group clashes of one sort or another. This is the difficult explaining we have to do, if it is so obvious to the human intellect that it would be better to live peacefully. We describe these killings as men behaving 'like animals', but if we could find a wild animal that showed signs of acting in this way, it would be more precise to describe it as behaving like men. The fact is that we cannot find such a creature. We are dealing with another of the dubious properties that make modern man a unique species.

Biologically speaking, man has the inborn task of defending three things: himself, his family and his tribe. As a pair-forming, territorial, group-living primate he is driven to this, and driven hard. If he or his family or his tribe are threatened with violence, it will be all too natural for him to respond with counter-violence. As long as there is a chance of repelling the attack, it is his biological duty to attempt to do so by any means at his disposal. For many other animals the situation is the same, but under natural conditions the amount of actual physical violence that occurs is limited. It is usually little more than a threat of violence answered by a counter-threat of counter-violence. The more truly violent species all appear to have exterminated themselves – a lesson we should not overlook.

This sounds straightforward enough, but the last few thousand years of human history have over-burdened our evolutionary inheritance. A man is still a man and a family is still a family, but a tribe is no longer a tribe. It is a super-tribe. If we are ever to understand the unique savageries of our national, idealistic and racial conflicts, we must once again examine the nature of this super-tribal condition. We have seen some of the tensions it has set up inside itself – the aggressions of the status battle; now we must look at the way it has created and magnified tensions outside itself, between one group and another.

It is a story of piling on the agony. The first important step was taken when we settled down in permanent dwellings. This gave us a definite object to defend. Our closest relatives, the monkeys and apes, live typically in nomadic bands. Each band keeps to a general home range but constantly moves about inside it. If two groups meet and threaten one another, there is little serious development of the incident. They simply move off and go about their business. Once early man became more strictly territorial, the defence system had to be tightened up. But in the early days there was so much land and so few men that there was plenty of room for all. Even when the tribes grew bigger, the weapons were still crude and primitive. The leaders were themselves much more personally involved in the conflicts. (If only today's leaders were forced to serve in the front lines, how much more cautious and 'humane' they would be when making their initial decisions. It is perhaps not too cynical to suggest that this is why they are still prepared to wage 'minor' wars, but are frightened of major nuclear wars. The range of nuclear weapons has accidentally put them back in the front lines again. Perhaps, instead of nuclear disarmament, what we should be demanding is the destruction of the deep concrete bunkers they have already constructed for their own protection.)

As soon as farming man became urban man, another vital step was taken towards more savage conflict. The division of labour and the specialization that developed meant that one category of the population could be spared for full-time killing – the military was born. With the growth of the urban super-tribes, things began to move more swiftly. Social growth became so rapid that its development in one area easily got out of phase with its progress in another. The more stable balance-of-tribal-power was replaced by the serious instability of super-tribal inequalities. As civilizations flourished and could afford to expand, they frequently found themselves faced, not with equal rivals who would make them think twice and indulge in the ritualized threat of bargaining and trade, but with weaker, more backward groups that could be invaded and assaulted with ease. Flicking through the pages of an historical atlas one can see at a glance the whole sorry story of waste and inefficiency, of construction followed by destruction, only to be followed again by more construction and more destruction. There were incidental advantages, of course, inter-minglings that brought the pooling of knowledge, the spread of new ideas. Ploughshares may have been turned into swords, but the impetus for research into better weapons did lead eventually to better implements as well. The cost, however, was heavy.

As the super-tribes became bigger and bigger, the task of ruling the sprawling, teeming populations became greater, the tensions of over-crowding grew, and the frustrations of the super-status race became more intense. There was more and more pent-up aggression, looking for an outlet. Inter-group conflict provided it on a grand scale.

For the modern leader, then, going to war has many advantages that the Stone Age leader did not enjoy. To start with, he does not have to risk getting his face bloody. Also, the men he sends to their deaths are not personal acquaintances of his: they are specialists, and the rest of society can go about its daily life. Trouble-makers who are spoiling for a fight, because of the super-tribal pressures they have been subjected to, can have their fight without directing it at the super-tribe itself. And having an outside enemy, a villain, can make a leader into a hero, unite his people and make them forget the squabbles that were giving him so many headaches.

It would be naive to think that leaders are so super-human that these factors do not influence them. Nevertheless, the major factor remains the urge to maintain or improve inter-leader status. The out-of-phase progress of the different super-tribes that I mentioned earlier is undoubtedly the greatest problem. If, because of its natural resources or its ingenuity, one super-tribe gets one jump ahead of another, then there is bound to be trouble. The advanced group will impose itself on the backward group in one way or another and the backward group will resent it in one way or another. An advanced group is, by its very nature, expansive, and simply cannot bear to leave things alone and mind its own business. It tries to influence other groups, either by dominating them or by 'helping' them. Unless it dominates its rivals to the point where they lose their identities and are absorbed into the advanced super-tribal body (which is often geographically impossible), the situation will be unstable. If the advanced super-tribe helps other groups and makes them stronger, but in its own image, then the day will dawn when they are strong enough to revolt and repel the super-tribe with its own weapons and its own methods.

While all this is going on, the leaders of other powerful, advanced super-tribes will be watching anxiously to make sure that these expansions are not too successful. If they are, then *their* inter-group status will begin to slip.

All this is done under a remarkably transparent but nevertheless persistent cloak of ideology. To read the official documents, one would never guess that it was really the pride and status of the leaders that were at stake. It is always, apparently, a matter of ideals, moral principles, social philosophies or religious beliefs. But to a soldier staring down at his severed legs, or holding his entrails in his hands, it means only one thing: a wasted life. The reason why it was so easy to get him into that position was that he is not only a potentially aggressive animal, but also an intensely co-operative one. All that talk of defending the principles of his super-tribe got through to him because it became a question of helping his friends. Under the stress of war, under the direct and visible threat from the out-group, the bonds between him and his battle companions became immensely strengthened. He killed, more not to let them down than for any other reason. The ancient tribal loyalties were so strong that, when the final moment came, he had no choice.

Given the pressures of the super-tribe, given the global overcrowding of our species, and given the inequalities in progress of the different super-tribes, there is little hope that our children will grow up to wonder what war was all about. The human animal has got too big for its primate boots. Its biological equipment is not strong enough to cope with the unbiological environment it has created. Only an immense effort of intellectual restraint will save the situation now. One sees a sign of this here and there, now and then, but as fast as it grows in one place, it shrivels in another. What is more, we are so resilient as a species that we always seem to be able to absorb the shocks, to make up for the waste, so that we are not even forced to learn from our brutal lessons. The biggest and bloodiest wars we have ever known have done no more, in the long run, than make a tiny, untidy kink in the soaring growth curve of the total world population. There is always a 'post-war bulge' in the birth rate, and the gaps are quickly filled. The human giant regenerates itself like a mutilated flatworm, and slides swiftly on.

What is it that makes a human individual one of 'them', to be destroyed like a verminous pest, rather than one of 'us', to be defended like a dearly beloved brother? What is it that puts him into an out-group and keeps us in the in-group? How do we recognize 'them'? It is easiest, of course, if they belong to an entirely separate super-tribe, with strange customs, a strange appearance and a strange language. Everything about them is so different from 'us' that it is a simple matter to make the gross over-simplification that they are *all* evil villains. The cohesive forces that helped to hold their group together as a clearly defined and efficiently organized society also serve to set them apart from us and to make them frightening by virtue of their unfamiliarity. Like the Shakespearean dragon, they are 'more often feared than seen'.

Such groups are the most obvious targets for the hostility of our group. But supposing we have attacked them and defeated them, what then? Supposing we dare not attack them? Supposing we are, for whatever reason, at peace with other super-tribes for the time being: what happens to our in-group aggression now? We may, if we are very lucky, remain at peace and continue to operate efficiently and constructively within our group. The internal cohesive forces, even without the assistance of an out-group threat, may be sufficiently strong to hold us together. But the pressures and stresses of the super-tribe will still be working on us, and if the internal dominance battle is fought too ruthlessly, with extreme subordinates experiencing too much suppression or poverty, then cracks will soon begin to show. If severe inequalities exist between the sub-groups that inevitably develop within the super-tribe, their normally healthy competition will erupt into violence. Pent-up sub-group aggression, if it cannot combine with the pent-up aggression of other sub-groups to attack a common, foreign enemy, will vent itself in the form of riots, persecutions and rebellions. Examples of this are scattered throughout history. When the Roman

Empire had conquered the world (as it then knew it), its internal peace was shattered by a series of civil wars and disruptions. When Spain ceased to be a conquering power, sending out colonial expeditions, the same thing happened. There is, unhappily, an inverse relationship between external wars and internal strife. The implication is clear enough: namely that it is the same kind of frustrated aggressive energy that is finding an outlet in both cases. Only a brilliantly designed super-tribal structure can avoid both at the same time.

It was easy to recognize 'them' when they belonged to an entirely different culture, but how is it done when 'they' belong to our own culture? The language, the customs, the appearance of the internal 'them' is not strange or unfamiliar, so the crude labelling and lumping is more difficult. But it can still be done. One sub-group may not look strange or unfamiliar to another sub-group, but it does look *different*, and that is often enough.

The different classes, the different occupations, the different age-groups, they all have their own characteristic ways of talking, dressing and behaving. Each sub-group develops its own accents or its own slang. The style of clothing also differs strikingly, and when hostilities break out between sub-groups, or are about to break out (a valuable clue), dressing habits become more aggressively and flamboyantly distinctive. In some ways they begin to resemble uniforms. In the event of a full-scale civil war, of course, they actually become uniforms, but even in lesser disputes the appearance of pseudo-military devices, such as arm-bands, badges and even crests and emblems, becomes a typical feature. In aggressive secret societies they proliferate.

These and other similar devices quickly serve to strengthen the sub-group identity and at the same time make it easier for other groups inside the super-tribe to recognize and lump together the individuals concerned as 'them'. But these are all temporary devices. The badges can be taken off when the trouble is over. The badge-wearers can quickly blend back into the main population. Even the most violent animosities can subside and be forgotten. An entirely different situation exists, however, when a sub-group possesses distinctive *physical* characteristics. If it happens to exhibit, say, dark skin or yellow skin, fuzzy hair or slant eyes, then these are badges that cannot be taken off, no matter how peaceful their owners. If they are in a minority in a super-tribe they are automatically looked upon as a sub-group behaving as an active 'them'. Even if they are a passive 'them' it seems to make no difference. Countless hair-straightening sessions and count-less eye-skin-fold operations fail to get the message across, the message that says, 'We are not deliberately, aggressively setting ourselves apart.' There are too many conspicuous physical clues left.

Rationally, the rest of the super-tribe knows perfectly well that these physical 'badges' have not been put there on purpose, but the response is not a rational one. It is a deep-seated in-group reaction, and when pent-up aggression seeks a target, the physical badge-wearers are there, literally ready-made to take the scapegoat role.

A vicious circle soon develops. If the physical badge-wearers are treated, through no fault of their own, as a hostile sub-group, they will all too soon begin to behave like one. Sociologists have called this a 'self-fulfilling prophecy'. Let me illustrate what happens, using an imaginary example. These are the stages:

1. Look at that green-haired man hitting a child.
2. That green-haired man is vicious.
3. All green-haired men are vicious.
4. Green-haired men will attack anyone.
5. There's another green-haired man – hit him before he hits you. (The green-haired man, who has done nothing to provoke aggression, hits back to defend himself.)
6. There you are – that proves it: green-haired men *are* vicious.
7. Hit all green-haired men.

This progression of violence sounds ridiculous when expressed in such an elementary manner. It is, of course, ridiculous, but nevertheless it represents a very real way of thinking. Even a dimwit can spot the fallacies in the seven deadly stages of mounting group prejudice that I have listed, but this does not stop them becoming a reality.

After the green-haired men have been hit for no reason for long enough, they do, rather naturally, become vicious. The original false prophecy has fulfilled itself and become a true prophecy.

This is the simple story of how the out-group becomes a hated entity. There are two morals to this tale: do not have green hair; but if you do, make sure you are known personally to people who do not have green hair, so that they will realize that you are not actually vicious. The point is that if the original man seen hitting a child had had no special features potentially setting him apart, he would have been judged as an individual, and there would have been no damaging generalization. Once the harm has been done, however, the only possible hope of preventing a further spread of in-group hostility must be founded on personal interchange and knowledge of the other green-haired individuals *as individuals*. If this does not happen, then the inter-group hostility will harden and the green-haired individuals – even those who are excessively non-violent – will feel the need to club together, even live together, and defend one another. Once this has occurred, then real violence is just around the corner. Less and less contact will take place between members of the two groups and they will soon be acting as if they belonged to two different tribes. The green-haired people will soon start to proclaim that they are proud of their green hair, when in reality it never had the slightest significance for them before it became singled out as a special signal.

The quality of the green-hair signal that made it so potent was its visibility. It had nothing to do with true personality. It was merely an accidental label. No out-group has ever been formed, for example, of people who belong to blood group O, despite the fact that, like skin colour or hair pattern, it is a distinct and genetically controlled factor. The reason is simple enough – you cannot tell who *is* group O, simply by looking at them. So, if a known group O man hits a child, it is difficult to extend antagonism towards him to other group O people.

This sounds so obvious, and yet it is the whole basis of the irrational in-group/out-group hatreds we usually refer to as 'racial intolerance'. For many it is hard to grasp that, in reality, this phenomenon has nothing whatsoever to do with significant racial differences in personality, intelligence, or emotional make-up (which have never been proved to exist), but only with insignificant and nowadays meaningless differences in superficial racial 'badges'. A white child or a yellow child, reared in a black super-tribe and given equal opportunities, would undoubtedly do as well and behave in the same way as the black children. The reverse is also true. If this does not appear to be so, then it is simply the result of the fact that they probably would *not* be given equal opportunities. To understand this we must take a brief look at the way the different races came into being in the first place.

To start with, the word 'race' is unfortunate. It has been abused too often. We speak of the human race, the white race and the British race, meaning respectively the human species, the white sub-species and the British super-tribe. In zoology, a species is a population of animals that breed freely amongst themselves, but cannot or do not breed with other populations. A species tends to split into a number of distinguishable sub-species as it spreads over a wider and wider geographical range. If these sub-species are artificially mixed up, they still breed freely with one another and can blend back into one overall type, but normally this does not happen. Climatic and other differences influence the colour, shape and size of the different sub-species in their various natural regions. A group living in a cold region, for example, may become heavier and stockier; another, inhabiting a forest region, may evolve a spotted coat that camouflages it in the dappled light. The physical differences help to tune the sub-species in to their environments so that each one is better off in its own particular area. Where the regions meet there is no hard line between the sub-species; they merge gradually into one another. If, as time goes on, they become increasingly different from one another, they may eventually cease to inter-breed at the borders of their range and a sharp dividing line develops. If later they spread and overlap they will no longer mix. They will have become true species.

The human species, as it began to spread out over the globe, started to form distinctive sub-species, just like any other animal. Three of these, the (white) Caucasoid group, the (black) Negroid group and the (yellow) Mongoloid group, have been highly successful. Two of them have not, and exist today as only remnant groups, shadows of their former selves. They are the Australoids – the Australian Aborigines and their relatives – and the Capoids – the southern African bushmen. These two sub-species once covered a much wider range (the bushmen at one time owning most of Africa), but they have since been exterminated in all but limited areas. A recent survey of the relative sizes of these five sub-species estimated their present world populations as follows:

Caucasoid: 1,757 million
Mongoloid: 1,171 million
Negroid: 216 million
Australoid: 13 million
Capoid: 126 thousand

Of the total world population of just over 3,000 million human animals, this gives the white sub-species the lead with over 55 per cent, the yellow sub-species close on their heels with 37 per cent, and the Negroid sub-species nearly 7 per cent. The two remnant groups together make up less than $^1/_2$ per cent of the total.*

These figures are inevitably approximations, but they give some idea of the general picture. They cannot be precise because, as I explained earlier, the characteristic of a sub-species is that it blends into its neighbours at the places where their ranges meet. An additional complication has arisen in the case of the human species as a result of the increased efficiency of transportation. There has been an enormous amount of migration and shifting about of sub-specific populations, so that in many regions complex mixtures have arisen and a further blending process has taken place. This has occurred despite the formation of in-group/out-group antagonisms and bloodshed because, of course, the different sub-species can still inter-breed fully and efficiently.

* These figures were calculated in the late 1960s. Today, in the 1990s, the total world population has risen to over 5,000 million. The percentages of the three major groups have altered only slightly, from 55, 37 and 7 to 55, 34 and 10.

Had the different human sub-species remained geographically separated for a longer period of time they might well have split up into distinct species, each physically adapted to its special climatic and environmental conditions. That was the way things were going. But man's increasingly efficient technical control over his physical environment, coupled with his great mobility, has made nonsense of this particular evolutionary trend. Cold climates have been subdued by everything from clothing and log fires to central heating; hot environments have been tamed by refrigeration and air-conditioning. The fact, for instance, that a Negro has more cooling sweat glands than a Caucasoid is rapidly ceasing to have any adaptive meaning.

In time it is inevitable that the sub-species differences, the 'racial characters', will blend completely and disappear altogether. Our distant successors will stare in wonder at the old photographs of their extraordinary ancestors. Unfortunately this will take a very long time indeed, because of the irrational misuse of these characters as badges for mutual hostility. The only hope of rapidly speeding up this valuable and ultimately inevitable process of remixing would be international obedience to a new law forbidding inter-breeding with a member of your own sub-species. Since this is pure fantasy, the solution we must rely on is an increasingly rational approach to what has hitherto been an immensely emotional subject. That this will come easily can soon be refuted by a brief study of the incredible extremes of irrationality that have prevailed on so many occasions. It will suffice to select only one example: the repercussions of the Negro slave trade to America.

Between the sixteenth and the nineteenth centuries a grand total of nearly fifteen million Negroes were captured in Africa and shipped as slaves to the Americas. There was nothing new about slavery, but the scale of the operation and the fact that it was carried out by super-tribes professing the Christian faith made it exceptional. It required a special attitude of mind – one that could only stem from a reaction to the physical differences between the sub-species involved. It could only be done if the African Negroes were looked upon as virtually a new form of domestic animal.

It had not begun like this. The first travellers to penetrate black Africa were astonished by the grandeur and organization of the Negro empire. There were great cities, scholarship and learning, complex administration and considerable wealth. Even today, for many people, this is hard to believe. There is so little evidence of it left, and the propaganda picture of the naked, indolent, murderous savage persists all too well. The glory of the Benin bronzes is easily overlooked. The early reports of Negro civilization have been comfortably hidden away and forgotten.

Let us take just one glimpse at an ancient Negro city in West Africa, as it was seen over three and a half centuries ago by an early Dutch traveller. He wrote:

> The town seemeth to be very great; when you enter into it, you go into a great broad street . . . seven or eight times broader than Warmoes street in Amsterdam . . . you see many streets on the sides thereof, which also go right forth . . . The houses in the town stand in good order one close and even with the other, as the houses in Holland stand . . . The King's Court is very great, within it having many great four-square plains, which round about them have galleries . . . I was so far within the court that I passed over four such great plains, and wherever I looked, still I saw gates upon gates to go into other places . . .

Hardly a crude mud-hut village. Nor could the inhabitants of these ancient West African civilizations be described as ferocious, spear-waving savages. As early as the middle of the fourteenth century a sophisticated visitor remarked on the ease of travel and the reliable availability of food and good lodgings for the night. He commented: 'There is complete security in their country. Neither traveller nor inhabitant in it has anything to fear from robbers or men of violence.'

After the early travellers, the later contacts rapidly turned to commercial exploitation. As the 'savages' were attacked, pillaged, subdued and exported, their civilization crumbled. The remnants of their shattered world began to fit the picture of a barbarous, disorganized race. Reports were more frequent now and they left no doubt as to the inferior nature of the Negroid culture. The fact that this cultural inferiority had been initially caused by white brutality and greed was conveniently overlooked. Instead, the Christian conscience found it easier to accept the idea that the black skin (and the other physical differences) represented outward signs of mental inferiorities. It was then a simple matter to argue that the culture was inferior *because* the Negroes were mentally inferior, and for no other reason. If this was so, then the exploitation did not appear to involve degradation because the 'breed' was already inherently degraded. As the 'proof' rolled in that the Negroes were little better than animals, the Christian conscience could relax.

The Darwinian theory of evolution had not yet arrived on the scene. There were two Christian attitudes towards the existence of Negroid humans: the monogenist and the polygenist. The monogenists believed that all types of men had sprung from the same original source, but that Negroes had long ago undergone a gross physical and moral decline, so that slavery was a proper role for them. Writing in the middle of the last century, an American priest put the position very clearly:

> The Negro is a striking variety, and at present permanent, as the numerous varieties of domestic animals. The Negro will remain what he is, unless his form is altered by intermixture, the simple idea of which is revolting; his intelligence is greatly inferior to that of the Caucasians, and he is consequently, from all we know of him, incapable of governing himself. He has been placed under our protection. The vindication of slavery is contained in the scriptures . . . It determines the duties of masters and slaves . . . we can effectively defend our institutions from the word of God.

With these words he taunted the early Christian reformers. How dare they go against the Bible?

This statement, coming several centuries after the start of the exploitation, makes it clear just how completely the original knowledge of the ancient civilization of the African Negroes had been suppressed. If it had not been suppressed, then the 'incapable of governing himself' lie would have been exposed and the whole argument, the whole justification, would have fallen to the ground.

Opposing the monogenists were the polygenists. They believed that each 'race' had been created separately, each with its own special properties, its strengths and its weaknesses. Some polygenists believed that there were as many as fifteen different species of man inhabiting the world. They put in a good word for the Negro:

> The polygenist doctrine assigns to the inferior races of humanity a more honourable place than in the opposite doctrine. To be inferior to another man either in intelligence, vigour, or beauty, is not a humiliating condition. On the contrary, one might be ashamed to have undergone a physical or moral degradation, to have descended the scale of beings, and to have lost rank in creation.

This too was written in the middle of the nineteenth century. Despite the difference in attitude, the polygenists' approach still automatically accepts the idea of racial inferiorities. Either way the Negroes lost out.

Even after the American slaves were given their official freedom, the old attitudes still persisted in one form or another. Had the Negroes not been saddled with their physical out-group 'badges', they would have been rapidly assimilated into their new super-tribe. But their appearance set them apart and the old prejudices were able to persist. The original lie – that their culture had always been inferior and that therefore *they* were inferior – still lurked at the back of white minds. It biased their behaviour and continued to aggravate relations. It influenced even the most intelligent and otherwise enlightened men. It continued to create black resentment, a resentment that was now backed by official social freedom. The outcome was inevitable. Since his inferiority was only a myth, invented by distorting history, the American Negro naturally failed to continue to behave as if he were inferior, once the chains had been removed. He began to rebel. He demanded actual equality as well as official equality.

His efforts were met with staggeringly irrational and violent responses. Real chains were replaced with invisible ones. Segregations, discriminations, and social degradations were heaped upon him. This had been anticipated by the early reformers and, at one point in the last century, it was seriously suggested that the whole American Negro population should be 'handsomely rewarded' for their trouble and returned to their native Africa. But repatriation would hardly have returned them to their original civilized condition. That had been smashed long ago. There was no turning back. The damage had been done. They stayed, and tried to gain what was due to them. After repeated frustrations they began to lose their patience, and during the last half-century their revolts have not only persisted but have increased in vigour. Their numbers have risen to around the twenty million mark. They are a force to be reckoned with and Negro extremists have now been driven to a policy, not of simple equality, but of black domination. A second American Civil War seems to be imminent.

Thoughtful white Americans struggle desperately to overcome their prejudice, but the cruel indoctrinations of childhood are difficult to forget. A new kind of prejudice creeps in, an insidious one of over-compensation. Guilt produces an over-friendliness, an over-helpfulness that creates a relationship as false as the one it replaces. It still fails to treat Negroes as individuals. It still persists in looking at them as members of an out-group. The flaw was neatly pin-pointed by an American Negro entertainer who, on being over-enthusiastically applauded by a white audience, chided them by pointing out that they would feel rather foolish if he turned out to be a white man who had blackened his face.*

Until human sub-species stop treating other human sub-species as if

---

* Since this was written, the word 'Negro' has been used less and less. In the 1960s it had replaced 'coloured' but was then swept away by 'black' in the 1970s. Now, in the U.S., 'African American' has become more 'correct', but this is slightly condescending because 'European Americans' still refer to themselves simply as 'Americans'. It also overlooks the fact that the population of Africa includes over 120 million Caucasians.

their physical differences denoted some kind of mental difference, and until they stop reacting to skin colour as if it were being deliberately worn as the badge of a hostile out-group, there will be pointless and wasteful bloodshed. I am not arguing that there can be a world-wide brotherhood of men. That is a naive utopian dream. Man is a tribal animal and the great super-tribes will always be in competition with one another. In well-organized societies these struggles will take the form of healthy, stimulating competition and the aggressive rituals of commercial trading and sport, helping to prevent communities from becoming stagnant and repetitive. The natural aggressiveness of men will not become excessive. It will take the acceptable form of self-assertion. Only when the pressures become too great will it boil over into violence.

At either level of aggression – the assertive or the violent – the ordinary (non-racial) in-groups and out-groups will face one another on their own terms. The individuals concerned will not be there by accident. But the situation is entirely different for the individual who, because of the colour of his skin, finds himself accidentally, permanently and inevitably trapped into a particular group. He cannot decide to enter a sub-species group, or leave it. Yet he is treated exactly as though he has become a member of a club, or joined an army. The only hope for the future, as I have said, is that the world-wide mixing up of the originally geographically distinct sub-species, which has been increasingly taking place, will lead to a greater and greater blending of characteristics until the strikingly visible differences have vanished. In the meantime the perpetual need for out-groups on to which in-group aggression can be vented will continue to confuse the issue and will continue to cast alien sub-species in unwarranted roles. Our irrational emotions fail to make the proper distinctions; only the imposition of our rational, logical intellects will help us.

I have selected the example of the American Negro dilemma because it is particularly relevant at the present moment. There is, unhappily, nothing unusual about it. The same pattern has been repeated all over the globe, ever since the human animal became really mobile. Even where there have been no sub-specific differences to fan the flames and keep them alight, extraordinary irrationalities have been widespread. The key error of assuming that a member of another group must possess certain special *inherited* character traits typical of his group is constantly arising. If he wears a different uniform, speaks a different language, or follows a different religion, it is illogically assumed that he also has a *biologically* different personality. Germans are said to be laboriously, obsessively methodical, Italians to be excitedly emotional, Americans to be expansive and extrovert, British to be stiff and retiring, Chinese to be devious and inscrutable, Spaniards to be haughty

263

and proud, Swedes to be bland and mild, French to be querulous and argumentative, and so on.

Even as superficial assessments of *acquired* national characters these generalizations are gross over-simplifications, but they are taken much further: for many people they are accepted as inborn traits of the out-groups concerned. It is really believed that in some way the 'breeds' have come to differ, that there has been some genetical change; but this is nothing more than the illogical wishful thinking of the in-grouping tendency. Confucius put it very well, over two thousand years ago, when he said: 'Men's natures are alike; it is their habits that carry them far apart.' But habits, being mere cultural traditions, can be changed so easily, and the in-grouping urge hopes for something more permanent, more basic, to set 'them' apart from 'us'. Being an ingenious species, if we cannot find such differences, we do not hesitate to invent them. With astonishing aplomb, we airily overlook the fact that nearly all the nations I have mentioned above are complex mixtures of a whole collection of earlier groupings, repeatedly cross-bred and re-fused. But logic has no place here.

The whole human species has a wide range of basic behaviour patterns in common. The fundamental similarities between any one man and any other man are enormous. One of these, paradoxically, is the tendency to form distinct in-groups and to feel that you are somehow different, really deep-down different, from members of other groups. This feeling is so strong that the view I have expressed in this chapter is not a popular one. The biological evidence, however, is overwhelming and the sooner it is appreciated, the more tolerant we can hope to become in our inter-group dealings.

Another of our biological characteristics, as I have already stressed, is our inventiveness. It is inevitable that we shall be constantly trying out new ways of expressing ourselves, and that these new ways will differ from group to group and from epoch to epoch. But these are superficial properties, easily gained and easily lost. They can come and go in a generation, whereas it takes hundreds of thousands of years to evolve a new species like ours and to build its basic biological features. Civilization is only ten thousand years old. We are fundamentally the same animals as our hunting ancestors. We all stemmed from that stock, all of us, regardless of our nationality. We all carry the same basic genetic properties. We are all naked apes beneath the wild variety of our adopted costumes. It is as well for us to remember this when we start playing our in-grouping games and when, under the tremendous pressures of super-tribal living, they begin to get out of hand and we find ourselves about to shed the blood of people who, beneath the surface, are exactly like ourselves.

Having said this, I am nevertheless left with an uneasy feeling. The reason is not hard to find. On the one hand I have pointed out that the in-grouping urge is illogical and irrational; on the other hand I have emphasized that conditions are so ripe for inter-group strife that our only hope is to apply rational, intelligent control. In urging the rational control of the deeply irrational, it could be argued that I am being unduly optimistic. It is not perhaps asking too much that rational processes should be brought to bear as an *aid* to the problem, but on the present evidence it does seem to be beyond hope that they alone will solve it. One only has to observe the most intellectual of protestors beating policemen over the heads with placards reading 'Stop this violence', or listen to the most brilliant of politicians supporting war 'to ensure peace', to realize that rational restraint in such matters is an elusive quality. Something else is needed. In some way we must tackle, at the roots, those conditions I referred to that are ripening us so effectively for inter-group violence.

I have already discussed these conditions, but it will help to summarize them briefly. They are:

1. The development of fixed human territories.
2. The swelling of tribes into over-crowded super-tribes.
3. The invention of weapons that kill at a distance.
4. The removal of leaders from the front line of battle.
5. The creation of a specialized class of professional killers.
6. The growth of technological inequalities between the groups.
7. The increase of frustrated status aggression within the groups.
8. The demands of the inter-group status rivalries of the leaders.
9. The loss of social identity within the super-tribes.
10. The exploitation of the co-operative urge to aid friends under attack.

The one condition I have deliberately omitted from this list is the development of differing ideologies. As a zoologist, viewing man as an animal, I find it hard to take such differences seriously in the present context. If one assesses the inter-group situation in terms of actual behaviour, rather than verbalized theorizing, differences in ideology fade into insignificance alongside the more basic conditions. They are merely the excuses, desperately sought for to provide reasons high-sounding enough to justify the destruction of thousands of human lives.

Examining the list of the ten more realistic factors it is difficult to see where one can begin to seek improvement in the situation. Taken together, they appear to offer an absolutely cast-iron guarantee that man will for ever be at war with man.

Remembering that I have described the present state as being that of a human zoo, perhaps there is something we can glean from looking inside the cages of an animal zoo. I have already made the point that wild animals in their natural environment do not habitually slaughter large numbers of their own kind; but what of the caged specimens? Are there massacres in the monkey house, lynchings in the lion house, pitched battles in the bird house? The answer, with obvious qualifications, is in the affirmative. The status struggles between established members of overcrowded groups of zoo animals are bad enough, but, as every zoo-man knows, the situation is even worse when one tries to introduce newcomers to such a group. There is a great danger that the strangers will be jointly set upon and relentlessly persecuted. They are treated as invading members of a hostile out-group. There is little they can do to stem the onslaught. If they huddle unobtrusively in a corner, rather than flaunt themselves in the middle of the cage, they are nevertheless hounded out and attacked.

This does not happen in all instances; where it is most prevalent, the species involved are usually those suffering from the most unnatural degree of spatial cramping. If the established cage owners have more than enough room, they may attack the newcomers initially and drive them away from the favoured spots, but they will not continue to persecute them with undue violence. The strangers are eventually permitted to take up residence in some other part of the enclosure. If the space is too small this stabilization of the relationship can never develop, and bloodshed inevitably ensues.

It is possible to demonstrate this experimentally. Sticklebacks are small fish that hold territories in the breeding season. The male builds a nest in the water-weeds and defends the area around it against other males of the species. Being solitary in this case, a single male represents the 'in-group' and each of his territory-owning rivals represents an 'out-group'. Under natural conditions, in a river or stream, each male has enough room, so that hostile encounters with rivals are restricted largely to threats and counter-threats. Prolonged battles are rare. If two males are encouraged to build nests, one at either end of a long aquarium tank, then, as in nature, they meet and threaten one another at a roughly mid-tank boundary line. Nothing more violent occurs. However, if the water-weeds in which they have nested have been experimentally planted in small, movable pots, it is possible for the experimenter to shift these pots closer together and artificially cramp the territories. As the pots are gradually brought nearer to one another, the two territory-owners intensify their threat displays. Eventually the system of ritualized threat and counter-threat breaks down, and serious fighting erupts. The males endlessly bite and tear at one another's fins, their nest-building duties forgotten, their world suddenly a riot of violence and savagery. The moment that their nest-pots are drawn apart again, however, peace returns and the battleground subsides once more into an arena for harmless, ritualized threat displays.

The lesson is obvious enough: when the small human tribes of early man became swollen into super-tribal proportions, we were, in effect, performing the stickleback experiment on ourselves, with much the same result. If the human zoo is to learn from the animal zoo, then, it is this second condition to which we should pay particular attention.

Viewed with the brutally objective eye of the animal ecologist, the violent behaviour of an over-populated species is an adaptive self-limiting mechanism. It could be described as being cruel to the individual in order to be kind to the species. Each type of animal has its own particular population 'ceiling'. If the numbers rise above this level, some sort of lethal activity intervenes and the numbers sink again. It is worth considering human violence in this light for a moment.

It may sound cold-blooded to express it in this way, but it is almost as if, ever since we first started to become overcrowded as a species, we have been frantically searching for a means to correct this situation and to reduce our numbers to a more suitable biological level. This has not been restricted merely to undertaking bulk slaughter in the form of wars, riots, revolts and rebellions. Our resourcefulness has known no bounds. In the past we have introduced a whole galaxy of self-limiting factors. Primitive societies, when they first began to experience over-crowding, employed practices such as infanticide, human sacrifice, mutilation, head-hunting, cannibalism, and all kinds of elaborate sexual taboos. These were not, of course, deliberately planned systems of population control, but they helped to control the population never-theless. They failed, however, to put a complete brake on the steady increase in human numbers.

As technologies advanced, the individual human life became more strongly protected and these earlier practices were gradually sup-pressed. At the same time, disease, drought and starvation came under heavy attack. As the populations began to soar, new self-limiting devices appeared on the scene. When the old sexual taboos vanished, strange new sexual philosophies emerged that had the effect of reduc-ing group fecundity; neuroses and psychoses proliferated, interfering with successful breeding; certain sexual practices increased, such as contraception, masturbation, oral and anal intercourse, homosexuality, fetishism and bestiality, which provided sexual consummation without the chance of fertilization. Slavery, imprisonment, castration and voluntary celibacy also played their part.

In addition we terminated individual lives by widespread abortion,

murder, the execution of criminals, assassination, suicide, duelling and
the deliberate pursuit of dangerous and potentially lethal sports and
pastimes.

All these measures have served to eliminate large numbers of human
beings from our over-crowded populations, either by the prevention of
fertilization, or by extermination. Assembled together in this way they
make a formidable list. Yet in the last analysis they have proved, even
in combination with mass warfare and rebellion, to be hopelessly
ineffectual. The human species has survived them all and has persisted
in over-breeding at an ever-increasing rate.

For years there has been a stubborn resistance to interpreting these
trends as indications that something is biologically wrong with our
population level. We have repeatedly refused to read them as danger
signals, warning us that we are heading for a major evolutionary
disaster. Everything possible has been done to outlaw these practices
and to protect the breeding and living rights of all human individuals.
Then, as the groups of human animals have swollen to increasingly
unmanageable proportions, we have applied our ingenuity to advancing technologies that help to make these unnatural social conditions
bearable.

As each day passes (adding, as it does so, another 150,000 to the
world population), the struggle becomes more difficult. If present
attitudes persist, it will soon become impossible. *Something* will
eventually arrive to reduce our population level, no matter what we
do. Perhaps it will be heightened mental instability leading to reckless
utilization of weapons of uncontrollable power. Perhaps it will be
mounting chemical pollution, or wildfire diseases of plague intensity.
We have a choice: either we can leave matters to chance, or we can
attempt to influence the situation. If we take the former course, then
there is a very real danger that, when a major population-control
factor does break through our defences and start to operate, it will be
like the bursting of a dam and will carry away our whole civilization.
If we take the latter course, we may be able to avert this disaster; but
how do we set about selecting our method of control?

The idea of enforcing any particular anti-breeding or anti-living
device is unacceptable to our fundamentally co-operative nature. The
only alternative is to encourage voluntary controls. We could, of
course, promote and glamorize increasingly dangerous sports and
pastimes. We might popularize suicide ('Why wait for disease? – Die
now, painlessly!'), or perhaps create a sophisticated new celibacy cult
('purity for kicks'). Advertising agencies throughout the world might
be employed to pour out persuasive propaganda extolling the virtues
of instant dying.

Even if we took such extraordinary (and biologically wasteful) steps, it is doubtful if they would lead to a significant level of population control. The method more generally favoured today is advanced contraception, with the added secondary measure of legalized abortion in the case of unwanted pregnancies. The argument favouring contraception, as I pointed out in an earlier chapter, is that preventing life is better than curing it. If something has to die, it is better that it should be human eggs and sperm, rather than thinking, feeling human beings, cared for and caring, who have already become an integral and interdependent part of society. If the argument of repugnant waste is applied to contraceived eggs and sperm, it can be pointed out that nature is already remarkably wasteful where these products are concerned, the human female being capable of producing around four hundred eggs during her lifetime, and the adult male literally millions of sperm every day.

There are drawbacks, nevertheless. Just as dangerous sports are likely to eliminate selectively the more adventurous spirits in society, and suicides the more highly strung and imaginative, so contraception may favour a bias against the more intelligent. At their present stage of development, contraceptive devices require a certain level of intelligence, thoughtfulness and self-control if they are to be utilized efficiently. Anyone below that level will be more liable to conceive. If their low level of intelligence is in any way governed by genetic factors, these factors will be passed on to their offspring. Slowly but surely these genetic qualities will spread and increase in the population as a whole.

For modern contraception to work effectively and without bias, therefore, it is essential for urgent progress to be made in the direction of finding less and less demanding techniques; techniques which require the absolute minimum of care and attention. Coupled with this must be a major assault on social attitudes towards contraceptive practices. Only when there are 150,000 fewer fertilizations per day than there are at present will we be holding the human population steady at its already overgrown level.

Furthermore, although this is difficult enough in itself to achieve, we must add to it the problem of ensuring that the increase in control is suitably spread around the world, rather than concentrated in one or two enlightened regions. If contraceptive advances are unevenly distributed geographically they will inevitably lead to the de-stabilizing of already strained inter-regional relationships.

It is difficult to be optimistic when contemplating these problems, but supposing for the moment they are magically solved and the world population of human animals is holding steady at around its present level of roughly 3,000 million. This means that if we take the whole land surface of the earth and imagine it evenly populated, we are already at a level of more than five hundred times the population density of primitive man. Should we manage to stop the increase and somehow contrive to spread people out more thinly over the globe, we must therefore not delude ourselves that we shall be achieving anything remotely resembling the social condition in which our early ancestors evolved. We shall still require tremendous efforts of self-discipline if we are to prevent violent social explosions and conflicts. But at least we might stand a chance. If, on the other hand, we wantonly allow the population level to go on rising, we shall soon have forfeited that chance.

As if this were not enough, we must also remember that being five hundred times over our natural primitive level is only one of the ten conditions contributing to our present warlike state. It is a frightening prospect, and the danger that we shall completely destroy civilization as we now know it is becoming daily more real.

It is intriguing to contemplate what will happen if we do go. We are making such great strides in the development of ever more efficient techniques of chemical and biological warfare that nuclear weapons may soon become quaintly old-fashioned. Once this has happened, these nuclear devices will gain the respectability of being dubbed conventional weapons and will be tossed recklessly back and forth between the major super-tribes. (With more and more groups joining the nuclear club, the 'hot line' will, of course, by then have become a hopelessly tangled 'hot network'.) The resultant radioactive cloud that will then circle the earth will dispense death to all forms of life in areas that experience rainfall or snowfall. Only the African bushmen and a few other remote groups living in the centres of the most arid desert regions will stand a chance of surviving. Ironically, the bushmen have, to date, been the most dramatically unsuccessful of all human groups and are living still in the primitive hunting condition typical of early man. It looks like being a case of back-to-the-drawing-board, or a supreme example, as someone once predicted, of the meek inheriting the earth.

5

IMPRINTING AND MAL-IMPRINTING

LIVING IN a human zoo we have a lot to learn and a lot to remember: but as biological learning machines go, our brain is easily the best in existence. With 14,000 million intricately connected cells churning away, we are capable of assimilating and storing an enormous number of impressions.

In everyday use the machinery runs very smoothly, but when something exceptional occurs in the outside world we switch on to a special emergency system. It is then that, in our super-tribal condition, things can go astray. There are two reasons for this. On the one hand, the human zoo in which we live shields us from certain experiences. We do not regularly kill prey – we buy meat. We do not see dead bodies – they are covered by a blanket or hidden in a box. This means that when violence does break through the protective barriers, its impact on our brains is greater than usual. On the other hand, the kinds of super-tribal violence that do break through are frequently of such unnatural magnitude that they are painfully impressive, and our brains are not always equipped to deal with them. It is this type of emergency learning that deserves more than a passing glance here.

Anyone who has ever been involved in a serious road accident will understand what I mean. Every tiny, nasty detail is, in a flash, burned into the memory and stays there for life. We all have personal experiences of this sort. At the age of seven, for example, I was nearly drowned, and to this day I can recall the incident as vividly as if it were yesterday. As a result of this childhood experience, it was thirty years before I was able to force myself to conquer my irrational fears of swimming. Like all children I had many other unpleasant experiences during the course of growing up, but the vast majority of them left no lasting scars.

It seems, then, that as we go through our lives we encounter two different kinds of experience. In one type, brief exposure to a situation makes an indelible and unforgettable impact; in the other, it produces only a mild and easily forgotten impression. Using the terms rather loosely, we can refer to the first as involving traumatic learning, the second as involving normal learning. In traumatic learning the effect produced is out of all proportion to the experience that caused it. In normal learning, the original experience has to be repeated over and over again to keep its influence going. Lack of reinforcement of ordinary learning leads to a fading of the response. In traumatic learning it does not.

Attempts to modify traumatic learning meet with enormous difficulties and can easily make matters worse. In normal learning this is not so. My drowning incident illustrates the point. The more I was shown the pleasures of swimming, the more intense my hatred of it became. If the early incident had not had such a traumatic effect I would have responded more and more positively instead of more and more negatively.

Traumas are not the main subject of this chapter, but they make a useful introduction to it. They show clearly that the human animal is capable of a rather special kind of learning, a kind that is incredibly rapid, difficult to modify, extremely long-lasting and requires no practice to keep it perfect. It is tempting to wish that we could read books in this way, remembering their entire contents for ever after only a single, brief scanning. However, if all our learning worked like this, we would lose all sense of values. Everything would have equal importance and we would suffer from a serious lack of selectivity. Rapid, indelible learning is reserved for the more vital moments of our lives. Traumatic experiences are only one side of this coin. I want to turn it over and examine the other side, the side that has been labelled 'imprinting'.

Whereas traumas are concerned with painful, negative experiences, imprinting is a positive process. When an animal experiences imprinting it develops a positive attachment to something. As with traumatic experiences, the process is quickly over, almost irreversible and needs no later reinforcement. In human beings it happens between a mother and her child. It can happen again when the child grows up and falls in love. Becoming attached to a mother, a child, or a mate are three of the most vital bits of learning that we can undergo in our entire lives and it is these that have been singled out for the special assistance that the phenomenon of imprinting gives. The word 'love' is, in fact, the way we commonly describe the emotional feelings that accompany the imprinting process. But before we go deeper into the human situation, a brief look at some other species will be helpful.

Many young birds, when they hatch from the egg, must immediately form an attachment to their mother and learn to recognize her. They can then follow her around and keep close to her for safety. If newly hatched chicks or ducklings did not do this, they might easily become lost and perish. They are too active and mobile for the mother to be able to keep them together and protect them without the assistance of imprinting. The process can take place in literally a matter of minutes. The first large moving object that the chicks or ducklings see automatically becomes 'mother'. Under normal conditions, of course, it really is their mother, but in experimental situations it can be almost anything. If the first large moving object that incubator chicks see happens to be an orange balloon, pulled along on a piece of string, then they will follow that. The balloon quickly becomes 'mother'. So powerful is this imprinting process that if, after some days, the chicks are then given a choice between their adopted orange balloon and their real mother (who has previously been kept out of sight), they will prefer the balloon. No more striking proof of the imprinting phenomenon can be provided than the sight of a batch of experimental chicks eagerly pattering along in the wake of an orange balloon and completely ignoring their genuine mother near by.

Without experiments of this kind, it could be argued that the young birds become attached to their natural mother because they are rewarded by being with her. Staying close to her means keeping warm, finding food, water and so on. But orange balloons lead to no such rewards and yet they easily become powerful mother-figures. Imprinting, then, is not a matter of responding to rewards, as in ordinary learning. It is simply a matter of exposure. We could call it 'exposure learning'. Also, unlike most normal learning, it has a critical period. Young chicks and ducklings are sensitive to imprinting for only a very brief period of days after hatching. As time passes they become frightened of large moving objects and, if not already imprinted, find it difficult to become so.

As they grow up, the young birds become independent and cease to follow the mother. But the impact of the early imprinting has not been lost. It has not only told them who their mother was, it has also told them what species they belong to. As adults, it helps them to select a sexual partner from their own species rather than from some alien species.

Again, this has to be proved by experiments. If young animals of one species are reared by foster-parents of another species, then, when they mature, they may try to mate with members of the foster species instead of with their own kind. This does not always occur, but there are many examples of it. (We still do not know why it occurs in some cases but not in others.)

Amongst captive animals this susceptibility to becoming fixated on the wrong species can lead to some bizarre situations. When doves reared by pigeons become sexually mature, they ignore other doves and try to mate with pigeons. Pigeons reared by doves try to mate with doves. A zoo peacock, reared on its own in a giant tortoise enclosure, displayed persistently to the bewildered reptiles, refusing to have anything to do with newly arrived peahens.

I have called this phenomenon 'mal-imprinting'. It occurs widely in the world of man/animal relationships. When certain animals, isolated from their own kind from birth, are hand-reared by human beings, they may later respond, not by biting the hand that fed them, but by copulating with it. Doves have often been found to react like this. It is not a new discovery. It has been known from ancient times, when Roman ladies kept small birds to amuse themselves in this way. (Leda, it seems, was more ambitious.) Pet mammals sometimes clasp and attempt to copulate with human legs, as certain dog-owners are painfully aware. Zoo keepers also have to keep a wary eye open during the breeding season. They must be ready to resist the advances of everything from an amorous emu to a rutting deer, when members of these species have been isolated and hand-reared. I myself was once the embarrassed recipient of sexual advances by a female giant panda. It occurred in Moscow, where I had arranged for her to be taken to be mated with the only male giant panda outside China. She ignored his persistent sexual attentions, but when I put my hand through the bars and patted her on the back, she responded by raising her tail and directing a full sexual invitation posture at me, while the male panda was only a few feet away. The difference between the two animals was that she had been isolated from other pandas at a much earlier age than the male. He had matured as a panda's panda, but she was now a people's panda.

Sometimes a 'humanized' animal may appear to be able to tell the difference between a human male and a human female, when making sexual advances to them, but this can be deceptive. A mal-imprinted male turkey, for instance, tried to mate with men, but attacked women. The reason was an intriguing one. Women wear skirts and carry handbags. Aggressive male turkeys display with drooping wings and with wattles. In the eyes of the mal-imprinted male turkey, the skirts became drooping wings and the handbags became wattles. It therefore saw women as rival males and attacked them, reserving its sexual advances for men.

Zoos are full of animals that, with misguided human kindness, have been painstakingly nurtured and hand-reared and then returned to the company of their own kind. But as far as tame isolates are concerned, their own kind are now aliens, members of some frightening, strange, 'other' species. There is an adult male chimpanzee at one zoo that has been caged with a female for over ten years. Medical tests show that he is sexually healthy and she is known to have bred before she was put with him. But because he was a hand-reared isolate, he ignores her completely. He never sits with her, grooms her, or attempts to mount her. As far as he is concerned, she belongs to another species. Years of exposure to her have not changed him.

Such animals may become extremely aggressive towards their own species, not because they are treating them as rivals, but because they see them as foreign enemies. The usual rituals, that under normal circumstances make for bloodless settlements, break down. A female mongoose, hand-reared and tame, was given a wild-caught male in the hope that they would breed, but she attacked him from the moment he entered the cage. Eventually they appeared to reach a state of fairly stable mutual disagreement, but the male must have been under considerable stress because he soon developed ulcers and died. The female immediately became her old friendly self again.

A hand-reared tigress was placed in a cage next to a wild-caught male for the first time in her life. She could see him and smell him, but they could not yet meet. This was just as well. She was so 'humanized' that as soon as she detected his presence, she fled to the far side of her cage and refused to move. This was an abnormal reaction for a tigress, but a much more normal one for a member of her adopted (human) species on encountering a tiger. She went further: she stopped eating, and continued to refuse food for several days, until the male was taken away. It took several weeks in her case to bring her back to her normally friendly, active self again, rubbing up against the bars to be patted and fondled by her keepers.

Sometimes the rearing conditions are such that the animal develops a dual sexual personality. If it is reared by humans in the presence of other members of its own species it may, as an adult, try to mate both with humans and with its own kind. The mal-imprinting is only partial, there being some degree of normal imprinting as well. This would be unlikely with a very rapid imprinter such as a duckling or a chick, but mammals tend to become socialized more slowly. There is time for a dual imprint to occur. Careful American studies with dogs have shown this very clearly. The socialization phase for domestic dogs lasts from the age of twenty days to sixty days. If domestic puppies are completely isolated from man (being fed by remote control) throughout this period, they emerge at the other end as virtually wild animals. If, however, they are reared in the presence of both dogs and men, then they are friendly towards both.

Monkeys reared in total isolation, both from other monkeys and other species including man, find it almost impossible to adapt in later life to any kind of social contact. Placed with sexually active members of their own species, they do not know how to respond. Most of the time they are terrified of making any social contact and sit nervously in a corner. They are so un-imprinted that they are virtually non-social animals, even though they belong to a highly sociable species. If they are reared with other young animals of their own kind, but without mothers, they do not suffer in this way, so that there also appears to be a kind of companion-imprinting as well as a parental one. Both processes can play a part in attaching an animal to its species.

The world of the mal-imprinted animal is a strange and frightening place. Mal-imprinting creates a psychological hybrid, performing patterns of behaviour belonging to its own species, but directing them to its adopted foster species. Only with enormous difficulty, and sometimes not even then, can it re-adapt. For some species the sexual signals of its own kind are strong enough, the responses to them instinctive enough, to enable it to survive its abnormal upbringing, but for many the power of imprinting is such that it overrides everything.

Animal lovers would do well to remember this when indulging in the 'taming' of young wild animals. Zoo officials have long been perplexed by the great difficulties they have encountered in breeding many of their animals. Sometimes this has been due to inadequate housing or feeding, but all too frequently the cause has been mal-imprinting before the animals concerned arrived at the zoo.

Turning to the human animal, the significance of imprinting is clear enough. During the early months of its life a human baby passes through a sensitive socialization phase when it develops a profound and long-lasting attachment to its species and especially to its mother. As with animal imprinting, the attachment is not totally dependent on physical rewards obtained from the mother, such as feeding and cleaning. The exposure learning typical of imprinting also takes place. The young baby cannot keep close to the mother by following her like a duckling, but it can achieve the same end by the use of the smiling pattern. Smiling is attractive to the mother and encourages her to stay with the infant and play with it. These playful, smiling interludes help to cement the bond between the child and its mother. Each becomes imprinted on the other and a powerful reciprocal attachment develops, a persistent bond that is extremely important for the later life of the child. Infants that are well fed and cleaned, but are deprived of the 'loving' of early imprinting, can suffer anxieties that stay with them for the rest of their lives. Orphans and babies which have to live in institutions, where personal attention and bonding are unavoidably limited, all too frequently become anxious adults. A strong bond cemented during the first year of life will mean a capacity for making strong bonds during the adult life that follows.

Good early imprinting opens a large emotional bank account for the child. If expenses are heavy later on, it will have plenty to draw upon.

If things go wrong with its parental care as it grows up (such as parental separation, divorce or death), its resilience will depend on the attachment quality of that first vital year. Later troubles will, of course, take their toll, but they will be minor compared with troubles in the early months. A five-year-old child, evacuated from London during the last war and separated from its parents, when asked who he was, replied: 'I'm nobody's nothing.' The shock clearly was damaging, but whether in such cases it will cause lasting harm will depend to a large extent on whether it is confirming or contradicting earlier experiences. Contradiction will cause bewilderment that can be rectified, but confirmation will tend to harden and strengthen earlier anxieties.

Passing on to the next great attachment phase, we come to the sexual phenomenon of pair-bonding. 'Falling in love at first sight' may not happen to all of us, but it is far from being a myth. The act of falling in love has all the properties of an imprinting process. There is a sensitive period (early adult life) when it is most likely to occur; it is a relatively rapid process; its effect is long-lasting in relation to the time it takes to develop; and it is capable of persisting even in the conspicuous absence of rewards.

Against this it might be argued that for many of us the earliest pair-bondings are unstable and ephemeral. The answer is that during the years of puberty and immediate post-puberty, the capacity to form a serious pair-bond takes some time to mature. This slow maturation provides a transition phase during which we can, so to speak, test the water before jumping in. If it were not so, we would all become completely fixated on our first loves. In modern society the natural transition phase has been artificially lengthened by the undue persistence of the parental bond. Parents tend to cling on to their offspring at a time when, biologically speaking, they should be releasing them. The reason is straightforward enough: the complex demands of the human zoo make it impossible for a fourteen- or fifteen-year-old individual to survive independently. This inability imparts a childlike quality which encourages the mother and father to continue to respond parentally despite the fact that their offspring is now sexually mature. This in turn prolongs many of the offspring's infantile patterns, so that they overlap unnaturally with the new adult patterns. Considerable tensions arise as a result and there is often a clash between the parent/offspring bond and the freshly developing tendency in the young to form a new sexual pair-bond.

It is not the parents' fault that their children cannot yet fend for themselves in the super-tribal world outside; nor is it the children's fault that they cannot avoid transmitting infantile signals of helplessness to their parents. It is the fault of the unnatural urban environment, which requires more years of apprenticeship than the biological growth rate of the young human animal provides.

Despite this interference with the development of the new pair-bond relationship, sexual imprinting soon forces its way to the surface. Young love may be typically ephemeral, but it can also be extremely intense – so much so that permanent fixations on 'childhood sweethearts' do occur in a number of cases, regardless of the socio-economic impracticability of the relationships. Even if, under pressure, these early pair-bonds collapse, they can leave their mark. Frequently it seems as if the later search for a sexual partner, at the fully independent adult phase, involves an unconscious quest to re-discover some of the key characteristics of the very first sexual imprint. Ultimate failure in the quest may well be a hidden factor helping to undermine an otherwise successful marriage.

This phenomenon of *bond confusion* is not confined to the 'childhood sweetheart' situation. It can occur at any stage, and is particularly likely to harass second marriages, where silent and sometimes not-so-silent comparisons with earlier mates are frequently being made. It can also play another important and damaging role when the parent/offspring bond is confused with the sexual pair-bond. To understand this it is necessary to look again at what the parent/offspring bond does to the infant. It tells it three things: 1. This is my particular, personal parent. 2. This is the species I belong to. 3. This is the species I shall mate with in later life.

The first two instructions are straightforward; it is the third that can go wrong. If the early bond with the parent of the opposite sex has been particularly persistent, some of their *individual* characteristics can also be carried over to influence the later sexual bonding of the offspring. Instead of taking in the message as 'This is the species I shall mate with in later life,' the child reads it as 'This is the type of person I shall mate with in later life.'

A limiting influence of this kind can become a serious problem. Interference with the sexual pair-bonding process, stemming from a persistent parental image, can lead to a particular mate-selection which, in all other respects, is highly unsuitable. Conversely, an otherwise thoroughly compatible mate can fail to achieve a full relationship because he or she lacks certain trivial but key characteristics of the partner's parent. ('My father would never do that.' – 'But I'm not your father.')

This troublesome phenomenon of bond confusion appears to be caused by the unnatural levels of family-unit isolation that so often develop in the crowded world of the human zoo. The 'strangers-in-our-midst' phenomenon tends to clamp down on the tribal-sharing, social-mixing atmosphere typical of smaller communities. Defensively, the families turn in on themselves, boxing themselves off from one another in neat rows of terraced or semi-detached cages. Unhappily there is no sign of the situation improving: rather the reverse.

Leaving the question of bond confusion, we now have to consider another, stranger aberration of human imprinting: our own version of mal-imprinting. Here we enter the unusual world of what has been termed sexual fetishism.

For a minority of individuals the nature of the first sexual experience can have a psychologically crippling effect. Instead of becoming imprinted with the image of a particular mate, this type of individual becomes sexually fixated on some inanimate object present at the time. It is not at all clear why so many of us escape these reproductively abnormal fixations. Perhaps it depends on the vividness or violence of certain aspects of the occasion of our first major sexual discovery. Whatever it is, the phenomenon is a striking one.

Judging by the case-history records that are available, it appears that the attachment to a sexual fetish occurs most frequently when the initial sexual consummation takes place spontaneously or when the individual is alone. In many instances it can be traced to the first ejaculation of a young adult male, which often occurs in the absence of a female and without the usual pair-bonding preliminaries. Some characteristic object that is present at the moment of ejaculation instantly takes on a powerful and lasting sexual significance. It is as if the whole imprinting force of pair-bonding is accidentally channelled into an inanimate object, giving it, in a flash, a major role for the rest of the person's sexual life.

This striking form of mal-imprinting is probably not quite so rare as it seems. Most of us develop a primary pair-bond with a member of the opposite sex, rather than with fur gloves or leather boots, and we are happy to advertise our pair-bonds openly, confident that others will understand and share our feelings; but the fetishist, firmly imprinted with his unusual sexual object, tends to remain silent on the subject of his strange attachment. The inanimate object of his sexual imprinting, which has such enormous significance for him, would mean nothing to others and, for fear of ridicule, he keeps it secret. It not only means nothing to the vast majority of people, the non-fetishists, but it also means little to other fetishists, each having their own particular speciality. Fur gloves have as little significance for a leather-boot fetishist as they do for a non-fetishist. The fetishist therefore becomes isolated by his own, highly specialized form of sexual imprinting.

Against this, it can be said that there are certain kinds of objects that do crop up with striking frequency in the fetish world. Rubber goods are particularly common, for instance. The significance of this will become clearer if we examine a few specific cases of fetish development.

A twelve-year-old boy was playing with a fox-fur coat when he experienced his first ejaculation. In adult life he was only able to achieve sexual satisfaction in the presence of furs. He was unable to copulate with females in the ordinary way. A young girl experienced her first orgasm when clutching a piece of black velvet as she masturbated. As an adult, velvet became essential to her sexually. Her whole house was decorated with it and she only married in order to obtain more money to buy more velvet. A fourteen-year-old boy had his first sexual experience with a girl who was wearing a silk dress. Later, he was incapable of making love to a naked female. He could only become aroused if she was wearing a silk dress. Another young boy was leaning out of a window when his first ejaculation occurred. As it happened, he saw a figure moving past in the road outside, walking on crutches. When he was married he could only make love to his wife if she wore crutches in bed. A nine-year-old boy was playing with a soft glove against his penis at the moment of his first ejaculation. As an adult he became a glove-fetishist with a collection of several hundred gloves. All his sexual activities were directed to these gloves.

There are many examples of this kind, clearly linking the adult fetish to the first sexual experience. Other common fetish objects include: shoes, riding boots, stiff collars, corsets, stockings, underclothes, leather, rubber, aprons, handkerchiefs, hair, feet, and special costumes such as nursemaids' uniforms. Sometimes these become the essential elements necessary for a successful (and otherwise normal) copulation. Sometimes they completely replace the sexual partner. Texture appears to be an important feature of most of them, often because pressures and frictions of various kinds are significant in causing the first sexual arousal in an individual's lifetime. If some substance with a highly characteristic tactile quality is involved, then it seems to stand a strong chance of becoming a sexual fetish. This could account for the high frequency of rubber, leather and silk fetishes, for example.

Shoe, boot and foot fetishes are also common and it seems likely that here, too, a pressure against the body could easily be involved. There is one classic case of a fourteen-year-old boy who was playing with a twenty-year-old girl who was wearing high-heeled shoes. He was lying on the ground and she playfully stood on him and trampled him. When her foot rested on his penis he experienced his first ejaculation. As an adult this became his only form of sexual activity. During his life he managed to persuade over a hundred women to trample on him wearing high-heeled shoes. Ideally the partner had to be of a particular weight and the shoes of a particular colour. The original encounter had to be recreated as precisely as possible to produce a maximum reaction.

This last case shows very clearly how masochism can develop. Another young boy, for example, had his first sexual experience spontaneously while wrestling with a much larger girl. In later life he was fixated on heavy, aggressive women who were prepared to hurt him during sexual encounters. It is not difficult to imagine how certain forms of sadism could develop in a similar way.

The attachment to a sexual fetish differs from the process of ordinary conditioning in several ways. Like imprinting (or the traumatic experiences I mentioned at the beginning of the chapter) it is very rapid, has a lasting effect and is extremely difficult to reverse. It also appears at a sensitive period. Like mal-imprinting, it fixes the individual on to an abnormal object, channelling sexual behaviour away from the biologically normal object, namely a member of the opposite sex. It is not so much the positive acquiring of sexual significance of an object such as a rubber glove that causes the damage; it is the total elimination of all other sexual objects that creates the problem. The mal-imprinting is so powerful in the cases I have mentioned that it 'uses up' all the available sexual interest. Just as the experimental duckling will follow only the orange balloon and completely ignore its real mother, so the glove fetishist will only mate with a glove and completely ignores potential mates. It is the exclusivity of the imprinting process that causes the difficulties when the mechanism misfires. We all find various textures and pressures stimulating as accessories to sexual encounters. There is nothing strange about responding to soft silks and velvets. But if we become exclusively fixated on them, so that we develop what amounts to a pair-bond with them (like the shoe-fetishist who, when he was alone with girls' shoes, 'blushed in their presence as if they were the girls themselves'), then something has gone drastically wrong with the imprinting mechanism.

Why should a small, but nevertheless considerable, number of human animals suffer from this kind of mal-imprinting? Other animals, under natural conditions in the wild, do not appear to do so. For them it only occurs when they are caught and hand-reared under highly artificial conditions, or when they are kept in enclosures with alien species, or when special experiments are carried out. Perhaps this gives the clue. As I have already emphasized, in a human zoo social conditions are highly artificial for our simple tribal species. In many of our super-tribes sexual behaviour is severely restricted at the critical stage of puberty. But although it becomes hidden and cloaked with all kinds of unnatural inhibitions, nothing can hold it back completely. It soon

bursts through. If, when it does so, there are certain highly characteristic objects present, then they may become over-impressive. Had
the developing adolescent become gradually more experienced in sexual matters at an earlier stage and had his initial sexual explorations
been richer and less constricted by the artificialities of the super-tribe,
then the later mal-imprinting could perhaps have been avoided. It
would be interesting to know how many of the extreme fetishists were
solitary children without brothers and sisters, or, as young adolescents,
were timid and shy of making personal contacts, or lived in a rather
strict household. Future research is needed here but my guess is that
the proportion would be high.

One important form of mal-imprinting that I have not mentioned is
homosexuality. I have left it until now because it is a more complex
phenomenon and because mal-imprinting is only part of the story.
Homosexual behaviour can arise in one of four ways. Firstly, it can
occur as a case of mal-imprinting in much the same way as fetishism.
If the earliest sexual experience in an individual's life is a powerful one
and occurs as a result of an intimate encounter with a member of the
same sex, then a fixation on that sex can rapidly develop. If two
adolescent boys are wrestling together or indulging in some form of
sex-play, and ejaculation occurs, this can lead to mal-imprinting. The
strange thing is that boys often share early sexual experiences of one
kind or another and yet the majority survive to become adult hetero-
sexuals. Again we need to know much more about what it is that
fixates a few but not the majority. As with the fetishists, it probably
has something to do with the degree of the richness of the boy's social
experience. The more restricted he has been socially, the more cut off
from personal interactions, the blanker will be his sexual canvas. Most
boys have, as it were, a sexual blackboard on which things are lightly
sketched, rubbed out and re-drawn. But the inward-living boy keeps
his sexual canvas virginally white. When finally something does get
drawn on it, it will have a much more dramatic impact and he will
probably keep the picture for life. Rough-and-tumble, extrovert boys
may become involved in homosexual activities, but they will simply
put them down to experience and pass on, adding more and more
experiences as they progress with their socializing explorations.

This leads me to the other causes of persistent homosexual behaviour. I say 'persistent' because, of course, brief and fleeting homosexual activities occur for the vast majority of both sexes at some point
in their lives, as part of general sexual explorations. For most people, like
the rough-and-tumble boys, they are mild experiences and are usually
confined to childhood. But for some, homosexual patterns persist
throughout life, frequently to the almost total, or total, exclusion of

heterosexual activities. Mal-imprinting of the type I have been discussing does not explain all these cases. A second, very simple cause is that the *opposite* sex behaves in an exceptionally unpleasant way towards a particular individual. A boy terrorized by girls may well come to regard other males as more attractive sexual partners, despite the fact that, as mates, they are sexually inadequate objects. A girl terrorized excessively by boys may react in the same way and turn to other girls as sexual partners. Terrorizing is not the only mechanism, of course: betrayal, and other forms of social or physical punishment from the opposite sex, can work just as effectively. (Even if the opposite sex is not directly hostile, cultural pressures placing powerful restrictions on heterosexual activities may lead to the same result.)

A third major influence in the creation of a persistent homosexual is the childhood assessment of the roles of his or her parents. If a child has a weak father who is dominated by the mother, it is particularly likely to get the masculine and feminine roles confused and reversed. This then tends to lead to a choice of the wrong sex as a pair-bond partner in later life.

The fourth cause is a more obvious one. If members of the opposite sex are totally absent from the environment for a long period of time, then members of the same sex become the next best thing for sexual encounters. A male isolated from females in this way, or a female isolated from males, may persistently indulge in homosexuality without any of the other three factors I have mentioned having any influence at all. A male prisoner, for instance, may have escaped mal-imprinting, may be fond of the opposite sex, and may have had a father who dominated his mother in a completely masculine way, and yet he may still become a long-term homosexual if he is confined in an all-male prison community, where the nearest thing to a female body is another male body. If, in prisons, in boarding-schools, on naval vessels, or in army barracks, the uni-sexual condition lasts for some years, the opportunist homosexual may eventually become conditioned to the rewards of his enforced sexual patterns and may persist in them even after he has returned to a heterosexual environment.

Of these four influences leading to persistent homosexual behaviour, only the first one is relevant to the present chapter, but it was important to discuss them all here in order to explain the partial role that mal-imprinting plays in this particular sexual phenomenon.

Homosexual behaviour in other animals is usually of the next-best-thing variety, and disappears in the presence of sexually active members of the opposite sex. There are a few cases of persistently homosexual animals, however, in instances where special social experiments have been carried out. If young mallard ducklings, for example, are kept in all-male groups of from five to ten individuals for the first seventy-five days of their lives, and never encounter a female of their species during that time, they become permanently homosexual. When released on to a pond as adults, now with both males and females present, they completely ignore the females and set up homosexual pair-bonds between themselves. This situation persists for many years, probably the whole lifetime of the homosexual ducks, and nothing the females can do will alter it. Doves kept in homosexual pairs are well known to copulate with one another and may form complete pair-bonds. Two males that became sexually imprinted on one another in this way went through the whole breeding cycle together, co-operating to build a nest, incubate eggs and rear the young. The fertile eggs, of course, had to be provided from the nest of a true pair, but they were quickly accepted, each of the homosexual males reacting as if they had been laid by his partner. If a real female had been introduced after the homosexual pair-bond had launched the two males into their pseudo-reproductive cycle, it is doubtful if they would have taken any notice of her. By that stage the homosexuality would have become persistent, at least for the duration of that complete breeding cycle.

Mal-imprinting in the human animal is not confined to sexual relationships. It can also occur in the parent-offspring relationships. As far as human infants becoming imprinted on parents of the wrong species is concerned, good evidence is lacking. The famous cases of so-called 'wolf-children' (abandoned or lost babies being suckled and reared by wolf bitches) have never been fully substantiated and must remain for the time being in the realm of fiction. If such a thing could occur, however, there is little doubt that the wolf-children would become fully mal-imprinted on their foster-parents.

The reverse process, by contrast, is encountered almost every day. When a young animal is hand-reared by a human foster-parent, it is not only the pet animal that becomes mal-imprinted. The human foster-parent also often becomes intensely mal-imprinted and responds to the young animal as if it were a human baby. The same kind of emotional devotion is lavished on it and the same kind of heartbreaks occur if something goes wrong.

Just as a pseudo-parent, such as the duckling's orange balloon, has certain key qualities that make it suitable for mal-imprinting (it is a large moving object), so the pseudo-infant becomes more suitable if it possesses certain qualities typical of the human infant. Human babies are helpless, soft, warm, rounded, flat-faced, big-eyed, and they cry. The more of these properties a young animal possesses, the more likely it is to encourage the setting up of a parent-offspring bond with a mal-imprinted human foster-parent. Many young mammals have nearly all these properties and it is extremely easy for a human being to become mal-imprinted with them in a matter of minutes. A soft, warm, big-eyed fawn bleating for its mother, or a helpless, rounded puppy crying for a missing bitch, projects a powerful infantile image which few human females can resist. Since some of the childlike properties of such animals are even stronger than those of a real human baby, the exaggerated stimuli from the pseudo-infant can frequently become more powerful than the natural ones, and the mal-imprinting becomes intense.

Animal pseudo-infants have one big drawback: they grow up too quickly. Even slow developers become active adults in only a fraction of the time it takes for a real human infant to mature. When this happens they often become unmanageable and lose their appeal. But the human animal is an ingenious species and has taken steps to deal with this unfortunate development. By selective breeding over a period of centuries, it has managed to make its domestic pets more infantile, so that adult cats and dogs, for example, are rather juvenile versions of their wild counterparts. They remain more playful and less independent, and continue to fulfil their roles as child substitutes.

With some breeds of dogs (the lap dogs or 'toy' dogs) this process has been taken to extremes. They not only behave in a more juvenile way, they also look, feel and sound more juvenile. Their whole anatomy has been altered to make them fit more closely to the image of a human baby, even when they are adult. In this way they can act as a satisfying pseudo-infant, not just for a few months as puppies, but for ten years or longer, a time span that begins to match that of human childhood. What is more, they go one better than the real baby, because they remain baby-like throughout the whole period.

The Pekinese is a good example. The wild ancestor of the Pekinese (as of all domestic dogs) is the wolf, a creature that can weigh up to 150 pounds or more. The average weight of an adult European human is much the same, about 155 pounds. The weight of a newborn human baby is roughly between five and ten pounds, the average being slightly over seven pounds. So, to convert the wolf into a good pseudo-infant, it has to be reduced in size to about one-fifteenth of its original, natural weight. The Pekinese is a triumph of this process, weighing today between seven and twelve pounds, with an average of about ten pounds. So far, so good. It matches the baby in weight and, even as an adult, has the first of the vital pseudo-infant properties: it is a small object. But some other improvements are needed. The legs of a typical dog are too long in relation to its body. Their proportion is more reminiscent of the human adult than the short-limbed human baby. So, off with their legs! By careful selective breeding it is possible to produce strains with shorter and shorter legs until they can only waddle along. This not only corrects the proportions, but as a bonus it also renders the animals more clumsy and helpless. Again, valuable infantile features. But something is still missing. The dog is warm enough to the touch, but not soft enough. Its natural wild-type hair is too short, stiff and coarse. So, on with the hair! Selective breeding again comes to the rescue, producing long, soft, flowing silky hair, creating the essential feel of infantile super-softness.

Further modifications are necessary to the natural wild shape of the dog. It has to become plumper, bigger-eyed, and shorter-tailed. One only has to look at a Pekinese to see that these changes have also been successfully imposed. Its ears stuck up and were too pointed. By the device of making them bigger, floppier and covered in long flowing hair, it was possible to convert them into a reasonable semblance of a growing infant's hair-style. The voice of the wild wolf is too deep, but the reduction in body size has taken care of that, producing a higher-pitched, more infantile tone. Finally, there is the face. A wild dog's face is far too pointed, and a little genetic plastic surgery is needed here, too. No matter if it deforms the jaws and makes feeding difficult, it has to be done. And so the Pekinese has its face squashed flat and childlike. Again there is an added bonus, because this also makes it more helpless and more dependent on its pseudo-parent for providing suitably pre-pared food, another essential parental activity. And there sits our Pekinese pseudo-infant, softer, rounder, more helpless, bigger-eyed and flatter-faced, ready to set up a powerful mal-imprinted bond in any susceptible adult human who happens along. And it works. It works so well that they are not only mothered, but also live with humans, travel with them, have their own (veterinary) doctors, and are frequently buried in graves like humans and even left money in wills like real human offspring.

As I have said before, on other topics, this is a description, not a criticism. It is difficult to see why so many people criticize such activities when they so obviously fulfil a basic need that often cannot be satisfied in the normal way. It is even harder to see why some people can accept this kind of imprinting, but not other kinds. Many humans are repelled by sexual mal-imprinting, for example, and revolted by the idea of a man making love to a fetish object, or copulating with another male, yet they happily accept parental mal-imprinting where a human adult is fondling a pet lap-dog or feeding a baby monkey from a bottle. But why do they make the distinction? Biologically speaking, there is virtually no difference between the two activities. They both involve mal-imprinting and they are both aberrations of normal human relationships. But although, in the biological sense, they must both be classed as abnormalities, neither of them causes any harm to bystanders, to individuals outside the relationships. We may feel that it would be more gratifying for the fetishist or the childless animal lover if they could enjoy the rewards of a full family life, but it is their loss, not ours, and we have no cause to be hostile to either of them.

We have to face the fact that, living in a human zoo, we are inevitably going to suffer from many abnormal relationships. We are bound to be exposed in unusual ways to unusual stimuli. Our nervous systems are not equipped to deal with this and our patterns of response will sometimes misfire. Like the experimental or zoo animals, we may find ourselves fixated with strange and sometimes damaging bonds, or we may suffer from serious bond confusion. It can happen to any of us, at any time. It is merely another of the hazards of existing as an inmate of a human zoo. We are all potential victims, and the most appropriate reaction, when we come across it in someone else, is sympathy rather than cold intolerance.

6

THE STIMULUS STRUGGLE

WHEN A MAN is reaching retiring age he often dreams of sitting quietly in the sun. By relaxing and 'taking it easy' he hopes to stretch out an enjoyable old age. If he manages to fulfil his sun-sit dream, one thing is certain: he will not lengthen his life, he will shorten it. The reason is simple – he will have given up the Stimulus Struggle. In the human zoo this is something we are all engaged in during our lives and if we abandon it, or tackle it badly, we are in serious trouble.

The object of the struggle is to obtain the optimum amount of stimulation from the environment. This does not mean the maximum amount. It is possible to be over-stimulated as well as under-stimulated. The optimum (or happy medium) lies somewhere between these two extremes. It is like adjusting the volume of music coming from a radio: too low and it makes no impact, too high and it causes pain. At some point between the two there is the ideal level, and it is obtaining this level in relation to our whole existence that is the goal of the Stimulus Struggle.

For the super-tribesman this is not easy. It is as if he were surrounded by hundreds of behaviour 'radios', some whispering and others blaring away. If, in extreme situations, they are all whispering, or monotonously repeating the same sounds over and over again, he will suffer from acute boredom. If they are all blaring, he will experience severe stress.

Our early tribal ancestor did not find this such a difficult problem. The demands of survival kept him busy. It required all his time and energy to stay alive, to find food and water, to defend his territory, to avoid his enemies, to breed and rear his young and to construct and maintain his shelter. Even when times were exceptionally bad, the challenges were at least comparatively straightforward. He can never have been subjected to the intricate and complex frustrations and conflicts that have become so typical of super-tribal existence. Nor is he likely to have suffered unduly from the boredom of gross under-stimulation that, paradoxically, super-tribal life can also impose. The advanced forms of the Stimulus Struggle are therefore a speciality of the urban animal. We do not find them amongst wild animals or 'wild' men in their natural environments. We do, however, find them in both urban men and in a particular kind of urban animal – the zoo inmate.

Like the human zoo, the animal zoo provides its occupants with the security of regular food and water, protection from the elements and freedom from natural predators. It looks after their hygiene and their health. It may also, in certain cases, put them under severe strain. In this highly artificial condition, zoo animals, too, are forced to switch from the struggle for survival to the Stimulus Struggle. When there is too little input from the world around them, they have to contrive ways of increasing it. Occasionally, when there is too much (as in the panic of a freshly caught animal), they have to try and damp it down.

The problem is more serious for some species than for others. From this point of view there are two basic kinds of animals: the specialists and the opportunists. The specialists are those which have evolved one supreme survival device on which they depend for their very existence, and which dominates their lives. Such creatures are the ant-eaters, the koalas, the giant pandas, the snakes and the eagles. So long as ant-eaters have their ants, koalas have their eucalyptus leaves, pandas have their bamboo shoots, and snakes and eagles have their prey, they can relax. They have perfected their diet specializations to such a pitch that, providing their particular requirements are met, they can accept a lazy and otherwise unstimulating pattern of life. Eagles, for instance, will thrive in a small empty cage for over forty years without so much as biting their claws, providing, of course, they can sink them daily into a freshly killed rabbit.

The opportunists are not so fortunate. They are the species – such as dogs and wolves, raccoons and coatis, and monkeys and apes – that have evolved no single, specialized survival device. They are jacks-of-all-trades, always on the look-out for any small advantage the environment has to offer. In the wild, they never stop exploring and investigating. Anything and everything is examined in case it may add yet another string to the bow of survival. They cannot afford to relax for very long and evolution has made sure that they do not. They have evolved nervous systems that abhor inactivity, that keep them constantly on the go. Of all species, it is man himself who is the supreme opportunist. Like the others, he is intensely exploratory. Like them, he has a biologically built-in demand for a high stimulus input from his environment.

In a zoo (or a city) it is clearly these opportunist species that will suffer most from the artificiality of the situation. Even if they are provided with perfectly balanced diets and are immaculately sheltered and protected, they will become bored and listless and eventually neurotic. The more we have come to understand the natural behaviour of such animals, the more obvious it has become, for example, that zoo monkeys are little more than distorted caricatures of their wild counterparts.

But opportunist animals do not give up easily. They react to the unpleasant situation with remarkable ingenuity. So, too, do the inmates of the human zoo. If we compare the animal zoo reactions with those we find in the human zoo, it will serve to bring home to us the striking parallels that exist between these two highly artificial environments.

The Stimulus Struggle operates on six basic principles and it will help if we look at them one by one, examining in each case first the animal zoo and then the human zoo. The principles are these:

1. If stimulation is too weak, you may increase your behaviour output *by creating unnecessary problems which you can then solve.*

We have all heard of labour-saving devices, but this principle is concerned with labour-wasting devices. The Stimulus Struggler deliberately makes work for himself by elaborating patterns that could otherwise be performed more simply, or that need no longer be performed at all.

In its zoo cage, a wild cat may be seen to throw a dead bird or a dead rat up into the air and then leap after it and pounce on it. By throwing the prey, the cat can put movement and therefore 'life' back into it, giving itself the chance to perform a 'kill'. In the same way, a captive mongoose can be seen 'shaking to death' a piece of meat.

Observations of this kind extend to domestic animals as well. A pet dog, pampered and well fed, will drop a ball or a stick at its master's feet and wait patiently for the object to be thrown. Once it is moving through the air or across the ground, it becomes 'prey' and can be chased after, caught, 'killed' and brought back again for a repeat performance. The domestic dog may not be hungry for food, but it is hungry for stimulation.

In its own way, a caged raccoon is equally ingenious. If there is no food to search for in a near-by stream, the animal will search for it anyway, even if there is no stream. It takes its food to its water-dish, drops it in, loses it, and then searches for it. When it finds it, it scrabbles with it in the water before eating it. Sometimes it even destroys it by this process, pieces of bread becoming a hopeless mush. But no matter, the frustrated food-searching urge has been satisfied. This, incidentally, is the origin of the long-standing myth that raccoons wash their food.

There is a large rodent, looking like a guinea-pig on stilts, called an agouti. In the wild it peels certain vegetables before eating them. It holds the vegetables in its front feet and pares them with its teeth as we might pare an orange. Only when it has completely skinned the object does it start to eat. In captivity, this peeling urge refuses to be frustrated. If a perfectly clean apple or potato is given to an agouti, the animal still peels it fastidiously and, after eating it, devours the peel as well. It even attempts to 'peel' a piece of bread.

Turning to the human zoo, the picture is strikingly similar. When we are born into a modern super-tribe, we are thrust into a world where human brilliance has already solved most of the basic survival problems. Just like the zoo animals, we find that our environment emanates security. Most of us have to do a certain amount of work, but thanks to technical developments, there is plenty of time left over for participating in the Stimulus Struggle. We are no longer totally absorbed in the problems of finding food and shelter, rearing our offspring, defending our territories, or avoiding our enemies. If, against this, you argue that you never stop working, then you must ask yourself a key question: could you do less work and still survive? The answer in many cases would have to be 'yes'. Working is the modern super-tribesman's equivalent of hunting for food and, like the animal zoo inmates, he frequently performs the pattern much more elaborately than is strictly necessary. He creates problems for himself.

Only those sectors of the super-tribe that are enduring what we would call severe hardship are working totally for survival. Even they, however, will be forced to indulge in the Stimulus Struggle when they can spare a moment, for the following special reason: the primitive, hunting tribesman may have been a 'survival-worker', but his tasks were varied and absorbing. The unfortunate subordinate super-tribesman who is a 'survival-worker' is not so well off. Thanks to the division of labour and industrialization, he is driven to carry out intensely dull and repetitive work – the same routine thing day after day, year after year – making a mockery of the giant brain housed inside his skull. When he does get a few moments to himself, he needs to indulge in the Stimulus Struggle as much as anyone else in our modern world, for the problem of stimulation is concerned with variety as well as amount, with quality as well as quantity.

For the others, as I have said, much of the activity is work for work's sake and, if it is exciting enough, the struggler – a businessman, for instance – may find that he has scored so many points during his working day that, in his spare time, he can allow himself to relax and indulge in the mildest of activities. He might doze at his fireside with a soothing drink, or dine out at a quiet restaurant. If he dances when he dines, it is worth observing how he does it. The point is that our survival-worker may also go dancing in the evening. At first sight there appears to be a contradiction here, but closer examination reveals that there is a world of difference between the two kinds of dancing. Big-businessmen do not go in for strenuous competitive ballroom dancing, or wild abandoned folk-dancing. Their clumsy shuffling on the night-club floor (the small size of which has been tailored to their low-stimulus demands) is far from being competitive or wild. The unskilled workman is likely to become a skilled dancer; the skilled businessman is likely to be an unskilled dancer. In both cases the individual achieves a balance which is, of course, the goal of the Stimulus Struggle.

In over-simplifying to make this point I have made the difference between the two types sound too much like a class distinction, which it is not. There are plenty of bored businessmen, suffering from repetitive office tasks that are almost as monotonous as packing boxes at a factory bench. They too will have to seek more stimulating forms of recreation in their spare time. Also, there are many simple labouring jobs where the work is rich and varied. The more fortunate labourer, in the evening, is more like the successful businessman, relaxing with a quiet drink and a chat.

The under-stimulated housewife is another interesting phenomenon. Surrounded by her modern labour-saving devices, she has to invent labour-wasting devices to occupy her time. This is not as futile as it sounds. She can at least *choose* her activities: therein lies the whole advantage of super-tribal living. In primitive tribal life there was no choice. Survival made its own demands. You had to do this, and this, and this, or die. Now you can do this, or that, or the other – anything you like, so long as you realize that you have to do *something*, or break the golden rules of the Stimulus Struggle. And so the housewife, her washing spinning automatically away in the kitchen, must busy herself with something else. The possibilities are endless and the game can be a most attractive one. It can also go astray. Every so often it suddenly seems to the under-stimulated player that the compensating activity he or she is pursuing so relentlessly is really rather meaningless. What *is* the point of rearranging the furniture, or collecting postage stamps, or entering the dog for another dog-show? What does it prove? What does it achieve? This is one of the dangers of the Stimulus Struggle. Substitutes for real survival activity remain substitutes, no matter how you look at them. Disillusionment can easily set in, and then it has to be dealt with.

There are several solutions. One is a rather drastic one. It is a variation of the Stimulus Struggle called Tempting Survival. The disillusioned teenager, instead of throwing a ball on a playing field, can throw it through a plate-glass window. The disillusioned housewife, instead of stroking the dog, can stroke the milkman. The disillusioned businessman, instead of stripping down the engine of his car, can strip down his secretary. The ramifications of this manœuvre are dramatic. In no time at all the individual is involved in the true survival struggle of fighting for his social life. During such phases there is a characteristic loss of interest in furniture rearranging and postage-stamp collecting. After the chaos has died down, the old substitute activities suddenly seem more appealing again.

A less drastic variant is Tempting Survival by Proxy. One form this takes consists of meddling in other people's emotional lives and creating for them the sort of chaos that you would otherwise have to go through yourself. This is the malicious gossip principle: it is extremely popular because it is so much safer than direct action. The worst that can happen is that you lose some of your friends. If it is operated skilfully enough, the reverse may occur: they may become substantially *more* friendly. If your machinations have succeeded in breaking up their lives, they may have a greater need of your friendship than ever before. So, providing you are not caught out, this variation can have a double benefit: the vicarious thrill of watching their survival drama, and the subsequent increase in their friendliness.

A second form of Tempting Survival by Proxy is less damaging. It consists of identifying yourself with the survival drama of fictional characters in books, films, plays and on television. This is even more popular, and a giant industry has grown up to meet the enormous demands it creates. It is not only harmless and safe, but it also has the distinction of being remarkably inexpensive. The straight game of Tempting Survival can end up costing thousands, but this variant, for no more than a few shillings, can permit the Stimulus Struggler to indulge in seduction, rape, adultery, starvation, murder and pillage, without so much as leaving the comfort of his chair.

2. If stimulation is too weak, you may increase your behaviour output *by over-reacting to a normal stimulus.*

This is the over-indulgence principle of the Stimulus Struggle. Instead of setting up a problem to which you then have to find a solution, as in the last case, you simply go on and on reacting to a stimulus that is already to hand, although it no longer excites you in its original role. It has become merely an occupational device.

In zoos where the public are permitted to feed the animals, certain bored species with nothing else to do will continue to eat until they become grossly over-weight. They will have already eaten their complete zoo diet and are no longer hungry, but idle nibbling is better than doing nothing. They get fatter and fatter, or become sick, or both.

Goats eat mountains of ice-cream cartons, paper, almost anything they are offered. Ostriches even consume sharp metal objects. A classic case concerns a female elephant. She was observed closely for a single typical zoo day and during that period (in addition to her normal, nutritionally adequate zoo diet) she devoured the following objects offered to her by the public: 1,706 peanuts, 1,330 sweets, 1,089 pieces of bread, 811 biscuits, 198 segments of orange, 17 apples, 16 pieces of paper, 7 ice-creams, 1 hamburger, 1 boot-lace and 1 lady's white leather glove. There are cases on record of zoo bears dying of suffocation caused by the enormous pressure of food in their stomachs. Such are the sacrifices made to the Stimulus Struggle.

One of the strangest examples of this phenomenon concerns a large male gorilla which regularly ate, regurgitated, and then re-ate his food, performing his own version of a Roman banquet. This process was taken a stage further by a sloth bear which was frequently observed to regurgitate its food more than a hundred times, each time eating it up again with the gurgling and sucking sounds typical of its species.

If the possibilities of over-indulging in feeding behaviour are limited and there is nothing else to do, an animal can always clean itself excessively, extending the performance until long after its feathers or its fur are perfectly cleansed and groomed. This, too, can lead to trouble. I recall one sulphur-crested cockatoo that had only a single feather left, a long yellow crest-plume, the rest of its body being as naked as an oven-chicken's. That was an extreme case, but not an isolated one. Mammals can scratch and lick bare patches until sores develop and set up their own vicious circle of irritation and picking.

For the human Stimulus Struggler, the unpleasant forms that this principle takes are well known. In infancy there is the example of prolonged thumb-sucking, which results from too little contact and inter-action with the mother. As we grow older we can indulge in occupational eating, nibbling aimlessly away at chocolates and biscuits to pass the time, and getting fatter and fatter as a result, like the zoo bears. Or we can groom ourselves into trouble, like the cockatoo. For us it will probably take the form of nail-biting or scab-picking. Occupational drinking, if the drinks are long and sweet, can lead again to fatness; if short and alcoholic they can lead to addiction and possibly liver damage. Smoking can be another time-killer and this, too, has its dangers.

Clearly there are pitfalls if the Stimulus Struggle is tackled badly. The snag with these over-indulgence time-killers is that they are so limited that they make development impossible. All one can do with them is repeat them over and over again, to stretch them out. To be effective in a major way they must be indulged in for long periods and that means trouble. Harmless enough in the ordinary course of events, as minor time-killers, they become damaging when carried to excess.

3. If stimulation is too weak, you may increase your behaviour output *by inventing novel activities.*

This is the creativity principle. If familiar patterns are too dull, the intelligent zoo animal must invent new ones. Captive chimpanzees, for instance, will contrive to introduce novelty into their environment by exploring the possibilities of new forms of locomotion, rolling over and over, dragging their feet along, and performing a variety of gymnastic patterns. If they can find a small piece of string, they will thread it through the cage roof, hang on to both ends with their teeth or their hands and spin round in the air, suspended like circus acrobats.

Many zoo animals use visitors to relieve the boredom. If they ignore the people who walk by their cages they are liable to be ignored in return, but if they stimulate them in some way, then the visitors will stimulate them back. It is surprising what you can get zoo visitors to do, if you are an ingenious zoo animal. If you are a chimpanzee or an orang-utan and you spit at them, they scream and rush wildly about. It helps to pass the day. If you are an elephant, you can flick spittle at them with the tip of your trunk. If you are a walrus, you can splash water over them with your flipper. If you are a magpie or a parrot, you can entice them with ruffled head feathers to preen you and then nip their fingers with your beak.

One particular male lion perfected his audience-manipulation in a remarkable way. His usual method of urination (as with tom-cats) was to squirt a jet of urine horizontally backwards at a vertical landmark, depositing his personal scent upon it. When he did this against one of the vertical bars of his cage-front he found that the spray reached his visitors and created an interesting reaction. They leapt back, shouting. As time passed, he not only improved his aim, but also added a new trick. After the first spraying, when the front row of his audience had retreated, the second row quickly took its place to get a better look. Instead of loosing his jet in one stream, he saved some of it for a second spraying and in this way managed to excite the new front row as well.

Food-begging (as distinct from food-nibbling) is a less drastic measure, but equally rewarding, and is practised by a wide variety of species. All that is necessary is to invent some peculiar action or posture that appeals to the passers-by and makes them believe that you are hungry. Monkeys and apes find that an outstretched palm is adequate, but bears have proved more inventive. Each has its own speciality: one will stand on its hind legs and wave a paw; another will sit on its rump in a curved posture, clasping its hind paws with its front feet; another will sit up and hook one of its front paws on to the lower jaw of its open mouth; another will stand up and nod or make come-hither movements with its head. It is amazing how easy it is to train zoo visitors to react to these displays if you are an intelligent zoo bear. The trouble is that in order to keep the visitors' interest, you have to reward them every so often by eating the objects they throw at you. If you fail to comply, they soon move away and the exciting stimulation of the social interaction you have invented is lost. The result of this we have already observed: you have to switch to the less satisfactory 'over-indulgence principle' and you get fat and sick.

The essential point about these zoo gymnastics and begging routines is that the motor patterns involved are not found in nature. They are inventions geared to the special conditions of captivity.

In the human zoo this creativity principle is carried to impressive extremes. I have already pointed out that disillusionment can set in when the survival-substitute activities of the Stimulus Struggle begin to seem pointless, often because the activities chosen are rather limited in their scope. In avoiding these limitations, men have sought for more and more complex forms of expression, forms which become so absorbing that they carry the individual on to such high planes of experience that the rewards are endless. Here we move from the realms of occupational trivia to the exciting worlds of the fine arts, philosophy and the pure sciences. These have the great value that they not only effectively combat under-stimulation, but also at the same time make maximum use of man's most spectacular physical property – his gigantic brain.

Because of the vast importance these activities have assumed in our civilizations, we tend to forget that they are in a sense no more than devices of the Stimulus Struggle. Like hide-and-seek or chess, they help to pass the time between the cradle and the grave, for those who are lucky enough not to be totally bound up in the struggle for crude survival. I say lucky, because, as I mentioned earlier, the great advantage of the super-tribal condition is that we are comparatively free to choose the forms that our activities take, and when the human brain can devise such beautiful pursuits as these, we must count ourselves fortunate to be amongst the Stimulus Strugglers, rather than the strugglers for survival. This is man the inventor playing for all he is worth. When we study the researches of science, listen to symphonies, read poetry, watch ballets, or look at paintings, we can only marvel at the lengths to which mankind has pushed the Stimulus Struggle and the incredible sensitivity with which he has tackled it.

4. If stimulation is too weak, you may increase your behaviour output *by performing normal responses to sub-normal stimuli.*

This is the overflow principle. If the internal urge to perform some activity becomes too great, it can 'overflow' in the absence of the external objects that normally provoke it.

Objects which in the wild state would never rate a reaction are given the full treatment in the bleak zoo environment. With monkeys this may take the form of coprophagy: if there is no food to chew, then faeces will do. If there is no territory to patrol, then stunted cage-pacing will do. The animal ambles back and forth, back and forth, until it has worn a track by its rhythmic, sterile pacing. Again, it is better than nothing.

In the absence of a suitable mate, a zoo animal may attempt to copulate with virtually anything that is available. A solitary hyena, for example, managed to mate with its circular food-dish, tipping it up on its side and rolling it back and forth beneath its body so that it pressed rhythmically against its penis. A male raccoon living alone used its bed as a mate. It could be seen to gather up a tight bundle of straw, clasp it beneath its body and then make pelvic thrusts into it. Sometimes, when an animal is kept with another of a different species, the alien companion can be used as a mate-substitute. A male brush-tailed porcupine living with a tree-porcupine repeatedly tried to mount it. The two species are not closely related and the arrangement of the spines differs markedly, with the result that it was an extremely painful affair for the frustrated male. In another cage a little squirrel monkey was housed with a large kangaroo-shaped rodent called a springhaas, which was about ten times its size. Undaunted, the diminutive monkey used to leap on to the sleeping rodent's back and attempt to copulate. The result of its desperate frustrations were reported in the local press, but totally misunderstood. It was recorded as having indulged in a charming game, 'riding on the big animal's back like a little furry jockey'.

The sexual examples are reminiscent of fetishism, but must not be confused with it. In the case of 'overflow activities', as soon as the natural stimulus is introduced into the environment, the animal reverts to normality. In the instances I have mentioned, the males immediately switched their attentions to females of their own species when these became available. They were not 'hooked' on their female-substitutes, like the true fetishists I discussed in the last chapter.

An unusual mutual overflow activity occurred when a female sloth and a small douroucouli monkey were housed together. In nature this monkey makes a snug den for itself inside a hollow tree, where it sleeps during the day. The female sloth, had she given birth in the wild, would have carried her offspring on her body for a considerable period. In the zoo, the monkey lacked a warm, snug bed and the sloth lacked an offspring. The problem was neatly solved for both of them by the simple expedient of the monkey sleeping clamped tightly on to the sloth's body.

The operation of this fourth principle of the Stimulus Struggle is not so much a case of any-port-in-a-storm as any-port-when-becalmed and, despite the many winds that blow through the human zoo, the human animal frequently finds itself in this sort of situation. The emotional patterns of the super-tribesman are constantly being blocked for one reason or another. In the midst of material plenty there is much behavioural deprivation. Then he, like the zoo animals, is driven to respond to sub-normal stimuli, no matter how inferior these may be.

In the sexual sphere, man is better equipped than most species to solve the absent-mate problem by masturbating, and this is the most common human solution. Despite this, zoophilia, or the act of copulation performed between a human being and some other animal species, does occur from time to time. It is rare, but less rare than most people imagine. A recent American survey revealed that in that country, among boys raised on farms, about 17 per cent experience orgasm as a result of 'animal contacts' at least once during their lives. There are many more that indulge in milder forms of sexual interaction with farm animals, and in certain districts the total figure has been put as high as 65 per cent of farm boys. The animals favoured are usually calves, donkeys and sheep, and occasionally some of the larger birds such as geese, ducks and chickens.

Zoophilic activities are much rarer among human females. Out of nearly six thousand American women only twenty-five had experienced orgasm as a result of stimulation by another animal species, usually a dog.

To most people such activities seem bizarre and revolting. The fact that they occur at all reveals the extraordinary lengths to which Stimulus Strugglers will go in avoiding inactivity. The parallel with the zoo world is inescapable.

Other forms of sexual behaviour, such as certain cases of 'better-than-nothing' homosexuality, also fall into this category. In the absence of normal stimulation, the sub-normal object becomes adequate. Starving men will chew wood and other nutritionally worthless objects, rather than chew nothing. Aggressive individuals with no enemies to attack will violently smash inanimate objects or mutilate their own bodies.

5. If stimulation is too weak, you may increase your behaviour output *by artificially magnifying selected stimuli*.

This principle concerns the creation of 'super-normal stimuli'. It operates on the simple premise that if natural, normal stimuli produce normal responses, then super-normal stimuli should produce super-normal responses. This idea has been put to great use in the human zoo, but it is rare in the animal zoo. Students of animal behaviour have devised a number of super-normal stimuli for experimental animals, but the accidental occurrence of the phenomenon is limited to only a few examples, one of which I will describe in detail.

It stems from my own research. For some time I had been keeping a mixed collection of birds in a large aviary on the roof of a research department. At one point they became troubled by nocturnal visits from a predatory owl which attempted to attack them through the wire of the aviary. Investigating the problem led me to make a number of dusk watches. The owl never came while I was there, and in fact was never heard of again, but although I drew a blank in that respect, what I did see was some very strange behaviour going on inside the aviary itself.

Among the birds were some doves and some small finches called Java sparrows. These finches normally roost together, pressed closely up against one another on a branch. To my surprise, the finches in the aviary were ignoring one another, favouring the doves instead as roosting companions. Each dove had a tiny finch pressed tightly up against its plump body. The small birds were snuggling down contentedly for the night, and the doves, although somewhat startled at first by their strange sleeping partners, were too drowsy to do anything about them, and eventually they too settled down for the night's sleep.

I was completely at a loss to explain this peculiar pattern of behaviour. The two species had not been reared together, so there could be no question of mal-imprinting. The finches had not even been bred in captivity. They should, by all the rules, have roosted with other members of their own species. There was another problem. Why, out of all the other species in the aviary, did they choose the doves to sleep with?

Returning to my roost-time vigil on subsequent nights, I was able to observe even more curious behaviour. Before going to sleep, the tiny finches often preened their doves, again an action which under normal circumstances they would only direct towards one of their own kind. Stranger still, they began to play leapfrog over the backs of their huge companions. A finch would leap on to the back of its dove, then off again at the other side; then back again, and so on. The ultimate oddity came when I saw one of the small birds push up underneath the body of its dove and shove itself between the big bird's legs. The sleepy dove stretched high on its legs and stared down at the struggling form beneath its rounded breast. Once in position, the finch settled down and the dove subsided on to it. There they sat, with the finch's pink beak protruding from the bottom of the dove's chest.

Somehow I had to find an explanation for this extraordinary relationship. There was nothing odd about the doves, except perhaps their remarkable tolerance. It was the finches that demanded further study. I found that they had a special signal at roosting time that indicated to other members of their species that they were ready to go to sleep. When they were active they kept their distance from one another, but when it was time to clump together for the night, one finch, presumably the sleepiest, would fluff out its feathers and squat low on its perch. This was the signal to other members of its group that they could join it without being repulsed. A second finch would fly in and squat up against the first one, fluffing its feathers out as it did so; then a third, a fourth, and so on, until a row of roosting birds had been formed. Late-comers would often hop along the backs of the row and squeeze down into a warmer and more favourable position in the middle. Here were all the clues I needed.

The combined fluffing-and-squatting action made the finches look bigger and more spherical than when they were actively moving about. This was the key signal, saying, 'come roost with me.' A roosting dove was even bigger and more spherical, and therefore could not help sending out a much more powerful version of the same signal. Furthermore, unlike the other species in the aviary, the doves had the same greyish colour as the little finches. As they were so big, rounded and grey, they gave out a *super-normal signal* to the finches which the small birds simply could not resist. Being innately programmed to this combination of size, shape and colour, the finches automatically responded to the doves as super-normal stimuli for roosting, preferring them to their own species. The snag was that the doves did not form rows. A finch clumping with one found itself at the end of a 'row', jumped on to the dove's back, failed to find the middle of the 'row', and jumped off the other side. The dove was so big that it must have seemed like a whole row of finches, so the small bird tried again, but still without success. With great persistence, the finch eventually tried pushing up from underneath the dove and at last found a snug position in the 'middle of the row', between the bigger bird's legs.

As I said earlier, this is one of the few known instances of a non-human super-normal stimulus occurring without a deliberate experiment being carried out. Other, better known examples have always involved the use of an experimental dummy. Oystercatchers, for instance, are ground-nesting birds. If one of their eggs rolls out of the nest, it is pulled back in with a special action of the beak. If dummy eggs are placed near the nest, the birds will pull these in too. If offered dummy eggs of different sizes, they always prefer the biggest one. They will, in fact, try to heave in eggs many times the size of their own real eggs. Again, they cannot help reacting to a super-normal stimulus.

Herring-gull chicks, when they beg for food from their parents, peck at a bright red spot that is situated near the tip of the adult birds' bills. The parents respond to this pecking by regurgitating fish for their young. The red spot is the vital signal. It was discovered that the chicks would even peck at flat cardboard models of their parents' heads. By a series of tests it was found that the other details of the adult head were unimportant. The chicks would peck at a red spot by itself. Furthermore, if they were offered a stick with three red spots on it, they would actually peck *more* at that than at a complete and realistic model of their parents. Again, the stick with the three red spots was a super-normal stimulus.

There are other examples, but these will suffice. Clearly, it is possible to improve on nature, a fact which some have found distasteful. But the reason is simple: each animal is a complex system of compromises. The conflicting demands of survival pull it in different directions. If, for example, it is too brightly coloured, it will be detected by its predators. If it is too drably coloured, it will be unable to attract a mate, and so on. Only when the pressures of survival are artificially reduced will this system of compromises be relaxed. Domesticated animals, for instance, are protected by man and no longer need fear their predators. Without risk, their dull colours can be replaced by pure whites, gaudy piebalds and other vivid patterns. But if they were turned loose again in their natural habitat, they would be so conspicuous that they would quickly fall prey to their natural enemies.

Like his domesticated animals, super-tribal man can also afford to ignore the survival restrictions of natural stimuli. He can manipulate stimuli, exaggerate them and distort them to his heart's content. By increasing their strength artificially – by creating super-normal stimuli – he can give an enormous boost to his responsiveness. In his super-tribal world he is like an oystercatcher surrounded by giant eggs.

Everywhere you look you will find evidence of some kind of super-normal stimulation. We like the colours of flowers, so we breed bigger and brighter ones. We like the rhythm of human locomotion, so we develop gymnastics. We like the taste of food, so we make it spicier and tastier. We like certain scents, so we manufacture strong perfumes.

We like a comfortable surface to sleep on, so we construct super-normal beds with springs and mattresses.

We can start by examining our appearance – our clothes and our cosmetics. Many male costumes include padding of the shoulders. At puberty there is a marked difference in the growth rate of the shoulders in males and females, those of boys becoming broader than those of girls. This is a natural, biological signal of adult masculinity. Padding the shoulders adds a super-normal quality to this masculinity and it is not surprising that the most exaggerated trend occurs in that most masculine of spheres, the military, where stiff epaulets are added to further increase the effect. A rise in body height is also an adult feature, especially in males, and many an aggressive costume is crowned by some form of tall headgear, creating the impression of super-normal height. We would no doubt wear stilts, too, if they were not so cumbersome.

If males wish to appear super-normally young, they can wear toupees to cover their bald heads, false teeth to fill their ageing mouths, and corsets to hold in their sagging bellies. Young executives, who wish to appear super-normally old, have been known to indulge in artificial greying of their juvenile hair.

The adolescent female of our species undergoes a swelling of the breasts and a widening of the hips that mark her out as a developing sexual adult. She can strengthen her sexual signals by exaggerating these features. She can raise, pad, point, or inflate her breasts in a variety of ways. By tightening her waist she can throw into contrast the width of her hips. She can also pad out her buttocks and her hips, a trend that found its most super-normal development in the periods of bustles and crinolines.

Another growth change that accompanies the maturation of the female is the lengthening of the legs in relation to the rest of the body. Long legs can therefore come to equal sexuality and exceptionally lengthy legs become sexually appealing. They cannot, of course, become super-normal stimuli themselves, being natural objects (although high heels will help a little), but artificial lengthening can occur in erotic drawings and paintings of females. Measurements of drawings of 'pin-ups' reveal that the girls are usually portrayed with unnaturally long legs, sometimes almost one and a half times as long as the legs of the models on which they are based. The recent fashion for very short skirts owes its sexual appeal not simply to the exposure of bare flesh, but also to the impression of longer legs it gives when contrasted with the earlier longer-skirted styles.

A glittering array of super-normal stimuli can be found in the world of female cosmetics. A clear, unblemished skin is universally attractive sexually. Its smoothness can be exaggerated by powders and creams. At times when it has been important to show that a female did not have to toil in the sun, her cosmetics aided her by creating a super-normal whiteness for her visible skin. When conditions changed and it became important for her to reveal that she could afford the *leisure* to lie in the sun, then tanning of the skin became an asset. Once again her cosmetics were there to provide her with super-normal browning. At other periods, in the past, it was important that she displayed her healthiness, and the super-normal flush of rouge was added. Another feature of her skin is that it is less hairy than that of the adult male. Here again, a super-normal effect can be achieved by various forms of depilation, the tiny hairs being shaved or stripped from the legs, or painfully plucked from the face. The eyebrows of the male tend to be bushier than those of the female, so super-normal femininity can be obtained by plucking here, too. Add to all this her super-normal eye make-up, lipstick, nail-varnish, perfume and occasionally even nipple-rouge, and it is easy to see how hard we work the super-normal principle of the Stimulus Struggle.

We have already observed in a previous chapter the lengths to which the male penis has gone in becoming a super-normal phallic symbol. In ordinary clothing it has not fared so well, except for a brief moment of glory during the epoch of the codpiece. Today we are left with little more than the super-normal pubic tuft of the Scotsman's sporran.

The strange world of aphrodisiacs is entirely devoted to the subject of super-normal sexual stimuli. For many centuries and in many cultures, ageing human males have attempted to boost their waning sexual responses by means of artificial aids. A dictionary of aphrodisiacs lists over nine hundred items, including such delightful potions as angel water, camel's hump, crocodile dung, deer sperm, goose tongues, hare soup, lion's fat, necks of snails and swan's genitals. Doubtless many of these aids proved successful, not because of their chemical properties, but because of the inflated prices paid for them. In the eastern world, powdered rhino horn has been so highly valued as a super-normal sexual stimulus that certain species of rhinoceros have nearly become extinct. Not all aphrodisiacs were swallowed. Some were rubbed on, others smoked, sniffed or worn on the body. Everything from aromatic baths to scented snuff seems to have been pressed into service in the frantic search for stronger and more violent stimulation.

The modern pharmacy is less sexually orientated, but it is bulging with super-normal stimuli of many kinds. There are sleeping pills to produce super-normal sleep, pep pills to produce super-normal alertness, laxatives to produce super-normal defecation, toilet preparations to produce super-normal body-cleaning, and toothpaste to produce a super-normal smile. Thanks to man's ingenuity there is hardly any natural activity which cannot be provided with some form of artificial boost.

The world of commercial advertising is a seething mass of super-normal stimuli, each trying to outpace the others. With competing firms marketing almost identical products, the super-normal Stimulus Struggle has become big business. Each product has to be presented in a more stimulating form than its rivals. This requires endless attention to subtleties of shape, texture, pattern and colour.

An essential feature of a super-normal stimulus is that it need not involve an exaggeration of *all* the elements of the natural stimulus on which it is based. The oystercatcher responded to a dummy egg that was super-normal in only one respect – its size. In shape, colour and texture it was similar to a normal egg. The experiment with the gull chicks went one step further. There, the vital red spots were exaggerated and, in addition, the other features of the parent figure, the unimportant ones, were eliminated. A double process was therefore taking place: magnification of the essential stimuli and, at the same time, elimination of the inessential ones. In the experiment this was done merely to demonstrate that the red spots alone were sufficient to trigger the reaction. Nevertheless, taking this step must also have helped in focusing more attention on the red spots by removing irrelevancies. With many human super-normal stimuli this dual process has been employed with great effect. It can be expressed as an additional, subsidiary principle for the Stimulus Struggle:

This states that when selected stimuli are magnified artificially to become super-normal stimuli, the effect can be further enhanced by reducing other (non-selected or irrelevant) stimuli. By simultaneously creating sub-normal stimuli in this way, the super-normal stimuli appear relatively stronger. This is the principle of *stimulus extremism*.

If we wish to be entertained by books, plays, films, or songs, we automatically subject ourselves to this procedure. It is the very essence of the process we call dramatization. Everyday actions performed as they happen in real life would not be exciting enough. They have to be exaggerated. The operation of the stimulus extremism principle ensures that irrelevant detail is suppressed and relevant detail is heightened and made more extravagant. Even in the most realistic schools of acting, or, for that matter, in non-fiction writing and documentary filming, the negative process still operates. Irrelevancies are pared away, thus producing an indirect form of exaggeration. In the more stylized performances, such as opera and melodrama, the direct forms of exaggeration are more important and it is remarkable to see how far the voices, the costumes, the gestures, the actions and the plot can stray away from reality and yet still make a powerful impact on the human brain. If this seems strange, it is worth recalling the case of the experimental birds. The gull chicks were prepared to respond to a substitute for their parents that consisted of something as remote from an adult gull as a stick with three red spots on it. Our reactions to the highly stylized rituals of an opera are no more outlandish.

Children's toys, dolls and puppets illustrate the same principle very vividly. A rag doll's face, for example, has certain important features magnified and others omitted. The eyes become huge black spots, while the eyebrows disappear. The mouth is shown in a vast grin, while the nose is reduced to two small dots. Enter a toyshop and you enter a world of contrasting super-normal and sub-normal stimuli. Only the toys for the older children become less contrasted and more realistic.

The same is true of the children's own drawings. In portrayals of the human body, those features that are important to them are enlarged; those that are unimportant are reduced or omitted. Usually the head, eyes and mouth receive the most disproportionate magnification. These are the parts of the body that have most meaning for a young child, because they form the area of visual expression and communication. The external ears of our species are inexpressive and comparatively unimportant and they are therefore frequently left out altogether.

Visual extremism of this kind is also prevalent in the arts of primitive peoples. The size of heads, eyes and mouths is usually super-normal in relation to the dimensions of the body and, as with children's drawings, other features are reduced. The stimuli selected for magnification do vary from case to case, however. If a figure is shown running, then its legs become super-normally large. If a figure is simply standing and is doing nothing with either its arms or its legs, they may become mere stumps or disappear altogether. If a prehistoric figurine is concerned with representing fertility, its reproductive features may become super-normalized to the exclusion of all else. Such a figure may boast a huge pregnant belly, enormous protruding buttocks, wide hips and vast breasts, but have no legs, arms, neck or head.

Graphic manipulations of subject-matter in this way have often been referred to as the creation of ugly deformities, as if the beauty of the human form were somehow being subjected to malicious damage and insult. The irony is that if such critics examined their own bodily adornments they would find that their own appearance was not exactly 'as nature intended'. Like the children and the primitive artists they are no doubt laden with 'deforming' super-normal and sub-normal elements.

The fascination of stimulus extremism in the arts lies in the way these exaggerations vary from case to case and place to place, and in the way the modifications develop new forms of harmony and balance. In the modern world, animated cartoon films have become major purveyors of this type of visual exaggeration, and a specialized form of it is to be found in the art of caricature. The expert caricaturist picks out the naturally exaggerated features of his victim's face and deftly super-normalizes these already existing exaggerations. At the same time he reduces the more inconspicuous features. The magnification of a large nose, for example, can become so extreme that it ends up with its dimensions doubled or even tripled, without rendering the face unrecognizable. Indeed, it makes it even more recognizable. The point is that we all identify individual faces by comparing them in our minds with an idealized 'typical' human face. If a particular face has certain features that are stronger or weaker, bigger or smaller, longer or shorter, darker or fairer, than our 'typical' face, these are the items that we remember. In drawing a successful caricature, the artist has to know intuitively which features we have selected in this way, and he then has to super-normalize the strong points and sub-normalize the weak ones. The process is fundamentally the same as that employed in the drawings of children and primitive peoples, except that the caraciturist is concerned primarily with individual differences.

The visual arts, throughout much of their history, have been pervaded by this device of stimulus extremism. Super-normal and sub-normal modifications abound in almost all the earlier art forms. As the centuries passed, however, realism came more and more to dominate European art. The painter and the sculptor became burdened with the task of recording the external world as precisely as possible. It was not until the last century, when science took over this formidable duty (with the development of photography), that artists were able to return to a freer manipulation of their subject-matter. They were slow to react at first, and although the chains were broken in the nineteenth century, it was not until the twentieth century that they were fully shaken off. During the past sixty years wave after wave of rebellion has occurred as stimulus extremism has reasserted itself more and more powerfully. The rule once again has become: magnify selected elements and eliminate others.

When paintings of the human face began to be manipulated in this way by modern artists, there was an outcry. The pictures were scorned as decadent lunacies, as if they reflected some new disease of twentieth-century life, instead of a return to art's more basic business of pursuing the Stimulus Struggle. The melodramatic exaggerations of human behaviour in theatrical productions, ballets and operas, and the extreme magnifications of human emotions expressed in songs and poems, were happily accepted, but it took some time to adjust to similar stimulus extremisms in the visual arts. When totally abstract paintings began to appear they were attacked as meaningless by people who were perfectly willing to enjoy the total abstraction of any musical performance. But music had never been forced into the aesthetic strait-jacket of portraying natural sounds.

I have defined a super-normal stimulus as an artificial exaggeration of a natural stimulus, but the concept can also be applied in a special way to an invented stimulus. Let me take two clear-cut cases. The pink lips of a beautiful girl are, without any question, a perfectly natural, biological stimulus. If she exaggerates them by painting them a brighter pink, she is obviously converting them into a super-normal stimulus. There the issue is simple, and it is this sort of example I have been concentrating on up to now. But what about the sight of a shiny new motor car? This can be very stimulating, too, but it is an entirely artificial, invented stimulus. There is no natural, biological model against which we can compare it to find out if it has been super-normalized. And yet, as we look around at various motor cars, we can easily pick out some that seem to have the quality of being super-normal. They are bigger and more dramatic than most of the others. Manufacturers of motor cars are, in fact, just as concerned with producing super-normal stimuli as manufacturers of lipsticks. The situation is more fluid, because there is no natural, biological base-line against which to work; but the process is essentially the same. Once a new stimulus has been invented, it develops a base-line of its own. At any point in the history of motor cars it would be possible to produce a sketch of the typical, common and therefore 'normal' car of the period. It would also be possible to produce a sketch of the outstanding luxury motor car of the period which, at that time, was the super-normal vehicle. The only difference between this and the lipstick example is that the 'normal base-line' of the motor car changes with technical progress, whereas the natural pink lips stay the same.

The application of the super-normal principle is therefore widespread and penetrates almost all of our endeavours in one way or another. Freed from the demands of crude survival, we wring the last drop of stimulation out of anything we can lay our hands or eyes on. The result is that we sometimes get stimulus indigestion. The snag with making stimuli more powerful is that we run the risk of exhausting ourselves by the strength of our response. We become jaded. We begin to agree with the Shakespearean comment that

To gild refinèd gold, to paint the lily,
To throw a perfume on the violet . . .
Is wasteful and ridiculous excess.

But at the same time we are forced to admit, with Wilde, that 'Nothing
succeeds like excess.' So what do we do? The answer is that we bring
into operation yet another subsidiary principle of the Stimulus
Struggle:

This states that because super-normal stimuli are so powerful and
our response to them can become exhausted, we must from time to
time vary the elements that are selected for magnification. In other
words, we ring the changes. When a switch of this sort occurs it is
usually dramatic, because a whole trend is reversed. It does not,
however, stop a particular branch of the Stimulus Struggle from being
pursued, it merely shifts the points of super-normal emphasis. Nowhere is this more clearly illustrated than in the world of fashionable
clothing and body adornment.

In female costumes, where sexual display is paramount, this has
given rise to what fashion experts refer to as the Law of Shifting
Erogenous Zones. Technically, an erogenous zone is an area of the
body that is particularly well supplied with nerve endings responsive to
touch, direct stimulation of which is sexually arousing. The main areas
are the genital region, the breasts, the mouth, the ear-lobes, the buttocks and the thighs. The neck, the armpits and the navel are sometimes added to the list. Female fashions are not, of course, concerned
with tactile stimulation, but with the visual display (or concealment)
of these sensitive areas. In extreme cases all these areas may be displayed at once, or, as in female Arab costumes, all may be concealed.
In the vast majority of super-tribal communities, however, some are
displayed and others simultaneously concealed. Alternatively, some
may be emphasized, although covered, while others are obliterated.

The Law of Shifting Erogenous Zones is concerned with the way in
which concentration on one area gives way to concentration on another as time passes and fashions change. If the modern female emphasizes one zone for too long, the attraction wears off and a new
super-normal shock is required to re-awaken interest.

In recent times the two main zones, the breasts and the pelvis, have
remained largely concealed, but have been emphasized in various
ways. One is by padding or tightening the clothing to exaggerate the
shapes of these regions. The other is by approaching them as closely as
possible with areas of exposed flesh. When this exposure creeps up on
the breast region, with exceptionally low-cut costumes, it usually
creeps away from the pelvic region, the dresses becoming longer. When

the zone of interest shifts and the skirts become shorter, the neckline rises. On occasions when bare midriffs have been popular, exposing the navel, the other zones have usually been rather well covered, often to the extent of the legs being concealed with some sort of trousers.

The great problem for fashion designers is that their super-normal stimuli are related to basic biological features. As there are only a few vital zones, this creates a strict limitation and forces the designers into a series of dangerously repetitive cycles. Only with great ingenuity can they overcome this difficulty. But there is always the head region to play with. Ear-lobes can be emphasized with ear-rings, necks with necklaces, the face with make-up. The Law of Shifting Erogenous Zones applies here too, and it is noticeable that when eye make-up becomes particularly striking and heavy, the lips usually become paler and less distinct.

Male fashion cycles follow a rather different course. The male in recent times has been more concerned with displaying his status than his sexual features. High status means access to leisure, and the most characteristic costumes of leisure are sporting clothes. Students of fashion history have unearthed the revealing fact that practically everything men wear today can be classified as 'ex-sports clothes'. Even our most formal attire can be shown to have these origins.

The system works like this. At any particular moment in recent history there has always been a highly functional costume to go with the high-status sport of the day. To wear such a costume indicates that you can afford the time and money to indulge in such a sport. This status display can be super-normalized by wearing the costume as ordinary day clothes, even when not pursuing the particular sport in question, thus magnifying the display by spreading it. The signals emanating from the sports clothes say, 'I am very leisured,' and they can say this almost as well for a non-sporting man who cannot afford to participate in the sport itself. After a while, when they have become completely accepted as everyday wear, they lose their impact. Then a new sport has to be raided for its unusual costume.

Back in the eighteenth century, English country gentlemen were exhibiting their status by taking to the hunting field. They adopted a sensible manner of dress for the occasion, wearing a coat that was cut away in the front, giving it the appearance of having tails at the back. They abandoned big floppy hats and began to wear stiff top hats, like prototype crash helmets. Once this costume was fully established as a high-status-sport outfit, it began to spread. At first it was the young bloods (the young swingers of the day) who started using a modified hunting costume as everyday wear. This was considered the height of daring, if not downright scandalous. But little by little the trend spread (young swingers get older), and by the middle of the nineteenth century the costume of top hat and tails had become normal everyday wear.

Having become so accepted and traditional, the top hat and tails had to be replaced with something new by the more daring members of society who wished to display their super-normal leisure signals. Other high-status sports available for raiding were shooting, fishing and golf. Billycock hats became bowlers and shooting tweeds became check lounge suits. The softer sporting hats became trilby hats. As the present century has advanced, the lounge suit has become more accepted as formal day wear and has become more sombre in the process. 'Morning dress', with top hat and tails, has been shifted one step further towards formality, being reserved now for special occasions such as weddings. It also survives as evening dress, but there the lounge suit has already caught up with it and stripped it of its tails to create a dinner-jacket suit.

Once the lounge suit had lost its daring, it had to be replaced, in its turn, by something more obviously sporting. Hunting may have dropped out of favour, but horse-riding in general still retained a high-status value, so here we go again. This time it was the hacking jacket that soon became known as a 'sports jacket'. Ironically it only acquired this name when it lost its true sporting function. It became the new casual wear for everyday use and still holds this position at the present time. Already, however, it is creeping into the more formal world of the business executive. Amongst the most daring dressers, it has even invaded that holy of holies, the formal evening occasion, in the guise of a patterned dinner jacket.

As the sports jacket spread into everyday life, the polo-necked sweater spread with it. Polo was another very high-status sport, and wearing the typical round-necked sweater of the game imparted instant status to the lucky wearer. But already this characteristic garment has lost its daring charm. A silk version of it was recently worn for the first time with a formal dinner jacket. Instantly shops were bombarded by young males clamouring for this latest sports attack on formality. It may have lost its impact as day wear, but as evening wear it was still able to shock, and its range spread accordingly.

Other similar trends have occurred during the last fifty years. Yachting blazers with brass buttons have been worn by men who have never stepped off dry land. Skiing suits have been worn by men (and women) who have never seen a snow-capped mountain. Just so long as a particular sport is exclusive and costly, it will be robbed for its costume signals. During the present century, leisure sports have been replaced to a certain extent by the habit of taking off for the sea-shores of warmer climates. This began with a craze for the French Riviera.

Visitors there began copying the sweaters and shirts of the local fishermen. They were able to show that they had indulged in this expensive new status holiday by wearing modified versions of these shirts and sweaters back home. Immediately, a whole new range of casual clothes burst on to the market. In America, it became fashionable for wealthy, high-status males to own a ranch in the country, where they would dress in modified cowboy clothes. In no time at all, many a young ranchless city-dweller was striding along in his (further) modified cowboy suit. It could be argued that he took it straight from the Western movies, but this is unlikely. It would still have been fancy dress. However, once real, contemporary, high-status males are wearing it when they take their leisure, then all is well and a new take-over bid is on its way.

None of this, you may feel, explains the bizarre clothing of the way-out male teenager, who wears cravats, long hair, necklaces, coloured scarves, bracelets, buckled shoes, flared trousers and lace-cuffed shirts.

What kind of sport is *he* modifying? There is nothing mysterious about the micro-skirted female teenager. All she has done, apart from shifting her erogenous zone to her thighs, is to take an emancipated leaf out of the male's fashion book, and steal a sports costume for everyday wear. The tennis skirt of the 1930s and the ice-skating skirt of the 1940s were already full-blooded micro-skirts. It only remained for some daring designer to modify them for everyday wear. But the flamboyant young male, what on earth is he doing? The answer seems to be that, with the recent setting up of a 'sub-culture of youth', it became necessary to develop an entirely new costume to go with it, one that owed as little as possible to the variations of the hated 'adult sub-culture'. Status in the 'youth sub-culture' has less to do with money and much more to do with sex appeal and virility. This has meant that the young males have begun to dress more like females, not because they are effeminate (a popular jibe of the older group), but because they are more concerned with sex attraction displays. In the recent past these have been largely the concern of the females, but now both sexes are involved. It is, in fact, a return to an earlier (pre-eighteenth-century) condition of male dressing, and we should not be too surprised if the codpiece makes its reappearance any minute now. We may also see the return of elaborate male make-up. It is hard to say how long this phase will last because it will gradually be copied by older males who are already feeling disgruntled by the overt sex displays of their juniors. In returning to a peacock display, the young males of the 'youth sub-culture' have hit where it hurts most. The human male is in his sexually most potent condition at the age of sixteen to seventeen. By abandoning leisure status dress and replacing it with sex status dress, they have chosen the ideal weapon. However, as I said earlier, young bloods and young swingers grow older. It will be interesting to see what happens in twenty years' time, when there are bald Beatles in the board-room, and a new sub-culture of youth has arisen.*

Almost everything we wear today, then, is the result of this Stimulus Struggle principle of ringing-the-changes to produce the shock effect of sudden novelty. What is daring today becomes ordinary tomorrow and stuffy the next day, and we quickly forget where it has all come from. How many men, climbing into their evening dress clothes and putting on their top hats, realize that they are donning the costume of a late eighteenth-century hunting squire? How many sombrely lounge-suited businessmen realize that they are following the dress of early nineteenth-century country sportsmen? How many sports-jacketed young men think of themselves as horse-riders? How many open-neck-shirted, loose-knitted-sweatered young men think of themselves as Mediterranean fishermen? And how many mini-skirted young girls think of themselves as tennis-players or ice-skaters?

The shock is quickly over. The new style is quickly absorbed, and another one is then required to take its place and to provide a new stimulus. One thing we can always be sure of: whatever is today's most daring innovation in the world of fashion will become tomorrow's respectability, and will then rapidly fossilize into pompous formality as new rebellions crowd in to replace it. Only by this process of constant turn-over can the extremes of fashion, the super-normal stimuli of design, maintain their massive impact. Necessity may be the mother of invention, but where the super-normal stimuli of fashion are concerned it is also true to say that novelty is the mother of necessity.

Up to this point we have been considering the five principles of the Stimulus Struggle that are concerned with raising the behaviour output of the individual. Occasionally the reverse trend is called for. When this happens the sixth and final principle comes into operation:

6. If stimulation is too strong, you may reduce your behaviour output *by damping down responsiveness to incoming sensations.*

This is the cut-off principle. Some zoo animals find their confinement frightening and stressful, especially when they are newly arrived, moved to a fresh cage, or housed with hostile or unsuitable companions. In their agitated condition they may suffer from abnormal over-stimulation. When this happens and they are unable to escape or hide,

* The peacock fashions of the late-1960s faded quickly. Later styles included a contrasting swing to contrived, unshaven 'scruffiness' as a rebellion against adult neatness.

they must somehow switch off the incoming stimuli. They may do this simply by crouching in a corner and closing their eyes. This, at least, shuts off the visual stimuli. Excessive, prolonged sleeping (a device also used by invalids, both animal and human) also occurs as a more extreme form of cut-off. But they cannot crouch or sleep for ever.

While active, they can relieve their tensions to some extent by performing 'stereotypes'. These are small tics, repetitive patterns of twitching, rocking, jumping, swaying or turning, which, because they have become so familiar through being constantly repeated, have also become comforting. The point is that for the over-stimulated animal the environment is so strange and frightening that any action, no matter how meaningless, will have a calming effect so long as it is an old familiar pattern. It is like meeting an old friend in a crowd of strangers at a party. One can see these stereotypes going on all around the zoo. The huge elephants sway rhythmically back and forth; the young chimpanzee rocks its body to and fro; the squirrel leaps round and round in a tight circle like a wall-of-death rider; the tiger rubs its nose left and right across its bars until it is raw and bleeding.

If some of these over-stimulation patterns also occur from time to time in intensely bored animals, this is no accident, for the stress caused by gross under-stimulation is in some ways basically the same as the stress of over-stimulation. Both extremes are unpleasant and their unpleasantness causes a stereotyped response, as the animal tries desperately to escape back to the happy medium of moderate stimulation that is the goal of the Stimulus Struggle.

If the inmate of the human zoo becomes grossly over-stimulated, he too falls back on the cut-off principle. When many different stimuli are blaring away and conflicting with one another, the situation becomes unbearable. If we can run and hide, then all is well, but our complex commitments to super-tribal living usually prevent this. We can shut our eyes and cover our ears, but something more than blindfolds and ear-plugs are needed.

In extremis we resort to artificial aids. We take tranquillizers, sleeping pills (sometimes so many that we cut-off for good), over-doses of alcohol, and a variety of drugs. This is a variant of the Stimulus Struggle which we can call Chemical Dreaming. To understand why, it will help to take a closer look at natural dreaming.

The great value of the process of normal, night-time dreaming is that it enables us to sort out and file away the chaos of the preceding day. Imagine an over-worked office, with mountains of documents, papers and notes pouring into it all through the day. The desks are piled high. The office workers cannot keep up with the incoming information and material. There is not enough time to file it away neatly before the end of the afternoon. They go home leaving the office in chaos. Next morning there will be another great influx and the situation will rapidly get out of hand.

If we are over-stimulated during the day, our brains taking in a mass of new information, much of it conflicting and difficult to classify, we go to bed in much the same condition as the chaotic office was left in at the end of the working day. But we are luckier than the over-worked office staff. At night-time someone comes into the office inside our skull and sorts everything out, files it neatly away and cleans up the office ready for the onslaught of the next day. In the brain of the human animal this process is what we call dreaming. We may obtain physical rest from sleep, but little more than we could get from lying awake all night. But awake we could not dream properly. The primary function of sleeping, then, is dreaming rather than resting our weary limbs. We sleep to dream and we dream most of the night. The new information is sorted and filed and we awaken with a refreshed brain, ready to start the next day.

If day-time living becomes too frenzied, if we are too intensely over-stimulated, the ordinary dreaming mechanism becomes too severely tested. This leads to a preoccupation with narcotics and the dangerous pursuit of Chemical Dreaming. In the stupors and trances of chemically induced states, we vaguely hope that the drugs will create a mimic of the dream-like state. But although they may be effective in helping to switch off the chaotic input from the outside world, they do not usually seem to assist in the positive dream function of sorting and filing. When they wear off, the temporary negative relief vanishes and the positive problem remains as it was before. The device is therefore doomed to be a disappointing one, with the addition of the possible anti-bonus of chemical addiction.

Another variation is the pursuit of what we can call Meditation Dreaming, in which the dream-like state is achieved by certain thought disciplines, yogal or otherwise. The cut-off, trance-like conditions produced by yoga, voodoo, hypnotism, and certain magical and religious practices all have certain features in common. They usually involve sustained rhythmic repetition, either verbal or physical, and are followed by a condition of detachment from normal outside stimulation. In this way they can help to cut down the massive and usually conflicting input that is being suffered by the over-stimulated individual. They are therefore similar to the various forms of Chemical Dreaming, but as yet we have little information about the way they may, in addition, provide positive benefits of the kind we all enjoy when dreaming normally.

If the human animal fails to escape from a prolonged state of over-stimulation, he is liable to fall sick, mentally or physically. Stress diseases or nervous breakdowns may, for the luckier ones, provide their own cure. The invalid is forced, by his incapacity, to switch off the massive input. His sick-bed becomes his animal hiding-place.

Individuals who know they are particularly prone to over-stimulation often develop an early-warning signal. An old injury may start to play up, tonsils may swell, a bad tooth may throb, a skin rash may break out, a small twitch may reappear, or a headache may begin again. Many people have a minor weakness of this kind which is really more of an old friend than an old enemy, because it warns them that they are 'over-doing' things and had better slow down if they wish to avoid something worse. If, as often happens, they are persuaded to have their particular weakness 'cured', they need have little fear of losing the early-warning advantage it bestowed on them; some other symptom will in all probability soon emerge to take its place. In the medical world this is sometimes known as the 'shifting syndrome'.

It is easy enough to understand how the modern super-tribesmen can come to suffer from this over-burdened state. As a species, we originally became intensely active and exploratory in connection with our special survival demands. The difficult role our hunting ancestors had to play insisted on it. Now, with the environment extensively under control, we are still saddled with our ancient system of high activity and high curiosity. Although we have reached a stage where we could easily afford to lie back and rest more often and more lengthily, we simply cannot do it. Instead we are forced to pursue the Stimulus Struggle. Since this is a new pursuit for us, we are not yet very expert performers and we are constantly going either too far or not far enough. Then, as soon as we feel ourselves becoming over-stimulated and over-active or under-stimulated and under-active, we veer away from the one painful extreme or the other and indulge in actions that tend to bring us back to the happy medium of optimum stimulation and optimum activity. The successful ones hold a steady centre course; the rest of us swing back and forth on either side of it.

We are helped to a certain extent by a slow process of adjustment. The countryman, living a quiet and peaceful life, develops a tolerance to this low level of activity. If a busy townsman were suddenly thrust into all that peace and quiet, he would quickly find it unbearably boring. If the countryman were thrown into the hurly-burly of chaotic town life, he would soon find it painfully stressful. It is fine to have a quiet weekend in the country as a de-stimulator, if you are a townsman, and it is great to have a day up in town as a stimulator, if you are a countryman. This obeys the balancing principles of the Stimulus Struggle; but much longer, and the balance is lost.

It is interesting that we are much less sympathetic towards a man who fails to adjust to a low level of activity than we are to one who fails to adjust to a high level. A bored and listless man annoys us more than a harassed and over-burdened one. Both are failing to tackle the Stimulus Struggle efficiently. Both are liable to become irritable and bad-tempered, but we are much more prone to forgive the over-worked man. The reason for this is that pushing the level up a little too high is one of the things that keeps our cultures advancing. It is the intensely over-exploratory individuals who will become the great innovators and will change the face of the world in which we live. Those who pursue the Stimulus Struggle in a more balanced and successful way will also, of course, be exploratory, but they will tend to provide new variations on old themes rather than entirely new themes. They will also be happier, better adjusted individuals.

You may remember that at the outset I said the stakes of the game are high. What we stand to win or lose is our happiness, in extreme cases our sanity. The over-exploratory innovators should, according to this, therefore be comparatively unhappy and even show a tendency to suffer from mental illness. Bearing in mind the goal of the Stimulus Struggle, we should predict that, despite their greater achievements, such men and women must frequently live uneasy and discontented lives. History tends to confirm that this is so. Our debt to them is sometimes paid in the form of the special tolerance we show towards their frequently moody and wayward behaviour. We intuitively recognize that it is an inevitable outcome of the unbalanced way in which they are pursuing the Stimulus Struggle. As we shall see in the next chapter, however, we are not always so understanding.

7

THE CHILDLIKE ADULT

IN MANY RESPECTS the play of children is similar to the
Stimulus Struggle of adults. The child's parents take care of its
survival problems and it is left with a great deal of surplus energy. Its
playful activities help to burn up this energy. There is, however, a
difference. We have seen that there are various ways of pursuing the
adult Stimulus Struggle, one of which is the invention of new patterns
of behaviour. In play, this element is much stronger. To the growing
child, virtually every action it performs is a new invention. Its naivety
in the face of the environment more or less forces it to indulge in a
non-stop process of innovation. Everything is novel. Each bout of
playing is a voyage of discovery: discovery of itself, its abilities and
capacities, and of the world about it. The development of inventiveness
may not be the specific goal of play, but it is nevertheless its predomi-
nant feature and its most valuable bonus.

The explorations and inventions of childhood are usually trivial and
ephemeral. In themselves they mean little. But if the processes they
involve, the sense of wonder and curiosity, the urge to seek and find
and test, can be prevented from fading with age, so that they remain
to dominate the mature Stimulus Struggle, over-shadowing the less
rewarding alternatives, then an important battle has been won: the
battle for creativity.

Many people have puzzled over the secret of creativity. I contend
that it is basically no more than the extension into adult life of these
vital childlike qualities. The child asks new questions; the adult
answers old ones; the childlike adult finds answers to new questions.
The child is inventive; the adult is productive; the childlike adult is
inventively productive. The child explores his environment; the adult
organizes it; the childlike adult organizes his explorations and, by
bringing order to them, strengthens them. He creates.

It is worth examining this phenomenon more closely. If a young
chimpanzee, or a child, is placed in a room with a single familiar toy,
he will play with it for a while and then lose interest. If he is offered,
say, five familiar toys instead of only one, he will play first here, then
there, moving from one to the other. By the time he gets back to the
first one, the original toy will seem 'fresh' again and worthy of a little

further play attention. If, by contrast, an unfamiliar and novel toy is offered, it will immediately command his interest and produce a powerful reaction.

This 'new toy' response is the first essential of creativity, but it is one phase of the process. The strong exploratory urge of our species drives us on to investigate the new toy and to test it out in as many ways as we can devise. Once we have finished our explorations, then the unfamiliar toy will have become familiar. At this point it is our inventiveness that will come into action to utilize the new toy, or what we have learned from it, to set up and solve new problems. If, by re-combining our experiences from our different toys, we can make more out of them than we started with, then we have been creative.

If a young chimpanzee is put in a room with an ordinary chair, for example, it starts out by investigating the object, tapping it, hitting it, biting it, sniffing it, and clambering over it. After a while these rather random activities give way to a more structured pattern of activity. It may, for instance, start jumping over the chair, using it as a piece of gymnastic equipment. It has 'invented' a vaulting box, and 'created' a new gymnastic activity. It had learned to jump over things before, but not in quite this way. By combining its past experiences with the investigation of this new toy, it creates the new action of rhythmic vaulting. If, later on, it is offered more complex apparatus, it will build on these earlier experiences again, incorporating the new elements.

This developmental process sounds very simple and straightforward, but it does not always fulfil its early promise. As children we all go through these processes of exploration, invention and creation, but the ultimate level of creativity we rise to as adults varies dramatically from individual to individual. At the worst, if the demands of the environment are too pressing, we stick to limited activities we know well. We do not risk new experiments. There is no time or energy to spare. If the environment seems too threatening, we would rather be sure than sorry: we fall back on the security of tried and trusted, familiar routines. The environmental situation has to change in one way or another before we will risk becoming more exploratory. Exploration involves uncertainty and uncertainty is frightening. Only two things will help us to overcome these fears. They are opposites: one is disaster, and the other is greatly increased security. A female rat, for instance, with a large litter to rear, is under heavy pressure. She works non-stop to keep her offspring fed, cleaned and protected. She will have little time for exploring. If disaster strikes – if her nest is flooded or destroyed – she will be forced into panic exploration. If, on the other hand, her young have been successfully reared and she has built up a large store of food, the pressure relaxes and, from a position of greater security, she is able to devote more time and energy to exploring her environment.

There are, then, two basic kinds of exploration: panic exploration and security exploration. It is the same for the human animal. During the chaos and upheaval of war, a human community may be driven to inventiveness to surmount the disasters it faces. Alternatively, a successful, thriving community may be highly exploratory, striking out from its strong position of increased security. It is the community that is just managing to scrape along that will show little or no urge to explore.

Looking back on the history of our species it is easy to see how these two types of exploration have helped human progress to stumble on its way. When our early ancestors left the comforts of a fruit-picking, forest existence and took to open country, they were in serious difficulties. The extreme demands of the new environment forced them to be exploratory or die. Only when they had evolved into efficient, co-operative hunters did the pressure ease a little. They were at the 'scraping by' stage again. The result was that this condition lasted for a very long time, thousands upon thousands of years, with advances in technology occurring at an incredibly slow rate, simple developments in such things as implements and weapons, for example, taking hundreds of years to take a small new step.

Eventually, when primitive agriculture slowly emerged and the environment came more under our ancestors' control, the situation improved. Where this was particularly successful, urbanization developed and a threshold was passed into a realm of new and dramatically increased social security. With it came a rush of the other kind of exploration – security exploration. This, in its turn, led to more and more startling developments, to more security and more exploration.

Unfortunately this was not the whole story. Man's rise to civilization would be a much happier tale if only it had been. But, sadly, events moved too fast and, as we have seen throughout this book, the pendulum of success and disaster began to swing crazily back and forth. Because we unleashed so much more than we were biologically equipped to cope with, our magnificent new social developments and complexities were as often abused as they were used. Our inability to deal rationally with the super-status and super-power that our super-tribal condition thrust upon us, led to new, more sudden, and more challenging disasters than we had ever known. No sooner had a super-tribe settled down to a phase of great prosperity, with security exploration operating at full intensity, and wonderful new forms of creativity blossoming out, than something went wrong. Invaders, tyrants and aggressors smashed the delicate machinery of the intricate new social structures, and panic exploration was back on a major scale. For each new invention of construction, there was another of destruction, back and forth, back and forth, for ten thousand years, and it still goes on today. It is the horror of atomic weapons that has given us the glory of atomic energy, and it is the glory of biological research that may yet give us the horror of biological warfare.

In between these two extremes there are still millions of people living the simple lives of primitive agriculturalists, tilling the soil much as our early ancestors did. In a few areas primitive hunters survive. Because they have stayed at the 'scraping by' stage, they are typically non-exploratory. Like the left-over great apes – the chimpanzees, the gorillas and the orang-utans – they have the potential for inventiveness and exploration, but it is not called forth to any appreciable extent. Experiments with chimpanzees in captivity have revealed how quickly they can be encouraged to develop their exploratory potential: they can operate machines, paint pictures and solve all kinds of experimental puzzles; but in the wild state they do not even learn to build crude shelters to keep out the rain. For them, and the simpler human communities, the scraping-by existence – not too difficult and not too easy – has blunted their exploratory urges. For the rest of us, one extreme follows the other and we constantly explore from either an excess of panic or an excess of security.

From time to time there are those among us who cast an envious backward glance at the 'simple life' of primitive communities and start to wish we had never left our primeval Forest of Eden. In some cases, serious attempts have been made to convert such thoughts into actions. Much as we may sympathize with these projects, it must be realized that they are fraught with difficulties. The inherent artificiality of pseudo-primitive drop-out communities, such as those that have appeared in North America and elsewhere recently, is a primary weakness. They are, after all, composed of individuals who have tasted the excitements of super-tribal life as well as its horrors. They have been conditioned throughout their lives to a high level of mental activity. In a sense they have lost their social innocence, and the loss of innocence is an irreversible process.

At first, all may go well for the neo-primitive, but this is deceptive. What happens is that the initial return to the simple way of life throws up an enormous challenge to the ex-inmate of the human zoo. His new role may be simple in theory, but in practice it is full of fascinatingly novel problems. The establishment of a pseudo-primitive community by a group of ex-city-dwellers becomes, in fact, a major exploratory act. This, rather than the official return to pure simplicity, is what makes the project so satisfying, as any Boy Scout will testify. But what happens once the initial challenge has been met and overcome? Whether it is a remote, rural or cave-dwelling group, or whether it is a self-insulated, pseudo-primitive group set up in an isolated pocket inside the city itself, the answer is the same. Disillusionment sets in, as the monotony begins to assail the brains that have been irreversibly trained to the higher, super-tribal level. Either the group collapses, or it starts to stir itself into action. If the new activity is successful, then the community will soon find itself becoming organized and expanding. In no time at all it will be back in the super-tribal rat-race.

It is difficult enough in the twentieth century to remain as a genuine primitive community, like the Eskimos or Aborigines, let alone a pseudo-primitive one. Even the traditionally resistant European gypsies are gradually succumbing to the relentless spread of the human zoo condition.

The tragedy for those who wish to solve their problems by a return to the simple life is that, even if they somehow contrived to 'de-train' their highly activated brains, such individuals would still remain extremely vulnerable in their small rebel communities. The human zoo would find it hard to leave them alone. They would either be exploited as a tourist attraction, as so many of the genuine primitives are today, or, if they become an irritant, they would be attacked and disbanded. There is no escape from the super-tribal monster and we may as well make the best of it.

If we are condemned to a complex social existence, as it seems we are, then the trick is to ensure that *we* make use of *it*, rather than let *it* make use of *us*. If we are going to be forced to pursue the Stimulus Struggle, then the important thing is to select the most rewarding method of approach. As I have already indicated, the best way to do this is to give priority to the inventive, exploratory principle, not inadvertently like the drop-outs, who find themselves all too soon in an exploratory blind alley, but deliberately, gearing our inventiveness to the mainstream of our super-tribal existence.

Given the fact that each super-tribesman is free to choose which way he pursues the Stimulus Struggle, it remains to ask why he does not select the inventive solution more frequently. With the enormous exploratory potential of his brain lying idle and with his experience of inventive playfulness in childhood behind him, he should in theory favour this solution above all others. In any thriving super-tribal city *all* the citizens should be potential 'inventors'. Why, then, do so few of them indulge in active creativity, while the others are satisfied to enjoy their inventions second-hand, watching them on television, or are content to play simple games and sports with strictly limited possibilities for inventiveness? They all appear to have the necessary background for becoming childlike adults. The super-tribe, like a gigantic parent, protects them and cares for them, so why is it that they do not all develop bigger and better childlike curiosity?

Part of the answer is that children are subordinate to adults. Inevitably, dominant animals try to control the behaviour of their subordinates. Much as adults may love their children, they cannot help seeing them as a growing threat to their dominance. They know that with ultimate senility they will have to give way to them, but they do everything they can to postpone the evil day. There is therefore a strong tendency to suppress inventiveness in members of the community younger than oneself. An appreciation of the value of their 'fresh eyes' and their new creativeness works against this, but it is an uphill struggle. By the time the new generation has matured to the point where its members could be wildly inventive, childlike adults, they are already burdened with a heavy sense of conformity. Struggling against this as hard as they can, they in turn are then faced with the threat of another younger generation coming up beneath them, and the suppressive process repeats itself. Only those rare individuals who experience an unusual childhood, from this point of view, will be able to achieve a level of great creativity in adult life. How unusual does such a childhood have to be? It either has to be so suppressive that the growing child revolts against the traditions of its elders in a big way (many of our greatest creative talents were so-called delinquent children), or it has to be so un-suppressive that the heavy hand of conformity rests only lightly on its shoulder. If a child is strongly punished for its inventiveness (which, after all, is essentially rebellious in nature), it may spend the rest of its adult life making up for lost time. If a child is strongly rewarded for its inventiveness, then it may never lose it, no matter what pressures are brought to bear on it in later years. Both types can make a great impact in adult society, but the second will probably suffer less from obsessive limitations in his creative acts.

The vast majority of children will, of course, receive a more balanced mixture of punishment and reward for their inventiveness and will emerge into adult life with personalities that are both moderately creative and moderately conformist. They will become the adult-adults. They will tend to read the newspapers rather than make the news that goes into them. Their attitude to the childlike adults will be ambivalent; on the one hand they will applaud them for providing the much-needed sources of novelty, but on the other they will envy them. The creative talent will therefore find himself alternately praised and damned by society in a bewildering way, and will be constantly in doubt about his acceptance by the rest of the community.

Modern education has made great strides in encouraging inventiveness, but it still has a long way to go before it can completely rid itself of the urge to suppress creativity. It is inevitable that bright young students will be seen as a threat by older academics, and it requires great self-control for teachers to overcome this. The system is designed to make it easy, but their nature, as dominant males, is not. Under the circumstances it is remarkable that they manage to control themselves as well as they do. There is a difference here between the school level and the university level. In most schools the dominance of the master over his pupils is strongly and directly expressed, both socially and intellectually. He uses his greater experience to conquer their greater inventiveness. His brain has probably become more rigidified than theirs, but he masks this weakness by imparting large quantities of 'hard' facts. There is no argument, only instruction. (The situation is improving and there are, of course, exceptions, but this still applies as a general rule.)

At the university level the scene changes. There are many more facts to be handed down, but they are not quite so 'hard'. The student is now expected to question them and assess them, and eventually to invent new ideas of his own. But at both stages, the school and the university, there is something else going on beneath the surface, something that has little to do with the encouragement of intellectual expansion, but a great deal to do with the indoctrination of supertribal identity. To understand this we must look at what happened in simpler tribal societies.

In many cultures children at puberty have been subjected to impressive initiation ceremonies. They are taken away from their parents and kept in groups. They are then forced to undergo severe ordeals, often involving torture or mutilation. Operations are performed on their genitals, or their bodies may be scarred, burned, whipped, or stung with ants. At the same time they are instructed in the secrets of the tribe. When the rituals are over, they are accepted as adult members of the society.

Before we see how this relates to the rituals of modern education, it is important to ask what value these seemingly damaging activities have. In the first place they isolate the maturing child from its parents. Previously it could always run to them for comfort when in pain. Now, for the first time, the child must endure pain and fear in a situation in which its parents cannot be called upon for assistance. (The initiation ceremonies are usually performed in strict privacy by the tribal elders and the rest of the tribe is excluded.) This helps to smash the child's sense of dependence on its parents and shift its allegiance away from the family home and on to the tribal community as a whole. The fact that at the same time he is allowed to share the adult tribal secrets strengthens the process by giving a fabric to his new tribal identity. Secondly, the violence of the emotional experience accompanying his instruction helps to burn the details of the tribal teachings into his brain. Just as we find it impossible to forget the details of a traumatic experience, such as a motor accident, so the tribal initiate will remember to his dying day the secrets that were imparted to him on that frightening occasion. The initiation is, in a sense, contrived traumatic teaching. Thirdly, it makes it absolutely clear to the sub-adult that, although he is now joining the ranks of his elders, he is doing so very much in the role of a subordinate. The intense power which they exercised over him will also be vividly remembered.

Modern schools and universities may not sting their students with ants but, in many ways, the educational system today shows striking similarities to the earlier tribal initiation procedures. To start with, the children are taken away from their parents and put in the hands of super-tribal elders – the academics – who instruct them in the 'secrets' of the super-tribe. In many cultures they are still made to wear a separate uniform, to set them apart and strengthen their new allegiance. They may also be encouraged to indulge in certain rituals, such as singing school or college songs. The severe ordeals of the tribal initiation ceremony no longer leave physical scars. (German duelling scars never really caught on.) But physical ordeals of a less vicious sort have persisted almost everywhere until very recently, at least at the school level, in the form of buttock-caning. Like the genital mutilations of the tribal ceremonies, this form of punishment has always had a sexual flavour and cannot be dissociated from the phenomenon of Status Sex.

In the absence of a more violent form of ordeal stemming from the teachers, the older pupils frequently take over the role of 'tribal elders' and administer their own tortures on 'new boys'. These vary from place to place. At one school, for instance, newcomers are 'grassed', having bundles of grasses stuffed inside their clothing. At another they are 'stoned', being bent over a large stone and spanked. At another they are forced to run down a long corridor lined with older pupils who kick them as they pass. At yet another they are 'bumped', being held by arms and legs and banged on the ground as many times as their age in years. Alternatively on the day a new pupil wears his first school uniform, he may have his flesh pinched once for each new article of clothing, by each senior pupil. In rare cases, the ordeal is much more elaborate and may almost approach a full-scale tribal initiation ceremony. Even today, occasional deaths are reported as a result of these activities.

Unlike the primitive tribal situation, there is nothing to stop a tortured boy from complaining to his parents, but this hardly ever happens because it would bring shame to the boy in question. Many parents are not even aware of the trials their children are undergoing. The ancient practice of alienating a child from its family home has already begun to work its strange magic.

Although these unofficial initiation rites have persisted here and there, the official punishment of caning by teachers has recently lost ground, due to the pressure of public opinion and the revised ideas of certain teachers. But if official ordeal by physical means is disappearing, there always remains the alternative of the mental ordeal. Throughout virtually the whole of the modern educational system there now exists one powerful and impressive form of super-tribal initiation ceremony, which goes under the revealing name of 'examinations'. These are conducted in the heavy atmosphere of high ritual, with the pupils cut off from all outside assistance. Just as in the tribal ritual, no one can help them. They must suffer on their own. At all other times in their lives they can make use of books of reference, or discussions over difficult points, when they are applying their intelligence to a problem, but not during the private rituals of the dreaded examinations.

The ordeal is further intensified by setting a strict time limit and by crowding all the different examinations together in the short space of a few days or weeks. The overall effect of these measures is to create a considerable amount of mental torment, again recalling the mood of the more primitive initiation ceremonies of simple tribes.

When the final exams are over, at university level, the students who have 'passed the test' become qualified as special members of the adult section of the super-tribe. They don elaborate display robes and take part in a further ritual called the degree ceremony, in the presence of the academic elders wearing their even more impressive and dramatic robes.

The university student phase usually lasts for three years, which is a long time, as initiation ceremonies go. For some, it is too long. The isolation from parental assistance and the comforting social environment of the home, coupled with the looming demands of the examination ordeal, often proves too stressful for the young initiate. At British universities roughly twenty per cent of the undergraduates seek psychiatric assistance at some time during their three-year course of study. For some, the situation becomes unbearable and suicides are unusually frequent, the university rate being three to six times higher than the national average for the same age group. At Oxford and Cambridge universities, the suicide rate is seven to ten times higher.

Clearly the educational ordeals I have been describing have little to do with the business of encouraging and expanding childhood playfulness, inventiveness and creativity. Like the primitive tribal initiation ceremonies, they have instead to do with the indoctrination of super-tribal identity. As such they play an important cohesive role, but the development of the creative intellect is another matter altogether.

One of the excuses given for the ritual ordeals of modern education is that they provide the only way of ensuring that the students will absorb the huge mass of facts now available. It is true that detailed knowledge and specialist skills are necessary today, before an adult can even begin to be confidently inventive. Also, the examination ceremonies prevent cheating. Furthermore, it could be claimed that students should deliberately be subjected to stress to test their stamina. The challenges of adult life are also stressful, and if a student cracks up under the strain of educational ordeals, then he probably was not equipped to withstand the post-educational pressures either. These arguments are plausible, and yet one still senses the crushing of creative potentials under the heavy boot of educational ritual procedures. It is undeniable that the present system is a considerable advance over earlier educational methods, and that for those who survive the ordeals there is a great deal of exploratory nourishment to be gained. Our super-tribes today contain more successful childlike adults than ever before. Yet despite this, in many spheres there still exists an oppressive atmosphere of emotional resistance to radically new, inventive ideas. Dominant individuals encourage minor inventiveness in the form of new variations on old themes, but resist major inventiveness that leads to entirely new themes.

To give an example: it is astonishing the way we go on and on trying to improve something as primitive as the engine used in present-day motor vehicles. There is a strong chance that by the twenty-first century it will have become as obsolete as the horse and cart is today. That this is only a strong chance and not a complete certainty is due entirely to the fact that at this moment all the best brains in the profession are busily absorbed by the minor inventive problems of how to achieve minute improvements in the performance of the existing machinery, rather than searching for something entirely new.

This tendency towards short-sightedness in adult exploratory behaviour is a measure of the insecurity of a peaceful society. Perhaps, as we move farther into the atomic age, we shall reach such peaks of super-tribal security, or hit such depths of super-tribal panic, that we shall become increasingly exploratory, inventive and creative.

It will not be an easy struggle, however, and recent events at universities all over the world bear this out. The improved educational systems have already been so effective that many students are no longer prepared to accept without question the authority of their elders. The community was not ready for this and has been taken by surprise. The result is that when groups of students indulge in noisy protest, society is outraged. The educational authorities are horrified. The ingratitude of it all! What has gone wrong?

If we are ruthlessly honest with ourselves, the answer is not hard to find. It is contained in the official doctrines of these same educational authorities. As they face the upheaval, they must contemplate the uncomfortable fact that they have brought it on themselves. They literally asked for it. 'Think for yourselves,' they said, 'be resourceful, be active, be inventive.' Contradicting themselves in the same breath, they added: 'But do it on *our* terms, in *our* way, and above all abide by *our* rituals.'

It should be obvious, even to a senile authority, that the more the first message is obeyed, the more the second will be ignored. Unfortunately the human animal is remarkably good at blinding itself to the obvious if it happens to be particularly unappealing, and it is this self-blinding process that has caused so many of the present difficulties.

When they called for increased resourcefulness and inventiveness, the authorities did not anticipate the magnitude of the response that was to follow, and it quickly got out of control. They did not seem to realize that they were encouraging something which already had a strong biological backing. They mistakenly treated resourcefulness and a sense of creative responsibility as properties alien to the young brain, when in fact they were hidden there all the time, only waiting to burst out if they got the chance.

The old-fashioned educational systems, as I have already pointed out, had done their best to suppress these properties, demanding a much greater obedience to the established rules of the elders. They had rigorously imposed parrot-fashion learning of rigid dogmas. Inventiveness had been forced to fight its own battles, pushing its way to the surface only in exceptional, isolated individuals. When it did manage to break through, however, its value to society was indisputable, and this led eventually to the present-day movement on the part of the establishment actively to encourage it. Approaching the matter rationally, they saw inventiveness and creativity as immense aids to greater social progress. At the same time, the deeply ingrained urge of these super-tribal authorities to retain their vice-like grip on the social order still persisted, making them oppose the very trend which they were now officially supporting. They entrenched ever more firmly, moulding society into a shape that was guaranteed to resist the new waves of inventiveness that they themselves had unleashed. A collision was inevitable.

The initial response of the establishment, as the mood of experimentation grew, was one of tolerant amusement. Cautiously viewing the younger generation's increasingly daring assaults on the accepted traditions of the arts, literature, music, entertainment and social custom, they kept their distance. Their tolerance collapsed, however, as this trend spread into the more threatening areas of politics and international affairs.

As isolated, eccentric thinkers grew into a massive, querulous crowd, the establishment switched hurriedly to its most primitive form of response – attack. The young intellectual, instead of being patted tolerantly on the head, found himself struck on the skull by a police baton. The lively brains that society had so carefully nurtured were soon suffering not from strain but from concussion.

The moral for the authorities is clear: do not give creative liberties unless you expect people to take them. The young human animal is not a stupid, idle creature that has to be driven to creativity; it is a fundamentally creative being that in the past has been made to appear idle by the suppressive influences exerted on it from above. The establishment's reply is that dissenting students are bent, not on positive innovation, but on negative disruption. Against this, however, it can be argued that these two processes are very closely related and that the former only degenerates into the latter when it finds itself blocked.

The secret is to provide a social environment capable of absorbing as much inventiveness and novelty as it sets out to encourage in the first place. As the super-tribes are constantly swelling in size and the human zoo is becoming ever more cramped and crowded, this requires careful and imaginative planning. Above all, it calls for considerably more insight into the biological demands of the human species, on the part of politicians, administrators and city planners, than has been evident in the recent past.

The more closely one looks at the situation, the more alarming it becomes. Well-meaning reformers and organizers busily work towards what they consider to be improved living conditions, never for one moment doubting the validity of what they are doing. Who, after all, can deny the value of providing more houses, more flats, more cars, more hospitals, more schools and more food? If perhaps there is a

degree of sameness about all these bright new commodities, this cannot be helped. The human population is growing so fast that there is
not enough time or space to do it any better. The snag is that while,
on the one hand, all those new schools are bursting with pupils,
inventiveness at the ready, dead set to *change* things, the other
new developments are conspiring to render startling new innovations
more and more impossible. In their ever-expanding and highly
regimented monotony, these developments unavoidably favour widespread indulgence in the more trivial solutions to the Stimulus
Struggle. If we are not careful, the human zoo will become increasingly
like a Victorian menagerie, with tiny cages full of twitching, pacing
captives.

Some science-fiction writers take the pessimistic view. When depicting the future, they portray it as an existence in which human individuals are subjected to a suffocating degree of increased uniformity,
as if new developments have brought further invention almost to a
standstill. Everyone wears drab tunics and automation dominates the
environment. If new inventions do occur, they only serve to squeeze
the trap tighter around the human brain.

It could be argued that this picture merely reflects the paucity of the
writers' imaginations, but there is more to it than that. To some extent
they are simply exaggerating the trend that can already be detected in
present-day conditions. They are responding to the relentless growth
of what has been called the 'planner's prison'. The trouble is that as
new developments in medicine, hygiene, housing and food-production
make it possible to cram more and more people efficiently into a given
space, the creative elements in society become more and more side-
tracked into problems of quantity rather than quality. Precedence is
given to those inventions that permit further increases in repetitive
mediocrity. Efficient homogeneity takes precedence over stimulating
heterogeneity.

As one rebel planner pointed out, a straight path between two
buildings may be the most efficient (and the cheapest) solution, but
that does not mean that it is the *best* path from the point of view of
satisfying human needs. The human animal requires a spatial territory
in which to live that possesses unique features, surprises, visual oddities, landmarks and architectural idiosyncrasies. Without them it can
have little meaning. A neatly symmetrical, geometric pattern may be
useful for holding up a roof, or for facilitating the prefabrication of
mass-produced housing-units, but when such patterning is applied at
the landscape level, it is going against the nature of the human animal.
Why else is it so much fun to wander down a twisting country lane?
Why else do children prefer to play on rubbish dumps or in derelict

buildings, rather than on their immaculate, sterile, geometrically arranged playgrounds?

The current architectural trend towards austere design-simplicity can easily get out of hand and be used as an excuse for lack of imagination. Minimal aesthetic statements are only exciting as contrasts to other, more complex statements. When they come to dominate the scene, the results can be extremely damaging. Modern architecture has been heading this way for some time, strongly encouraged by the human zoo planners. Huge tower blocks of repetitive, uniform apartments have proliferated in many cities as a response to the housing demands of the swelling super-tribal populations. The excuse has been slum clearance, but the results have all too often been the creation of the super-slums of the immediate future. In a sense they are worse than nothing because, since they falsely give the impression of progress, they create complacency and deaden the chance of a genuine advance.

The more enlightened animal zoos have been getting rid of their old monkey houses. The zoo directors saw what was happening to the inmates, and realized that putting more hygienic tiles on the walls and improving the drainage was no real solution. The directors of human zoos, faced with mushrooming populations, have not been so far-sighted. The outcome of their experiments in high-density uniformity is now being assessed in the juvenile courts and psychiatrists' consulting rooms. On some housing estates it has even been recommended that prospective tenants of upper-storey apartments should undergo psychiatric examination *before* taking up residence, to ascertain whether, in the psychiatrist's opinion, they will be able to stand the strain of their brave new way of life.

This fact alone should provide sufficient warning to the planners, revealing to them clearly the enormity of the folly they are committing, but as yet there is little sign that they are heeding such warnings. When confronted with the shortcomings of their endeavours, they reply that they have no alternative; there are more and more people and they have to be housed. But somehow alternatives must be found. The whole nature of city-complexes must be re-examined. The harassed citizens of the human zoo must in some way be given back the 'village-community' feeling of social identity. A genuine village, seen from the air, looks like an organic growth, not a piece of slide-rule geometry, a point which most planners seem studiously to ignore. They fail to appreciate the basic demands of human territorial behaviour. Houses and streets are not primarily for looking at, like set pieces, but for moving about in. The architectural environment should make its impact second by second and minute by minute as we travel along our territorial tracks, the pattern changing subtly with each new line of vision. As we turn a corner, or open a door, the last thing our navigational sense wants to be faced with is a spatial configuration that drearily duplicates the one we have just left. All too frequently, however, this is precisely what happens, the architectural planner having loomed over his drawing-board like a bomber-pilot sighting a target, rather than attempting to project himself down as a small mobile object travelling around *inside* the environment.

These problems of repetitive monotony and uniformity do, of course, permeate almost all aspects of modern living. With the ever-increasing complexity of the human zoo environment, the dangers of intensified social regimentation are mounting daily. While organizers struggle to encase human behaviour in a more and more rigid framework, other trends work in the opposite direction. As we have seen, the steadily improving education of the young and the growing affluence of their elders both lead to a demand for more and more stimulation, adventure, excitement and experimentation. If the modern world fails to permit these trends, then tomorrow's super-tribesmen will fight hard to change that world. They will have the training and the time and the exploratory energy to do so, and somehow they will manage it. If they feel themselves trapped in a planner's prison they will stage a prison riot. If the environment does not permit creative innovations, they will smash it in order to be able to start again. This is one of the greatest dilemmas our societies face. To resolve it is our formidable task for the future.

Unfortunately we tend to forget that we are animals with certain specific weaknesses and certain specific strengths. We think of ourselves as blank sheets on which anything can be written. We are not. We come into the world with a set of basic instructions and we ignore or disobey them at our peril.

The politicians, the administrators and the other super-tribal leaders are good social mathematicians, but this is not enough. In what promises to be the ever more crowded world of the future, they must become good biologists as well, because somewhere in all that mass of wires, cables, plastics, concrete, bricks, metal and glass which they control, there is an animal, a human animal, a primitive tribal hunter, masquerading as a civilized, super-tribal citizen and desperately struggling to match his ancient inherited qualities with his extraordinary new situation. If he is given the chance he may yet contrive to turn his human zoo into a magnificent human game-park. If he is not, it may proliferate into a gigantic lunatic asylum, like one of the hideously cramped animal menageries of the last century.

For us, the super-tribesmen of the twentieth century, it will be interesting to see what happens. For our children, however, it will be more than merely interesting. By the time they are in charge of the situation, the human species will no doubt be facing problems of such magnitude that it will be a matter of living or dying.

CHAPTER REFERENCES

It is impossible to list all the many works that have been of assistance in writing *The Human Zoo*. I have therefore included only those which have either been important in providing information on a specific point, or are of particular interest for further reading. They are arranged below on a chapter-by-chapter and topic-by-topic basis. From the names and dates given, it is possible to trace the full references in the bibliography.

1 TRIBES AND SUPER-TRIBES

Home range of prehistoric man: Washburn and DeVore, 1962.
Prehistoric man: Boule and Vallois, 1957. Clark and Piggott, 1965. Read, 1925. Tax, 1960. Washburn, 1962.
Farming origins: Cole, 1959. Piggott, 1965. Zeuner, 1963.
Urban origins: Piggott, 1961, 1965. Smailes, 1953.
Mourning dress: Crawley, 1931.

2 STATUS AND SUPER-STATUS

Behaviour of baboons: Hall and DeVore, 1965.
Dominance patterns: Caine, 1960.
Status seekers: Packard, 1960.
Mimicry: Wickler, 1968.
Suicide: Berelson and Steiner, 1964. Stengel, 1964. Woddis, 1957.
Re-direction of aggression: Bastock, Morris and Moynihan, 1953.
Cruelty to animals: Jennison, 1937. Turner, 1964.

3 SEX AND SUPER-SEX

Sexual behaviour: Beach, 1965. Ford and Beach, 1952. Hediger, 1965. Kinsey *et al.*, 1948, 1953. Morris, 1956, 1964, 1966 and 1967.
Masturbation: Kinsey *et al.*, 1948.

Religious ecstasy: Bataille, 1962.
Boredom: Berlyne, 1960.
Displacement activities: Tinbergen, 1951.
Monkey prostitutes: Zuckerman, 1932.
Feline display: Leyhausen, 1956.
Sexual mimicry: Wickler, 1967.
Status sex: Russell and Russell, 1961.
Phallic symbols: Knight and Wright, 1957. Boullet, 1961.
Maltese cross: Adams, 1870.

4 IN-GROUPS AND OUT-GROUPS

Aggression and War: Ardrey, 1963, 1967. Berkowitz, 1962. Carthy and Ebling, 1964. Lorenz, 1963. Richardson, 1960. Storr, 1968.
Races of man: Broca, 1864. Coon, 1963, 1966. Montagu, 1945. Pickering, 1850. Smith, 1968.
Racial conflict: Berelson and Steiner, 1964. Segal, 1966.
Population levels: Fremlin, 1965.

5 IMPRINTING AND MAL-IMPRINTING

Imprinting in animals: Lorenz, 1935. Sluckin, 1965.
Mal-imprinting in animals: Hediger, 1950, 1965 (zoo animals). Morris, 1964 (zoo animals). Scott, 1956, 1958 (dogs). Scott and Fuller, 1965 (dogs). Whitman, 1919 (pigeons).
Social isolation in monkeys: Harlow and Harlow, 1962.
Human infant bonding: Ambrose, 1960. Brackbill and Thompson, 1967.
Pair-bonding: Morris, 1967.
Fetishism: Freeman, 1967. Hartwich, 1959.
Homosexuality: Morris, 1952, 1954, 1955. Schutz, 1965. West, 1968.
Pet-keeping: Morris and Morris, 1966.

6 THE STIMULUS STRUGGLE

Zoo animals: Appelman, 1960. Hediger, 1950. Inhelder, 1962. Lang, 1943. Lyall-Watson, 1963. Morris, 1962, 1964, 1966.
Boredom and stress: Berlyne, 1960.
Aesthetics: Morris, 1962.
Bestiality: Kinsey *et al.*, 1948, 1953.

CHAPTER REFERENCES

Super-normal stimuli: Morris, 1956. Tinbergen, 1951, 1953.
Children's drawings: Morris, 1962.
Costume: Laver, 1950, 1952, 1963.
Cut-off: Chance, 1962.

7   THE CHILDLIKE ADULT

Chimpanzee curiosity: Morris, 1962. Morris and Morris, 1966.
Initiation ceremonies: Cohen, 1964.
School rituals: Opie and Opie, 1959.

INTIMATE BEHAVIOUR

CONTENTS

INTRODUCTION

To be intimate means to be close, and I must make it clear at the outset that I am treating this literally. In my terms, then, the act of intimacy occurs whenever two individuals come into bodily contact. It is the nature of this contact, whether it be a handshake or a copulation, a pat on the back or a slap in the face, a manicure or a surgical operation, that this book is about. Something special happens when two people touch one another physically, and it is this something that I have set out to study.

My method has been that of the zoologist trained in ethology, that is, in the observation and analysis of animal behaviour. I have limited myself, in this case, to the human animal, and have given myself the task of observing what people do – not what people say, or even what they say they do, but what they actually do.

The method is simple enough – merely to use one's eyes – but the task is not as easy as it sounds. The reason is that, despite the self-discipline, words persist in filtering through and preconceived ideas repeatedly get in the way. It is hard for an adult human being to look at a piece of human behaviour as if he were seeing it for the very first time, but that is what the ethologist must attempt to do if he is to bring new understanding to the subject. The more familiar and commonplace the behaviour, of course, the worse the problem becomes; in addition, the more intimate the behaviour, the more emotionally charged it becomes, not only for the performers, but also for the observer.

Perhaps this is why, despite their importance and interest, so few studies have been made of commonplace human intimacies. It is far more comfortable to study something as remote from human involvement as, say, the territorial scent-marking behaviour of the giant panda, or the food-burying behaviour of the green acouchi, than it is to tackle scientifically and objectively something as 'well known' as the human embrace, the mother's kiss or the lover's caress. But in a social environment that is ever more crowded and impersonal, it is becoming increasingly important to reconsider the value of close personal relationships, before we are driven to ask the forlorn question, 'Whatever happened to love?' Biologists are often wary of using this word 'love', as if it reflected no more than some kind of culturally inspired romanticism. But love is a biological fact. The subjective, emotional rewards and agonies associated with it may be deep and mysterious, and difficult to deal with scientifically, but the outward signs of love – the actions of loving – are readily observable, and there is no reason why they should not be examined like any other type of behaviour.

It has sometimes been said that to explain love is to explain it away, but this is quite unjustified. In a way, it is an insult to love, implying that, like an ageing, cosmetic-caked face, it cannot stand scrutiny under a bright light. But there is nothing illusory about the powerful process of the formation of strong bonds of attachment between one individual and another. This is something we share with thousands of other animal species – in our parent-offspring relationships, our sexual relationships and our closest friendships.

Our intimate encounters involve verbal, visual and even olfactory elements, but, above all, loving means touching and body contact. We often talk about the way we talk, and we frequently try to see the way we see, but for some reason we have rarely touched on the way we touch. Perhaps touch is so basic – it has been called the mother of senses – that we tend to take it for granted. Unhappily, and almost without our noticing it, we have gradually become less and less touchful, more and more distant, and physical untouchability has been accompanied by emotional remoteness. It is as if the modern urbanite has put on a suit of emotional armour and, with a velvet hand inside an iron glove, is beginning to feel trapped and alienated from the feelings of even his nearest companions.

It is time to take a closer look at this situation. In doing so, I shall endeavour to keep my opinions to myself, and to describe human behaviour as seen through the objective eyes of a zoologist. The facts, I trust, will speak for themselves, and will speak loudly enough for the reader to form his own conclusions.

I

THE ROOTS OF INTIMACY

A S AN ADULT human being, you can communicate with me in a variety of ways. I can read what you write, listen to the words you speak, hear your laughter and your cries, look at the expressions on your face, watch the actions you perform, smell the scent you wear and feel your embrace. In ordinary speech we might refer to these interactions as 'making contact', or 'keeping in touch', and yet only the last one on the list involves bodily contact. All the others operate at a distance. The use of words like 'contact' and 'touch' to cover such activities as writing, vocalization and visual signalling is, when considered objectively, strange and rather revealing. It is as if we are automatically accepting that bodily contact is the most basic form of communication.

There are further examples of this. For instance, we often refer to 'gripping experiences', 'touching scenes' or 'hurt feelings', and we talk of a speaker who 'holds his audience'. In none of these cases is there an actual physical grip, touch, feel or hold, but this does not seem to matter. The use of physical-contact metaphors provides a satisfying way of expressing the various emotions involved in the different contexts.

The explanation is simple enough. In early childhood, before we could speak or write, body contact was a dominant theme. Direct physical interaction with the mother was all-important and it left its mark. Still earlier, inside the womb, before we could see or smell, leave alone speak or write, it was an even more powerful element in our lives. If we are to understand the many curious and often strongly inhibited ways in which we make physical contact with one another as adults, then we must start by returning to our earliest beginnings, when we were no more than embryos inside our mothers' bodies. It is the intimacies of the womb, which we hardly ever consider, that will help us to understand the intimacies of childhood, which we tend to ignore because we take them so much for granted, and it is the intimacies of childhood, re-examined and seen afresh, that will help us to explain the intimacies of adult life, which so often confuse, puzzle and even embarrass us.

The very first impressions we receive as living beings must be sensations of intimate body contact, as we float snugly inside the protective wall of the maternal uterus. The major input to the developing nervous system at this stage therefore takes the form of varying sensations of touch, pressure and movement. The entire skin surface of the unborn child is bathed in the warm uterine liquid of the mother. As the child grows and its swelling body presses harder against the mother's tissues, the soft embrace of the enveloping bag of the womb becomes gradually stronger, hugging tighter with each passing week. In addition, throughout this period the growing baby is subjected to the varying pressure of the rhythmic breathing of the maternal lungs, and to a gentle, regular swaying motion whenever the mother walks.

Towards the end of pregnancy, in the last three months before birth, the baby is also capable of hearing. There is still nothing to see, taste or smell, but things that go bump in the night of the womb can be clearly detected. If a loud, sharp noise is made near to the mother's belly, it startles the baby inside and makes it jump. The movement can easily be recorded by sensitive instruments and may even be strong enough for the mother to feel it herself. This means that during this period before birth the baby is undoubtedly capable of hearing the steady thump of the maternal heartbeat, 72 times every minute. It will become imprinted as the major sound-signal of life in the womb.

These, then, are our first real experiences of life – floating in a warm fluid, curling inside a total embrace, swaying to the undulations of the moving body and hearing the beat of the pulsing heart. Our prolonged exposure to these sensations in the absence of other, competing stimuli leaves a lasting impression on our brains, an impression that spells security, comfort and passivity.

This intra-uterine bliss is then rudely and rapidly shattered by what must be one of the most traumatic experiences in our entire lives – the act of being born. The uterus, in a matter of hours, is transformed from a cosy nest into a straining, squeezing sac of muscle, the largest and most powerful muscle in the whole human body, athlete's arms included. The lazy embrace that became a snug hug now becomes a crushing constriction. The newly delivered baby displays, not a happy, welcoming grin, but the strained, tightly contorted facial expressions of a desperate torture victim. Its cries, which are such sweet music to the anxiously waiting parents, are in reality nothing short of the wild screams of blind panic, as it is exposed to the sudden loss of intimate body contact.

At the moment of birth the baby appears floppy, like soft, wet rubber, but almost at once it makes a gasping action and takes its first breath. Then, five to six seconds later, it starts to cry. Its head, legs and arms begin to move about with increasing intensity and for the next thirty minutes it continues to protest in irregular outbursts of limb-thrashing, grasping, grimacing and screaming, after which it usually subsides exhausted into a long sleep.*

The drama is over for the moment, but when the baby reawakens it is going to need a great deal of maternal care, contact and intimacy to compensate it for the lost comforts of the womb. These post-uterine substitutes are provided by the mother, or those who are helping her, in a number of ways. The most obvious one is the replacement of the embrace of the womb by the embrace of the mother's arms. The ideal maternal embrace enfolds the baby, bringing as much of its body surface into contact with the mother as is possible without restricting breathing. There is a great difference between embracing the baby and merely holding it. An awkward adult who holds the infant with a minimum of contact will soon discover how dramatically this reduces the comforting value of the action. The maternal chest, arms and hands must do their best to re-create the total engulfment of the lost womb.

Sometimes the embrace alone is not enough. Other womb-like elements have to be added. Without knowing quite why, the mother starts to rock her child gently from side to side. This has a strong soothing effect, but if it fails she may get up and start slowly walking back and forth with the baby cradled in her arms. From time to time she may joggle it up and down briefly. All these intimacies have a comforting influence on a restless or crying child, and they do so, it seems, because of the way they copy certain of the rhythms experienced earlier by the unborn baby. The most obvious guess is that they succeed by re-creating the gentle swaying motions felt inside the womb whenever the mother walked about during her pregnancy. But there is a catch to this. The speed is wrong. The rate of rocking is considerably slower than the rate of normal walking. Furthermore, 'walking the baby' is also done at a pace that is much slower than an average walk of the ordinary kind.

Experiments were carried out recently to ascertain the ideal rocking speed for a cradle. At very low or very high speeds the movements had little or no soothing effect, but when the mechanically operated cradle was set at between sixty and seventy rocks per minute there was a striking change, the babies under observation immediately becoming much calmer and crying much less. Although mothers vary somewhat in the speed at which they rock their babies when they hold them in their arms, the typical maternal rocking rate is much the same as that in the experiments, and the pace when 'walking the baby' is also in that

---

* New birthing techniques that leave the naked newborn in close, gentle contact with its mother's body for some time after delivery have been shown to reduce dramatically the struggling and screaming described here.

region. An average walking rate under ordinary circumstances, however, usually exceeds a hundred paces per minute.

It seems, therefore, that although these comforting actions may well soothe by virtue of the way they copy the swaying motions felt inside the womb, the speed at which they are carried out requires some other explanation. There are two rhythmic experiences available to the unborn child, apart from the mother's walking: the steady rise and fall of her chest as she breathes, and the steady thump of her heartbeat. The breathing rate is much too slow to be considered, being roughly between ten and fourteen respirations per minute, but the heartbeat speed, at 72 beats per minute, looks like the ideal candidate. It appears as if this rhythm, whether heard or felt, is the vital comforter, reminding the baby vividly of the lost paradise of the womb.

There are two other pieces of evidence that support this view. First, the recorded heartbeat sound, if played experimentally to babies at the correct speed, also has a calming effect, even without any rocking or swaying movement. If the same sound is played faster, at over a hundred beats per minute – that is, at the speed of normal walking – it immediately ceases to have any calming effect. Second, as I reported in *The Naked Ape*, careful observations have revealed that the vast majority of mothers hold their babies in such a way that their infants' heads are pressed to the left breast, close to the maternal heart. Even though these mothers are unaware of what they are doing, they are nevertheless successfully placing their babies' ears as close as possible to the source of the heartbeat sound. This applies to both right-handed and left-handed mothers, so that the heartbeat explanation appears to be the only one that fits.

Clearly this is susceptible to commercial exploitation by anyone who takes the trouble to manufacture a cradle that can be mechanically rocked at heartbeat speed, or that is equipped with a small machine that plays a non-stop amplified recording of the normal heartbeat sound. A de luxe model incorporating both devices would no doubt be even more effective, and many a harassed mother could simply switch on and relax as it automatically and relentlessly calmed her baby to sleep, just as her washing machine so efficiently deals with the baby's dirty clothing.

Inevitably it is only a matter of time before such machines do appear on the market, and undoubtedly they will do a great deal to assist the busy mother of modern times, but there is an inherent danger in their use if it becomes excessive. It is true that mechanical calming is better than no calming, both for the mother's nerves and for the baby's well-being, and where demands on the mother's time are so heavy that she has no other choice, then mechanical calming will certainly be advantageous. But old-fashioned maternal calming is always going to be better than its mechanical replacement. There are two reasons for this. First, the mother does more than the machine could ever do. Her comforting actions are more complex and contain special features that we have yet to discuss. Second, the intimate interaction between mother and child that occurs whenever she comforts it by carrying, embracing and rocking it provides the important foundation for the strong bond of attachment that will soon grow between them. True, during its first months after birth the baby will respond positively to any friendly adult. It accepts any intimacies from any individual who offers them, regardless of who they are. After a year has passed, however, the child will have learned its own mother and will have started to reject intimacies from strangers. This change is known to occur around the fifth month in most babies, but it does not happen overnight and there is great variability from child to child. It is therefore difficult to predict with certainty the exact moment at which the infant will begin to respond selectively to its own mother. It is a critical time, because the strength and quality of the later bond of attachment will depend on the richness and intensity of the body-contact behaviour that occurs between mother and infant at just this threshold phase.

Obviously, excessive use of mechanical mothers during this vital stage could be dangerous. Some mothers imagine that it is the provision of food and other similar rewards that make the baby become attached to them, but this is not so. Observations of deprived children and careful experiments with monkeys have shown conclusively that it is instead the tender intimacies with the soft body of the mother which are vital in producing the essential bond of attachment that will be so important for successful social behaviour in later life. It is virtually impossible to give too much body-loving and contact during these critical early months, and the mother who ignores this fact will suffer for it later, as will her child. It is difficult to comprehend the warped tradition which says that it is better to leave a small baby to cry so that it does not 'get the better of you', which is encountered all too often in our civilized cultures.

To counterbalance this statement, however, it must be added that when the child is older, the situation changes. It then becomes possible for the mother to be over-protective and to hold her child back just when it should be striking out and becoming more independent. The worst twist that can happen is for a mother to be under-protective, strict and disciplinary with a tiny infant, and *then* over-protective and clinging with an older child. This completely reverses the natural order of bond development, and sadly it is a sequence that is frequently observed today. If an older child, or adolescent, 'rebels', this twisted pattern of rearing is very likely to be found lurking in the background.

Unfortunately, by the time this happens it is a little late in the day to correct the early damage that has been done.

The natural sequence I have described here – love first, freedom later – is basic not only to man, but also to all the other higher primates. Monkey mothers and ape mothers keep up the body-contact intimacies non-stop from the moment of birth for many weeks. They are greatly aided, of course, by the fact that baby monkeys and apes are strong enough to cling to them for long periods of time without assistance. In the great apes, such as the gorilla, the babies may take a few days to get started as active clingers, but after that, despite their weight, they manage it with remarkable tenacity. The smaller monkeys cling from the moment of birth, and I have even seen one half-born baby clutching tightly with its hands to the body of its mother during delivery, while the rear part of it was still inside the womb.

The human baby is much less athletic. Its arms are weaker and its short-toed feet are clingless. It therefore poses a much greater problem for the human mother. Throughout the early months it is she who must perform all the physical actions that serve to keep her and her baby in bodily contact. Only a few fragments of the infant's ancestral clinging pattern remain, rudimentary reminders of its ancient, evolutionary past, and even these are of no practical use today. They last for little more than two months after birth and are known as the grasp reflex and the Moro reflex.

The grasp reflex arrives early, the sixth-month-old foetus already having a strong grip. At birth, stimulation of the palm results in a tight clasping of the hand, and the grasp is strong enough for an adult to lift the whole weight of the baby's body in this way. Unlike the young monkey, however, this clinging cannot be sustained for any length of time.

The Moro reflex can be demonstrated by sharply and swiftly lowering the baby through a short distance – as if it were being dropped – while supporting it under its back. The infant's arms quickly fling outwards, with the hands opening and the fingers spreading wide. Then the arms come together again, as if groping for an embrace to steady itself. Here we can see clearly the ghost of the ancestral primate clinging action that is employed so effectively by every healthy young monkey. Recent studies have shown this even more clearly. If the baby feels itself dropping down while at the same time its hands are held and allowed to grasp, its reaction is not first to fling its arms wide before making the embracing action, but simply to go straight into the tight clinging response. This is precisely what a startled young monkey would do if it were grasping lightly on to the fur of its resting mother, who then suddenly leapt up in alarm. The infant monkey would instantly tighten its grip in this way, ready to be carried off rapidly by the mother to safety. The human baby, until it is eight weeks old, still has enough of the monkey left in it to show us a remnant of this response.

From the human mother's point of view, however, these 'monkey' reactions are of no more than academic interest. They may intrigue zoologists, but they do nothing in practical terms to lighten the parental burden. How, then, can she deal with the situation? There are several alternatives. In most so-called primitive cultures, the baby is almost constantly in touch with the mother's body during its early months. When the mother is resting, the baby is held all the time, either by herself or by someone else. When she sleeps, it shares her bed. When she works or moves about, it is carried strapped firmly to her body. In this way she manages to provide the almost non-stop contact typical of other primates. But modern mothers cannot always go to such lengths.

An alternative is to swathe the unheld baby in a sheath of clothes. If the mother cannot offer the baby the snug embrace of her arms or close contact with her body, day and night, hour after hour, she can at least provide it with the snug embrace of a wrapping of smooth, soft clothing, as an aid to replacing the lost engulfment of the womb. We usually think of clothing the baby merely as an act of keeping it warm, but there is more to it than this. The embrace of the material, as it wraps around and makes contact with the body surface of the infant, is equally important. Whether this wrapping should be loose or firm, however, remains a hotly debated point. Cultures vary considerably in their attitude towards the ideal tightness of this post-natal cloth-womb.

In the Western world today, tight swaddling is generally frowned upon, and the baby, even when new-born, is wrapped only lightly, so that it can move its body and limbs freely if it feels so inclined. Experts have expressed the fear that 'it might cramp the child's spirit' to wrap it more firmly. The vast majority of Western readers would instantly agree with this comment, but it does bear looking at more closely. The ancient Greeks and Romans swaddled their babies, yet even the most fanatical anti-swaddler would have to admit that there were quite a few uncramped spirits among their number. British babies, up to the end of the eighteenth century, were swaddled, and many Russian, Yugoslav, Mexican, Lapp, Japanese and American Indian babies are still swaddled at the present day. Recently the matter was examined scientifically, with both swaddled and unswaddled babies checked for discomfort by sets of sensitive instruments. The conclusions were that swaddling did in fact make the babies less fretful, as was shown by reduced heartbeat rate, breathing rate and crying frequency. Sleep, on the other hand, was increased. Presumably this is because the tight wrapping is more reminiscent of the tight womb-hug experienced by the advanced foetus during the final weeks of its gestation.

If this appears to be entirely in favour of the swaddlers, it should nevertheless be remembered that even the bulkiest, most belly-swelling foetus is never so tightly hugged by the womb that it cannot manage occasional kicks and struggles. Any mother who has felt these movements inside her will be aware that she is not 'swaddling' her unborn infant to the point of total immobility. A moderately close swaddling after birth is therefore probably more natural than the really tight binding that is applied in some cultures. Furthermore, swaddlers tend to prolong unnecessarily the close wrapping of their infants, well beyond the point where it is advisable. It may be helpful during the earliest weeks, but if it is extended over a period of months it may start to interfere with the healthy growing processes of muscular and postural development. Just as the foetus has to leave the real womb, so the new-born must soon leave the cloth-womb, if it is not to become 'overdue' at its next stage of maturation. We normally speak of babies being premature or overdue only in relation to the moment of their birth, but it is useful to apply this same concept to the later stages of childhood development as well. At each phase, from infancy to adolescence, there are relevant forms of intimacy, of bodily contact and caring, that should occur between parent and child if the offspring is to pass through the various stages successfully. If the intimacy offered by the parent at any particular stage is either too advanced or too retarded for that particular stage, then trouble may ensue later.

Up to this point we have been looking at some of the ways in which the mother helps her baby to relive some of its womb-time intimacies, but it would be wrong to give the impression that comforts during the early post-birth phase are no more than prolongations of foetal comforts. Such extensions provide only one part of the picture. Other interactions are taking place at the same time. The baby stage has its own, new forms of comfort to be added. They include fondling, kissing and stroking by the mother, and the cleaning of the baby's body surface with gentle, tactile manipulations such as rubbing, wiping and other mild frictions. Also, there is more to the embrace than merely embracing. In addition to offering the overall enveloping pressure of her arms, the mother frequently does something else. She pats the baby rhythmically with one of her hands. This patting action is restricted largely to one region of the baby's body, namely the back. It is delivered at a characteristic speed, and with a characteristic strength, neither too weak nor too strong. To call it a 'burping' action is misleading. It is a much more widespread and basic response of the mother's, and is not limited to that one specific form of infantile discomfort. Whenever the baby seems to need a little extra comforting, the mother embellishes her simple embrace by the addition of back-patting. Frequently she adds swaying or rocking movements at the same time, and often coos or croons softly with her mouth held near to the baby's head. The importance of these early comforting actions is considerable, for, as we shall see later, they reappear in many forms, sometimes obvious, sometimes heavily disguised, in the various intimacies of adult human life. They are so automatic to the mother that they are seldom thought about or discussed, with the result that the transformed roles they play in later life are usually overlooked.

In origin, the patting action is what students of animal behaviour refer to as an intention movement. This can best be illustrated by giving an animal example. When a bird is about to fly, it bobs its head as part of the action of taking off. During evolution this head-bobbing may become exaggerated as a signal to other birds that it is about to depart. It may perform vigorous, repeated head-bobs for some time before actually taking off, giving its companions the warning message that it is about to leave and enabling them to be ready to accompany it. In other words, it signals its intention to fly, and the head-bobbing is referred to as an intention movement. Patting by human mothers appears to have evolved as a special contact signal in a similar way, being a repeated intention movement of tight clinging. Each pat of the mother's hand says, 'See, this is how I will cling tightly to you to protect you from danger, so relax, there is nothing to worry about.' Each pat repeats the signal and helps to soothe the baby. But there is more to it than this. Again, the example of the bird can help. If the bird is mildly alarmed, but not sufficiently to fly away, it can alert its companions simply by giving a few mild head-bobs, but without actually taking off. In other words, the intention-movement signal can be given by itself, without being carried through into the full action of flight. This is what has happened with the human patting action. The hand pats the back, then stops, then pats again, then stops again. It is not carried through into the full danger-protection clinging action. So the message from the mother to the baby reads not only 'Don't worry, I will cling to you like this if danger threatens,' but also 'Don't worry, there is no danger, or I would be clinging to you tighter than this.' The repeated patting is therefore doubly soothing.

The signal of the softly cooing or humming voice soothes in another way. Again, an animal illustration may help. When certain fish are in an aggressive mood, they indicate this by lowering the head-end of the body and raising the tail-end. If the same fish are signalling that they are definitely not aggressive, they do the exact opposite, that is, they raise the head and lower the tail. The soft cooing of the mother works on this same antithesis principle. Loud, harsh sounds are alarm signals for our species, as they are for many others. Screams, shouts, snarls and roars are widespread mammalian messages of pain, danger, fear and aggression. By employing tonal qualities which are the antithesis of these sounds, the human mother can, as it were, signal the opposite of these messages, namely, that all is well. She may use verbal messages in her cooing and crooning, but the words are, of course, of little importance. It is the soft, sweet, smooth tonal qualities of the cooing that transmit the vital, comforting signal to the infant.

Another important new, post-womb pattern of intimacy is the presentation of the nipple (or bottle-teat) for sucking by the baby. Its mouth experiences the intrusion of a soft, warm, rubbery shape from which it can squeeze a sweet, warm liquid. Its mouth senses the warmth, its tongue tastes the sweetness, and its lips feel the softness. Another very basic comfort – a primary intimacy – has been added to its life. Again, it is one that will reappear in many disguises later, in adult contexts.

These, then, are the most important intimacies of the baby phase of the human species. The mother embraces her offspring, carries, rocks, pats, fondles, kisses, strokes, cleans and suckles it, and coos, hums and croons to it. The baby's only really positive contact action at this early stage is to suck, but it does have two vital signals that act as invitations to intimacy and encourage the mother to perform her close-contact actions. These signals are crying and smiling. Crying initiates contact and smiling helps to maintain it. Crying says 'come here' and smiling says 'please stay'.

The act of crying is sometimes misunderstood. Because it is used when the baby is hungry, uncomfortable or in pain, it is assumed that these are the only messages it carries. If a baby cries, a mother often automatically concludes that one of these three problems must exist, but this is not necessarily the case. The message says only 'come here'; it does not say why. If a baby is well fed, comfortable and in no pain, it may still cry, simply to initiate intimate contact with the mother. If the mother feeds it, makes sure there is no discomfort, and then puts it down again, it may immediately restart its crying signal. All this means, in a healthy baby, is that it has not had its quota of intimate bodily contact, and it will go on protesting until it gets it. In the early months the demand is high, and the baby is fortunate in having a powerful attraction-signal in its happy smile, to reward the mother for her labours.

Amongst primates, the smile of the human baby is unique. Monkey and ape babies do not have it. They simply do not need it, because they are strong enough to cling on to their mother's fur and stay close to her in this way by virtue of their own actions. The human baby cannot do this and somehow has to make itself more appealing to the mother. The smile has been evolution's answer to the problem.

Crying and smiling are both backed up by secondary signals. Human crying starts out in a monkey-like way. When a baby monkey cries, it produces a series of rhythmic screaming sounds, but sheds no tears. During the first few weeks after birth, the human baby cries in the same tearless way, but after this initial period, weeping is added to the vocal signal. Later, in adult life, the weeping can occur separately, by itself, as a silent signal, but for the infant it is essentially a combined act. For some reason, man's uniqueness as a weeping primate has seldom been commented on, but clearly it must have some specific significance for our species. Primarily it is, of course, a visual signal and is enhanced by the hairlessness of our cheeks, on which tears can glisten and trickle so conspicuously. But another clue comes from the mother's response, which is usually to 'dry the eyes' of her infant. This entails a gentle wiping away of the tears from the skin of the face – a soothing act of intimate body contact. Perhaps this is an important secondary function of the dramatically increased secretion of the lachrymal glands that so often floods the face of the young human animal.

If this seems far-fetched, it is worth remembering that the human mother, as in many other species, has a strong basic urge to clean the body of her offspring. When it wets itself with urine, she dries it, and it is almost as if copious tears have evolved as a kind of 'substitute urine' serving to stimulate a similar intimate response at times of emotional distress. Unlike urine, tears do not help to remove waste products from the body. At low levels of secretion they clean and protect the eyes, but during full weeping their only function appears to be that of transmitting social signals, and they then justify a purely behavioural interpretation. As with smiling, the encouragement of intimacy seems to be their main business.

Smiling is supported by the secondary signals of babbling and reaching. The infant grins, gurgles and stretches out its arms towards its mother in an intention movement of clinging to her, inviting her to pick it up. The mother's response is to reciprocate. She smiles back, 'babbles' to the baby, and reaches out her own arms to touch it or lift it. Like weeping, the smiling complex does not appear until about the second month after birth. In fact, the first month could well be called the 'monkey phase', the specifically human signals appearing only after these first few weeks have passed.

As the baby moves on into its third and fourth months, new patterns of body contact begin to appear. The early 'monkey' actions of the grasp reflex and the Moro reflex disappear and are replaced by more sophisticated forms of *directed* grasping and clinging. In the case of the primitive grasp reflex, the baby's hand automatically took hold of any object pressed into it, but now the new, selective grasping becomes a positive action in which the infant co-ordinates its eyes and its hands, reaching out for and grabbing a particular object that catches its attention. Frequently this is part of the mother's body, especially her hair. Directed grasping of this kind is usually perfected by the fifth month of life.

In a similar way, the automatic, undirected clinging movements of the Moro reflex make way for orientated hugging in which the baby clings specifically to the mother's body, adjusting its movements to her position. Directed clinging is normally established by the sixth month.

Leaving the baby stage behind and passing on to the later period of childhood, it becomes obvious that there is a steady decline in the extent of primary bodily intimacy. The need for security, which extensive body contact with the parent has satisfied so well, meets a growing competitor, namely the need for independent action, for discovering the world, for exploring the environment. Clearly this cannot be done from the encompassing circle of the mother's arms. The infant must strike out. Primary intimacy must suffer. But the world is still a frightening place, and some form of indirect, remote-control intimacy is needed to maintain the sense of security and safety while independence is asserting itself. Tactile communication has to give way to increasingly sensitive visual communication. The baby has to replace the confining, hampering security of the embrace and the cuddle with the less restricting device of exchanged facial expressions. The shared hug makes way for the shared smile, the shared laugh, and all the other subtle facial postures of which the human being is capable. The smiling face, which earlier was an invitation to an embrace, now replaces it. The smile, in effect, becomes itself a symbolic embrace, operating at a distance. This enables the infant to function more freely and yet, at a glance, to re-establish emotional 'contact' with the mother.

The next great phase of development arrives when the child begins to speak. In the third year of life, with the acquisition of a basic vocabulary, verbal 'contact' is added to the visual. Now the child and its mother can express their 'feelings' towards one another in words.

As this phase progresses, the early, primary intimacies of direct body contact are inevitably restricted still further. It becomes babyish to be cuddled. The strongly growing need for exploration, independence and separate, individual identity increasingly dampens down the urge to be held and fondled. If parents over-express these primary body contacts at this stage, the child does not feel protected, but mauled. Being held now becomes being held back, and the parents have to adjust to the new situation.

For all this, body contact does not disappear completely. In times of pain, shock, fear or panic, the embrace will still be welcomed or sought, and even at less dramatic moments some contact will still occur. There are considerable changes, however, in the form which it takes. The all-embracing hugs shrink to small fragments of their former selves. The semi-embrace, the arm around the shoulder, the pat on the head and the hand-clasp begin to appear.

The irony of this phase of later childhood is that, with all the stresses that exploration brings, there is still a great inner need for the comfort of bodily contact and intimacy. This need is not so much reduced as suppressed. Tactile intimacy spells infancy and has to be relegated to the past, but the environment still demands it. The conflict that results from this situation is resolved by the introduction of new forms of contact that provide the required bodily intimacy without giving the impression of being babyish.

The first sign of these disguised intimacies appears early and takes us almost back to the baby stage. It starts in the second half of the first year of life and has to do with the use of what have been called 'transitional objects'. These are, in effect, inanimate mother-substitutes. There are three common ones: a favourite sucking-bottle, a soft toy and a piece of soft material, usually a shawl or a particular piece of bedding. In the early baby phase they were experienced by the infant as part of the intimate contacts it had with its mother. They were not preferred to the mother then, of course, but were nevertheless strongly associated with her bodily presence. In her absence they become substitutes for her, and many children will refuse to go to sleep without their comforting proximity. The shawl or the soft toy has to be in the cot at bedtime, or there is trouble. And the demands are quite specific – it has to be *the* toy or *the* shawl. Similar but unfamiliar ones will not do.

At this stage the objects are employed only when the mother herself is not available, which is why they become so important at bedtime, when contact with the mother is being broken. But then, as the child grows, a change takes place. Now, as the infant becomes more independent of the mother, the favourite objects become more, rather than less, important as comforters. Some mothers misinterpret this and imagine that the infant is feeling unnaturally insecure for some reason. If the child becomes desperate for contact with its 'Teddy', or 'shawly' or 'cuddly' – these objects always seem to acquire a special nickname – the mother may look upon it as a backward step. In fact, it is just the reverse. What the child is doing, in effect, is saying, 'I want body contact with my mother, but that would be too babyish. I am too independent for that now. Instead I will make contact with this object, which will make me feel secure without throwing me back into my mother's arms.' As one authority has put it, the transitional object 'is reminiscent of the pleasurable aspects of mother, is a substitute for mother, but is also a defense against re-envelopment by mother'.

As the infant grows older and the years pass, the comforter may persist with remarkable tenacity, surviving in some cases even into middle childhood. In rare cases it may last into adulthood. We are all familiar with the nubile girl sprawled on her bed clutching her giant Teddy bear. I say 'rare cases', but that requires qualification. It is certainly rare for us to retain an obsession for precisely the same transitional object that we employed in infancy. For most of us the nature of the act would become too transparent. Instead, we find substitutes for the substitutes – sophisticated adult substitutes for the childish mother-body substitutes. When the baby-shawl becomes transformed into a fur coat we treat it with more respect.

Another form of disguised intimacy in the growing child is found embedded in the development of rough-and-tumble play. If it is baby-ish to embrace and cuddle but the need is still there, then the problem can be solved by embracing the parent in such a way that the body contact involved does not look like an embrace. The loving hug becomes a mock-aggressive bear-hug. The embrace becomes a wres-tling-match. In play-fighting with the parent, the child can once again relive the close intimacy of babyhood whilst hiding behind the mask of aggressive adulthood.

This device is so successful that mock body-fighting with the parent can be seen even in later adolescence. Still later, between adults, it is usually restricted to a friendly punch on the arm or thump on the back. Admittedly, the play-fighting of childhood involves more than merely disguised intimacy. There is a great deal of body-testing as well as body-touching, of exploring new physical possibilities as well as reliv-ing old ones. But the old ones are certainly there and they are import-ant – much more so than is usually recognized.

With the arrival of puberty, a new problem asserts itself. Bodily contact with the parent is now further curtailed. Fathers find that their daughters suddenly become less playful. Sons become body-shy with their mothers. In the post-baby phase, the need for independent action began to express itself, but now, at puberty, the need intensifies and introduces a powerful new demand: privacy.

If the baby's message was 'hold me tight', and the child's was 'put me down', that of the adolescent is 'leave me alone'. One psychoanalyst has described the way in which, at puberty, 'The young person tends to isolate himself; from this time on, he will live with the members of his family as though with strangers.' This statement makes its point, of course, by over-emphasis. Adolescents do not go around kissing strangers, but they do continue to kiss their parents. True, the actions have now become more formalized – the smacking kiss has become a peck on the cheek – but brief intimacies do still take place. As with adults, however, they are at this stage largely limited to greetings, farewells, celebrations and disasters. In fact, the adolescent is already adult – sometimes super-adult – as far as his or her family intimacies are concerned. With unconscious ingenuity, loving parents overcome this problem in a variety of ways. A typical example is the 'clothing-adjustment' pattern. If they cannot perform a direct loving touch, they can nevertheless make body contact by disguising it as 'let me straighten your tie' or 'let me brush your coat'. If the reply is 'stop fussing, mother' or 'I can do it myself', then this means that the adolescent has, equally unconsciously, seen through the trick.

When post-adolescence arrives and the young adult moves outside the family, then – from the point of view of body intimacy – it is as if a second birth has been experienced, the womb of the family being abandoned as was the womb of the mother two decades before. The primary sequence of changing intimacies – 'hold me tight/put me down/leave me alone' – now goes back to the beginning again. The young lovers, like the baby, say 'hold me tight'. Occasionally they will even call one another 'baby' as they say it. For the first time since babyhood, intimacies are once again extensive. As before, the body-contact signals begin to weave their magic and a powerful bond of attachment starts to form. To emphasize the strength of this attachment, the message 'hold me tight' becomes amplified by the words 'and never let me go'. Once the pair-formation is completed, however, and the lovers have set themselves up as a new two-piece family unit, the second babyhood phase is over. The new intimacy sequence moves relentlessly on, copying the first, primary one. The second babyhood gives way to the second childhood. (This is the true second childhood, and is not to be confused with the senility phase, which appears much later and has sometimes been mistakenly referred to as the second childhood.)

Now the enveloping intimacies of courtship begin to weaken. In extreme cases, one or both members of the new pair start to feel trapped, their independence of action threatened. It is normal enough, but it feels unnatural and so they decide it was all a mistake and split up. The 'put me down' of the second childhood is replaced by the 'leave me alone' of the second puberty, and the primary family separation of adolescence becomes the secondary family separation of divorce. But if divorce creates a second adolescence, then what is this neo-adolescent doing alone, without a lover? So, after the divorce, each of the ex-pair members finds a new lover, goes through the second babyhood stage once more, remarries and, with a crash, is back at the second childhood. To their astonishment, the process has repeated itself.

This description may be a cynical over-simplification, but it helps to make the point. For the lucky ones, and even today there are many of them, the second adolescence never comes. They accept the conversion of the second babyhood into the second childhood. Augmented with the new intimacies of sex and the shared intimacies of parenthood, the pair-bond survives.

In later life the loss of the parental factor is softened by the arrival of new intimacies with the grandchildren, until eventually the third and final babyhood appears on the scene with the approach of senility and the helplessness of old age. This third run-through of the intimacy sequence is short-lived. There is no third childhood, at least in earthly terms. We end as babies, snugly tucked in our coffins, which are softly padded and draped – just like the cradles of our first babyhood. From the rock of the cradle we have passed on to the rock of ages.

For many, it is difficult to contemplate the third great intimacy-sequence ending there. They refuse to accept that the third babyhood does not continue into a third childhood, to be found in heaven. There the condition is ideal and permanent, and there is no fear of being over-mothered, for God the Father has no wife.

In tracing these patterns of intimacy from the womb to the tomb, I have dwelt at greater length on the early phases of life, passing more swiftly over the later stages of adulthood. With the roots of intimacy exposed in this way, we can now look more closely at adult behaviour in the chapters that follow.

2

INVITATIONS TO SEXUAL INTIMACY

E VERY HUMAN BODY is constantly sending out signals to its social companions. Some of these signals invite intimate contact and others repel it. Unless we are accidentally thrown against someone's body, we never touch one another until we have first carefully read the signs. Our brains, however, are so beautifully tuned in to the delicate business of assessing these invitation signals that we can often sum up a social situation in a split second. If we unexpectedly spot a loved one amongst a crowd of strangers, we can be embracing them within a few moments of setting eyes on them. This does not imply carelessness; it simply means that the computers inside our skulls are brilliant at making rapid, almost instantaneous calculations concerning the appearance and mood of all the many individuals we encounter during our waking hours. The hundreds of separate signals coming from the details of their shape, size, colour, sound, smell, posture, movement and expression, crowd at lightning speed into our specialized sense organs, the social computer whirrs into action, and out comes the answer, to touch or not to touch.

As babies, our small size and our helplessness act as powerful stimuli encouraging adults to reach out and make friendly contact with us. The flat face, the large eyes, the clumsy movements, the short limbs and the generally rounded contours, all contribute to our touch-appeal. Add to this the broad smile and the alarm signal of crying and screaming, and the human infant is clearly a massive invitation to intimacy.

If, as adults, we send out similar signals of helplessness or pain, as, for instance, when we are sick or the victims of an accident, we provoke a pseudo-parental response of much the same kind. Also, when we are performing the first tentative body contact, in the form of shaking hands, we nearly always accompany it with a smiling facial expression.

These are the basic invitations to intimacy; but, with sexual maturity, the human animal moves into a whole new sphere of contact signals – the signals of sex appeal – which serve to encourage a male and female to start touching one another with more than mere friendliness in mind.

Some of the sex signals are universal and apply to all adult human beings; others are cultural variations on these biological themes. Some concern our adult appearances as males and females, and others have to do with our adult behaviour – our postures, gestures and actions. The simplest way to survey them is to make a tour of the human body, pausing briefly at each of the main points of interest in turn.

The crotch. Since we are dealing with sexual signals, it is logical to start with the primary genital region, and work our way outwards. The crotch is the major taboo zone, and this is not merely because it is the site of the external reproductive organs. Packed into this one small area of the body are all the main taboo subjects: urination, defecation, copulation, fellation, ejaculation, masturbation and menstruation. With this array of activities it is little wonder that it has always been the most concealed area of the human body. To expose it directly to view as a visual invitation to intimacy is much too strong a sexual signal to be employed as a preliminary device, before a relationship has progressed through the earlier stages of body contact. Ironically, by the time that a relationship has reached the more advanced phase of genital intimacy, it is usually too late for visual displaying, so that the first experience of the partner's genitals is normally tactile. The act of looking directly at the genitals of the opposite sex therefore plays a comparatively small role in modern human courtship. There is, nevertheless, considerable interest in this region of the body, and if direct exposure of the genitals is not possible, alternatives can still be found.

The first is to employ articles of clothing which underline the nature of the organs hidden beneath them. For the female this means wearing trousers, shorts or bathing costumes which are one size too small for comfort, but which, by their tightness, cut into the genital cleft and thereby reveal its shape to the attentive male eye. This is an exclusively modern development, but the male equivalent has a longer history. For a period of nearly two hundred years (from roughly 1408 to 1575) many European male genitals were indirectly displayed in all their glory by the wearing of a codpiece. This began life in a modest way as a front flap which formed a small pouch at the fork of the extremely tight trousers, or hose, worn by the males of the day. The hose were, in fact, so tight that they had little choice. 'Cod' is an old word for scrotum, hence the name 'codpiece'. As the years passed, it grew dramatically in size until it became a blatant phallus-piece, rather than a mere scrotum-piece, and gave the impression that the wearer was the possessor of a permanently erect penis. To emphasize the point further, it was often differently coloured from the surrounding material, and even decorated with gold and jewels. Eventually the codpiece went so far that it began to become something of a joke, so that when Rabelais described the one worn by his hero, he was able to comment that 'For the codpiece sixteen yards of material was required. The shape was that of a triumphal arch, most elegant and secured by two gold rings fastened to enamel buttons, each the size of an orange and set with heavy emeralds. The codpiece protruded forward as much as three feet.'

Today such extravagant penile displays are missing from the world of fashion, but an echo can still be found, for the young male of the 1960s and 1970s has once again started to wear his 'hose' too tight. Like the modern female, he squeezes himself into closely fitting jeans and bathing costumes, in which he is forced to change the resting position of his penis. Unlike the older males, who still fill the generation gap between their legs with a loosely hanging penis beneath their slacker slacks, the younger male of today walks with his penis in the upright position. Held firmly in its vertical posture by the snugly clinging material, it presents a mild but distinctly visible genital bulge to the interested female eye. In this way the young male's costume once again permits him to display a pseudo-erection and, as with the earlier codpiece, this has surprisingly happened without any undue criticism from puritan quarters. It remains to be seen whether the codpiece itself will make a come-back in the years ahead, and how far the trend will have to go before this male embellishment once again becomes so sexually obtrusive that it falls into disrepute.

Other modern costume-genital displays are of a frankly exotic nature and not in wide use. They include female bathing costumes and panties trimmed with fur in the pubic area, or with lace-up fronts which imitate the shape of the genitals. Another form of indirect genital display that has, without comment, survived the passage of time is the Scotsman's sporran, a symbolic genital pouch worn in the scrotal region and frequently covered with symbolic pubic hair.

A less direct way of transmitting visual genital signals is to use some other part of the body as a 'genital-echo', or copy. This enables a primary sexual message to be transmitted, whilst keeping the real genitals completely obscured. There are several ways in which this is done and to understand them we must look again at the anatomy of the female sex organs. For purposes of symbolism, they consist of an orifice – the vagina – and paired flaps of skin – the minor and major labia. If these are covered up, then any other body organs or details that resemble them in some way are likely to become employed as 'genital-echoes', for signalling purposes.

As orifice-substitutes, candidates available are the navel, the mouth, the nostrils and the ears. All have mild taboos associated with them. It is impolite to pick the nose or clean the ear with a finger in a public place. This contrasts with a cleaning action such as wiping the forehead, or rubbing the eyes, which we permit without comment. The mouth is frequently covered for some reason, if not with a veil, at least when we are yawning, gasping or giggling. The navel is even more taboo, having frequently been air-brushed out of photographs completely in past decades, to protect our eyes from its suggestive shape. Of these four types of 'aperture', only the mouth and the navel appear to have been specifically employed as genital-orifice substitutes.

The mouth is by far the most important of these, and transmits a great deal of pseudo-genital signalling during amorous encounters. I suggested in *The Naked Ape* that the unique development of everted lips in our species may well be part of this story, their fleshy pink surfaces having developed as a labial mimic at the biological, rather than the purely cultural, level. Like the true labia, they become redder and more swollen with sexual arousal, and like them they surround a centrally placed orifice. Since the earliest historical times, the lip signals of the female have been heightened by the application of artificial colouring. Lipstick is today a major cosmetic industry and, although the colours vary from fad to fad, they always return before long to something in the pink-red range, thereby copying the flushing of the labia during the advanced stages of sexual arousal. This is not, of course, a conscious imitation of the genital signals; it is merely 'sexy' or 'attractive', with no further questions asked.

The lips of the adult female are typically slightly larger and fleshier than those of the male, which is what one would expect if they are playing a symbolic role, and this size difference has sometimes been exaggerated by spreading the lipstick over a wider area than the lips themselves. This also mimics the enlargement they undergo when sexually engorged with blood.

Many authors and poets have seen the lips and mouth as a powerfully erotic region of the body, with the male tongue inserting itself, penis-like, into the female mouth during deep kissing. It has also been suggested that the structure of the lips of a particular female reflects the structure of the (as yet unseen) genitals below. A woman with fleshy lips is supposed to be the possessor of fleshy labia; one with tight, thin lips to have tight, thin genitals. Where this is in fact the case, it does not, of course, reflect the precision of the body-mimicry, but merely the overall somatotype of the female in question.

The navel has excited far less comment than the mouth, but some curious things have been happening to it in recent years which reveal that it does have a distinct role to play as a genital-echo. Not only was it painted out of early photographs, but its exposure was also expressly forbidden by the original 'Hollywood Code', so that harem dancing-girls in pre-war films were always obliged to appear with ornamental navel-covers of some kind. No real explanation was ever given for this taboo, except for the lame excuse that the exposure of the navel might start children asking what it was for and thereby precipitate an awkward 'facts of life' interlude for the parent. In adult contexts, however, this is clearly nonsensical, and the true reason was obviously that the navel is strongly reminiscent of a 'secret orifice'. Since harem girls are likely, at the drop of a veil, to start wriggling their abdomens about in an Eastern belly-dance, with the result that this pseudo-orifice starts gaping, stretching and writhing in a sexually inviting manner, Hollywood decided that this indelicate piece of anatomy would have to be masked. Ironically, as we started to move into the second half of the twentieth century and the grip of the Hollywood Code was beginning to relax in the West, the Arab world, with its new-found republican spirit, set about reversing the trend. Egyptian belly-dancers were officially informed that it was improper and undignified to expose the navel during their 'traditional folklore' performances. In future, their new government insisted, the midriff would be suitably covered by some kind of light cloth. And so, as the European and American navels fought their way back, in the cinemas and on the beaches, the blind orifices of North Africa retreated into a new obscurity.

Since their reappearance, the naked navels of the Western world have undergone a curious modification. They have started to change shape. In pictorial representations, the old-fashioned circular aperture is tending to give way to a more elongated, vertical slit. Investigating this odd phenomenon, I discovered that contemporary models and actresses are six times more likely to display a vertical navel than a circular one, when compared with the artist's models of yesterday. A brief survey of two hundred paintings and sculptures showing female nudes, and selected at random from the whole range of art history, revealed a proportion of 92 per cent of round navels to 8 per cent of vertical ones. A similar analysis of pictures of modern photographic models and film actresses shows a striking change: now the proportion of vertical ones has risen to 46 per cent. This is only partly due to the fact that the girls have become generally much slimmer, for although it is true that a fat, sagging female cannot offer the contemplator of navels a vertical slit, it is equally true that a skinny female need not do so either. Modigliani's slender girls display navels that are quite as round as those of Renoir's plump models. Furthermore, two young girls of the 1970s with similar figures can easily display the two distinct navel shapes.

How this change has come about, and whether it has been unconsciously arrived at or knowingly encouraged by modern photographers, is not entirely clear. It appears to have something to do with a subtle alteration in the trunk posture of the displaying model, possibly partly connected with an exaggerated intake of breath. The ultimate significance of the new navel shape is, however, reasonably certain. The classical round navel, in its symbolic orifice role, is rather too reminiscent of the anus. By becoming a more oval, vertical slit, it automatically assumes a much more genital shape, and its quality as a sexual symbol is immensely increased. This, apparently, is what has been happening since the Western navel came out into the open and began to operate more overtly as an erotic signalling device.

The buttocks. Leaving the crotch and its substitute echoes and passing round to the back of the pelvic region, we come to the paired, fleshy hemispheres of the buttocks. These are more pronounced in the female than the male and are a uniquely human feature, being absent in other species of primates. If a human female were to bend down and present her buttocks visually to a male, as if she were adopting the typical primate invitation-to-copulation posture, her genitals would be seen framed by the two hemispheres of smooth flesh. This association makes them an important sexual signal for our species, and one that probably has a very ancient biological origin. It is our equivalent of the 'sexual swellings' of other species. The difference is that, in our case, the condition is permanent. In other species the swellings rise and fall with the menstrual cycle, reaching their maximum size when the female is sexually receptive, around the time of ovulation. Naturally, as the human female is sexually receptive at virtually all times, her sexual 'swellings' remain permanently 'inflated'. As our early ancestors became more upright in their standing posture, the genitals came to be displayed more from the front than the rear, but the buttocks still retained their sexual significance. Even though copulation itself became an increasingly frontal affair, the female could still send out sexual signals by emphasizing her rump in some way. Today, if a girl slightly increases the undulations of her buttocks when she walks, it acts as a powerful erotic signal to the male. If she adopts a posture in which they are 'accidentally' protruded a little more than usual, this has the same effect. Sometimes, as in the famous 'bottoms up' cancan dance posture, the full version of the ancient primate rump-presentation display is still to be seen, and jokes about the man who wants to slap or caress the bottom of the girl who has innocently bent down to pick up a dropped object are commonplace.

From earlier days, there are two buttock phenomena that deserve comment. The first is the natural condition known as 'steatopygia', and the second is the artificial device of the bustle. Literally, steatopygia means fat-rumped, and refers to a remarkably pronounced protuberance of the buttocks found amongst certain human cultures, notably the Bushmen of Southern Africa. It has been suggested that this is a case of fat-storage similar to that of the camel's hump, but since it is much more exaggerated in females than males, it seems more likely that it is a specialization of the sexual signals that emanate from this region of the body. It looks as if the

Bushman females went further in the development of this signal than other races. It is even possible that this condition was typical of most of our early ancestors, but that later on it was reduced in favour of a more athletically adaptable compromise in the shape of the less ambitious female buttocks we see around us today. Certainly the Bushmen were once much more common than they are now, and owned much of Africa before the later Negro expansion.

It is also curious that so many prehistoric female figurines, from Europe and elsewhere, frequently show a similar condition, with huge protruding buttocks which are completely out of proportion with the general obesity of the bodies depicted. There are only two explanations. Either prehistoric women were endowed with huge rumps, sending out massive sexual signals to their males, or the prehistoric sculptors were so obsessed with the erotic nature of buttocks that, like many cartoonists of the present day, they permitted themselves a considerable degree of artistic licence. Either way, the prehistoric buttocks reigned supreme. Then, strangely, in region after region, as the art forms advanced, the big-buttocked females began to disappear. In the prehistoric art of every locality where they occur, they are always the earliest figures to be found. Then they are gone, and more slender females move in to take their place. Unless fat-rumped females really were common in earlier days and then gradually died out, the reason for this widespread change in prehistoric art remains a mystery. The male interest in female buttocks survived, but with a few exceptions they now became reduced to the natural proportions we observe on the cinema screens of the twentieth century. The dancing girls depicted on the murals of ancient Egypt could easily find work in a modern night-club and, had Venus de Milo lived, she would have had a hip measurement of no more than 38 inches.

The exceptions to this rule are intriguing, since, in a sense, they demonstrate a throw-back to prehistoric times and show man's revived interest in gross exaggeration of the female rump region. Here we move from the fleshy phenomenon of steatopygia to the artificial device of the bustle. The effect is the same in both cases – namely, a vast enlargement of the buttock region – but with the bustle this was done by inserting thick padding, or some kind of framework, beneath the female costume. In origin it was a sort of crinoline-in-retreat. The habit of padding the hips all round the pelvic region occurred frequently in European fashions, and all that was needed to permit a new rump-display to appear on the scene was to remove the padding at the front and sides of the body. This made the creation of the bustle a 'reduction' rather than an exaggeration and permitted its introduction into high fashion to take place without undue comment. Having arrived negatively in this way, it managed to avoid its obvious sexual implications. The hooped and padded bustle of the 1870s vanished after a few years, but returned in triumph in an even more exaggerated form in the 1880s. Now it became a great shelf sticking out at the back, kept in place by wire netting and steel springs, and created an impression that even a tired Bushman would have responded to. By the 1890s, however, this too was gone, and the increasingly athletic female of the twentieth century has never wished to reinstate it. Instead, the enlarged buttocks of modern times have been confined to rarely used 'bottom-falsies', to provocative 'come-hither' posturing and to the drawing-boards of the cartoonists.

The legs. Moving downwards from the pelvic region, the female legs have also been the subject of considerable male interest as sexual signalling devices. Anatomically, the outsides of the female thighs have a greater fat deposit than in the male, and at certain periods a plump leg has therefore been an erotic one. At other times, the mere exposure of leg flesh has been sufficient to transmit sexual signals. Needless to say, the higher the exposure goes, the more stimulating it becomes, for the simple reason that it then approaches more closely the primary genital zone. Artificial aids to the legs have included 'calf-falsies', to be worn under opaque stockings, but these have been as rarely used as 'bottom-falsies'. High-heeled shoes have been much more commonly adopted, the tilt of the foot supposedly enhancing the leg contour and also increasing the apparent length of the legs, which relates to the fact that elongation of the limbs is a characteristic of adolescent maturing. 'Legginess' equals sexual maturity, and therefore sexiness.

The feet themselves have often been squeezed into shoes which are much too tight for them, a tendency that results from the fact that the foot of the adult female is slightly smaller than that of the adult male. It therefore follows that to increase this difference a little will make the female foot more feminine, as a sexual signal to the male. The petite foot has often been praised by male admirers, and many a female has suffered tortures to obtain it. Byron's words sum up the traditional masculine attitude, when he writes of 'feet so small and sylph-like, suggesting the more perfect symmetry of the fair forms which terminate so well'. This view of the female foot is reflected in the perennial story of Cinderella – a tale that can be traced back at least two thousand years – where her ugly sisters have feet too big to fit into the tiny glass slipper. Only the beautiful heroine has small enough feet to win the heart of the Prince.

In China the trend towards small female feet was once taken to horrifying lengths, young girls often having to undergo such tight binding that they suffered serious deformity. The bound foot, or 'golden lily', which looked so attractive in its tiny decorated shoe, had more the appearance of a deformed pig's trotter when it was viewed unshod. So important was this painful practice that a girl's commercial value was measured by the smallness of her feet, her shoes being displayed to establish her worth during the bartering over bride-price. The modern woman whose 'feet are killing her' is merely echoing in a mild way this ancient phenomenon. The official reason given to explain the custom of the 'golden lilies' was that it showed that the female in question did not need to work – because she was so crippled that she could not. But her husband did not need to work either, yet his feet remained un-crushed, so that the exaggerated sex-difference element provides a more basic explanation. This rule applies in many other cases as well. A particular distortion, or exaggeration, is officially performed for reasons of 'high fashion' or status, but the deeper explanation is that the modification in question in some way emphasizes a female (or male) biological characteristic. The artificial tightening of the female waist is another case in point.

In addition to their anatomy, the posture of the legs is capable of transmitting sexual signals. In many cultures, girls are told that it is improper to stand or sit with the legs wide apart. To do this is to 'open' the genitals, and even if they are not visible, the message remains fundamentally the same. With the advent of female trousers and the disappearance of strict codes of etiquette, the legs-apart posture has become much more common in recent years, and is used with increas-ing frequency by model girls in advertisements. What was once a too-powerful signal has now become merely challenging; what was shocking is now no more than tantalizing. But nevertheless, a girl in a skirt still obeys the old rules. To expose the opened crotch, clad only in panties, remains too strong an invitation signal in most cases.

The traditionally 'polite girl' therefore keeps her legs together, but there is a danger, also, if she goes too far in this direction and presses them too tightly against one another. If she does this, or crosses them so vigorously that her thighs are squeezed together, she begins to 'protest too much' and thereby makes a new kind of sexual comment. As with all puritan statements, she reveals that she has sex very much on her mind. In fact, the girl who tries to protect her genitals unduly draws almost as much attention to them as the one who exposes them to view. Similarly, if a skirt rides up slightly when a girl sits down and exposes more leg than she intended, she only enhances the sexuality of the situation by making attempts to tug it down again. The only non-sexual signal is the one that avoids both extremes.

For the male, the opening of the legs carries much the same signal as with the female, because here, too, it says 'I am exposing my genitals to you.' Sitting with the legs wide apart is the gesture of a dominant, confident male (unless, of course, he is so fat that he cannot get them together).

The belly. Moving up above the genital region now, we come to the belly, which has two characteristic shapes: flat and 'pot'. Lovers tend to be flat-bellied, while pot-bellies are most commonly seen in starving children and overfed men. The adult female is less likely to become pot-bellied than the adult male, even if she becomes equally overweight. This is because, as a female, her tissues are hungrier for fat in the thigh and hip region than around the belly. If a male and female go far enough, of course, they will both end up with much the same spheroid shape, but at milder levels of overeating the difference in fat-distribution is marked. Many comparatively skinny men manage to develop moderate-sized pot-bellies in middle and old age. What is the explanation for this?

Sometimes a visual joke says more than the artist intended or even understood. A cartoon of a pot-bellied, middle-aged man standing on a beach shows him being approached by a beautiful young girl in a bikini. As she comes near, he spots her and starts to contract his sagging stomach, so that when she is abreast of him his chest is swollen out and his belly held tightly in. As she passes by, the stomach slowly begins to sag again until, as she disappears in the distance, it is completely back to its gross, original shape. This joke was clearly intended to reflect a conscious control of the man's body contour – and his sexual image – but what it also depicts is something that happens unconsciously and semi-permanently, as part of the male human sexual display. For sexual excitement, or prolonged sexual interest, has the automatic effect of tightening up the stomach muscles. Regardless of individual variations, this is revealed in the general difference in the body contour of young males and old males. Young males are more sexually potent than older ones, and their general body shape goes in more as it goes down. They have the typical masculine display shape of our species, with broadened shoulders, expanded chests, and narrow hips. The flat belly is part of this general body-tapering. The older male tends towards sagging, rounded shoulders, flattened chest, and heavier hips. The swollen belly again is part of this now *inverted* tapering of the body contour. With this shape, the older male says plainly, 'I am past the pair-formation stage.'

In modern times, older males, who have made youthfulness and sexual potency into high-status cults, struggle desperately to stem the tide of this almost inevitable contour inversion. They diet ruthlessly, they indulge in physical training, they sometimes even wear tight corsets, and they consciously hold in their loosening stomach muscles as best they can. It would, of course, be much simpler if they could keep on falling in love over and over again. Belly-wise, they would find that an affair is as good as a diet and a corset combined, and that the physical exercise comes built-in. Under the influence of their passionate emotions, their stomach muscles would automatically contract and stay contracted, for by the act of falling in love they would be genuinely and biologically returning to a youthful condition, and their body would do its best to match their mood. Many men do, of course, take steps in this direction from time to time, but unless the process is a more or less continuous one, irreversible contour inversion will have started to take its toll, and their body success will be limited. Needless to say, such steps also play havoc with the older male's true biological role, which is by now that of the parental head of an established family unit.

The situation has not always been like this. Long ago, before the miracles of modern medicine stretched out our lives to such an unnatural extent, most older males soon vanished beneath the turf. To judge by our primate body weight and various other features of our life cycle, the natural life-span for man is probably somewhere between forty and fifty years, no more. Beyond that everything is a bonus. Also, in previous historical periods, a dominant, ageing male has often maintained his status more by his social power than by his youthfulness. An attractive young female has frequently been bought rather than wooed. A fat lord, or a fat master of the harem, cared little about his gross shape or the anti-sexual signals he transmitted. In the harem, this situation gave birth to the phenomenon of belly-dancing. Originally this consisted of the performance of pelvic thrusts by the female on the podgy, incapacitated form of her lord and master. Unable to make the thrusting movements himself, he had to be serviced by trained girls who would be able to take over the masculine role in the encounter, inserting his immobile penis into their vaginas, and then undulating and jolting their pelvises to stimulate it to a climax, in what amounted to little more than an act of fertile masturbation. The clever and varied movements developed by such females to arouse their fat, dominant males formed the basis of the famous Eastern belly-dance and, as a visual preliminary, this became more and more elaborate, until it grew into the display we see so often today in night-clubs and cabarets.

For the modern male, sexual conquests with no thought given to the masculine invitation signals are largely limited to brief visits to prostitutes. For his long-term relationships, he now has to rely much more heavily on his personal sexual appeal. In this respect he has returned to a much more natural situation for the human species, but at the same time his life-span has been artificially elongated. It is this situation that has led to the new concern about 'youthfulness and vigour' for the male who, as he leaves his twenties behind, inevitably starts to feel the decline of his sexual powers. Coasting to a natural death at forty, this would not have posed such a problem, leaving just enough time to rear the offspring and be off. But now, with something more like half a century confronting the post-parental male, the problem has become acute, and we have all the diet books, health farms and other paraphernalia of contemporary life to prove it.

The waist. Here we return again to the sexual-signalling world of the female. The waist is narrower in the female than the male, or appears to be so because of her widened, childbearing hips and her swollen, rounded breasts. Narrowness of waist has therefore become an important female sexual signal, available for artificial exaggerations of the type we have met before. The signal can be made more powerful directly or indirectly, either by tightening the waist or by enlarging the bust and hips. The maximum signal can be transmitted by doing both at once. Busts can be magnified by holding or pushing them up with tight clothing, by padding them out, or by cosmetic surgery. The hips can be enlarged by padding or by the wearing of stiff clothing that curves out from the body. The waist itself can be reduced by tight lacing or the wearing of a belt.

Female waist corsets have a long and sometimes unhappy history. In previous eras they have occasionally become so severe that they have damaged the rib and lung development of young girls and interfered with their healthy respiration. In late Victorian times, an attractive girl was one whose waist measurement in inches was the same as her age at her last birthday. To achieve this goal many young ladies were forced to sleep in their tightly laced corsets, as well as wear them throughout the day. In historical periods when the crinoline was fashionable, the restrictions on the waist could be eased considerably, because, of course, any waist looked small by comparison with the enormous apparent hip width provided by the huge skirts.

Twentieth-century waists have suffered much less from the artificial compressions of the corset, and frequently have been completely free of them, relying instead on the 'compressions' of a rigid diet. The average British woman of today has a waist measurement of $27\frac{3}{4}$ inches. The model girl Twiggy, the typical *Playboy* 'playmate' and the average Miss World all have a 24-inch waist. Modern female athletes, making more masculine demands on their bodies, tend towards a 29-inch waist.

These figures acquire more meaning when they are related to the bust and hip measurements, thereby revealing the 'waist indentation' factor that transmits the essential female contour signal. Twiggy ($30\frac{1}{2}$–24–33) and Miss World (36–24–36) now part company, the latter sending out a much more powerful waist signal.

There is a further waist factor that requires comment. Indentation comes from both above and below, and one can be greater than the other. Miss World is perfectly balanced – from both bust to waist and hip to waist, she indents by 12 inches. The average British female, however ($37-27\frac{3}{4}-39$), goes in more from hip to waist than she does from bust to waist. The fact that her hips are two inches larger than her bust gives her what is called a 2-inch 'drop'. This is common in the average female of other Western countries as well. In Italy it is also 2 inches; in Germany and Switzerland it is $2\frac{3}{8}$ inches; in Sweden and France, $3\frac{1}{8}$ inches.

These figures show a marked and significant difference from the *Playboy* 'playmate', a typical example being 37-24-35: in other words, a 2-inch 'rise', rather than a 'drop'. That she is referred to as being 'big-bosomed' is therefore not merely a reflection of her breast size. Her bust measurement is exactly the same as that of the average British female. Her apparent 'bustiness' is more the result of the fact that, although her bust measurement is no bigger, both her waist and hip measurements are roughly four inches smaller, giving her a top-heavy contour which magnetically draws attention to her breasts. To find such an unusual female is not easy. Since, for the purposes of the magazine, the girl is to be photographed with her breasts naked, the problem is a strictly biological one – artificial aids cannot help her. To consider this more fully, we can now leave the waist behind and concentrate more specifically on the chest region.

The breasts. The adult female of the human species is unique amongst primates in possessing a pair of swollen, hemispherical mammary glands. These remain protuberant and swollen even when she is not producing milk and are clearly more than a mere feeding device. I have suggested that, in their shape, they can better be thought of as another mimic of a primary sexual zone; in other words, as biologically developed copies of the hemispherical buttocks. This gives the female a powerful sexual signal when she is standing vertically, in the uniquely human posture, and facing a male.

There are two other echoes of the basic buttock shape, but they are less powerful than the breast signals. One is the smooth, rounded shoulder of the female, which, if it is 'only just' exposed by the pulling down of a blouse or sweater, presents a suitably curved hemisphere of flesh. This is a common erotic device in periods of low-cut dresses. The other echo is found in the smooth, rounded knees of the female, which, when the legs are bent and pressed together, present another pair of feminine hemispheres to the eyes of the male. Again, knees have frequently been referred to in an erotic context. As with the shoulders, they make their greatest impact when they are 'only just' exposed by the skirt. If the whole leg is visible, some of their impact is lost because they then become merely the rounded ends of the thighs, rather than a pair of hemispheres in their own right. But these are much milder buttock-echoes, and it is with the breasts that the major impact is achieved.

It is important to distinguish here between the childhood reaction to the female breast and the adult sexual one. Most men see their interest in the female bosom as purely sexual. By contrast, some scientific theorists have viewed it as purely infantile. Both views are one-eyed, because both factors are operating. The male lover who kisses the nipple of his female's breast may well be harking back to the pleasures of infancy, rather than kissing a pseudo-buttock, but the amorous male who ogles or fondles a female's breasts may well be responding primarily to their hemispherical buttock shape, rather than reliving the feeling of his mother's breast with his infantile hand. To the tiny hand of an infant, the mother's breast is too vast an object to be cupped or clasped in the palm, but to the adult hand it presents a rounded surface remarkably reminiscent of the buttock hemisphere. Visually the same is true, a pair of breasts offering an image that is much closer to that of a pair of buttocks than to the looming shape seen at close quarters by a baby being suckled.

This sexual significance of the female breasts is therefore of primary importance to our species, and although this is not the whole story, it plays a major role in society's perennial preoccupation with the feminine bosom. For the early English Puritans it meant flattening the breasts completely with a tight bodice. In seventeenth-century Spain, the measures taken were even more severe, young ladies having lead plates pressed into their swelling bosoms in an attempt to prevent their development. Such steps do not, of course, indicate a lack of interest in the female breast, which could only be demonstrated by completely ignoring it. Rather, they show acceptance of the fact that sexual signals do come from this region and, for cultural reasons, have to be stopped.

A much more widespread and frequent tendency has been for the breasts to be emphasized in some way. This emphasis has nearly always been to make them not so much larger as more upstanding. In other words, the tendency has been to improve their hemispherical, pseudo-buttock appearance. They are pushed up by tight dresses, so that they bulge above them, or they are pushed together to make the cleavage between them more like that between the real buttocks, or they are clasped in stretched brassières so that they protrude forward instead of sagging down. At some points in history, even greater attention has been paid to the problem, an ancient Indian love manual advising that 'Continuous treatment with antimony and rice water will cause the breasts of a young girl to become large and prominent so that they will steal the heart of the connoisseur as a robber steals gold.'

In a few primitive cultures, however, sagging or drooping breasts do find favour, and young girls are encouraged to pull at them regularly in order to hasten their downward turn. Also, nearer home, the small-breasted, or even flat-chested, female has her ardent followers, and these exceptions to the general rule require explanation. The social anthropologist would probably merely put them down to 'cultural variations' and leave the matter there. Every culture and every period has its own special standards of beauty, he would say, and virtually anything goes, providing it becomes the accepted fashion for a particular tribe or society. There is no basic biological theme with variations, but merely a wide range of equally valid alternatives, and each one must be looked at on its own merits. But to take such a view is to beg the fundamental issue of why the adult male and female human animals have evolved so many different body details, differences which are typical for our species as a whole. The typical female *does* have swollen breasts which are lacking in the male, she *does* display them regardless of the presence or absence of milk production for her offspring, and other species of primate *do* lack this prominent visual feature. They do therefore present a basic biological theme for *Homo sapiens*, variations from which must be viewed as unusual and requiring special explanation, rather than simply as equally valid cultural alternatives that require none, except to say that they are 'differing tribal customs'.

To understand the exceptions, it helps to look at the 'life cycle' of a typical female breast. In the child it starts out as no more than a nipple on a flat chest. Then, at puberty, it swells into a breast-bud. At this stage the swelling points straight forward. As it grows and becomes heavier, its weight begins to pull it down, its underside becoming slightly more curved than its upper surface. The nipples, however, still project straight out in front. This is the condition of the older teenage girl. Then, as she moves into her twenties and the breast continues to swell, it slowly begins to turn downwards until, in middle age, a full-breasted female shows a marked sag if she does not employ some artificial booster device. There are, then, three basic stages: small for the immature girl, pointed and protuberant for the young adult, and drooping for the older adult.

Viewed in this light, the cultural variations begin to make more sense. If, for some reason, immature girls are thought of as sexually appealing, then small breasts will be favoured. If older women are preferred, then drooping breasts will be in fashion. For the vast majority, the intermediate stage will be preferred, since it represents the true phase of first sexual activity in the human female. Under-developed females will imitate the pointed, protuberant condition by padding their small bosoms, the older ones who wish to give the impression of being in the first bloom of sexual life will imitate it by adding an artificial up:hrust.

In cases where pseudo-immature girls are preferred, there are several possible explanations. For the male living in a sexually repressed, puritan culture, flattening the female breast helps to damp down the strong sexual signal. For the male who has a strong inclination to play the 'father' role to a 'daughter' bride, the little-girl look of small breasts will be appealing. For the latent homosexual, small breasts give a more boyish look, which will be strongly attractive. Going to the other extreme, in societies where the maternal role of the female has become culturally much more important than the sexual one, the sagging breasts of the older female will be more appealing, even in younger girls. The latter must then 'age' their breasts by pulling at them to make them hang down.

For the majority of human beings, however, the maximum breast-appeal will come at the point where the hemispheres have reached their fullest protrusion *before* becoming so large that they begin to droop downwards. This explains the dilemma of the *Playboy* photographer, for as one breast quality (increased size) grows, the other (non-droop) diminishes. To take a picture of a super-breast, he has to search for a rare girl who has retained the firmness of her young breast past the point where it has already swollen to a full adult size. It is interesting that this limits him to a narrow age range in the late teens. Clearly this is the most vital point in the life cycle for this type of sexual signalling, and is the phase which older females are attempting to imitate and artificially prolong with their various breast-supporting techniques.

The super-breast effect is indirectly enhanced by selecting girls with modest-sized waists and hips. Here we come back to the question of the more general change in female body contour that occurs with age. Tests have shown that the average adult female puts on three pounds in weight every five years. The small proportion of this that goes on the breasts is what continues to pull them down as the years pass. The hips and thighs get an unfair share of this increase, giving the middle-aged female her characteristic 'hippy' (in the old sense) appearance. This is what accounts for the 'drop' mentioned earlier – the slightly larger size of hip than bust measurement. In some parts of the world, especially the Mediterranean region, this change can occur with startling rapidity as girls pass through their twenties. At one moment they are slim and slender and then, almost overnight, they start swelling out in the pelvic region, to assume the typical 'maternal' shape of the older woman. In other regions the change is more gradual, but the basic trend is the same. Not until very old age is there a reversal, when the body starts to shrink again.

To many Western women who wish to stay young-looking, this natural biological tendency of the species presents a severe challenge and demands the constant agonies of diet control. It is not merely that they are fighting gluttony, they are fighting nature as well. They must not merely eat 'normally', they must deliberately undereat, if they are to retain their girlish figures. The situation has not always been as extreme as it is today. In the past, a plumply rounded adult female figure has been perfectly acceptable sexually. There is nothing unfeminine about ample curves. They do, however, signal the maternal rather than the virginal phase, and the modern woman, under the influence of the contemporary youth-cult, wishes to remain a flesh-virgin, even while she copulates and bears children.

That the plump curves of the adult female are essentially linked to the maternal, rather than the courting, phase of life, is borne out by the fact that for every seven pounds a married woman with a child puts on, an unmarried woman only gains two pounds. The moral of this is that if a female wishes to remain girl-shaped, she should retain girl-status. Being unmarried, regardless of age, means that, biologically speaking, she is still displaying to a potential mate and therefore tends to stay the shape that has evolved to fit such a context. Once she becomes married, she begins to slip into the more 'comfortable' maternal shape, and her body contour starts to signal this new condition.

Although this trend is viewed by most modern women as merely a nuisance, it is too basic to be an accident. In biological terms it must have some value. One reason often given for it is that the plump, broad-hipped woman is a better child-bearer, but there seems to be little evidence to support this, especially as most of the added pelvic width consists not of a broader spacing of the bones that surround the birth-passage, but of heavy layers of fat. (The average human female body contains 28 per cent fat, the average male only 15 per cent.) There is another, more sexual explanation that seems to make better sense. The slender-girl shape is the one men enjoy visually – stare at, touch lightly, kiss, and fall in love with. The fuller, more womanly shape is the one they spend years copulating with. Perhaps, then, what happens is that the ideal visual shape changes to become the ideal tactile body, the 'dancing gazelle' becoming a cushion of 'pneumatic bliss'. Such a change would certainly explain the difference between the bony fashion model, who is to be seen but not touched, and the full-bodied female, who is to be embraced and hugged, and has already completed her biological task of attracting and becoming bonded with an adult male.

I am dealing here, of course, with extremes. In the case of the average female, the girlish body is not so bony as to be unpleasant to copulate with, and the womanly body is not so full as to be uneasy on the eye. The change is only a minor one, and both stages are successful in both a visual and a tactile sense. What has gone wrong is that modern society has swallowed the romantic myth that young lovers stay dreamily in love for ever, year in and year out, with the maximum intensity of the pair-formation stage lasting permanently, even after the pair has become fully bonded. Instead of accepting that the state of being crazily 'in love' must inevitably mature into a condition of deep but less violent 'loving', the married couple struggle to maintain the breathless ardour of their first contacts, and the physical contours that went with them. When, as must happen, they sense the initial intensity slackening, they imagine that something must have gone wrong and feel let down. In retrospect, those early Hollywood movies may have a lot to answer for.

Body skin. For both sexes and in all cultures, a smooth, clean, disease-free skin has a major sexual significance. Wrinkles, dirt and skin diseases are always anti-erotic. (The deliberately scarred or tattooed skin that occurs in some cultures is a different matter and adds to rather than detracts from the owner's sex appeal.)

In addition, the female body and limb skin is less hairy than that of the male, so that she not only attempts to increase its smoothness by the use of oils, lotions and massage, but also depilates it in a variety of ways, to exaggerate the sexual difference. Depilation has been used in many cultures and for thousands of years. It was practised, not only by certain 'primitive' tribes, but also in particular by the ancient Greeks, where women went so far as to remove much of the pubic hair as well, this being done either by tugging the hairs out manually – 'myrtle bunches pulled out by the hand', to quote one classical author – or by singeing them off with a burning lamp or hot ashes.

In modern times, females have depilated themselves with electric or safety razors and, more recently, by chemical means. Beauty experts claim that 80 per cent of women in Britain are the possessors of 'unwanted' body hair, which, although far more sparse and light than the male equivalent, nevertheless makes them feel a shade too masculine for comfort. In addition to shaving and the application of hair-dissolving creams, lotions and aerosol sprays, the beauty advisers recommend several other alternatives, such as waxing, rubbing, plucking and electrolysis. Waxing involves heating a special wax until it can be applied to the skin as a soft mould which then hardens. Once it is firm, it is stripped away, pulling the small body hairs with it. This is essentially the same device as that used from early times by Arab women, except that in their case they used a thick syrup of equal parts of water and sugar, with lemon juice added. This was poured on to the offending skin and allowed to harden, then ripped off in a similar way.

At first sight it is surprising that modern males, so many of whom daily endure a long struggle to depilate their stubble-strewn faces, have never risked experimenting with anything beyond the traditional razor shave. At second sight, however, a hidden factor emerges. This is not cowardice, or lack of initiative, but the result of a paradoxical desire to appear bearded even when beardless. Shaving always leaves a masculine blue sheen to the lower face, a tell-tale ghost of the beard that was. If, with some new technique, the adult male beard could be banished for ever, so would that virile blue sheen, and the treated face would begin to look too feminine to please its owner. Instead, by the time he dies, the average clean-shaven man will have spent well over two thousand hours scraping and rubbing at his face – a high price to pay for so contradictory a signal.

During the intense phases of pre-copulatory and copulatory activity, the whole of the body-skin surface of both the male and the female undergoes considerable change in textural quality. It glows with heat and, at the point of orgasm, there may be profuse sweating. In erotically posed photographs, models are sometimes shown displaying these conditions as visual signals. The skin may be oiled or greased to make it shine, or sprayed with water to give the impression of profuse sweating. In such cases the water is not meant to be consciously thought of as sweat, the water is plainly water and the model can often be seen to be rising from a pool or a bath to prove it. It would be too direct a comment to make the skin look obviously sweaty. Instead, the dampened surface makes its impact by unconscious association. The same is true of the habit of using a high level of red in printing colour pictures. This gives the girl's skin an erotic, flushed appearance, as though she is sexually overheated, and is a commonly used device in many magazines. Editorial requests to 'have the red brought up' do not, however, go so far that the viewer becomes consciously aware of what is being done.

Products have recently been marketed to produce an erotic skin glow artificially on a more private basis. Lovers can now cover themselves with various strange substances that make them look (and feel) as though they are in an advanced arousal condition before they have even begun to perform pre-copulatory contacts. For example, 'Love Foam' comes in aerosol cans. When it is sprayed on it looks rather like shaving cream, but when rubbed into the skin, according to the manufacturers, it makes the body 'take on a magical glow'. For even more exotic tastes, there is a substance which goes by the colourful name of 'Orgy Butter'. Called 'The Luxury Lubricant', it is advertised as 'A bold, red warm body rub . . . Gives a slippery, sensual effect. Works into the skin with rubbing, providing an after-glow.' Here, quite clearly, are the vital arousal signals – red/slippery/glow – once again mimicking the vaso-dilation and sweating of the real arousal state of the human body skin.

The shoulders. The rounded female shoulder has already been mentioned, but the larger male shoulder also deserves a comment. Width of shoulders becomes an important secondary sexual characteristic that starts to develop at puberty. The shoulders of the male adolescent widen out more extensively than those of the female, and by the time the young adult stage is reached the male is decidedly more broad-shouldered than his female companion. Like the other contour differences, this one has also been exaggerated artificially in various ways. Throughout history, male clothing has repeatedly involved an extra padding of the shoulder region, making it appear even larger as a masculine signal. The extreme situation is reached with the military epaulet, which not only makes the shoulders appear wider but also makes them much more angular. In this way they provide a double contrast with the narrower and more rounded shoulders of the female, and completely lose their hemispherical visual quality.

The jaw. There are several important sexual differences in the head region, and the first of these concerns the jaw and chin. The average human male has a slightly heavier jaw and chin region than the average female. For some reason this fact is seldom commented on, and yet it remains the one certain give-away clue in the case of an otherwise perfectly disguised male transvestite or female impersonator. Such men can pad out their body contours, depilate all visible areas of skin, cover their faces in heavy make-up, have wax injections to give themselves artificial breasts, and generally present such an alluring female appearance that occasionally a sailor in a foreign port has found himself, rather late in the transaction, involved with a 'female' prostitute who is not all that 'she' at first seemed to be. But even the very best of transvestites can do nothing about his jaw and chin, short of major surgery. Unless, by chance, he happens to be an abnormally small-jawed male, he will always present a tell-tale, heavy-jowled appearance to a discerning eye.

In some races, especially in the Far East, the extra heaviness of the male jaw and chin is less pronounced, and it is significant that in these same races there is typically a much less marked beard growth. It seems as if there is a connection between these two features. Jutting the jaw forward, by either sex, is an aggressive act – an intention movement of thrusting forward into the attack. It is the opposite of the submissive lowering of the head that occurs during a meek bowing action. The male, by having a more powerful jaw, is, so to speak, performing a permanent assertive jut. That this is important as a masculine trait is borne out by the fact that males who have receding chins are sometimes sneered at as 'chinless wonders', implying that they lack the normal male assertiveness.

Since one of the most obvious masculine characteristics of our species is the possession of a beard, it seems highly probable that this evolved in company with the more jutting jaw, the heavier bony structure providing a better base for the hairs, and the two features together producing a maximum virile jut. The peculiar chin of our species is important here. Unlike other primates, we have an outward protuberance of bone in the chin region, a protuberance which anatomists have now decided has no internal mechanical function. In the past, many theories have been put forward to explain this unique human feature, relating it to special properties of the jaw muscles and the tongue, but all of these arguments have recently been demolished. Instead, our chin jut is now accepted as primarily a signal feature and, as such, it must be seen as one that underlies the assertive jutting of the male beard.

The cheeks. Moving up the face and avoiding the mouth, which was dealt with earlier, we come to the cheeks. Here the most important signal is the blush, a reddening of the skin caused by vaso-congestion. This always starts in the cheek region, where it is most obvious, and may then spread to cover the whole of the face, the neck and, in some cases, the upper part of the trunk. Blushing is more common in females than males, and in young females than older ones. Accompanying the reddening there is a turgidity of the skin which takes on a surface glow that is visible even in a Negro blush. Blushing occurs in all human races and even in the deaf/blind, so that it appears to be a basic biological characteristic of the species. Darwin devoted a whole chapter to the subject of blushing and concluded that it reflected shyness, shame, or modesty, and indicated 'self-attention to personal appearance'. Its sexual significance is illustrated by the fact that the records show that girls who blushed freely when being offered for sale in ancient slave-markets for use in harems fetched higher prices than those that did not. As an invitation to intimacy, whether desired or not, the blush seems to have been a powerful signal.

The eyes. The most important of human sense organs, these not only see all the various signals we have been examining, but also transmit some on their own account. We all make and break eye contacts repeatedly during face-to-face encounters, looking at people to check their changing moods and then looking away to avoid the threat of staring at them. Between lovers, however, the stare can become more prolonged without being disconcerting or aggressive. Lovers 'gaze deeply into one another's eyes' for a particular reason. Under the influence of strong emotions of a pleasant kind, our pupils dilate to an unusual degree, the small dark spot in the centre of the eye becoming a great black disc. Unconsciously, this transmits a powerful signal to the loved one, indicating the intensity of the love felt by the dilator. This fact has only recently been studied scientifically, but it has been known for centuries, Italian beauties of earlier days having placed drops of belladonna into their eyes in order to create the effect artificially. In modern times a similar device has been used by advertisers, who, using black ink instead of belladonna, have touched up photographs of model girls, enlarging their pupils to make them more appealing.

Another change that takes place in the eyes when emotions run high is a slight increase in tear production. In an intensely loving condition this does not usually go to the extreme of producing tears that actually start to run from the eye, but merely gives the eye surface an increased glistening quality. These are the shining eyes of love, and combine with the pupil dilation to leave no doubt about the condition of the signaller.

Eye movements of various kinds also invite intimacy. Apart from the well-known wink, the rolling of the eyes is also reported to be a direct invitation to copulation in certain cultures. A demure dropping of the eyes also transmits its message in the female, while a slight narrowing of them can indicate interest on the part of a male. When making a first encounter, the holding of a glance slightly longer than is usual can also make an impact, acting as a hint, so to speak, of the deep gazing that may develop later.

A wide-eyed, or magnified, stare is sometimes employed by a female inviting intimacy, and associated with this is the feminine device of fluttering the eyelashes or batting the eyelids. The word 'bat' used in this connection is a modification of 'bate', which means a beating of wings, and, in our culture at least, is a decidedly non-masculine action, so much so that it is sometimes used by a male when he is mockingly imitating a female gesture. Perhaps because eyelash movements of this kind are so essentially feminine, a great deal of lash exaggeration is practised by females at the present time. This began with the use of mascara, to make the eyelashes look heavier and more conspicuous, then moved on to eyelash curlers, and finally, in the 1960s, culminated with the development of sets of long artificial lashes which were added to the real ones. Today, one company alone offers no less than fifteen different styles of artificial eyelashes, including 'Wispy-tipped Starry Lashes' that 'open up eyes' and 'Raggedy Lashes' that 'enlarge small eyes'. These are fixed to the upper eyelids, as are such exotic concoctions as 'Cluster-lashes', 'Natural-fluff Lashes' and 'Super-sweepers'. For the lower eyelid, there are 'Winged Under-lashes' to 'widen, brighten eyes'. As in so many other parts of the body, when the female has something that sends out an important feminine signal, she makes the most of it. This new trend to exaggerate the eyelash region would certainly provide a feast for an amorous male Trobriander, who, as an important part of his love-making, regularly bites off the lashes of his loved one. Luckily for the latter, eyelashes grow remarkably quickly, each one only lasting for three to five months even in the unbitten condition.

The eyebrows. Above the eyes the human animal possesses two unique patches of hair at the bottom of an otherwise hairless forehead. The eyebrows were once thought to operate as devices for stopping the sweat running down into the eyes, but their basic function is that of signalling changing moods. They are raised in fear and surprise and lowered in anger, knitted together in anxiety and cocked in a questioning glance. At the moment of a friendly acknowledgment they are flicked quickly once up and down.

The eyebrows of the female are less thick and bushy than those of the male, so, once again, they are available for exaggeration to make a female condition more feminine. They have frequently been plucked to make them thinner, and in the 1930s they were reduced to a mere pencil line. Even this extreme was exceeded in earlier days, however, Japanese brides once going so far as to shave them off altogether at the time of their marriage.

The sexual nature of this comparatively trivial modification to the female appearance is well illustrated by the fact that, in 1933, a girl who applied for a post as a nurse in a London hospital was warned by the matron that, amongst other things, she would not be permitted to continue plucking her eyebrows. A complaint was made, her case was taken up, and the London County Council was urged to give the matron an official rebuke, but the request was refused. The hospital patients were therefore spared the undue stimulation of a sensuously plucked eyebrow, and unmodified female shapes continued to glide through the long white corridors.

The face. Before leaving the facial region, it is worth taking a last glance at it as a whole, rather than as a set of smaller details. It is, without any doubt, the most expressive region of the entire human body, capable of transmitting incredibly varied and subtle emotional messages by means of its complex expressions. By contracting and relaxing the special muscles, particularly those around the mouth and eyes, we can signal everything from joy and surprise to sadness and anger. As a device for inviting intimacy, this is of major importance. A face that is soft and smiling, or alert and excited, attracts us strongly. One that is forlorn and helpless, or in agony, may also stimulate us to approach and console. A tense, hard or grumpy face has the opposite effect. This is common enough knowledge, but there is an interesting long-term effect that operates on the human face, and which deserves brief comment.

Where facial expressions are concerned, we can speak of an 'on-face' and an 'off-face'. The on-face is the one we use during social en-counters. We speak of 'putting on a happy face', or 'putting a good face on it', and we try to avoid 'losing face' in public. If we want to appear friendly, we adopt a soft, smiling expression. Alternatively, we set our faces into a grim or pompous look to cope with more serious occasions. When we are alone and unseen, however, we let our faces go off-duty. When this happens they arrange themselves into the posture typical of our overall long-term mood. The basically anxiety-ridden man, who tried his best to look happy at the party, now tenses up his solitary face, revealing his true emotional condition, but revealing it, of course, to no one but himself. (Unless he catches sight of himself in a mirror, even he may not be aware of it.) The basically happy and contented man, who tried his best to look sad and serious at the funeral, now relaxes his solitary face, with his lips softened and the furrows in his brow smoothed out.

Most of us change our long-term moods from time to time, so that our facial muscles do not experience prolonged exposure to one particular off-face condition. We may feel depressed in the morning but happy again by the evening, and in our solitary moments our facial postures will vary accordingly. For individuals who live in a state of more or less permanent private anxiety, depression or anger, the situation is differ-ent. For them, there is a danger that their off-faces will become com-pletely set. In such instances, the facial musculature seems to become moulded in one basic expression. The crease lines on the forehead, around the mouth and at the side of the nose become almost permanent.

Such people find it hard to switch to the appropriate on-face during social encounters. The anxious person still looks anxious even when smiling a greeting. The grumpy man still looks grumpy even when laughing at a joke. The set of the muscles somehow survives and the on-face becomes superimposed on the off-face, rather than replacing it. In this way, facial expressions can tell us something about a person's past as well as his present emotional condition.

It is not clear how long these off-face wrinkles will last after there has been a fundamental change in a way of life. If someone who has been anxious and worried all his life suddenly achieves a contented condi-tion, the wrinkles will not vanish overnight. If the person in question is elderly when this welcome change occurs, they may never fade. Certainly, in all such cases there will be a period of time when the old set-face will be lingering on despite the fact that its message is no longer relevant, but I know of no studies in which the length of this period has been measured.

These comments also apply, incidentally, to the general body posture of the human being. There are slumped bodies and alert bodies, stiffly tense bodies and softly lithe ones. Again, we are capable of changing our body-muscle tensions to match our social moods and occasions, but, as with the face, a prolonged, extreme condition can give us a set posture that is difficult to shake off even when we want to. Rounded shoulders can develop into a permanent hunch that even winning a million cannot straighten out, and a tense stiff-legged walk can become a companion for life.

The hair. Finally we reach man's crowning glory, his densely packed mop of roughly 100,000 head-hairs. In some races these are woolly or frizzy, in others they hang down long and straight, or flow in the wind. They grow at a rate of nearly five inches a year, and each one lasts up to six years before it falls out and is replaced. This means that the average uncut head of straight hair would reach down to the hips, and must have given an unbarbered, primitive human being an extraordinary appearance, when compared with any other species of primate. While our body hair became stunted and almost invisible at a distance, our head hair ran amok.

Apart from the fact that older males, but not older females, tend to become bald in many cases, there is no sex difference in the head hair. Biologically, both men and women are long-haired, and this characteristic has developed as a species recognition signal, rather than a sexual one. Culturally, however, it has frequently been modified as a gender signal. Sometimes men have worn longer hair than women, but usually the reverse has been the case. In recent centuries, masculine hair has been cropped basically as an anti-parasite device, aggressive army sergeants referring to long male hairs as 'louse-ladders'. Women have nearly always maintained a moderate hair length; it is the men who have fluctuated wildly from one extreme to the other. In the past this has sometimes led to the wearing of huge, drooping male wigs, a practice still seen today in the case of British judges. In general, however, the longer locks of recent times have become so firmly associated with the female sex that for a man to wear his hair in anything remotely approaching its natural length has been quite wrongly viewed as essentially feminine. During the last decade this situation has changed dramatically amongst young males, and hair length seems once again to be re-establishing itself in its truly non-sexual role. It is perhaps ironic that, although it is modern hygiene that has made this so much less of a parasite risk, it is the anti-hygiene hippie movement that has taken the lead.

Cleaning, grooming, washing and oiling the hair has always been an important accompaniment to its cultural use as a sexual signal. Ancient urbanites, like their modern counterparts, were prepared to go to great lengths to obtain the desired effect. The oldest known hair tonic consisted of 'Paws of a dog, one part; kernels of dates, one part; hoof of a donkey, one part. Cook thoroughly with oil and anoint.' Today, glistening, gleaming, swishing hair remains almost every girl's ideal and, as the advertisers repeatedly tell us, hair that is 'dull and lifeless' ruins its owner's chances of inviting intimacy.

In this grand tour of the signalling human body the different parts have been taken one by one, but there is still the whole person to be considered. The isolated parts are displayed not singly but all at once, in a general combination and in a specific context. It is the tremendously varied way in which they can combine with one another, and the great range of contexts in which they can be presented, that makes social interaction so complex and so fascinating. Every time we enter a room, or walk down a street, we are transmitting a mass of signals, some purely biological and others culturally modified, and we are always unconsciously aware of this fact and adjusting them in a hundred subtle ways as we move through our many kinds of social encounter. Nearly always we are striving to send out a balanced set of signals, some attracting intimacy, others repelling it. Just occasionally we go much further in one direction or the other, either displaying our invitations blatantly, or presenting ourselves in a hostile, rejecting manner to those around us.

Throughout this chapter, in surveying the various visual invitations to sexual intimacy, I have tended to dwell on extremes. I have selected the most vivid examples in order to underline the points I have been making. Codpieces, corsets and epaulets may seem remote from the ordinary signals of sex appeal used by the average adult of today, but they nevertheless help to draw our attention to the less exaggerated devices – the tight trousers, the belts and the padded shoulders – that are more widely and less obviously employed. Similarly, belly-dancing may be no more than an exotic form of entertainment, but again its inclusion in this survey helps us to appreciate the less extreme dance movements used nightly by hundreds of thousands of ordinary girls at parties and discotheques.

Whether, as adults, we go to great lengths to improve and display our visual signals of sex appeal, or whether we treat the whole business in a more off-hand manner, whether we resort to artificial aids (and there are few of us who do not employ *some*), or whether we scorn them and prefer a more 'natural' approach, we are all of us constantly transmitting a complex set of visual signals to our companions. Many of these signals inevitably have to do with our adult sexual qualities, and even when we are totally unaware of what we are doing, we never stop 'reading the signs'. In this way we prepare ourselves for a major social step – the step that leads us to initiate the first tentative contact with a potential sexual partner and carries us over a vital threshold into the whole complex world of sexual intimacy itself.

3

SEXUAL INTIMACY

IN DISCOVERING HIS personal identity, the growing child must reject the soft embrace of his mother's arms. At last, as a young adult, he stands alone. As a baby, his trust in his mother was unlimited, his intimacy total. Now, in maturity, both his relationships with other adults and his intimacies with them are severely limited. Like them, he keeps his distance. Blind trust is replaced by alert manœuvring, dependency by interdependency. The gentle intimacies of infancy that gave way to the joyful games of childhood have become the tough social transactions of adulthood.

There is no denying the excitement. There are things to explore, goals to achieve and status to be raised. But where did all the loving go? Loving was an act of giving, giving oneself without question to another person, but adult relationships are not like that . . .

Up to this point my words could apply as much to a growing monkey as to a growing human. The pattern is the same. But now comes a difference. If the monkey is a male, it will never again, as an adult, know the total intimacy of a loving bond. Until the day it dies, it will continue to exist in the loveless world of rivalries and partnerships, of competition and co-operation. If it is a female, it will eventually regain the loving condition, as a mother with an infant of its own, but like the male it will know no such bond with another adult monkey. Close friendships, yes; partnerships, yes; brief sexual encounters, yes; but total intimacy, no.

For the adult human, however, there is such a possibility. He is capable of forming a powerful and lasting bond of attachment for a member of the opposite sex that is much more than a mere partnership. To say that 'marriage is a partnership', as is so often done, is to insult it, and to completely misunderstand the true nature of a bond of love. A mother and her baby are not 'partners'. The baby does not trust the mother because she feeds and protects him; he loves her because she is who she is, not because of what she does. In a partnership one merely exchanges favours; the partner does not give for the sake of giving. But between a pair of adult human lovers there develops a relationship like that between mother and child. A total trust develops and, with it, a total bodily intimacy. There is no 'give and take' in true loving, only giving. The fact that it is 'two-way giving' obscures this, but the 'two-way receiving' that inevitably results from it is not a condition of the giving, as it is in a partnership; it is simply a pleasing adjunct to it.

For the cautious, calculating adult, the entry into such a relationship appears a hazardous affair. The resistance to 'letting go' and trusting is enormous. It breaks all the rules of bargaining and dealing that he is so used to in all his other adult relationships. Without some help from the lower centres of his brain, his higher centres would never permit it. But in our species that help is forthcoming and, often against all reason, we fall in love. For some, the natural process is suppressed, and if they do enter into a state of marriage or its equivalent, they do so as if it were a business transaction: you rear child, I earn money. This 'baby-buying', or 'status-buying', has sadly become commonplace in our crowded human zoos, but it is fraught with dangers. The mated pair are held together, not by internal bonds of attachment, but by the external pressures of social convention. This means that the couple's natural potential for falling in love still lies waiting inside their brains and can leap into action without warning at any time, to create a true bond somewhere outside their official one.

For the lucky ones this sequence does not take place. As young adults, they find themselves falling in love uncontrollably and forming a true bond of attachment. The process is a gradual one, although it does not always appear to be. 'Love at first sight' is a popular concept. It is, however, usually a retrospective judgment. What occurs is not 'total trust at first sight', but 'powerful attraction at first sight'. The progress from first attraction to final trust is nearly always a long and complex sequence of gradually increasing intimacies, and it is this sequence that we must now examine.

To do so, the simplest method is to take a pair of 'typical lovers', as seen in our Western culture, and follow them through the process of pair-formation from first glimpse to ultimate copulation. In doing this we must always remember that there is in reality no such thing as the 'typical lover', any more than there is the 'average citizen' or 'man in the street'. But it helps if we start out by trying to imagine one, and then, afterwards, consider the variations.

All animal courtship patterns are organized in a typical sequence, and the course taken by a human love affair is no exception. For convenience we can divide the human sequence up into twelve stages, and see what happens as each threshold is successfully passed. The twelve (obviously over-simplified) stages are these.

1. *Eye to body*. The most common form of social 'contact' is to look at people from a distance. In a fraction of a second it is possible to sum up the physical qualities of another adult, labelling them and grading

them mentally in the process. The eyes feed the brain with immediate information concerning the sex, size, shape, age, colouring, status and mood of the other person. Simultaneously a grading takes place on a scale from extreme attractiveness to extreme repulsiveness. If the signs indicate that the individual in view is an attractive member of the opposite sex, then we are ready to move on to the next phase in the sequence.

2. *Eye to eye.* While we view others, they view us. From time to time this means that our eyes meet, and when this happens the usual reaction is to look away quickly and break the eye 'contact'. This will not happen, of course, if we have recognized one another as previous acquaintances. In such cases, the moment of recognition leads instantly to mutual greeting signals, such as sudden smiling, raising of the eyebrows, changes in body posture, movements of the arms and eventually vocalizations. If, on the other hand, we have locked eyes with a stranger, then the rapid looking away is the typical reaction, as if to avoid the temporary invasion of privacy. If one of the two strangers does continue to stare after eye contact has been made, the other may become acutely embarrassed and even angry. If it is possible to move away to avoid the staring eyes, this will soon be done, even though there was no element of aggression in the facial expressions or gestures accompanying the stare. This is because to perform prolonged staring is in itself an act of aggression between unfamiliar adults. The result is that two strangers normally watch one another in turn, rather than simultaneously. If, then, one finds the other attractive, he or she may add a slight smile to the next meeting of glances. If the response is returned, so is the smile, and further, more intimate contact may ensue. If the response is not returned, a blank look in reply to a friendly smile will usually stop any further development.

3. *Voice to voice.* Assuming there is no third party to make introductions, the next stage involves vocal contact between the male and female strangers. Invariably the initial comments will concern trivia. It is rare at this stage to make any direct reference to the true mood of the speakers. This small-talk permits the reception of a further set of signals, this time to the ear instead of the eye. Dialect, tone of voice, accent, mode of verbal thinking and use of vocabulary permit a whole new range of units of information to be fed into the brain. Maintaining this communication at the level of irrelevant small-talk enables either side to retreat from further involvement, should the new signals prove unattractive despite the promise of the earlier, visual signals.

4. *Hand to hand.* The previous three stages can all occur in seconds, or they may take months, with one potential partner silently admiring the other from a distance, not daring to make vocal contact. This new

stage, the hand to hand, may also take place quickly, in the form of the introduction handshake, or it is likely to be delayed for some considerable time. If the formalized, non-sexual handshake does not come into operation, then the first actual body contact to occur is likely to be disguised as an act of 'supporting aid', 'body protection' or 'directional guidance'. This is usually performed by the male towards the female and consists of holding her arm or hand to help her cross a street, or climb over an obstruction. If she is about to walk into an obstacle or danger spot, then the hand of the male can quickly take the opportunity to reach out swiftly and take her arm to alter her course or check her movement. If she slips or trips, a supporting action with the hands may also facilitate the first body contact. Again, the use of acts which are irrelevant to the true mood of the encounter is important. If the body of the girl has been touched by the man in the act of assisting her in some way, either partner can still withdraw from further involvement without loss of face. The girl can thank the man for his help and leave him, without being forced into a position where she has to deliver a direct rebuff. Both parties may be well aware that a behaviour sequence is just beginning, and that it is one that may lead eventually to greater intimacies, but neither as yet does anything which openly states this fact, so that there is still time for one to back out without hurting the other's feelings. Only when the growing relationship has been openly declared will the action of hand-holding or arm-holding become prolonged in duration. It then ceases to be a 'supportive' or 'guiding' act and becomes an undisguised intimacy.

5. *Arm to shoulder*. Up to this point the bodies have not come into close contact. When they do so, another important threshold has been passed. Whether sitting, standing or walking, physical contact down the side of the body indicates a great advance in the relationship from its earlier hesitant touchings. The earliest method employed is the shoulder embrace, usually with the man's arm placed around the girl's shoulders to draw the two partners together. This is the simplest introduction to trunk contact because it is already used in other contexts between mere friends as an act of non-sexual companionship. It is therefore the smallest next step to take, and the least likely to meet rebuff. Walking together in this posture can be given the air of slight ambiguity, half way between close friendship and love.

6. *Arm to waist*. A slight advance on the last stage occurs with the wrapping of the arm around the waist. This is something the man will not have done to other men, no matter how friendly, so that it becomes more of a direct statement of amorous intimacy. Furthermore, his hand will now be in much closer proximity to the genital region of the female.

7. *Mouth to mouth*. Kissing on the mouth, combined with the full frontal embrace, is a major step forward. For the first time there is a strong chance of physiological arousal, if the action is prolonged or repeated. The female may experience genital secretions and the male's penis may start to become erect.

8. *Hand to head*. As an extension of the last stage, the hands begin to caress the partner's head. Fingers stroke the face, neck and hair. Hands clasp the nape and the side of the head.

9. *Hand to body*. In the post-kissing phase, the hands begin to explore the partner's body, squeezing, fondling and stroking. The major advance here is the manipulation by the male of the female's breasts. Further physiological arousal occurs with these acts and reaches such a pitch that, for many young females, this is a point at which a temporary halt is called. Further developments mean increasing difficulty in breaking off the pattern without continuing to completion, and if the bond of attachment has not reached a sufficient level of mutual trust, more advanced sexual intimacies are postponed.

10. *Mouth to breast*. Here the threshold is passed in which the interactions become strictly private. For most couples this will also have applied to the last stage, especially where breast manipulations are concerned, but advanced kissing and body-fondling does occur frequently in public places under certain circumstances. Such actions may cause reactions of disapproval in other members of the public, but it is rare in most countries for serious steps to be taken against the embracing couple. With the advance to breast-kissing, however, the situation is entirely different, if only because it involves the exposure of the female breast. Mouth-to-breast contacts are the last of the pre-genital intimacies and are the prelude to actions which are concerned not merely with arousal, but with arousal to climax.

11. *Hand to genitals*. If the manual exploration of the partner's body continues, it inevitably arrives at the genital region. After tentative caressing of the partner's genitals, the actions soon develop into gentle, rhythmic rubbing that simulates the rhythm of pelvic thrusting. The male repeatedly strokes the labia or clitoris of the female and may insert his finger or fingers into the vagina, imitating the action of the penis. Manual stimulation of this kind can soon lead to orgasm for either sex, and is a common form of culmination in advanced pre-copulatory encounters between lovers.

12. *Genitals to genitals*. Finally, the stage of full copulation is reached and, if the female is a virgin, the first irreversible act of the entire sequence occurs with the rupture of the hymen. There is also, for the first time, the possibility of another irreversible act, namely that of fertilization. This irreversibility puts this concluding act in the se-

quence on to an entirely new plane. Each stage will have served to tighten the bond of attachment a little more, but, in a biological sense, this final copulatory action is clearly related to a phase where the earlier intimacies will already have done their job of cementing the bond, so that the pair will want to stay together after the sex drive has been reduced by the consummation of orgasm. If this bonding has failed, the female is liable to find herself pregnant in the absence of a stable family unit.

These then are twelve typical stages in the pair-formation process of a young male and female. To some extent they are, of course, culturally determined, but to a much greater extent they are determined by the anatomy and sexual physiology common to all members of our species. The variations imposed by cultural traditions and conventions, and by the personal peculiarities of certain unusual individuals, will alter this major sequence in a number of ways, and these can now be considered against the background of the typical sequence through which we have just passed.

The variations take three main forms: a reduction of the sequence, an alteration in the order of the acts and an elaboration of the pattern.

The most extreme form of reduction is forcible mating, or rape. Here the first stage runs as quickly as is physically possible to the last stage, with all the intervening phases condensed to the absolute minimum. After eye-to-body contact has been made by the male, he simply attacks the female, omitting all arousal stages, and goes as rapidly to genital-to-genital contact as her resistance will allow. The non-genital body contacts are limited purely to those necessary to overpower her and strip her genital region of clothing.

Viewed objectively, rape in the human species lacks two important ingredients: pair-formation and sexual arousal. The rapist male, by omitting all the intermediate stages of the sexual sequence, clearly does not allow a bond of attachment to grow between himself and the female in question. This is obvious enough, but it is relevant in biological terms, because our species does require this personal attachment to develop, as a means of safeguarding the successful rearing of the offspring that may result from the copulation. There are other species, with little or no parental responsibilities, where, in theory, rape would create no problems. That it is rare is due, amongst other things, to the physical difficulty of achieving the goal of rape. A man without a pair of grasping hands and verbalized threats would find it virtually impossible, and this is a dual advantage which other species lack. Even where animal rape does seem to occur, appearances can be deceptive. For example, carnivores can, and most carnivore species do, grab their females by the scruff of the neck with their jaws during the act of mating, as if to prevent their mates from escaping, but there is still the problem of successfully inserting the penis into the vagina of a writhing female form. If the female is unresponsive, the male stands little chance. The truth is that the superficially savage act of neck-gripping in male carnivores is a rather specialized movement. Although it looks like the act of a rapist male, it is in fact the carnivore equivalent of a gentle parental embrace in our own species. The bite is strongly inhibited, so that the teeth do not harm the female. This is the pattern of behaviour a parent carnivore employs towards its young when carrying them from place to place. The male is, in effect, treating the female like a cub or a kitten, and if she is sexually responsive to him, she reacts like one, going limp in his jaws as she once did when being protectively transported by her mother in early life.

For the human male animal, rape is comparatively easy. If physical force is not enough, he can add threats of death or injury. Alternatively he can contrive to render the female unconscious or semi-unconscious, or can enlist the aid of other males to hold her still. If the absence of the female's sexual arousal makes penis insertion difficult or painful, he can always resort to the use of some alternative form of lubrication to replace the missing natural secretions.

For the female in question these proceedings are, to say the least, both unsatisfying and unsatisfactory, and, to say the most, may result in severe trauma and psychological damage. Only in cases of rape where the partners are already known to one another and where the female has a strong masochistic streak is there any chance at all of an emotional attachment developing as a result of this violent reduction of the normal sexual sequence of our species.

I have gone into this question of rape at some length because it bears a close relationship to another form of sexual reduction which is much more widespread and important in our culture. To contrast it with the violent rape we have been discussing, we might call it 'economic rape'. Unlike the violent kind, it occurs, not in derelict building-sites or damp hedgerows, but in decorously draped boudoirs and cosy bedrooms. It is the loveless mating act of the marriage of economic convenience, the act of couples who marry and copulate with only the barest hint of a true bond of attachment.

In past centuries, the parentally controlled status marriage was commonplace. Today it is becoming increasingly rare, but the psychological scar it left on the children it spawned is more lasting. As growing, developing witnesses to this loveless relationship, the offspring of the match were themselves in danger of becoming sexually crippled, so that they were incapable of expressing the full amatory sequence typical of our species. Their sexual anatomy was in perfect working order; their physiological arousal mechanisms were operating efficiently; but their ability to relate these biological features to a deep and lasting bond of attachment will have been stunted by the atmosphere in which they matured. They, in turn, will find it difficult to make successful pair-bonds, but social pressures will encourage them to try, and once again the next generation of children will suffer. Such a reverberation is difficult to eliminate, and the untold damage caused by past cultural interferences with the natural human process of falling deeply in love is still with us today, even though the parentally controlled marriage is fading into history.

The pattern of 'economic rape' is not, of course, as extreme as that of violent rape, in the way in which it condenses the full twelve-stage sequence. Superficially it may appear very similar to the full pattern, with the partners 'going through the motions' of the different stages, one by one, until copulation is reached. But if we examine the actions in detail we find that they are all reduced in intensity, duration and frequency.

Take first the classic case of the young couple who are pushed together to satisfy the economic or status relationship of their two families. In earlier centuries, their pre-marital courtship might typically involve no more than a few abbreviated embraces and kisses, following prolonged verbal exchanges. Then, with little knowledge of one another's bodies or sexual emotions, they are thrust together into a marriage bed. The bride is advised that nasty but necessary things will have to be done to her by her bridegroom in order to ensure the future population of the nation, and that while this is going on, she is to 'lie still and think of England'. The male is given rudimentary instruction concerning female anatomy and told to be gentle with his bride because she will bleed when he penetrates her. With this information the young couple perform their sexual duties as simply and quickly as possible, with a minimum of pleasure and a minimum of pair-bonding. For the female, there is rarely if ever an orgasm. For the male, there is an unresponsive object in his bed, which happens to be his wife, but which sexually is little more than an object, the vagina of which he uses like a hand to masturbate his penis. In their public, social lives the young couple will, naturally, be provided with a set of rules enabling them to simulate a loving relationship. Each of the public intimacies, severely restricted in the form it can take, is precisely described and defined by the books of polite etiquette, so that it will become almost impossible to tell the true loving couple from the false. Almost, but not quite, and for the children it will be painfully easy, for they have not yet had their heads crammed with detailed rules of conduct, and they will intuitively detect the degree of lovingness or lovelessness in their parents' relationship. And so the damage will go on.

If this description seems bizarre today, in the latter half of the twentieth century, then this is not because such marriages no longer occur, but rather because they are less blatantly organized than they once were. A much greater show of loving is put on nowadays to mask such relationships, but a show it remains nevertheless. Parents are less involved than they were, which also disguises the pattern. Now it is one or both members of the pair who set out themselves to construct an economically based marriage. Behind her wedding veil the bride's lips are moving, but she is not overcome with emotion; she is busy calculating her alimony rating. The man at her side with a faraway expression on his face is not lost in a romantic dream, he is working out the impact his socially efficient bride will have on his business colleagues. Admittedly, the brides no longer lie still and think of wherever-it-is on their wedding nights. Instead, they check their orgasm frequencies against the national average for their age group, educational level, and racial and urban background. If they fall below the required level, they employ a firm of private investigators to find out where the husband is relocating the additional 1.7 orgasms per week that they should be getting. Meanwhile the husbands are trying to estimate how many early evening drinks they can consume before running the risk of alcoholically impairing their ability to achieve an insertable erection later that night. All too often it is these that are the sweet mysteries of life in modern urbania.

In looking at reductions in the sexual sequence, we have progressed from rape, to the parentally controlled marriage of the past, to the so-called 'bitch/bastard' marriage of modern times. The obsession with orgasm frequency in the last of these is an important new development that appears to lead us away from the reduction and compression of the full sexual sequence that we were discussing. Indeed, it looks like a swing in the other direction, towards elaboration rather than reduction. But the matter is not quite so simple. Basically, what has happened in the new 'sexual freedom' is that the later stages have been given much greater emphasis. The elaboration of the sequence has all been concentrated at one end of it – the copulatory end. The earlier courtship patterns, so important for pair-formation, instead of being elaborated, have been reduced and simplified. It is worth trying to discover the way this has come about.

In earlier centuries, the courtship stages were prolonged in time, but severely restricted in intensity. Insistence on obeying formal rules of procedure, down to the last detail, reduced their emotional impact. Then, after marriage, the later pre-copulatory and copulatory patterns were strongly curtailed by ignorance and anti-erotic propaganda. The males solved this problem with brothels and mistresses. The females, by and large, did not solve it. In the first half of the present century the situation changed. Parental control was loosened and serious attempts were made at sexual education, with the publication of books on 'married love'. The result was that young couples were much freer to search for a partner that suited them and to indulge in less formalized courtship activities. The phenomenon of the chaperone vanished. The rules of conduct for body contact were relaxed, so that virtually all stages of the sexual intimacy sequence were permitted, excluding only the final, genital ones. These pre-marital activities were, however, still expected to cover a considerable period of time. Eventually, when marriage did take place, the couple were able to take to their mating bed with a much greater knowledge of one another's bodies and emotional personalities. Efficient contraception had appeared on the scene, combining with the new sexual knowledge to make marital pleasures less restricted and more satisfying.

During this phase there was a tendency for young couples, before marriage, to indulge in prolonged 'petting sessions'. The idea that they were now allowed to go so far, but no further, seemed good in theory, but was difficult in practice. The reason is obvious enough. Unlike the young couples of earlier days, they were now permitted to go past the first stages of courtship – the ones which helped to form a bond of attachment, but which did not produce strong physiological arousal of the sex organs – and on into the stages which were really concerned with pre-copulatory stimulation. The half-way mark between the two is the act of the mouth-to-mouth kiss. Simply performed this is a pleasant bonding act, but energetically repeated it is also the starting point of pre-copulatory arousal.

This led to a new type of crisis for the young lovers. Prolonged petting resulted in prolonged erections for the males and prolonged sexual secretions for the females. One of three things then happened: they broke off the sequence in accordance with 'official ruling' and remained intensely frustrated, or they continued by non-copulatory means to achieve mutual orgasm, or they broke the rules and copulated. If they used the second method and masturbated or fondled one another to orgasm, and if they continued to do this over a long pre-marital period, there was always the danger that this pattern of climax would begin to assume too great a significance in their sexual relationship, making for difficulties when, eventually, they were able to complete the act of copulation after marriage. If they chose the third alternative and broke the rules, there was the problem of guilt and secrecy. Nevertheless, despite these difficulties, the prolonged pre-marital phase was strongly conducive to the formation of a power-ful bond of attachment, so that it had a great deal to commend it over the previous situation, when the couple were severely restricted in their actions.

Moving on now to more recent times, a new change has occurred. Although official attitudes may still be the same, they are less strictly enforced. With further improvements in contraception, virginity has lost its significance for many young girls. The non-copulation rule that was once reluctantly broken is now commonly ignored. Virginity, far from being prized, has almost become a stigma, an indication of some kind of sexual inadequacy. Pre-marital copulation is an accepted pat-tern by the young lovers, if not by their parents. The result of this development, which is more widespread than some people like to admit, is that the lovers no longer experience the intense 'petting frustrations' of their predecessors, nor are they in danger of becoming fixated on masturbatory activities. Instead, their courtship grows natur-ally, and without undue postponement, into the full twelve-stage sex-ual sequence.

Assuming the existence of adequate venereal hygiene and efficient, readily available contraception, what, if any, is the danger of this new situation? The answer, as some have seen it, is the 'tyranny of the orgasm', the need created by the social pressures of the new permissive convention to achieve maximum copulatory performance. This is seen as a threat to the person who truly falls in love, but who is incapable of sufficiently impressive orgasmic achievements.

There is something curiously short-sighted about this criticism. Ear-lier I did mention the obsession with orgasm frequency, but that was in connection with the loveless marriage, the modern equivalent of the economic or status marriage. There, as I pointed out, the female may feel she has failed if her sexual athleticism is below standard, because she is concerned, even in matters of sex, with the business of status. But if two young people are in love today, they will laugh at the desperate athleticism of the copulating non-lovers. For them, as for true lovers at all points in history, a fleeting touch on the cheek from the one they adore will be worth more than six hours in thirty-seven positions with someone they do not. This has always been so, but for these new lovers there is the added advantage that, circumstances permitting, they are no longer restricted to a mere cheek touch. They can do anything they wish with one another's bodies, as much *or* as little as they wish. If the powerful bond of attachment has formed, then it will be the quality of the sexual behaviour that counts, not its mere quantity. For them, the new conventions simply permit, they do not insist, as some critics seem to think.

Another point which the critics seem to miss is that when a couple have started to fall in love, they will not want to omit the earlier stages of the sexual sequence. They will not give up holding hands merely because they are permitted to copulate. Furthermore, they are not the ones who will be likely to have much trouble in quite naturally enjoying maximum orgasmic pleasure when they reach the final stages of the sequence. The emotional intensity of their personal relationship will ensure this, and they will happily reach climax after climax without having to resort to the contorted wrestlers' postures of the ubiquitous modern sex manuals.

Perhaps the greatest danger that the permissive young lovers face today is an economic one, for they still inhabit a complex, economically structured society, and it was no accident that economics came to figure so largely in the marriages of yesterday. Previously this aspect was safeguarded by all the rigid restrictions placed on early sexual behaviour. Sexual intimacy suffered, but social status was ensured and organized. Now the social status of the young pair has become a problem, as sexual intimacy blossoms. How can a pair of seventeen-year-old lovers, who are fully sexually mature, who have developed a powerful bond of attachment, and who are enjoying a full sexual life, set up home in our modern economy? Either they have to wait in a kind of social limbo, or they have to 'drop out' of the accepted social pattern. The choice is not an easy one, and the problem has yet to be solved.

We arrived at this discussion by considering the ways in which the sexual sequence becomes reduced from its full expression. We must now leave the young lovers, fully expressing it, but enduring considerable social problems in so doing, and return to the reductions once more. What of the sexually active non-lover? We have dealt with the rapist and the sexually inhibited marriage partners who reduced their copulatory stages to a child-spawning minimum, but what about the loveless sexual athletes of today? How do they reduce the sexual sequence of the typical lovers? For them, the later genital stages are not expressed as the culmination of a pattern, but as a replacement for it. In earlier times this was precisely what occurred when a man visited a prostitute. There was no holding of hands, no cuddling or murmuring of sweet nothings, merely a quick business transaction and then straight to direct genital contact. What we might call 'commercial rape'. In previous generations this was often a young man's first introduction to copulation, but today such professional services are hardly necessary. Instead they are replaced by what is referred to as 'sleeping around'. In such cases there is frequently a massive reduction in the early stages of the sequence, just as there was with the visit to the prostitute. The situation is summed up by the girl in the cartoon who says 'Boys simply don't want to kiss any more', as she returns late at night to her room in an obvious post-copulatory condition, exhausted and with her clothes dishevelled, but with her make-up still neatly intact.

The result of this type of reduction is to provide the maximum of copulatory activity with the minimum of pair-bond development. As a status device this may permit repeated ego-boosting, but as a source of intense pleasure it degrades sexual activity to the level of something like urinating. It is not surprising, therefore, that for the 'permissive', non-loving copulator some elaboration of the consummatory act becomes desirable. If it has lost its emotional intensity along with its powerful bond of personal attachment, then it must be given some increased physical intensity as a compensation. This is where the illustrated sexual manuals come on the scene, and it is worth analysing some of them to see what they recommend.

A sample taken at random from the large number now on sale included a combined total of several hundred photographs, each showing a young naked couple in the act of 'making love'. Of these illustrations, no more than 4 per cent showed any of the first eight stages in my twelve-stage sequence described earlier. By contrast, 82 per cent showed full copulation, each book including between thirty and fifty posture variations. This means that the vast majority of all the various sexual intimacies were illustrated as the final stage of genital/genital contact, and clearly demonstrates the great emphasis placed on this concluding element of the sequence. Whereas previous censorship limited the illustration of amorous activities to the earlier stages, the removal of this censorship, instead of enriching the situation, has simply had the effect of shifting concentration from one end of the scale to the other. The implied message is that the act of copulation should be made as complex and varied as possible, and forget about the rest. Many of the postures shown are obviously uncomfortable, if not actually painful to maintain for any length of time, except possibly for trained circus acrobats. Their inclusion can only reflect a desperate search for copulatory novelty as a means to obtaining further arousal. The emphasis is no longer on loving, but on sexual athletics.

There is, of course, nothing harmful about these patterns as amusing and playful additions to sexual behaviour, but if obsession with them replaces and excludes the personal emotional aspects of the interaction between the male and female involved, then their ultimate effect is to decrease the true value of the relationship. They may elaborate one element of the sexual sequence, but overall they reduce it.

Young lovers who *need* these copulation variations, rather than merely play with them in an exploratory fashion from time to time, are perhaps not young lovers at all. At a later point, when they have passed right through the intense pair-formation stage of their lives, and have arrived at the more mellow pair-maintenance phase, they may well find that some embellishments and novel additions to their sexual activities will provide a valuable way to re-heighten intensity, but if as young lovers they are truly in love it is surprising if this will be necessary.

Needless to say, this does not mean that any of these sexual intimacies, no matter how contrived or unusual, should be condemned or suppressed. Providing they are performed voluntarily in private by adults and cause no physical harm, there is no biological reason why they should be outlawed or attacked by society. In certain countries this is still, however, the case. One example concerns oral/genital contacts, which I omitted from my earlier twelve-stage list. The reason why they were not included is because they do not represent a clear-cut stage in the progress from first encounter between the lovers to final copulation. In the vast majority of cases they appear only after the first copulations have occurred, as a further embellishment of the genital intimacies. Later, when copulation has become a regular feature of a relationship, they are frequently included as a standard pre-copulatory pattern, and evidence from ancient art and history indicates that they have long been so employed.

Modern American surveys indicate that today oral/genital contacts are used by about half of all married couples as part of their pre-copulatory activities. The application of the male mouth to the female genitals was recorded in 54 per cent of cases, that of female mouth to male genitals in 49 per cent. Although this is well below the figures for the other pre-copulatory patterns in my twelve-stage list (mouth to mouth, hand to breast, mouth to breast, and hand to genitals all occur in over 90 per cent of cases), an average of roughly 50 per cent of the population for mouth-to-genital contact can hardly justify calling this activity 'abnormal'. Yet despite this, and despite the fact that it occurs widely in other mammalian species, it is often regarded as an 'unnatural' intimacy. It is condemned by the Judaeo-Christian religious codes, even between married partners, and in many places it is not only considered immoral but is also illegal. It is surprising to find that, right into the second half of the twentieth century, this has been the case in almost all the states in the United States of America. To be more specific, in only Kentucky and South Carolina can an American married couple privately engage in any kind of oral/genital contact without breaking the law. This means that in recent times 50 per cent of all other Americans have, technically speaking, been sexual law-breakers at some time in their married lives. In the states where it is outlawed, the act is rated as a felony everywhere except in New York, where it is classified as a mere 'misdemeanor'. In the states of Illinois, Wisconsin, Mississippi and Ohio, the law has applied a curious kind of sexual inequality, the act being legal when the husband makes oral contact with his wife's genitals, but a felony when she performs a similar intimacy with him.

These strange legal restrictions have seldom been applied in practice and have been rendered nonsensical in recent years, during which open sale and advertisement of flavoured vaginal douches has been permitted in America, but they do appear from time to time in divorce cases, where oral/genital acts have been cited as factors contributing to 'mental cruelty' in marriage. It has also been pointed out that, in theory, these laws could lead to instances of blackmail. Biologically, as I have already pointed out, there is no case against mouth-to-genital contacts. On the contrary, if they heighten the emotional intensity of pre-copulatory activities, they will merely serve to tighten the bonds of attachment between the mated pair and thereby strengthen the married condition which is so vigorously protected in so many other ways by the Church and the laws of the land.

If we examine the exact form which this type of intimacy takes in the human species, it is possible to detect a difference between man and the other mammals that employ mouth-to-genital contacts. Usually, in other species, the action begins as sniffing and nuzzling and then extends into licking. Rhythmic friction is less common. The significance of the act lies in the acquisition of detailed information concerning the state of the partner's genitals. Unlike man, other mammalian species only come into full sexual condition at certain times of the year, or at certain restricted phases of the menstrual cycle, and it is important for the partner, especially the male, to know as much as possible about the precise state of arousal before attempting copulation. The application of the nose and tongue to the genital region provides valuable clues concerning odour, taste and texture. The actual stimulation of the partner by these contacts is probably of secondary importance.

In man, the situation is reversed, the stimulation element becoming more important. The mouth is employed to arouse the partner rather than to learn about his or her sexual condition. It is for this reason that, in man, rhythmic friction takes on a more important role than mere touching or licking, the female using her mouth as a pseudo-vagina and simulating the actions of pelvic thrusting by movements of her neck. The male may also use his tongue as a pseudo-penis, but he is more likely to employ clitoral stimulation, using rhythmic tongue pressure. Again, he will imitate the repeated massage of the female organ that occurs during the pelvic thrusting of copulation itself. The great advantage to the male of this form of mimic-copulation is that he can provide prolonged stimulation for the female without himself becoming orgasmically satiated. In this way he can compensate for the longer time the average female takes to reach the orgasmic level.

This last fact no doubt explains why the extensive use of this type of sexual intimacy is more common amongst males than females. A contrast, however, has been found in the portrayal of these acts in 'blue movies'. A recent study of the history of films of this type made over the past half-century reveals that here the female action is portrayed much more frequently than the male. There is a special reason for this. These films have traditionally been made for use at all-male gatherings, or 'stag-parties', and are often referred to as 'stag movies'. Such occasions have little to do with love, being concerned instead with sex as a status device. If male status is involved, the historians of blue movies have pointed out that it 'belittles' the man to show him in a subordinate posture to the female, but enhances his feelings of dominance to portray him in the superior role of being 'served' by the female. Here we are back to basic animal behaviour and the postures of inferior submissions. When kneeling and bowing are performed as submissive acts, their biological significance lies in the lowering of the inferior's body in front of the superior. It is significant that the slang expression for oral/genital contact in man is 'going down' on the partner. In order to apply the mouth to the genitals, the active partner must considerably lower his or her body in relation to that of the passive one. This applies in any position, but is particularly clear when the act is performed with the passive partner standing. The active male or female must then kneel or crouch in front of the standing body to apply the mouth to the genital region, thus entailing the adoption of an almost medieval posture of servility. It is little wonder, then, that the performance of this act by the female to the male appeals so strongly to the status sex gatherings of males at stag-parties. Between lovers in private, of course, the situation is entirely different. Unless the encounter is one of loveless sexual gratification, the act will be one of giving pleasure and not of status-boosting, and will show no such bias. Because of the time-lag difference in reaching orgasm that exists between the sexes, it will then, as we have already seen, become more commonly extended as a male action rather than a female one.

In considering variations in the basic sexual sequence, we have so far been dealing with some of the ways in which it has been reduced and elaborated, but I also mentioned a third possibility, namely shifts in the order of appearance of the acts. Clearly these are many, and the fixed sequence I outlined is often altered in some way. As it stands, it is no more than a rough guide to the general trend which the various actions follow, from the moment of the first encounter to the final pattern of copulation. It remains a true picture of the average sequence of events, but the formalization of certain specific elements will have a marked effect on the order in which they follow one another in many cases. Some examples will clarify this.

The first three acts listed were: eye to body, eye to eye, and voice to voice. These pre-tactile 'contacts' rarely change position in the sequence. Today, exceptions can occur in instances where the initial encounter takes place over the telephone, and one sometimes hears someone say, 'It's pleasant to meet after having spoken on the phone so often.' This implies that vocal exchanges on the telephone do not by themselves constitute a 'meeting'. Combined with eye contact, however, they do. The phrase 'we met last year' need not signify any tactile contact, but merely a combination of visual and verbal exchanges. Nevertheless, a 'meeting' does usually include at least the minimal body contact of the handshake. To 'meet someone' it does seem important to have displayed some degree of physical touching. Since, in modern life, we encounter so many strangers, it is not surprising to find that this initial touching is rigidly stylized in form. A more variable type of body contact would involve too great an intimacy at such an early stage in the time-scale of the sequence of a developing relationship.

Because it has become so formalized, the handshake can frequently jump almost to the front of the whole sequence. A third party says simply, 'Here is someone I want you to meet'; and, within seconds of making eye contact, skin contact has taken place, the two hands immediately stretching out and joining. The action may even occur slightly before verbal contact has been made.

This basic rule, that the more formalized a contact action becomes, the more it can jump forward in the sequence, is also particularly well illustrated by the mouth-to-mouth kiss. Although, strictly speaking, this is the first of the pre-copulatory arousal actions, and should belong to the second half of the sequence rather than the first, it often jumps forward in time by virtue of the accepted convention of the 'goodnight kiss' between young lovers. It is significant that the first kiss usually takes place as an act of farewell. The device employed here, which enables the kiss, combined with a full frontal embrace, to jump ahead of the less intimate semi-embraces of arm-around-shoulder and arm-around-waist, and possibly even hand-holding, is the way in which it borrows 'innocence' from the non-sexual kissing of family greetings and farewells. The young couple, having met and talked for some hours, may, at the moment of parting, go straight into a brief, formal embrace-and-kiss, even though they have not touched one another's bodies previously in any way. This contrasts sharply with the situation where a man visits a prostitute, when the kiss may drop back in the sequence to a point where genital contact has already been made, or where kissing may even be totally omitted.

It will have become obvious that, in discussing these sexual variations, I have been thinking primarily of modern 'civilized' societies. In other cultures and tribes the patterns vary to a certain extent, but the general principles of a sequence of escalating intimacy still apply. An American survey of nearly two hundred different human cultures revealed that 'unless social conditioning imposes inhibitions upon active foreplay it is very likely to occur'. Nearly all the arousal actions occur in most societies, but sometimes they take a slightly different form. The nose, for example, occasionally takes over from the mouth as a contact organ, nose-rubbing or -pressing replacing mouth-kissing. In certain tribes, mutual nose-and-face-pressing appears at the point where, more usually, mouth-to-face or mouth-to-mouth contacts would occur. In others, mouth-to-mouth and nose-to-nose contacts are performed simultaneously. Some males employ nose-rubbing rather than lip contact, to stimulate the female breasts. In other tribes, kissing takes the form of placing the lips close to the partner's face and inhaling. In still others, it is more a question of reciprocal lip-and-tongue-sucking. These variations of detail are interesting in their own right, but to over-emphasize their importance, as has sometimes been done in the past, is to obscure the fact that, in more general terms, there is a great similarity in the courtship and pre-copulatory patterns of all human beings.

Having looked at the sequence of human sexual intimacies, we now come to the question of their frequency. I have gone on record as saying that man is the sexiest of all the primates, a comment which has met with some criticism. The biological evidence, however, is irrefutable, and the argument that the high level of sexual activity observed in certain quarters today is the artificial product of civilized life is nonsensical. If anything, it is the remarkably *low* level of sexual activity in certain other quarters that can best be ascribed to the artificiality of modern living. As anyone who has been under severe stress will know, anxiety is a powerful anti-sexual influence, and, since there is a great deal of stress involved in the high-pressure existence of our modern urban communities, the fact that so much sexual behaviour still occurs is a remarkable testimony to the sexuality of our species.

Let me be more specific. If I put my statement in a slightly different way, namely that man is *potentially* the sexiest of all the primates, then there can be no argument. In the first place, other primates are limited in their sexual activities to a brief section of the female's monthly sexual cycle. At these times her external sexual organs undergo a change that in most species is clearly visible to the male. This makes her sexually attractive to him. At other times she will have little or no appeal for him. In the human species the active phase is extended throughout almost the whole of the monthly cycle, more or less tripling the time when the female will be appealing to the male. Already, in this respect alone, the human animal has three times the sexual potential of his near relatives, the monkeys and apes.

Secondly, the female human remains both sexually appealing and responsive during most of her period of pregnancy, when other primates are not. Also, she becomes sexually active again much sooner after giving birth than do the other species. Finally, the average human animal of modern times can expect to enjoy roughly half a century of active sexual life, a range that few other mammals can match.

Not only is there this enormous potential for sexual activity, but in the vast majority of cases it is fully realized, so that I see no reason to modify my original statement. Most human beings express themselves sexually by finding partners and indulging in frequent sexual interactions, but even those who do not, or who are temporarily sexually isolated, do not normally become inactive. For them, it is typically the case that masturbation at fairly high frequency will be employed to compensate for the absence of a mate.

Above all, the human sexual pattern is complex. It involves not only vigorous copulation, but also all the gentle subtleties of courtship and the intense arousal actions of pre-copulatory behaviour. In other words, it not only occurs with high frequency over a long period of years, and with few interruptions from 'dead periods' in female reproductive cycles, but, when it does occur, it is prolonged and elaborate. This enlargement of the sexual life of the species is achieved by the addition to its primate inheritance of a great variety of sexual body contacts and intimacies, of the kind we have been discussing. Here the contrast with other species is striking and, to clarify this point, it is worth pausing to look at the way in which monkeys and apes perform sexually.

Monkeys do not form deep bonds of attachment to their mates and there is little courtship or pre-copulatory behaviour. During the few days of her monthly cycle when she is in full sexual condition, the female approaches the male or is approached by him, she turns her rump towards him, crouching down slightly with the front end of her body, he mounts her from behind, inserts his penis, makes a few quick pelvic thrusts, ejaculates, dismounts, and they move apart again. The whole encounter is usually over in a few seconds. A number of examples will give a clear idea of the widespread occurrence of this extreme brevity. In the bonnet monkey, the male makes only 5 to 30 pelvic thrusts. In the howler monkey, the figure is 8 to 28 thrusts, with an average of 17, taking 22 seconds, with a preceding 10 seconds for 'body adjustment'. The rhesus monkey performs 2 to 8 pelvic thrusts, lasting a total of no more than 3 or 4 seconds. Baboons, in one report, made up to 15 pelvic thrusts, lasting for a total of 7–8 seconds; in another report, an average of 6 thrusts, lasting 8–20 seconds; in a third, 5 to 10 thrusts, lasting 10–15 seconds. Two reports for chimpanzees give the average male performance as 4 to 8 thrusts, with a maximum of 15 in one instance, and 6 to 20 thrusts in a total copulating time of 7–10 seconds in another.

These details plainly indicate that our hairier relatives do not linger at the business of mating. To be fair, however, they do perform these 'instant copulations' with a very high frequency during the short period of days while the female is in sexual condition. In some species, remounting occurs within a few minutes, and may be repeated a number of times in quick succession. In the South African baboon, for instance, there are usually from 3 to 6 mountings one after the other, with only two-minute intervals between them. In the rhesus monkey this figure rises, and there may be from 5 to 25 mountings with intervals of only a minute or so between each. It seems as if the male only ejaculates after the final mounting, which is more vigorous and intense, so that here the pattern does appear to be more complex. In all these cases, however, the mating activity differs markedly from that of the human animal.

In the human species, not only is there more sexual foreplay, but the act of copulation itself takes longer. In the pre-copulatory phase, over 50 per cent of human couples spend longer than ten minutes indulging in a wide variety of arousal techniques. After this, pelvic thrusting for the male could, in most cases, lead to ejaculation in only a few minutes, but typically he prolongs this phase. He does so because, unlike the monkeys, his female is able to experience a sexual climax, similar to his own in emotional intensity, but she usually takes from ten to twenty minutes to do so. This means that, for an ordinary human couple, the whole pattern, including both foreplay and copulation, takes roughly half an hour, which is more than a hundred times as long as for the typical monkey pair. Again, to be fair to the monkeys, they will be likely to repeat their brief encounter much sooner than the human pair, but against this must be set the fact that the female monkey will be receptive only for the few days of her period of heat.

To compare the situation for the female monkey and the female human, the former comes into heat as ovulation time approaches and stays in heat for nearly a week. During this time, copulation neither arouses nor exhausts her sexually. She remains permanently aroused during the whole mating period. For the human female it is as if each mating pattern is a short period of heat, now unrelated to the time of ovulation, but related instead to the pre-copulatory stimulations of the male. She has, in effect, become a mate-responder, rather than an ovulation-responder. Her physiological arousal is geared to the shared sexual intimacies with her male, rather than to the rigidly fixed sequence of the monthly cycle of ovulation and menstruation. This vital step, which represents a fundamental change in the usual primate sexual system, leads inevitably to a far greater degree and complexity of body contact between the mated pair and forms the basis of human sexual intimacy.

This leads us to the question of the origins of the more complex human sexual acts. What are the sources of all the additional body contacts? Since monkeys do little more than mount and copulate, the act of mounting, making rhythmic pelvic thrusts and ejaculating is virtually all we share with them. So from where do we derive all the gentle, hesitant touchings and hand-holdings of the courtship period, and all the passionate arousal actions of sexual foreplay? The answer seems to be that in almost all cases they can be traced back to the intimacies of the mother-infant relationship that were described earlier. Hardly any of them appear to be 'new' actions, specifically evolved in connection with sexuality itself. In terms of the behaviour involved, falling in love looks very much like a return to infancy.

In tracing the way in which the primary embrace of our earliest years gradually becomes restricted as we mature, we watched the decline and fall of close body intimacy. Now, as we observe the young lovers, we see the whole process put into reverse. The first actions in the sexual sequence are virtually identical with those of any other kind of adult social interaction. Then, little by little, the hands of the behavioural clock start to turn backwards. The formal handshake and small-talk of the first introduction grow back into the protective hand-holding of childhood. The young lovers now walk hand in hand, as each once did with his or her parent. As their bodies come closer together with increasing trust, we soon witness the welcome return of the intimate frontal embrace, with the two heads touching and kissing. As the relationship deepens, we travel still further back, to the earlier days of gentle caresses. The hands once again fondle the face, the hair and the body of the loved one. At last, the lovers are naked again and, for the first time since they were tiny babies, the most private parts of their bodies experience the intimate touch of another's hands. And, as their movements travel backwards in time, so too do their voices, the words spoken becoming less important than the soft tonal quality with which they are delivered. Frequently even the phrases used become infantile, as a new kind of 'baby-talk' develops. A wave of shared security envelops the young couple and, as in babyhood, the hurly-burly of the outside world has little meaning. The dreamy expression of a girl in love is not the alert face of an active child; it is the almost blank face of a satisfied baby.

This return to intimacy, so beautiful to those who are experiencing it, is often belittled by those who are not. The epigrams tell the story: 'The first sigh of love is the last of wisdom'; 'Love is a sickness full of woes'; 'Love is blind'; 'We are easily duped by what we love'; 'Love's a malady without a cure'; ''Tis impossible to love and to be wise'; 'Lovers are fools, but nature makes them so'. Even in the scientific literature, the term 'regressive behaviour' takes on the flavour of an insult, instead of an impartial, objective description of what is taking place. Of course, to behave in an infantile manner in certain adult contexts is an inefficient way of coping with a situation, but here, in the case of young lovers forming a deep bond of personal attachment, it is exactly the opposite. Extensive, intimate body contacts are the very best way of developing such a bond, and those who resist them because they are 'babyish' or infantile will be the losers.

When courtship advances to the stage of pre-copulatory behaviour, the infantile patterns do not fade. Instead they grow still younger, and the clock ticks backwards to the sucking at the mother's breast. The simple kiss, in which lips are gently pressed to the lover's mouth or cheek, becomes a vigorous, moving pressure. With muscular actions of their lips and tongues, the partners work on one another's mouths as if to draw milk from them. They suck and squeeze rhythmically with their lips, explore and lick with their tongues, like hungry babies. This active kissing is no longer confined to the partner's mouth. It seeks other sites, as if searching for the long-lost mother's nipple. In its quest it travels everywhere, discovering the pseudo-nipples of the ear-lobes, the toes, the clitoris and the penis, and, of course, the lover's nipples themselves.

Earlier, I mentioned the reward these actions bring from the knowledge of the sexual pleasure they give, but clearly that was only part of the story. There is also this more direct reward of re-experiencing the strongly gratifying oral contact of the suckling interaction in infancy. The effect is heightened where the pseudo-breast can be made to produce pseudo-milk. This can be obtained from the increased salivation of the lover's mouth, from the increased sexual secretions of the female's genitals and from the seminal fluid of the male's penis. If the mouthing of the penis by the female is prolonged until ejaculation occurs, it is as if the action has finally succeeded in starting the 'milk-flow' from this pseudo-breast, a similarity which was recognized as long ago as the seventeenth century, when the slang expression 'milking' for this activity first came into common usage.

Even when the pre-copulatory patterns are terminated and copulation itself begins, the infantile actions do not completely disappear. For the mating monkeys, the only body contacts, apart from actual genital interaction, are the mechanical holding actions by the male's hands and feet. He grips the female's body, not as an amorous intimacy, but to steady himself while he makes the rapid pelvic thrusts. Such graspings also occur between the human partners, but in addition there are many contacts that have no 'body adjustment' function. The hands clasp or hold the partner, not for mechanical reasons of thrust facilitation, but as tactile signals of intimacy.

Turning again to the illustrated sexual manuals discussed earlier, and taking only the cases where the male and female were shown actually copulating, it is possible to score the frequency of these non-genital contacts that accompany the pelvic thrusting. In no less than 74 per cent of the copulation postures depicted, the hand (or hands) of one partner clasps or touches some part of the other's body, in a 'non-steadying' way. In addition, there are many embracing acts, kissing acts and non-kissing head-to-head contacts; also a number of hand-to-head and hand-to-hand contacts. All these actions are basically embraces, partial embraces or fragments of embraces. They indicate that, for the human animal, copulation consists of the adult primate mating act *plus* the returned infantile embracing act. The latter pervades the whole sexual sequence from its earliest courtship stages right through to its final moments. The human animal does not merely copulate with a set of genitals belonging to a member of the opposite sex; it 'makes love' – a significant phrase – to a complete and special individual. This is why, in our species, all stages of the sequence, including copulation, can serve to enhance the pair-bonding process, and why, presumably, the female evolved an extended period of sexual receptivity, stretching far beyond the limits of the ovulation period. It could even be said that we now perform the mating act, not so much to fertilize an egg as to fertilize a relationship. There is no reproductive danger in this, for even the small proportion of mating acts that do happen to coincide with the time of ovulation will still be sufficient to produce an adequate number of offspring, and there are more than three thousand million of us alive today to prove the point.*

* The figure has now risen to more than five thousand million.

4

SOCIAL INTIMACY

TO STUDY HUMAN sexual intimacy is to witness the rebirth of lavish bodily contact between adults, replacing the lost intimacies of infancy. To study human social intimacy is, by contrast, to observe the restraint of cautious, inhibited contact, as the conflicting demands of closeness and privacy, of dependence and independence, do battle inside our brains.

We all feel overcrowded from time to time, and over-exposed to the prying eyes and minds of others. The monkish idea of shutting ourselves away from it all becomes appealing. But for most of us a few hours will do, and the idea of lifelong monastic solitude is appalling. For man is a social animal, and the ordinary healthy human being finds prolonged isolation a severe punishment. Short of physical torture or death, solitary confinement is the worst agony that can be inflicted on a prisoner. He is driven eventually, in near-madness, to talk into his lavatory pan so that he can enjoy the echo of his voice coming back to him. It is the nearest thing to a social response he can get.

A shy single person living in a big city can find him- or herself in almost the same situation. For such people, if they have left the intimacies of home behind and are now existing by themselves in a small room or apartment, the loneliness can soon become intolerable. Too timid to make friends, they may ultimately prefer death by suicide to prolonged lack of close human contact. Such is the basic need for intimacy. For intimacy breeds understanding, and most of us, unlike the solitary monk, do want to be understood, at least by a few people.

It is not a question of being understood rationally or intellectually. It is a matter of being understood emotionally, and in that respect a single intimate body contact will do more than all the beautiful words in the dictionary. The ability that physical feelings have to transmit emotional feelings is truly astonishing. Perhaps it is their strength that is also their weakness. In tracing the sequence of intimacies that occur through life, from birth to death, we saw the way in which the two phases of massive body contact were also the two phases of powerful social bonding, first between parent and child and second between lovers. All the indications are that it is impossible to be lavish and uninhibited with one's body-to-body contacts and not become strongly bonded with the object of one's attentions. An intuitive understanding of this is perhaps what inhibits us so strongly from indulging in the pure pleasure of more widespread bodily intimacies. It is not enough to say that it is unconventional to hug and embrace one's business colleagues, for example. That does not explain how the convention of 'keeping oneself to oneself' or 'keeping one's distance' arose in the first place. We have to look deeper than that to understand the extraordinary lengths we go to to avoid touching one another in ordinary day-to-day affairs outside the close family circle.

Part of the answer has to do with the great overcrowding we experience in our modern urban communities. We encounter so many people in streets and buildings every day that we simply cannot embark on intimacies with them, or all social organization would come to a halt. Ironically this over-crowding situation has two completely incompatible effects on us. On the one hand it stresses us and makes us feel tense and insecure, and on the other hand it makes us cut down on the very exchanges of intimacies which would relieve this stress and tension.

Another part of the answer has to do with sex. It is not merely that we cannot afford the time or energy to start setting up the endless social bonds that would result from widespread indulgence in extensive body intimacies. There is also the problem that, between adults, body intimacy spells sex. This is an unhappy confusion, but it is not hard to see how it has arisen. Since copulation, short of artificial insemination, is impossible without body intimacy, it has become in some respects synonymous with it. To indulge in the mating act, even the most 'untouchable' adult has to touch and be touched. At almost all other times he can avoid it if he wishes to, but then he cannot. Some Victorians went as far as they could to reduce contact by wearing nightgowns with small slits in the front, but even they had to make contact to the extent of inserting a penis into a vagina if they were to populate their world with children. And so it came to be that, in the year 1889, the expression 'to be intimate' became a euphemism for having sexual intercourse. During the present century it has become increasingly difficult for any adult, of either sex and with either sex, to indulge in intimate body contact without giving the impression that there is a sexual element involved in what is taking place.

It would be wrong to give the idea that this is an entirely new trend. The problem, of course, has always been there, and adult intimacies have always been curtailed to some extent to avoid sexual implications. But one gets a distinct impression that the situation has tightened up more in recent years. We no longer seem to be quite so free at falling on one another's necks in joy, or weeping copiously in one another's arms. The basic urge to touch one another remains, however, and it is intriguing to study how we go about this in our day-to-day affairs outside the bosom of the family.

The answer is that we formalize it. We take the uninhibited intimacies of infancy and we reduce them to fragments. Each fragment becomes stylized and rigidified until it fits into a neat category. We set up rules of etiquette (a word taken from the French and meaning, literally, a label) and we train the members of our cultures to abide by them. No training is necessary where an embrace is concerned. This, as we have seen, is an inborn biological act that we share with all our primate relatives. But an embrace contains many elements, and which particular fragment we are to use at which particular social moment, and in what rigidly stylized form, is something that our genetical make-up cannot help us with. For the animal, there is either behaviour or no behaviour, but for us there is either behaviour or misbehaviour, good behaviour or bad behaviour, and the rules are complex. This does not mean, however, that we cannot study them biologically. No matter how culturally determined they may be, or how culturally variable, we can still understand them better if we view them as pieces of primate behaviour. This is because we can nearly always trace them back to their biological origins.

Before surveying the whole scene, let me give a single detailed example to illustrate what I mean. I will use an action that does not appear to have received much attention in the past, namely the pat on the back. You may think that this is too trivial a piece of behaviour to be of much interest, but it is dangerous to dismiss small actions in this way. Every twitch, every scratch, every stroke and every pat has the potential of changing a person's whole life, or even a nation's. The warm caress that was withheld at the vital moment when it was desperately needed can easily be the act, or rather the non-act, that finally destroys a relationship. The simple failure to return a smile, between two great rulers, can in the same way lead to war and destruction. So it is unwise to sneer at a 'mere' pat on the back. These little actions are the stuff that emotional life is made of.

If you have ever had a close personal relationship with a chimpanzee, you will know that back-patting is not a uniquely human activity. If your ape is particularly glad to see you, he is likely to come up, embrace you, press his lips warmly and wetly to the side of your neck, and then start rhythmically patting you on the back with his hands. It is a strange sensation, because in one way it is so human and yet in another it is subtly different. The kiss is not quite like a human kiss. I can best describe it as being more of a soft-open-mouth-press. And the patting is both lighter and faster than human back-patting, with the two ape hands alternating rhythmically. Nevertheless, the actions of embracing, kissing and patting are basically the same in the two species, and the social signals they transmit appear to be identical. We can therefore start out with a reasonable guess that back-patting is a biological feature of the human animal.

I have already, in the first chapter, explained the probable origin of the activity, as a repeated intention movement of clinging that says, 'I will cling to you like this if necessary, but at the moment it is not, so relax, all is well.' In infancy we only get patted as an embellishment of the embrace, but later on the friendly pat may be executed by itself, without any embracing. The patter's arm simply reaches out towards the companion and contact is made with the hand alone. Already, with this change, the process of formalization has started to take place. Seeing only an embraceless pat would make it impossible to guess its true origin. Another change occurs at the same time: the area of the body patted becomes less restricted. A baby is patted almost exclusively on the back, but an older child is patted almost anywhere, not only on the back but also on the shoulder, the arm, the hand, the cheek, the top of the head, the back of the head, the stomach, the buttocks, the thighs, the knees and the legs. The message of the pat also becomes extended. The comforting signal 'all is well' becomes the congratulatory signal 'all is *very* well' or 'you have done well'. Since the brain that has done well is housed inside the skull, it is natural that the pat on the head should be the action that comes to typify the congratulatory message. In fact, this particular form of the action becomes so strongly associated with childhood congratulation that it has to be abandoned later on when patting occurs between adults, because it then acquires the flavour of condescension.

Other changes also occur as we move from the context adult-pats-child to adult-pats-adult. In addition to the head, certain other areas now become taboo. Patting on the back, shoulder and arm is still unhampered, but patting the back of the hand, the cheek, the knee or the thigh begins to acquire a slightly sexual flavour, and patting on the buttocks a strongly sexual flavour. The situation is highly variable, however, and there are many exceptions to this rule. Handback-patting and thigh-patting between two women, for example, may occur without any hint of sexuality being involved. Also it is possible, by comic over-exaggeration, to pat almost any part of the body without causing offence. The patter then makes joking remarks, as he pats his victim on the head or cheek, such as 'there, there, little man', indicating that the contact he is making is not sexual but mock-parental, and is not to be taken seriously. There is an element of offensiveness involved, of course, but it has nothing to do with the breaking of sexual taboos on touching certain areas in certain ways.

To complicate matters further, there is an intriguing exception to this last exception. It works like this. One adult, say a male, wishes to make sexual body contact with another, say a female. He knows she will not accept a direct, undisguised contact, which she would find distasteful. He knows, in fact, that she finds him sexually unattractive in general, but his urge to touch her is strong enough to ignore the disencouraging signals she transmits to him. His strategy therefore becomes one of pretending to behave like a mock-parent. He makes a great joke of patting her on the knee and calling her a funny little girl. He hopes that she will accept the contact as a joke action, even though the true reward he is getting is sexual. Unhappily for him he is not usually good enough at concealing other sexual signals, especially from his facial expressions, and the girl in question can normally see through the device and respond in an appropriately negative way.

The least sexually loaded patting action is the primary pat on the back. This somehow manages to retain its original quality and may often be observed between complete strangers, in contexts of either condolence or congratulation. Two specific situations which illustrate this are the road accident and the sportsman's moment of triumph. Following a road accident, if one of those involved is sitting slumped by the roadside in a state of numbed shock, he will soon be approached closely by one of the helpers who arrive on the scene. Typically, the helper will peer closely at the shocked person and ask some quite nonsensical question, such as 'Are you all right?', when clearly he is not. Almost immediately the helper realizes how meaningless his words are in such a situation, and reverts instead to the more powerful and basic communication pattern of direct body contact. The most likely form that this contact will take is a comforting, gentle patting on the back of the victim. The same response can be observed, but with much more vigour, when a sportsman has just achieved some sort of athletic triumph. As he returns cheerfully from the field or arena, his fans will compete with one another to get close enough to pat him warmly on the back as he passes by.

We have already come a long way from the primary situation of the mother lovingly patting her tiny baby, but there is still further to go, for amongst adults the action of patting has extended its range even beyond the tactile. The basic touch signal has, in two important contexts, become a sound signal and a visual signal. Both human clapping, when an audience applauds a performer, and human waving, during greetings and farewells, have been derived from the primary patting action. Let us take clapping first.

For years I was puzzled by the widespread use of hand-clapping as a method of rewarding a performer. The sharp slapping of one hand against another seemed an almost aggressive action, as did the harsh sound it produced. Yet clearly it was the exact opposite of aggressive in its effect on the happy performer. For centuries, actors have craved the applause of a sea of clapping hands, and have devised many a 'trap to catch a clap', thereby giving the English language its word 'clap-trap'.

To understand the potent reward of hand-clapping, we must search for its origins in childhood. Careful studies of infants in the second half of the first year of life reveal that at this age a hand-clap may often be given as part of the greeting towards the mother, when she returns to her baby after a brief absence. The action may be performed either just before, or instead of, reaching out with its arms towards the mother. It appears at about the same time as the act of holding on to the mother with the arms. It is as if the baby, seeing the mother approaching, makes a movement in which the arms are brought out and round as if to hold her. But her body is not there yet to be held, and so the arms continue in the arc of embracing until the hands meet with a clap. At this stage the clap is performed, as it were, from the arm, not from the wrist as in the adult version of hand-clapping.

Detailed observations have indicated that this occurs where there is no evidence of the mother's having taught a clapping response previously. The baby clap, in other words, can best be interpreted as the audible culmination of a vacuum-embracing of the mother. The rhythmically repeated clapping-from-the-wrist that develops later can then be seen clearly as a kind of vacuum-patting added to the vacuum-embrace. When we applaud a performer, we are, in effect, patting him on the back from a distance. It is inconvenient or impossible for us to all rush up and actually make physical contact with him, giving him a real pat to show our approval, so instead we stay in our seats and repeatedly pat him in vacuo. If you test this by clapping your hands as though you were applauding, you will find that you do not bring both hands together with equal force. One tends to take the role of the performer's back and the other does the vigorous vacuum-back-patting on it. It is true that both hands move, but one makes a much stronger action than the other. In nine people out of ten it is the right hand, with the palm facing half-downwards, that takes the patting role, and the left hand, with the palm facing half-upwards, that takes the role of the back to be patted.

Occasionally one gets an unexpected glimpse, even in the adult world, of the basic relationship that exists between the primary embrace and the action of hand-clapping. When the first Russian cosmonaut returned in triumph to Moscow and stood in Red Square with the Russian leaders, a vast crowd filed past paying homage, reaching up and clapping their hands to him as they went by. A film of the event clearly shows one man in the crowd so overcome with emotion that he keeps on interrupting his prolonged hand-clapping with vacuum embraces. He claps his outstretched hands, embraces the air in front of him and hugs it to him, then reaches out, claps again and embraces again. When the power of the emotion breaks down the formalization of the conventional pattern in this way, it provides us with an eloquent confirmation of the origins of the adult act.

Russia happens to provide us with another interesting variation in hand-clapping. In that country it is usual for the performers to clap the audience as the audience claps them. This does not mean, as has sometimes been cynically suggested, that Russian performers are so narcissistic that they even applaud their own performances. They are simply returning the audience's formalized embrace, as they would do if they met them body to body. In the West we lack this convention, although a variant of it can sometimes be seen in the wide-stretched arms of the performers seeking applause at the end of an act. Circus performers and acrobats are particularly prone to adopt this posture. At the completion of a difficult stunt, they stand proudly poised, face the audience, and fling their arms out wide. The audience immediately breaks into ringing applause. The opening of the arms in this way is an example of an intention movement of embracing. The arms are positioned ready to embrace the audience, but do not go on to consummate the act in vacuo. Some cabaret singers who specialize in highly emotional songs use much the same action whilst actually singing, arousing emotion in the audience by offering them an imploring invitation to embrace as an accompaniment to the imploring words of the song.

Hand-clapping is also used occasionally as an act of summoning a servant. In the harem fantasy, it is the signal that says 'bring on the dancing girls'. In these cases the clapping is not the typical fast, rhythmically repeated action of the applause pattern, but simply one or two hard slaps of one hand against the other. In this respect it is much more like the infant clapping that was performed as a greeting to the mother. The message is also similar. The infant's request, 'come closer', to the mother, becomes the adult's demand for the same thing from the servant.

I said earlier that the basic touch signal of patting has been extended both as a sound signal, which we have been examining, and as a visual signal, in the form of waving. Like clapping, waving is usually taken for granted, but it, too, has some unexpected elements, and it is worth analysing in detail.

To start with, it seems obvious that we wave when greeting or saying goodbye because, from a distance, it makes us more conspicuous. This is true, but it is not the whole answer. If you watch people who are really desperate to make themselves conspicuous, say when hailing a taxi, or when trying to make visual contact with a person in a crowd who has not yet spotted them, they do not wave in the usual, conventional manner. Instead, they raise one arm stiffly, straight up, and start swinging it from side to side, moving it from the shoulder. Under even greater pressure, they may raise and wave both arms in this way at the same time. This is the action that makes for maximum conspicuousness at a distance. It is not, however, the way we wave to one another when we are already in visual contact. If we are waving goodbye to someone, or if we are waving a greeting to someone who has already seen us but is still out of reach, we do not normally wave our arms. We raise the arm, but what we wave is the hand. We do it in one of three ways. One is to wave it up and down, with the fingers pointing away from us. When the hand is in the up position, the palm faces outwards; when in the down position, the palm faces downwards. Here again is the ubiquitous patting action. The greeting arm reaches out to embrace and pat, but, as with the clapping, the intervening distance forces us to perform the action in vacuo. The difference is that whereas in hand-clapping the long-distance embrace-and-pat is elaborated into a sound signal, here it is modified into a visual signal. The arm goes upward instead of forward, as it would in a true contact embrace, because this increases the visibility of the action. Otherwise there is little difference.

A second form of hand-waving reveals a further modification towards visibility. Instead of moving the hand up and down, it is waved from side to side, with the palm kept in the out-facing position. The speed is much the same, but the action is now one step further removed from the primary patting motion. It is significant that this form of waving is favoured more by adults than by children, who seem to prefer the primitive up-and-down version.

The third type of hand-wave is one that will not be familiar to most Western readers. I have myself observed it only in Italy, but apparently it also occurs in Spain, China, India, Pakistan, Burma, Malaysia, East Africa, Nigeria, and amongst gypsies. (This is, to say the least of it, a most curious distribution, for which I can as yet find no explanation.) It is reminiscent of a hand-beckoning action, but one only has to see it performed as a goodbye signal to realize that this is not what it is. Like the first form of hand-waving I mentioned, it is an up-and-down action, but this time the hand starts out palm uppermost (as in a begging posture) and is then moved repeatedly upwards towards the body of the waver. Again, it is basically a patting movement, and in true back-patting one often sees the hand adopt this position, with the fingers pointing upwards on the back, when the embracing arm has the elbow in a low position.

Two rather specialized waves are related to this last one. They are the Papal wave and the British royal wave. In both cases, for some reason, they are performed neither from the shoulder, as in the conspicuous arm-wave, nor from the wrist, as in the usual patting-wave. Instead they are done from the elbow. The Pope normally uses both arms at once and brings his hands and forearms slowly, rhythmically and repeatedly towards himself, palms uppermost, making a series of embracing intention movements. But it is not as simple as that, because his arms do not bend directly towards his chest. He does not hug the crowd to his bosom. Instead, the arc the arms perform is half-inward and half-upward, as though his action is a compromise, embracing the crowds partly towards himself and partly towards the heavens above, where one day they presumably all hope to be welcomed.

The British royal wave is also typically stiff from the elbow, but is normally one-handed and with the fingers pointing straight up. The palm faces inwards towards the royal body, emphasizing the embracing nature of the action, and the forearm is slowly and rhythmically rotated, with the emphasis on the inward part of the rotation. In this way the Queen, in a highly stylized manner, embraces her subjects and reassures them with a rather formal pat on the back.

As in the case of clapping, one is sometimes lucky enough to see the formality of waving break down under emotional pressure in a revealing way, laying bare its basic origins. A specific instance of greeting will illustrate this. At a small airport where I was making observations of waving behaviour, there is a balcony from which friends and relatives can watch the new arrivals descend from the incoming aircraft and walk across the tarmac to the customs entrance. This entrance is just beneath the balcony, so that although the new arrivals cannot touch the excited figures waving frantically to them from above, they do come very close before finally disappearing into the airport building. This, then, is the setting, and the action usually takes place as follows. After the aircraft doors open and the passengers start piling out, there is a great deal of distant eye-scanning from both the arrivals and the greeters. If one makes eye contact before the other, he normally starts vigorous arm-waving, moving the arm from the shoulder in the maximum-conspicuousness manner. After mutual eye contact has been made, then both sides tend to adopt the arm-raised hand-wave action. This goes on for some time, but it is a long walk to the building and after a while they usually break off. Their waving and smiling urges have become temporarily exhausted (like the person being photographed who, as time passes, finds it hard to maintain a natural smile for the lens-fiddling cameraman), but they do not wish to appear 'ungreetful', so both sides now take a sudden interest in other aspects of the airport scene. The new arrival glances round to survey the landscape of the airfield, or rearranges that suddenly awkward piece of hand-luggage that is mysteriously slipping from his grip. The greeters for their part start to exchange comments about the appearance of the new arrival. Then, as the latter comes nearer and his facial details can be made out more clearly, both sides start up the vigorous hand-waving and smiling again until the arrivals have disappeared into the building below. Half an hour later, with customs inspection over, the first body contact is made, with handshakes, embraces, pattings, hugs and kisses.

This is the basic story. Naturally, there are many minor variations, and on one occasion the pattern was exceeded in a most revealing way. A man was returning to the bosom of his family after a long spell abroad. From the moment he stepped off the aircraft both he and his group of family greeters exploded into an orgy of arm- and hand-waving. As he reached the building and saw the details of their faces clearly above him, he found the convention of hand-waving inadequate for his emotional needs. With tears in his eyes and his mouth shaping unheard words of love, he had to do something with his arm that would better express the enormous intensity of his feelings at being reunited with his family. At this moment, as I watched, I saw his hand actions change. The ordinary waving movements ceased and were replaced by a perfect miming of a passionate series of back-pattings. The arm was now held out towards the family group, rather than raised upward, which fore-shortened it and reduced its conspicuousness. The hand curled round sideways and made vigorous, rapid patting actions in mid-air. The power of his emotions was so strong that all the secondary, conventionalized modifications of the primary embrace-and-pat actions, which help to improve the signal quality of this pattern from a distance by making it more clearly visible, were abandoned in the heat of the moment, and the original primary behaviour was once again laid bare.

The intensity of this encounter was confirmed by the tactile greeting behaviour that followed the customs interlude. When the man emerged into the body of the airport, he was embraced, hugged, kissed and patted with such vigour by all fourteen members of his family that, by the time they had finished, he was emotionally exhausted, his face flooded with tears, his body shaking. At one point a woman who appeared to be his mother augmented her embrace with vigorous face-kneading, taking both the man's cheeks in her hands and grappling with them as though she were in her kitchen kneading soft dough. While this was going on, the man's hands were embracing her and

418

patting her back at high intensity. However, after about the tenth passionate greeting, it seemed as if the emotional exhaustion of the occasion was beginning to tell on him. As this occurred, his patting actions changed significantly. Once again, a signal convention was breaking down under emotional pressure, and once again the origins of a formalized pattern were being laid bare. Whereas, before, the waving had reverted to vacuum-air-patting, this time it went one stage further back to its original source. The repeated patting gave way to brief, repeated clinging actions. Each pat became a tightening of the hand contact, a sort of clinging squeeze, that was repeatedly relaxed and then tightened again. Here, without doubt, was the original clinging intention movement. This was the 'ancestral' pattern, the one from which all the others had descended by a process of signal specialization: to a tactile signal by modification to on-off patting, to a sound signal by using the other palm as the more noisily patted object in the act of hand-clapping, and to a visual signal by patting in mid-air with a raised arm in the act of hand-waving. Such are the ramifications that occur with our so-called 'trivial' acts of human intimacy.

In following this one small human-contact action through all its various developments, I have tried to show how it is possible to see old familiar activities in a new light. The need that we, as adults, have to make body contact with one another is basic and powerful, but, as we have seen, it is rarely fully expressed. Instead, it appears in fragmented, modified or disguised forms in many of the signs, gestures and signals we make to one another in our daily lives. Often the true meaning of the actions is hidden from us and we have to trace them back to their origins to understand them fully. In the examples I have just been describing, the primary contact action frequently became remote, operating at a distance, but there are also many ways in which we do still make actual body contact with one another, and it is interesting to survey these and see what forms they take. To do this it will help to go back for a moment to the primary embrace itself. This is not commonly seen today between adults in public, but it still does occur from time to time and it is worth studying the situations in which it appears.

The full embrace. If we make careful observations of as many full embraces as possible, then it soon becomes clear that between adults this action falls into three distinct categories. The biggest group, as we might expect, consists of amorous contact between lovers. This accounts for about two-thirds of all public embraces today. The remaining third can be split up into the two types that we can call the 'relatives' reunion' and the 'sportsman's triumph'.

Young lovers perform the full embrace, not only when they meet and part, but also frequently during the times they are together. Amongst older couples, after marriage, it is rare to see a full embrace in public unless one partner is going away for some time, or returning from a separation of at least some days. At other times, if the embrace occurs at all, it will be publicly expressed merely as a token contact of a rather mild kind.

Between adult relatives, such as brothers and sisters, or parents and their adult children, the passionate embrace is even less common. It does occur with remarkable predictability, however, whenever there has been some major disaster from which one of the relatives has escaped. If he or she has been hijacked, kidnapped, taken captive, or trapped by some upheaval of nature, then you can be sure that the 'relatives' reunion' that follows safe delivery and return will exhibit full embraces of the most intense kind. Under such circumstances the action may even extend to close friends of either sex, who would normally give only a handshake or a kiss on the cheek. Such is the emotional intensity of the situation that the passionate embracing of a man by a man, or a woman by a woman, or a man-*friend* by a woman-*friend*, creates no difficulties with regard to sexual taboos. At a lower emotional intensity there would be more of a problem, but at moments of high drama the taboos are forgotten. In triumph, relief or despair it is acceptable even for two adult males in our culture to hug and kiss one another, but if, in a less dramatic situation, they were to perform a mere fragment of an embrace such as holding hands, or pressing cheek to cheek, they would immediately create a homosexual impression.

This difference is significant and requires explanation. It tells us something about the way in which basic body contacts are fragmented and formalized. To begin with, the full embrace is natural between parent and baby, and therefore also between parent and older child, even though it becomes less frequent. Between adults it is typical of lovers and mates. Other adults who, for a variety of reasons, feel the urge to embrace one another, must therefore somehow make it clear that there is no sexual element in their contact. This they do by employing some formalized fragment of the full embrace, a fragment which by conventional agreement is non-sexual. A man, for example, can put his arm around another man's shoulder without running any risk of sexual misinterpretation, either by his contact partner, or by others who see him perform the action. If, however, he were to perform other simple fragments, such as, say, kissing the man's ear, a sexual connotation would immediately be placed upon the act.

The situation is entirely different when two men are seen employing the full embrace, complete with hugging and kissing, in a situation of great triumph, disaster or reunion. No sexual interpretation is made here because it is recognized that the response is not formalized, but basic. The onlookers recognize the situation as being one where the usual conventions are overpowered by the intensity of the emotions involved. They know, intuitively, that what they are witnessing is the return to the primary, pre-sexual embrace of infancy, with all the layers of later, adult stylization stripped away, and they accept the contact as being perfectly natural. In fact, if two male homosexuals wished to make body contact in public without arousing the usual heterosexual hostility or curiosity, they would do better to indulge in a wild hug rather than a gentle kiss.

By studying the various formal fragments of the basic embrace, we should therefore be able to see how convention has pigeon-holed them into different categories, so that each one now signals something quite specific about the nature of the relationship between the 'contacters'.

Before doing this, however, there is still the third category of the full embrace to be mentioned, namely the 'sportsman's triumph'. The embrace of two men following a major disaster has been with us for a long time, but the passionate hugging of football players after a goal has been scored is something comparatively recent. How is it that this occasion has suddenly been elevated to the ranks of a major emotional experience? To find the answer we have to go much further afield than the football pitch. We do, in fact, have to go back many hundreds of years.

Two thousand years ago, when the world was less crowded and the relationships between members of a community were more clearly defined, the full embrace was much more commonly used as an ordinary form of greeting between equals. The embrace-and-kiss was employed between men and men as well as between women and women and between non-amorous men and women. In ancient Persia it was even common for men of equal rank to kiss one another on the mouth, reserving the kiss on the cheek for someone of slightly inferior status. It was more usual in other places, however, for the cheek kiss to be employed between equals. This situation persisted for many centuries, and was still to be found in medieval England, where valiant knights would kiss and embrace one another at times when their modern equivalents would do no more than nod and shake an outstretched hand.

Towards the end of the seventeenth century the situation in England began to change, and the non-sexual greeting embrace went into a rapid decline. This began in the towns and slowly spread to the country, as we know from a line in Congreve's *The Way of the World*: 'You think you're in the country, where great lubberly brothers slabber and kiss one another when they meet. 'Tis not the fashion here; 'tis not indeed, dear brother.'

Social life in the cities was becoming more crowded and personal relationships more complex and confusing, and with the arrival of the nineteenth century further restrictions were imposed. Now even the elaborate bowing and curtsying that had survived through the eighteenth century became more and more limited to formal occasions and lost their everyday role. By the 1830s that minimal body contact, the handshake, had arrived, and it has been with us ever since.

Similar trends were taking place elsewhere, but not always to the same extent. The Latin countries tended to restrict their body contacts far less than the British, and, even into the twentieth century, friendly embraces between adult males remained far more acceptable. They do so to this day, and this is where we return to the case of the 'footballer's hug'. Football, which began as a British sport, spread rapidly in the present century to many parts of the world. In Latin countries it became especially popular, and before long international matches of great emotional intensity were being played. When Latin teams visited England, the passionate embracing of their members, following a successful goal, was at first met with astonishment and derision, but the excellence of their playing soon changed all that. As the years passed, the 'Well done, old chap' of the British players when one of their members scored began to seem almost churlish. Back-patting gave way to mild embracing, and mild embracing grew into wild hugging, until today spectators have become quite accustomed to seeing a goal-scorer almost submerged beneath a dense cluster of passionately hugging congratulators.

In this one specific context, then, we have gone full circle back to the times of medieval knights and the ancient world beyond. It remains to be seen whether this trend will extend to other spheres. It may do so, but there is one limitation we must bear in mind. The players embracing on their football field are in a strictly non-sexual context. Their roles are clearly defined and their physical masculinity is being strongly demonstrated by the toughness of the game they are playing. In a social situation of a less clearly defined type, the situation would be different, and the usual restrictions of our complex society will probably then continue to apply. Only in spheres where the expression of powerful emotions is part of the ordinary stock-in-trade, such as the acting profession, can we expect to find major exceptions. If the rest of us find the social embracings of actors and actresses somewhat excessive, we must remember three things. Not only are they trained to show passions easily, but they are also put under severe emotional strain by the nature of their work and, in addition, theirs happens to be a particularly insecure occupation. They need all the mutual re-assurance they can get.

Leaving the full embrace now, we can start to look at its less intense forms of expression. So far we have been dealing with the maximal frontal embrace, in which the two partners are pressed close together with the sides of their heads in contact and their arms wrapped tightly around one another's bodies. When this action is performed at a lower intensity, there are usually three major changes. The bodies now make contact side to side instead of face to face; only one arm is wrapped around the partner's body instead of two; and the heads are usually apart instead of touching. My observations reveal that, between adults in public, a partial embrace of this type is six times as likely to occur as a full embrace.*

The shoulder embrace. The most common form of partial embrace is the shoulder embrace, in which one person wraps an arm around his partner so that the hand comes to rest on his far shoulder. This is twice as common as any other form of partial embrace.

The first difference to emerge, when comparing this with the full frontal embrace, is that it is predominantly a masculine act. Whereas the full embrace was given more or less equally by males and females, the shoulder embrace is five times more likely to be performed by a male than by a female. The reason is simple enough: men are taller than women, and women will always have to look up to them physic-ally, no matter what their attitudes may be in other respects. The consequence of this anatomical difference is that certain body contacts are much easier for men than for women, and the shoulder embrace is one of these.

This fact gives the shoulder embrace a special quality. Since, when it is performed between a man and a woman, it is nearly always done by the man, this means that there is nothing effeminate about it. This in turn means that it is also available for use between males in casual, friendly situations, without giving any sexual flavour to the contact. About one in four shoulder embraces do, in fact, occur between males, and this is the only form of body embrace that does occur commonly in an all-male context. The difference with the full frontal embrace is striking. There, if two males were involved, the situation was typically one of high drama, or intense emotion, but here the context can be much more relaxed, the usual scene being one of team-mates, 'old pals' or 'buddies'.

---

* This, and other similar quantitative statements, are based on personal observations backed up by a detailed analysis of 10,000 photographs taken at random from a wide variety of recent news magazines and newspapers.

This 'safely masculine' rule does not apply to other types of partial embrace, such as putting the arm around the partner's waist. Since this is easier for either sex to do and since it also brings the hand closer to the genital region, it is rarely if ever employed between males.

If we now move still further away from the full embrace, shifting our attention from partial embraces to mere fragments of the complete act, we find similar differences. Some embrace fragments have a non-sexual quality and can be used freely between males, while others retain a more amorous flavour and are largely restricted to use between lovers and mates.

The hand-on-shoulder. A common action is placing one hand on the partner's shoulder, without involving any actual embrace. This is a simple reduction of the shoulder embrace and, as we might expect, is employed in similar contexts. Being slightly less intimate, it is even more common between males. Whereas only one in four of the shoulder embraces was done between two males, here the figure is one in three.

The arm-link. If we watch the embrace disintegrate still further, to a mere linking of arms, then the situation shows a striking change. Here, instead of increasing, the figure for males sinks to one in twelve, and we have to ask the question why, with an even less intimate form of body contact, men should become so much less inclined to link arms with other men rather than with women. The answer is that this action is basically feminine. When it occurs between males and females, it is five times more likely to be done by a female to a male than the other way around. This exactly reverses the position we found in the shoulder embrace, and means that if this contact is to be made between members of the same sex, it is going to have an effeminate quality. This leads to a prediction that if it is going to occur between friends of the same sex, it should occur more between women than between men, and observations do, in fact, bear this out.

If we look for cases where one man does link his arm through that of another, then we find that they fall into two categories: the Latin and the elderly. Latin males, with their culturally freer body contacts, often do it, and in non-Latin Western countries it can also be observed as the support-seeking act of an elderly male who is past the sexually potent phase of life.

The hand-in-hand. Continuing our anatomical progress away from the full embrace, via the shoulder embrace, the hand-on-shoulder and the arm-in-arm, we come finally to the hand-in-hand (not to be confused with the handshake, which is discussed separately later). Although this is a more remote form of contact than the last three, with the two bodies usually kept separate from one another, it does have something in common with the full embrace that the others do not. It is a mutual act. I can place my hand on your shoulder, for example, without you doing anything, but if I hold your hand, you are also going to be holding mine. Since it occurs frequently between a male and a female, and since they are always *both* doing it, it acquires neither a masculine nor a feminine quality, but rather a heterosexual one. This makes it, in effect, like a tiny version of the full embrace, and it is not surprising, therefore, to find that it is seldom used today between two males in public.

This was not always the case. Back in the days when the full embrace was freely employed between men, they could also be seen holding hands as an act of non-sexual friendship. To give one example, on an occasion when two medieval monarchs met, it is recorded that they 'took each other by the hand, when the king of France led the king of England to his tent; the four dukes took each other by the hand and followed them.' Before long this custom faded and 'leading by the hand' became restricted to use between males and females. In modern times the action has become modified in two different directions. On formal occasions, such as when a male escorts a female into a banquet, or down the aisle of a church, it has grown into the more reassuring arm-link. On less formal occasions it has changed into the typical hand-holding of mutually grasped palms. And sometimes, when greater intimacy is required, a couple can be seen to perform both acts at the same time.

Despite this general trend, there are certain specialized occasions when males in our modern world do still hold hands. One example is the multiple hand-holding that occurs when a whole group of people are joining hands with one another to do such things as sing community songs, or take curtain calls in the theatre. Even here the tendency is to alternate the positions of the males and females, so that each person is flanked by members of the opposite sex, but if the numbers are not equal, or if it is too difficult to jostle everyone into the correct positionings, then it is permissible to hold hands with your own sex. This is because you are not in any sense forming a pair with them. The very size of the group eliminates the potential sexual flavour of the hand-holding.

Another highly stylized version of hand-holding that can occur between males is for one of them to take the other's hand in his own and raise it high in the air as a sign of victory. Although this originates from the world of boxing, it is perhaps more frequently met with today between pairs of male politicians, who appear to fantasize imaginary boxing-gloves on the hands of their victorious colleagues. The hand-holding is permissible in this context because of the primarily aggressive nature of the arm-raising gesture. In its more original form, prior to the hand-holding embellishment, this movement of holding the raised fist aloft was undoubtedly a winning fighter's signal that he was still capable of hitting, at a stage when his rival was not. It is the frozen intention movement of delivering an overarm blow and is the same gesture that has been adopted as the modern communist salute. Studies of fighting behaviour between young children have revealed that this form of hitting, bringing the arm downwards from above, is very basic to our species and does not have to be learned. So it is interesting to see that the modern boxer still uses the intention movement of this action as a victory gesture, even though, in his actual fighting, he no longer employs it, but uses instead highly stylized and 'unnatural' frontal punches. It is also intriguing to notice how, in more informal fighting, such as that which occurs during street rioting, both the police and the rioters revert largely to the more primitive form of overarm hitting.

Returning now to the question of males holding hands in public, there is one final special context in which this occurs. It concerns priests, and especially those of high rank in the Catholic Church. The Pope, for instance, can frequently be seen to hold hands with his followers, both male and female, and this exception illustrates the way in which a well-known public figure can set himself outside normal conventions. The Pope's image is so totally non-sexual that he can perform a whole variety of fragmented intimacies with perfect strangers, intimacies which ordinary citizens could never contemplate. Who else, for example, could reach out and hold the cheeks of a beautiful girl in a completely non-sexual way? The Pope is, in fact, able to behave very much like the 'holy father' that he is called, and can confidently make intimate body contacts with adult strangers in the same way as a real father would do with his real children. By adopting the role of a super-father, the pontiff can strip away the body-contact restrictions that others must use, and return to the more natural and primary intimacies typical of the early parent-offspring phase. If he still appears more inhibited with his followers than a real father would be with his children, this is not due to the sexual confusion that restricts the rest of us, but simply to the fact that, in the face of a family of 500 million children, he has to conserve his strength.

Up to this point we have been moving away from the full embrace by travelling, as it were, across the shoulders and down the arm to the hand, and in that direction we can go no further. Instead, we can look at what other parts of the body come into contact with one another during the full embrace, and see if there, too, there are sources of useful fragments that can be employed in day-to-day encounters.

The pressing together of the trunks and legs during the full frontal embrace does not appear to be a very rich source and it is not hard to see why. For adults to touch one another in these regions, in public, brings them too close to the forbidden sexual zones. But there is another important contact region involved in the full embrace, and that is the head. At high intensities, the sides of the heads are pressed together, caressed with the hands or touched with the lips, and from these actions we do find the development of three important fragments that are widely used in everyday life. They can be labelled as the head-to-head contact, the hand-to-head contact and the kiss.

The head contacts. Touching the partner's head with the hand, and putting the two heads together, are both specialities of young lovers. This is especially true of the former. Hand-to-head contacts are four times as common between young lovers as between older married couples. Head-to-head contacts are twice as common with young lovers, and in both cases this contrasts with an intimacy such as the shoulder embrace, which is more common between the older couples.

Males rarely employ these head contacts with other males. When men 'put their heads together', they do not usually do it literally, the function being to indulge in intimate conversation rather than true body intimacy. If a male hand touches a male head, it usually does so for one of three special reasons: giving first aid, bestowing a blessing or delivering an attack. If a male (or a female) encounters an accident victim, the helplessness of the wounded person transmits strong infantile signals that are hard to resist. Photographs of the victims of assassination attempts, for example, nearly always show someone cradling the victim's head in their hands. Medically this is a somewhat dubious procedure, but medical logic has no place here. This is not a trained act of assistance; it is a more basic response related to the primary parental care of a helpless child. It is much too difficult for an untrained person to pause and make a reasoned assessment of the physical injuries the victim has sustained before taking first-aid action. Instead, he will reach out and touch or lift as a primary act of comfort, with no thought to the further damage he may be causing. It is too painful to stand by and coldly calculate the best steps to be taken. The urge to make comforting body contact is overpowering, but we have to face the fact that it can sometimes prove fatal. Once, as a small boy, ignorant of what was happening, I watched a man killed in this way. Following an accident, his injured body was cradled in the loving arms of anxious helpers who lifted him into a car to drive him away. The loving act destroyed his life by pressing his splintered ribs through his lungs. Had he been 'callously' left lying where he was until a stretcher arrived, he might have lived. Such is the power of the urge to make body contact when tragedy strikes, and this applies to males and females equally, for disaster knows no sex.

The bestowing of priestly blessings is equally sexless, as in the laying on of hands of a bishop in ordination or confirmation. Here we are back again to a copy of the primary parent-child relationship.

The delivering of an attack by the hand of one male to the head of another requires little comment in itself, but it does provide one possible source of inter-male intimacy. If a man feels the friendly urge to touch another man's head but is inhibited about making it a friendly caress, he can employ the simple device of mock-aggression. Instead of fondling his partner's head, which would have too strong a sexual flavour, he can deliver a playful 'pretend-attack', such as ruffling the hair or squeezing the neck in a mock-grasp. Just as play-fighting helped the parent to prolong intimacies with his growing children, so many a fragment of play-assault can be observed between male friends, enabling them to be both manly and intimate at one and the same time.

The kiss. This brings us to the last of the important derivatives of the primary embrace, namely the kiss, an action with an intriguing and complex history. If you feel that a kiss is a simple enough act, think for a moment about the many ways in which you do it, even in today's supposedly informal society. You kiss your lover on the lips, an old friend of the opposite sex on the cheek, an infant on the top of the head; if a child hurts a finger, you kiss it 'to make it better'; if you are about to face danger, you kiss a mascot 'to bring you luck'; if you are a gambler, you kiss the dice before you roll them; if you are the best man at a wedding, you kiss the bride; if you are religious, you kiss the bishop's ring as a sign of respect, or the Bible when taking an oath; if you are bidding someone farewell and they are already out of reach, you kiss your own hand and blow the kiss towards them. No, the kiss is not a simple matter, and to understand it we must once again turn back the clock.

The most sensitive areas of skin on the human body are the tips of the fingers, the clitoris, the tip of the penis, the tongue and the lips. It is not surprising, therefore, that the lips should be used a great deal in intimate body contacts. Their role begins in the act of sucking at the mother's breast, which provides a major tactile reward in addition to the reward of obtaining milk. This has been proved by studying the behaviour of unfortunate babies who have been born with an abnormal, blocked oesophagus, and who have to be fed by artificial means. It was observed that if they were given rubber teats to suck, this helped to calm them down and stopped them crying. Since they had never taken any food through the mouth, the reward of having a teat between the lips could have nothing to do with the pleasure of the milk supply that normally results from such an action. It had to be a case of contact for contact's sake. So the touching of something soft with the mouth is an important and primary intimacy in its own right.

As the child grows and exchanges head-to-head contacts with the mother, feeling her lips pressed to his skin and his to hers, it is easy to see how this early mouth contact can develop into a potent act of friendly greeting. In the embrace of childhood, the usual position for the lips when touching the parent will be the cheek or the side of the head. As I have already mentioned, in ancient times, when the full embrace was given more freely between adults of either sex, the kiss on the cheek was the usual form of mouth contact employed between equals. This was, in a sense, the primitive greeting kiss taken straight from childhood with little modification, and it has persisted through the centuries right down to the present day. In our culture, male and female friends and relatives still kiss in this way when meeting or parting, and the act can be performed without any sexual implication whatever. The same is true of adult females with adult females. Between adult males the situation varies considerably from country to country, with France, for example, remaining much closer to the ancient system than England.

Direct mouth-to-mouth kissing has taken a different course. At various times and in various places it has been employed to some extent as a non-sexual greeting between close friends, but the joining of two body orifices in this way has usually seemed to be an act of too great an intimacy even for close friends, and generally speaking it has become more and more restricted to contacts between lovers and mates.

Since the female breasts are sexual signals as well as feeding devices, the kissing of a female breast by an adult male has also become totally sexual in context, despite the similarity of the act to the primitive action of sucking at the breast in infancy. Needless to say, kissing of the genitals is also exclusively sexual, and so indeed is the kissing of many other parts of the body, especially the trunk, thighs and ears. Certain specific parts of the body, however, have been formally set aside for a special kind of non-sexual kissing – what we might call the subordinate kiss, or the kiss of reverence. This differs categorically from both the friendly kiss and the sexual kiss, and to understand it we must look at the way in which a subordinate human being presents himself in front of a dominant.

It is well known from studies of animal behaviour that one way to appease the wrath of a dominant animal is to make yourself seem smaller and therefore less threatening to him. If you threaten him less, then he is less likely to see you as a challenge to his superiority and therefore less likely to take damaging action against you. He will simply ignore you as being beneath him, both metaphorically and literally, which, if you happen to be the weaker animal, is precisely what you want (at least for the moment). And so we see all manner of cringings and crouchings, grovellings and hunchings-up, downcast eyes and lowered heads, in a wide variety of animal species.

It is the same with man. Where there are no formalities, the response takes the animal form of cringing low on the ground, but in many situations the response of the inferior man has become highly stylized, and these stylizations have varied considerably from place to place and from time to time. This does not put them outside the realm of a biological analysis, however, for they all, without exception, still reveal basic features that clearly relate them to the submissive behaviour of other species of animals.

The most extreme form of submissive body-lowering ever seen in man is full prostration, in which the whole body lies flat on the ground, face down. You simply cannot get any lower than that without an act of burial. The dominant, on the other hand, can, and often did, enhance the lowering effect by viewing it from a raised platform or throne. This total act of servility was common and widespread in the ancient kingdoms, performed by prisoners to their captors, slaves to their masters, and servants to their rulers. Between it and the act of standing fully erect, there is a whole range of formalized submissions, and we can scan them briefly in ascending order.

Next to the full prostration comes the kowtow of the Eastern world, in which the body does not lie, but kneels down and then bends the trunk low until the forehead is touching the ground. Coming one stage up from this is the full kneel, with both knees on the ground, but without the forward bending of the body. This, too, was frequently used in the ancient world when confronting an overlord, but by medieval times it had already risen to the half-kneel, with only one knee lowered to the ground. Men were then specifically instructed to reserve the full kneel for God, who by this time had to be given a little more respect than the rulers of the day. In modern times it is rare for us to kneel to any man at any time, except on certain state occasions in the presence of royalty, but worshippers in church have not changed the ancient full-kneel custom to this day, God having maintained his dominant status rather more successfully than modern rulers.

Coming one stage further up, we reach the curtsy (or courtesy), which was no more than an intention movement of the half-kneel. One leg was drawn back slightly, as if its knee was going to go down and touch the ground, and then both knees started to bend, but neither reached ground level. The body was not bowed forward. Up until the time of Shakespeare both men and women made the curtsy. In this respect, at least, there was equality between the sexes. The male bow had not yet appeared. With the arrival of the curtsy, the act of servility was further reduced, and the half-kneel started to fade from the scene, being given only to royalty.

In the seventeenth century the sexes split up, the men now bowing from the waist while the women retained the curtsy. Both actions lowered the body in front of the dominant individual, but in entirely different ways. From that day to this the situation has remained much the same, except that the extent of the actions has been reduced. The flowery male bowing of the Restoration period gave way to the much simpler and stiffer bow of Victorian times, and the curtsy dwindled to little more than a bob down and up. Today, except in the presence of powerful rulers or royalty, the curtsy is rarely given by women, and the male bow, if given at all, is seldom more than a mere lowering and raising of the head.

The one exception to this rule occurs at the end of a theatrical performance, when, for some reason, the performers slip back several centuries and indulge in deep bows and elaborate curtsies. It is amusing that here we sometimes also see an entirely new trend, with female performers bowing deeply as if they were males. It looks as if this return to sexual equality in the act of subordination is mirroring the new trend towards female equality in all other matters, but if this is the case, the males can at least claim that it was the females who changed to the male action, rather than that the males had to return to their medieval curtsying. It is just possible that there is another reason altogether for the actress's bow which has nothing to do with the masculinizing of the modern female in our culture. It may instead have to do with the precise opposite, dating back to the earlier times of play-acting, when all the actors were male, and half the men were feminized to play the female roles. Perhaps the modern actress when she bows is by force of tradition merely imitating her male transvestite predecessors. Even allowing for the persistence of ancient traditions, however, this explanation appears improbable. It seems much more likely that she feels she is joining the men.

All the bowing and scraping of yesterday's everyday greetings has now been almost universally replaced by the far more forthright and upstanding handshake. In this act, at last, there is no lowering of the body. We greet erect and, in so doing, have travelled the full distance from flat prostration. Today all men are not only 'born equal', but are still considered to be so, in greetings at least, when they are fully adult.

I have gone into these formalities of greeting at some length, despite the fact that, until we reach the handshake, they do not in themselves involve the intimacies of body contact. This digression was necessary because of their importance in relation to the kiss of reverence. I began by saying that, in ancient times, two equals kissed one another on the cheek, that is, at equal body height. But for an inferior to kiss a superior in this way would have been unthinkable. If he was to show his friendship by a touch of the lips, then that touch had to be performed at a level low enough to match his inferiority. For the lowest subordinates this meant the act of kissing the dominant one's foot. For an abject prisoner even this was too good, and he was forced to kiss the ground near the dominant shoe. In modern times these actions are rare, rulers not being what they once were, but even now the ruler of Ethiopia, for example, may find himself accorded this honour by one of his subjects in a public place. And slang phrases such as 'kiss the dirt', 'bite the dust' and 'boot-licker' are still with us to remind us of the humiliations of yesterday.

For those of slightly less inferiority, it was permissible to kiss the hem of a garment, or to kiss the knee of the dominant individual. A bishop, for instance, was permitted to kiss the Pope's knee, but a lesser mortal had to be content with kissing the cross embroidered on his right shoe.

Coming one step further up the body, we arrive at the kiss on the hand. This, too, was performed to many a dominant male, but today, apart from high-ranking priests, we reserve it entirely as a mark of respect to a lady, and even then only in certain countries and on certain occasions.

There were therefore four points on the body which were licensed, so to speak, for non-sexual kissing: the cheek, for friendly equality; the hand, for deep respect; the knee, for humble submission; and the foot, for grovelling servility. The action of touching with the lips was the same in each case, but the lower the point of its application, the lower was the expression of relative status. Despite all the pomp and ritual, nothing could be closer to a sequence of typical animal appeasement gestures. When stripped in this way of the fussy details of cultural variation and viewed as a whole, even the most courtly of human behaviour patterns come remarkably close to the patterns of animal behaviour we see all around us.

Earlier I listed a number of forms of modern kissing, some of which I may not appear to have explained: for example, kissing the dice before rolling them, kissing a lucky mascot, or a hurt finger to make it well. These and other similar actions, all of which are basically concerned with bringing good luck, are related to the kiss of reverence I have been describing. It is impossible to kiss God, the most dominant one of all, and so worshippers have to make do with symbols of God, such as crosses, Bibles, and similar objects. Since kissing them symbolizes kissing God, the act brings good luck simply because it appeases God. Any lucky mascot, therefore, is being treated as a holy relic. It may be odd to think of a gambler in Las Vegas kissing God when he blows on his dice during a crap game, but that is in effect what he is doing, just as when he crosses his fingers for good luck he is doing no more and no less than making a reverential sign of the cross to protect him from God's wrath. When we kiss our hands in farewell and blow a kiss to our departing friends, we are performing another ancient act, for in earlier days it was more servile to kiss one's own hand than the hand of the dominant person. The hand-kiss at the modern airport is the only survivor of this custom, even though it is now distance rather than servility that makes us perform the movement.

The handshake. With that farewell kiss, we leave the world of the fragmented embrace, with all its complexities, and come to the last of the adult body contacts that is important enough to be examined in detail, namely the handshake. I have already mentioned that this did not gain wide usage until about a hundred and fifty years ago, but its precursor, the hand-clasp, was employed long before that. In ancient Rome it was used as a pledge of honour, and this was to remain its primary function for nearly two thousand years. In medieval times, for instance, a man might kneel and clasp hands with his superior as an act of swearing allegiance to him. The addition of a shaking movement to the clasping hands is mentioned as early as the sixteenth century. The phrase 'they shook hands and swore brothers' appears in Shakespeare's *As You Like It*, where again the function is the binding of an agreement.

In the early part of the nineteenth century the situation changed. Although the handshake was still employed after the making of a promise or a contract, as an act of sealing it, it was now for the first time used in ordinary greetings. The cause of this change was the industrial revolution and the massive expansion of the middle classes, forcing an ever widening wedge between the aristocracy and the peasants. These new middlemen, with their businesses and their trades, were forever 'doing deals' and 'making agreements', and sealing them with the inevitable handshakings. Dealing and trading was becoming the new way of life, and social relationships began more and more to revolve around them. In this way the contractual handshake invaded the social occasion. Its message became the bartering one of 'I offer you an exchange of friendly greeting'. Gradually it ousted the other forms of greeting until today it has become universally used as the principal act performed, not only when meeting equals, but also when meeting both subordinates and superiors. Whereas once we would have had a wide range of alternatives to suit each type of social encounter, today we have merely this one. What a president does to a farm worker is now the same as what a farm worker does to a president – they both offer a hand, clasp it and shake it, both smiling as they do so. Furthermore, when a president meets another president, or a farm worker meets another farm worker, they will behave in precisely the same way. In terms of body intimacies, times have certainly changed. But if the ubiquitous handshake has simplified matters in one way, it has complicated them in another. We may know that it is the thing to do, but when precisely do we do it? Who offers the hand to whom?

Modern books of etiquette are full of conflicting advice, clearly indicating the confusion that exists. One tells us that a man never holds out his hand to a woman for a handshake, while another informs us that in many parts of the world it is the man who takes the initiative. One tells us that a younger man should never offer his hand to an older one, while another advises that, whenever we are in doubt, we should offer the hand, rather than risk hurting someone's feelings. One authority insists that a woman should rise to shake hands, another that she should remain seated. Further complications exist in respect of whether we are hosts or guests, male hosts offering their hands to female guests, but male guests waiting for a female hand to be offered to them. There are also separate rules for business and social occasions. One book goes so far as to say that 'There are no rules as to when to shake hands,' but clearly this is a statement of despair, the truth being that there are in fact far too many rules.

Obviously there is some hidden complication in the superficially simple act of handshaking which we must unravel if we are to understand these confusions. To do this we must look back at the origins of the act. If we go back as far as our animal relatives, we find that a subordinate chimpanzee will often appease a dominant one by reaching out towards it with a limp hand, as if making a begging gesture. If the action is returned, the two animals briefly touch hands in a contact that looks remarkably like an abbreviated handshake. The initial signal reads, 'See, I am just a harmless beggar who dare not attack you,' and the reply is, 'I will not attack you either.' Developing into a friendly gesture between equals, the message then becomes simply, 'I will not hurt you, I am your friend.' In other words, the chimpanzee hand-offering can be done either by a subordinate to a dominant, as a submissive act, or by a dominant to a subordinate, as a reassuring one, or it can be performed between equals as an act of friendship. Nevertheless, in this respect, it is fundamentally an appeasement gesture and, translated into the modern terms of the etiquette books, we would expect more emphasis to be placed on the inferior individual offering the hand first, to the superior.

Advancing now to the ancient human hand-clasp, we can see this in a similar light. Specifically, the offering of an empty hand revealed that there was no weapon in it, which would explain why we always use the right, or weapon, hand. Showing the hand in this way could be done either submissively, by a weaker to a stronger, or reassuringly, by a stronger to a weaker, as in the chimpanzee case. Developed into a strong, mutual hand-clasp, it became a vigorous pact-making device with two men accepting one another, momentarily at least, as equals. Essentially, however, it remains an act in which neither performer is asserting his dominance but, regardless of his relative status, is temporarily displaying himself as harmless.

This is one likely origin for the modern handshake, but there is another that confuses the picture. One of the important greeting acts of a male to a female was the kissing of the hand. To do this, the man took her offered hand in his before applying his lips to it. As this action became more stylized, the actual kissing element declined in intensity to a point where the mouth only approached the back of the lady's hand and stopped before making contact, the lips then forming a kiss in mid-air. Becoming even more fossilized, the act then sometimes occurred as a mere holding and raising of the lady's hand, accompanied by a slight bow of the head towards it. In this modified form it is little more or less than a weak handshake, but with the vigorous pumping action omitted. One writer has seen this as the sole source of the modern handshake: 'As a contact salutation the handshake would appear to be a late derivative of the "face-kiss", with the "hand-kiss" as a connecting link.' In this respect, offering the hand is essentially the act of asserted dominance towards a subordinate, and therefore differs fundamentally from the display of the male-pact handshake.

The truth appears to be that both the hand-clasp theory and the hand-kiss theory are correct, and that this double origin is the cause of all the confusion in the modern etiquette books. The point is that we do not today shake hands for a single reason only. We do it as a greeting, as a farewell, to make a pact, to seal a bargain, to congratulate, to accept a challenge, to give thanks, to commiserate, to make up after a squabble and to wish one another luck. There are two elements here. In some cases it symbolizes a friendly bond, in others merely that we are friendly at the moment of shaking. If I shake hands with a man when I am introduced to him for the first time, it is merely a courtesy and says nothing about our past or even our future relationship.

To put it another way, we can say that the modern handshake is a double act masquerading as a single one. The 'pact handshake' and the 'greeting handshake' have different origins and different functions, but because they have come to have the same form, we think of them both

435

simply as the 'friendly handshake'. Hence all the confusion. Up to early Victorian times there was no problem. Then there was the pact hand-shake for males with males which said 'it's a deal', and the hand-kiss for males with females which said 'it's an honour to meet you'. But when the Victorians increasingly began to mix business with social life, the two became blended and muddled together. The vigorous pact handshake became softened and weakened, while the gentle clasping of the lady's hand in the already abbreviated hand-kiss became strengthened.

Although we accept this happily enough today, there was some resistance to it in nineteenth-century France, where the greeting shake was referred to as 'the American handshake', and frowned upon when it was done between visiting males and unmarried French girls. The reason for this was not so much the body contact involved, but simply that the French were still interpreting the handshake in its old masculine role. The visiting males were then seen to be 'making a pact' and establishing a bond of friendship with young girls they had only just met, which was considered highly improper. The foreign visitors, of course, thought that they were doing no more than offering a polite salutation.

This brings us back to the misunderstandings and confusions of the etiquette books. The big problem is who offers the hand to whom. Is it an insult to fail to offer the hand first and therefore to appear to be unfriendly, or to offer it first and therefore to appear to be demanding a modified hand-kiss? Careful observation of social occasions reveals that the confused greeters tend to solve the problem by watching for tiny clues. They look for the slightest sign of an intention movement of raising the arm on the part of the other person, and then try to make the contact appear to be simultaneous. Contributing to their confusion is the fact that with most other salutations the subordinate acts first to show his respect. The private salutes the officer before the officer salutes the private. In earlier days, it was always the junior who bowed first to the senior. But with kissing the hand it was different. The lady had to offer her hand first. No respectful male would grab it first without waiting for a sign from her. Because the hand-kiss is involved in the origin of handshake greetings, this rule still applies in most cases. The man waits for the woman to offer her hand for shaking, as if she were still offering it to be kissed. However, not to offer the hand to her first, now that the kissing has vanished from the act, is tanta-mount to saying that he is the officer and she the private, and that she must make the first sign of salute. Hence all the warnings and mum-blings of the etiquette experts.

The other origin of the handshake, concerning the making of pacts, further confuses the situation. The weaker male usually offers his hand first here, to show his eagerness to the stronger one. In a contest, it is usually the weaker loser who offers his hand to the stronger winner in the act of congratulation, to show that, despite his defeat, the bond of friendship is reconfirmed. So, for an eager young businessman to greet a senior colleague with an outstretched hand may be looked upon either as brashness ('you may kiss my hand'), or as humility ('you are the winner'). Again, as with the social occasions, the problem is usually solved by watching for small intention clues and trying to contrive a simultaneous act.

Given its complicated past and its confused present, one might expect to see the handshake on the decline in the increasingly informal world of today, and in certain contexts this appears to be the case. Social greetings are becoming increasingly verbal. Around the middle of the present century, the etiquette experts announced that 'Hand-shaking on introductions between men is nowadays on the wane in Great Britain.' Despite this, it is still much more common for a male/male shake to occur, rather than a male/female shake or a female/female shake. My observations suggest that two-thirds of all handshaking is done between males; in the remaining third, shaking between the sexes is three times as common as between females. These figures fit well with the history of the action, for men have inherited the handshake as a pact device and then added to it the greeting function, giving the all-male pattern double value, so to speak. Women with men have inherited it from the hand-kiss, but have not yet taken an equal role in business, so are poor on pact handshakes. And women with women never did hand-kiss, so are poor on both, and come at the bottom of the handshaking league.

A final point about this particular form of body contact, which may seem obvious but is significant nevertheless, is that lovers do not do it. Nor, in most countries, do married couples. Ask an Englishman who has been married for, say, twelve years when he last greeted his wife by shaking hands with her, and the answer is more likely to be twelve years than twelve days. This is, without any doubt, the least amorous of all friendly body contacts. In all the other cases mentioned in this chapter, from the full embrace to the kiss, there has always been a strong sexual element. They have all originated from the same primary source and they have all been performed more between lovers and mates than between adults in any other roles. When performed between males there have, in most cases, been special circumstances which make this possible. The handshake, by contrast, stemming originally not from a tender embrace but from a masculine act of pact-making, has avoided these difficulties. Even the later involvement of the hand-kiss in its history has not created any problems, because this was already a formalized, de-sexed kiss of reverence before it became incorporated. Strong men have therefore been able to shake one another's hands until they are blue in the palms without running the slightest risk of creating an amorous impression. The very act of shaking the clasped hands up and down in mid-air that typifies the action helps to make it more brusque and less gentle, and distinguishes it clearly, even at a distance, from the lovers' act of holding hands.

In this chapter we have looked at the way adults behave towards one another in public and we have seen how the all-enveloping, un-inhibited intimacies of infancy have become restricted, pigeon-holed and labelled. It can be argued that this has taken place because adults need greater independence of action and greater mobility than infants, and that more extensive body contacts would limit them in this respect. That would explain the reduction in the amount of time actually spent touching, but not the reduction of the intimacy of the touches that do still occur. It can be argued that it has taken place because adults do not need so much body contact; but if that is so, then why do they spend so much of their time indulging in second-hand intimacy in books, films, plays and television, and why is it that popular songs cry out the message hour after hour? It can be argued that our untouchability has to do with status, with not wishing to be touched by our inferiors and not daring to touch our superiors; but if this is so, then why are we not more intimate with our equals? It can be argued that we do not wish to have our intimate actions confused with those of lovers; but how does this explain the fact that lovers themselves, in public, restrict their intimacies so much more than they do in private?

All these arguments give partial answers, but there is something missing. This hidden factor seems to be the powerful bonding effect that close body intimacies have on those who perform them. We cannot be close physically without becoming 'close' emotionally. In our busy modern lives, we hold back from such involvements, even though we may need them. Our relationships are too profuse, too vague, too complex, and often too insincere for us to be able to risk the primitive bonding processes of body intimacy. In the ruthless world of business, we can dismiss a girl with whom we have only shaken hands, or we can betray a colleague with whom we have done no more than rest a hand on a shoulder; but what if the body contacts had been greater? What if, without any sexual involvement, we had experienced greater intimacies with them? Then, without a doubt, we would have seen our tough determination soften, and our competitiveness dwindle, when the moments for brutal decisions came. And if we dare not expose ourselves to these dangers, to these powerful reciprocal involvements that know no logic, then we certainly do not want to be reminded of them by seeing them flaunted in public by others. So the young lovers can keep to themselves and do it in private, and in case they ignore our request we will make it law. We will make it a crime to be intimate in public. And so it is that, even to this day, in certain sophisticated, civilized countries it remains a crime to kiss in public. A tender act of touching becomes immoral and illegal. A gentle intimacy becomes legally equated with an act of theft. So hide it away quickly, lest the rest of us see what we are missing!

It has sometimes been said that if only all the tight-lipped defenders of public morals would embrace one another lovingly, caress one another's faces, and kiss one another's cheeks, they might suddenly feel it was time to go home and leave the rest of society to go about its friendly, loving business without having to endure their desperate envy. But it is pointless to despise them, for society stitches its own strait-jacket. The teeming zoo in which we live is not the ideal setting for public intimacies. It suffers from people pollution; we bump into one another and apologize, when we should be reaching out to touch; we collide headlong and curse, when we should be embracing and laughing. There are strangers everywhere and so we hold ourselves back. There seems to be no alternative. Our only compensation is to indulge more heavily in private intimacies, but this we frequently fail to do. It seems as if our public restraint can spread to infect our conduct even in the bosom of our families. For many, the solution is to indulge in second-hand intimacies by spending the evening hours avidly watching the abandoned touchings and embracings of the professionals on our television or cinema screens, listening to endless words of love in our popular songs, or reading them in our novels and magazines. For some, there are other, more heavily disguised alternatives, as we shall see on the pages that follow.

5

SPECIALIZED INTIMACY

BY STUDYING THE behaviour of infants and lovers, it becomes clear that the degree of physical intimacy that exists between two human animals relates to the degree of trust between them. The crowded conditions of modern life surround us with strangers whom we do not trust, at least not fully, and we go to great pains to keep our distance from them. The intricate avoidance patterns of any busy street bear witness to this. But the frenzy of urban living creates stress, and stress breeds anxiety and feelings of insecurity. Intimacy calms these feelings, and so, paradoxically, the more we are forced to keep apart, the more we need to make body contact. If our loved ones are loving enough, then the supply of intimacy they offer will suffice, and we can go out to face the world at arm's length. But supposing they are not; supposing we have failed as adults to form close bonds with either friends or lovers, and have no children; what do we do then? Or supposing we have formed these bonds successfully, but then they have broken down, or become fossilized into the remoteness of indifference, with the 'loving' embrace and kiss becoming as formalized as a public handshake; what then? The answer for many is simply to grouse and bear it, but there *are* solutions, and one of these is the device of employing professional touchers, a measure which helps to some extent to compensate for the shortcomings of the amateur and amatory touchers who are failing to supply us with our much-needed quota of body intimacy.

Who are these professional touchers? The answer is that they are virtually any strangers or semi-strangers who, under the pretext of providing us with some specialist service, are required to touch our bodies. This pretext is necessary because, of course, we do not like to admit that we are insecure and need the comforting touch of another human body. That would be 'soft', immature, regressive; it would assail our image of ourselves as self-controlled, independent adults. And so we must get our dose of intimacy in some disguised form.

One of the most popular and widespread methods is being ill. Nothing serious, of course, merely some mild sickness that will stimulate in others the urge to perform comforting acts in intimacy. The majority of people imagine that when they fall prey to some minor ailment, they have simply been unlucky in accidentally encountering a hostile virus, bacterium, or some other form of parasite. If they come down with a nasty bout of influenza, for instance, they feel it could have happened to anyone – anyone who, like them, has been shopping in busy stores, standing on tightly packed buses, or jamming themselves into stuffy corners at overcrowded parties, where coughs and sneezes can be heard incessantly wafting their eager pathogens through the air. The facts, however, do not support this view. Even at the height of a flu epidemic, there are still many people – equally exposed to the infection – who do not succumb. How is it that *they* manage to avoid taking to their sick-beds? How, in particular, does the medical profession manage to remain so remarkably healthy? They, more than anyone, are massively exposed to infection all day and every day, but they do not seem to become proportionally sick.

Minor ailments, therefore, do not appear to be entirely a matter of unlucky accident. In a modern city there are hostile microbes everywhere. Almost every day, and in almost every place we walk and breathe, we are exposed to sufficient of them to bring us down with some sort of infection. If we defeat them, it is not so much that we manage to avoid them as that our bodies are equipped with a highly efficient defence system which slaughters them by the million, week in and week out. If we succumb, it is not so much that we have been accidentally exposed to them as that we have, for some reason, lowered our body defences. One way we do this (apart from excessive hygiene!) is to let ourselves become overstressed and overstrained by the pressures of urban life. In our weakened condition, we soon fall prey to one or other of the wide selection of unfriendly microbes that fill the world around us. Luckily for us, the disease is its own cure, for, in putting us to bed, it provides us with the very comfort we were lacking before. We might call this the 'instant-baby' syndrome.

The man who is feeling 'poorly' begins to look weak and helpless and starts to transmit powerful pseudo-infantile signals to his wife. She responds automatically as an 'instant-parent' and begins to mother him, insisting on tucking him up in his bed (cot) and bringing him soup, hot drinks and medicine (baby food). Her tone of voice becomes softer (the maternal coo) and she fusses over him, feeling his forehead and performing other intimacies of a kind that were missing before, when he was fit and well but equally in need of them. The curative effect of this comfort behaviour works wonders and he is soon back in action again, facing the hostile world outside.

This description does not imply malingering. It is essential for the patient to be truly and visibly sick in order to stimulate fully the necessary pseudo-parental care. This accounts for the high frequency of strongly debilitating, but comparatively pain-free, minor ailments in cases of emotionally induced illness. It is important, not only to be sick, but to be seen to be sick.

To some, these comments will seem cynical, but that is not my intention. If the stress of life demands that we shall obtain increased comfort and intimacy from our closest companions, and forces us once again to sink into the warm embrace of the soft bedding of our 'cots', then this is a valuable social mechanism and must not be sneered at.

It is, indeed, so useful a device that it has come to support a major industry. Despite all the impressive technological advances of modern medicine and our so-called conquest of the environment, we still get sick at an astonishingly high rate. The majority of the victims do not see the inside of a hospital ward. They may be out-patients, pharmacy clients, or merely self-treating in the home. They suffer from a great variety of common ailments such as coughs, colds, influenza, headaches, allergies, backaches, tonsillitis, laryngitis, stomach-aches, ulcers, diarrhoea, skin rashes, and the like. The fashions change from generation to generation – once it was 'the vapours', now it is 'a virus' – but basically the list remains much the same. In terms of simple frequency of occurrence, these cases account for the vast majority of present-day illnesses.

In Britain, for example, over 500 million pharmaceutical purchases are made every year to treat minor illnesses, which works out at roughly ten ailments per year, per head of the population. Some 100 million pounds is spent annually on these products. Over two-thirds of all illnesses are not serious enough to involve the services of a doctor.

The reason for this situation is simple enough. All the time, our populations are increasing in size and our communities becoming more and more overcrowded and overstressed. The larger numbers of people involved means that there is more and more money available for medical research, which finds better and better cures. In the meantime, however, the populations have grown again, the social stresses have become greater and the susceptibility to disease has increased. Therefore more medical research is needed, and so on, neck and neck, into an imaginary, disease-free future that will never come.

But suppose for a moment that I am being pessimistic; suppose that some medical miracle has eventually appeared on the scene and defeated and exterminated all the parasites. Will we then finally have arrived at a condition where the downtrodden, emotionally bruised urbanite can no longer collapse with impunity into the comforting arms of his sick-bed? The chances are more than remote that this miracle will ever happen, but even if it did there are still several alternatives open to the would-be 'instant-baby'. These are already in frequent use. In the absence of suitable viruses or bacteria, he can always have a 'nervous breakdown'. Minor mental illnesses have the advantage that they can operate in the absence of acquired microbes, and they are equally effective as comfort-producers. Indeed, they are so effective that even a murderer can plead 'temporary insanity' as an excuse for his actions and have his sentence modified on the grounds of 'diminished responsibility' – again being treated as if he were a 'temporary infant'. Pleading that he was suffering from a cold in the head at the time of the murder would provide less comfort to him, so there is clearly a lot to be said for the power of mental breakdown as a survival device when stress becomes extreme. The main disadvantage here is that many of the milder versions of mental illness are lacking in the external symptoms necessary to provoke the much-needed comforting reactions. The emotionally bruised individual is driven to extremes to produce the required response. Internal agonizing is not enough, but after a good bout of screaming hysterics, his collapsed body stands a very high chance of feeling itself snugly enveloped in the embracing arms of an earnest comforter. If the breakdown is more violent, he may instead find himself encased in forcibly restraining arms, but even then all is not lost, for he will have succeeded, albeit in a desperate way, in making some kind of intimate body contact with another human being. Only if he loses control completely will he fail and find himself condemned to the solitary self-embrace imposed by the canvas sleeves of a strait-jacket.

A second alternative in the absence of foreign parasites is the use of the patient's own endogenous microbes, the ones that he has been carrying on his body all his life. To explain how this works, we must take a close look, in fact a microscopic look, at the surfaces of our bodies.

Many people seem to imagine that *all* microbes are nasty and that they automatically mean disease or dirt, but this is not true. As any bacteriologist will testify, this is nothing more than the modern myth of the new hygiene religion, the religion whose aerosol prayers keep its worshippers 'free from all known germs', whose holy water is anti-septic solution, and whose god is totally sterile. Of course, there *are* vicious, deadly germs that we do well to destroy as ruthlessly as possible. There is no denying that. But what about the 'germs' whose main activity in life is killing off other germs? Do we really want to kill all known germs?

The fact is that we are each of us protected by a vast army of friendly microbes that do not hinder us but, on the contrary, actively help us to keep healthy. On our healthy, clean skin there is an average of five million of them to every square centimetre. Ordinary saliva, spat from the mouth, contains between ten million and 1,000 million bacteria in every cubic centimetre. Every time we defecate we lose 100,000 million microbes, but their numbers are soon made up again inside the body. This is the normal condition of the adult human animal. If we managed to live 'germ-free' of our own microbes all our lives, we would be at a grave disadvantage. Amongst other things, we would be less resistant to the foreign, and really vicious, microbes that we would encounter from time to time. We know this from careful experiments with germ-free laboratory animals. Our natural load of body microbes is therefore of great value to us, but now comes the catch. We have to pay a price for their good services, for even they can get out of hand when we become unduly stressed. Some of our diseases are caused, not by acquiring infections from others, but by a sudden eruption and 'over-crowding' of our own 'normal' microbes. The usual measures of public hygiene that assist in cutting down cross-infection from one person to another cannot help in such cases: we do not 'catch' the diseases; we carry the makings of them ourselves all the time. This is particularly true of many of the alimentary upsets so common in the emotionally stressed patient. If we have 'stomach trouble' we put it down to 'something bad' that we ate, but it is amazing what a healthy, happy person can devour and get away with. Probably nearly all the mild stomach and intestinal upsets we suffer from are due instead to emotional disturbances resulting from failures to adjust to the stresses and strains of modern living. To remind ourselves of this we have only to watch a natural history film of a flock of healthy vultures on the plains of Africa, gobbling down the putrid flesh of a decaying carcass, an event which is more likely to turn our own stomachs rather than those of the birds concerned.

A third alternative for the human individual in need of comfort is a more drastic one. Failing mental illness or endogenous illness, he can, with a little agitated carelessness, become dramatically accident-prone. If he trips over and breaks an ankle, he will soon be able to curse that he is 'as helpless as a baby', and in no time at all find himself helped and supported just like one. But surely accidents are accidental? Of course they can be, but nevertheless it is surprising how much people vary in their susceptibility to 'accidental' injuries. In a recent hospital investigation into the emotional backgrounds of disease patients, a number of accident patients were used as a control group, because it was assumed that they must be in their hospital beds 'by accident', in both senses of the word. Results showed that this was far from being the case, the accident victims proving to be, if anything, more emotionally disturbed than the disease patients.

Our stressed urban comfort-seeker, therefore, has several ways in which he can become suitably helpless and promote calming intimacies from those who attend him. There is a considerable advantage in being mildly ill from time to time, and if the advantage cannot be gained in one way, there is always another open. This method of increasing adult intimacies does, however, have its drawbacks. In all cases it involves the sick individual in the adoption of a submissive role. To obtain the comforting attention that his ailment provokes, he is forced to become genuinely inferior, either physically or mentally, to his comforters. This was not so with the young lovers, who went 'soft' on a reciprocal basis that did not lower their social status. Furthermore, the patient's comfort-bath soon grows cold when he regains his health and strength, and the tender intimacies of those who have been caring for him cease abruptly. The reward was temporary, and the only way to prolong it is to become a chronic invalid who, as the saying goes, 'enjoys bad health'. Apart from prolonging the inferior status, this also introduces a new danger, that of ailment escalation. The comforting fire that has been lit may get out of control and burn down the house. Even when used as a short-term measure, there is always the risk of long-term damage to the organism, as ulcer-sufferers know to their cost. But for many who find the tensions of modern living hard to bear, the risk is worth it. Temporary respite is better than no respite. If they are lucky, it gives them time to recharge their emotional batteries, and in so doing it can be said, in biological terms, to have considerable survival value in today's crowded human communities.

Although much of the comfort obtained in this way is provided by the close companions of the patient, whose intimacy quotient becomes dramatically increased in most cases, the phenomenon of 'going sick' also provides the additional reward of obtaining the intimate attentions of a group of people who are comparative strangers – the members of the medical profession. Doctors are 'licensed to touch', and to do so with a degree of intimacy forbidden between most adults. Intuitively aware of this important element in their work, they know well the curative value of the 'bedside manner'. The reassurance of the softly spoken word, the confident touch of the hand that takes the pulse, or taps the chest, or turns the head to examine eyes and mouth, these are the body-contact actions that, for some, are better than a hundred pills.

Sometimes a doctor will order the removal of a patient to a hospital bed on emotional grounds alone. For an individual whose source of stress lies solely in the outside world, such a move is unnecessary. By staying at home and taking to his bed, he escapes the tension that is damaging him. But if the tension lies in the home itself, there is no such escape. If the emotional pressures are coming from inside the family unit, then even his bedroom may not provide the necessary hiding-place, where he can curl up and find the comfort he needs so badly. Then the only solution is the hospital cot, and pray God for short visiting hours.

The medical solution for the adult intimacy-seeker is, as we have seen, a mixed blessing, and he would clearly do better to look else-where. If he is religious, he can perhaps enjoy an unmixed blessing from the hands of a priest, but, failing that, there are several other soothing contacts he can enjoy.

There is the whole lush world of body conditioning and beautifying to indulge in, where an army of professional touchers is waiting to rub, slap, stroke, smooth and pluck almost any part of your body you wish to indicate. This is like a kind of 'healthy medicine', the unpleasant stigma of sickness being replaced by a mood that is predominantly athletic or cosmetic. Or so it seems; but once again there is a powerful element of body-contact-for-contact's-sake underlying all these activities. To be massaged from head to toe by a young masseuse is, for a man, almost as intimate a procedure as if he were to make love to her. In some ways it is more so, since by the time she has finished she will have made active body contact with nearly every part of his body, applying to each section of it in turn a rich variety of pressures, touches and tactile rhythms. Herein, we might dare to say, lies the rub, for the interaction, although it involves no direct sexual contact, is far too close for some men's comfort.

Perhaps it would be more correct to say that it is too close for the comfort of Western society. Privately, the massaged body would no doubt enjoy itself greatly, but the public image of the massage parlour is, in our culture, not what it might be. One trend has been to reduce the imagined eroticism of the activity by introducing sexual segrega-tion, so that men massage men and women massage women. Even this step has failed to give this intrinsically harmless form of soothing body contact a widespread acceptance in modern society. In removing the heterosexual contact, the way was inevitably paved for dark murmur-ings about the homosexual element. Only intensely athletic males can, with ease, overcome this slur. For the boxer or the wrestler there is no problem. Like the triumphant footballers who could embrace passion-ately in public without criticism, because of their obviously aggressive, masculine role, the prize-fighter can luxuriate on his massage table without any adverse comment. In theory, the rest of the adult popula-tion could follow his lead and do so without the slightest sexual involvement, regardless of the sexes involved, but in practice this plainly has not happened, and so, for the unmassaged majority, we must look elsewhere for adult body intimacies.

One way in which the problem has been solved is to multiply the numbers involved and to eliminate the atmosphere of an intimate 'pair'. This is done in many gymnasiums and health farms, where groups of people gather to indulge in a variety of exercises which may include a great variety of body contacts without creating the flavour of two 'consenting adults in private'. Another method is to replace the human masseur or masseuse with a strictly sexless machine that embraces them, not with loving arms, but with an impersonal canvas belt which then proceeds to be mechanically intimate with them.

A more commonly employed solution is to restrict the body contacts to the less private parts of the human body. Here we move into the totally acceptable world of hairdressers and beauty experts, pausing only to cast a last sympathetic glance at the massage world, where some practitioners have attempted a similar restriction by coyly advertising that they provide only 'arm and leg massage'.

Since, in Western society, we all expose our heads to one another's public gaze, the hairdresser is automatically excused the stigma of increased nudity during the course of his professional body contacts. What he or she handles, we all see. Nevertheless, touching the head, as we saw in an earlier chapter, is normally reserved for only the closest of intimates and particularly characterizes the amorous contacts of young lovers. Between adult strangers it is almost taboo, and so the hairdresser, in the guise of a cosmetician, can fill an important gap for a contact-starved adult. This does not mean that the cosmetic role is unimportant, but merely that there is more to hairdressing than meets the mirror's eye.

Head-grooming, in the dual cosmetic/intimate role, has been with us for thousands of years. If we care to include our primate ancestors, we can safely put the figure at millions of years. The detailed and tender fingering that can be seen in any zoo monkey-house, as one monkey or ape works lovingly over the head-hair of its companion, leaves little doubt about the intimacy factor involved. Cleansing alone cannot account for the relaxed ecstasy of the primate groomee. And so it is with us, except that we, of course, cannot extend the interaction over our whole bodies like the furry monkey or ape. Where we cover our naked skins with clothing, we must rely on the deft, delicate touch of the tailor's fingers as he adjusts our new garments to rekindle – faintly, very faintly – the long-lost sensation of intimate body-grooming.

For the monkey, hair-grooming by another is an act of social bonding, so it is not surprising to find that in earlier periods of our history the professional hairdresser was a rarity. Hair was groomed by close intimates, rather than by comparative strangers. In the days when we lived in small tribes this was, of course, inevitable, since everyone in the social group knew everyone else on a personal basis. Later, when the urban revolution came and we found ourselves increasingly surrounded by strangers, the tendency was to restrict hairdressing and its associated activities to interactions between close personal contacts. Much later, with the increasingly complex coiffure that appeared after the Middle Ages, more expert attention was needed by the high-ranking members of society, and the professional hairdresser began to make his mark. At first, where ladies were concerned, his intimate operations were confined to the privacy of his clients' homes, but, gradually, more efficient public salons were opened and ladies of fashion began to flock to their doors. Even so, it was not until the second half of the last century that this became a common practice. Then the rush was on. In 1851 there were already 2,338 hairdressers operating in London, but fifty years later, in 1901, this figure had shot up to no less than 7,771, a dramatic rise that far outstripped the general increase in the city's population. Part of the reason for this change was undoubtedly economic, but perhaps, too, there was another factor, for the Victorian female was severely restricted in the other ways in which she could make adult body contacts. The rules of conduct were so tightened during this period that the caress of the hairdresser's hands must have provided a welcome intimacy in an era of such rigid restraint. Not only did more and more women venture forth, but they did so with increasing frequency. In the present century the pattern has spread out from the great cities and down to the smallest towns, involving almost the entire female population.

Aware that their modern clients craved more intimacy than the mere dressing of the hair could offer, this new army of professional touchers expanded the nature of their activities. Wherever there was exposed skin, they applied their delicate attentions. Manicures became popular. The 'facial' appeared on the scene. Mud-packs were applied, wrinkles smoothed and caressed, soft skin 'toned-up', the latest style of make-up demonstrated by a professional hand. 'Beauty', cried *Vogue* in 1923, 'is a full-time job.' There is no denying that the primary motive was visual, but the increased tactile intimacies involved in obtaining the desired visual effect were also undoubtedly of great importance. To visit a modern beauty parlour is nothing if not a touching experience.

By comparison, the modern male is poorly supplied with intimacies of this kind. Some men indulge in manicure and scalp massage, and a few still have their faces shaved occasionally, but for most the visit to the barber's is a quick snip-snip and home again to wash the hair themselves. It is interesting that the barber does his best to increase the intimacy of this simple snipping by employing a ritual device. If you are a male, the next time you visit your barber, listen to the snipping sound of his scissors and you will find that for every hair-snip there are a number of 'air-snips', the scissors snapping together rapidly in mid-air before closing in for the next actual cut. There is no mechanical function to these air-snips, but they create the impression of great activity in the proximity of the scalp and thereby effectively increase the impression of 'contact-complexity'.

For all that, the intimacy involved is a remarkably limited one, and it is surprising that today's males should accept such restrictions. Perhaps with the return of longer male hair-styles we shall witness some changes. So far, it must be admitted, there is little sign of any general increase, rather the reverse. If anything, long male hair now means that there is a rapid decline even in the simple snip-snip, with the hair-washing still done largely in the home. Only in the more sophisticated urban centres is there any indication that the new hair-styles are leading to greater barbering activities, and it remains to be seen whether this development will spread. But the fashion is new and, if it survives, it will take some time for it to regain the widespread respectability it once had. There is an unjustified stigma of 'effemi-nacy' attached to it by the older males in the population, who have still not woken up to the fact that their close-cropped styles arose primarily as anti-louse devices and that to insist on all males retaining such styles in a post-lousy era is the height of irrationality. As long as this slur persists, there will be a reluctance on the part of many of the younger males to continue their trend to its logical conclusion and once again luxuriate in more elaborate tonsorial intimacies.

Just about the only 'cosmetic' intimacy that the modern male enjoys more than the female is the use of a public shoe-shine and, as a trade, even that has been losing ground of late. In most big cities now it has become little more than a curiosity, found only at one or two special points. Apart from the oral-genital contacts discussed earlier, this is probably the only time in a modern man's life when he will see other human beings going on their knees before him to perform an act of body contact, and it is certainly the only time it will happen in public. (The shoe-shop attendant avoids the posture by sitting and leaning forward.) The kneeling posture of the shoe-shine creates such a strik-ing impression of servility that this feature has perhaps been his undoing. In the past, a man could more easily accept a display of humility of this kind, so that the humble intimacy performed was doubly rewarding, but with a growing respect for human equality such an overt submissiveness becomes almost embarrassing. A symbolic kissing of our feet is too much for us, and the shoe-shine is fast becoming a vanishing breed. It is not that we have ceased to be responsive to humiliating services – that would be too congratulatory a thought – but rather that we no longer wish to be *seen* to be so responsive to them.

In this survey of professional touchers, we have so far covered the doctor, the nurse, the masseur, the gymnastics and health-and-beauty instructors, the hairdresser, the tailor, the manicurist, the beautician, the make-up specialist, the barber, the shoe-shine and the shoe-shop attendant. To this list we could add many other related occupations such as those of the wig-maker, the hatter, the chiropodist, the dentist, the surgeon, the gynaecologist and a whole variety of medical and semi-medical specialists. Of these, few warrant special comment. The dentist usually causes too much stress for his oral intimacies to provide any contact reward. The surgeon, whose body intimacies go so very much deeper than those of even the most passionate lover, also has little impact on us emotionally, thanks to the use of anaesthetics.

The actions that must take place during a gynaecologist's examination of a patient are so similar, at a descriptive level, to the hand-to-genital contacts of a lover that here, too, paradoxically, there is no comfort in the intimacy. Embarrassment is reduced today by an intensely professional atmosphere, with both sides strictly on their guard against any misinterpretation of the anatomically sexual contact. Whereas holding a woman patient's hand while taking her pulse may provide the secondary benefits of soothing bodily intimacy, touching her genitals is inevitably *so* intimate that the emotional barriers clamp down immediately and no such benefits are possible.

In the past, the special nature of genital examinations has caused endless trouble for well-meaning gynaecologists. Extraordinary anti-intimacy procedures have been insisted on. Three hundred years ago he was even, on occasion, required to crawl into the pregnant woman's bedroom on his hands and knees to perform the examination, so that she would be unable to see the owner of the fingers which were to touch her so privately. At a later date, he was forced to work in a darkened room, or to deliver a baby by groping beneath the bed-clothes. A seventeenth-century etching shows him sitting at the foot of the labour bed with the sheet tucked into his collar like a napkin, so that he is unable to see what his hands are doing, an anti-intimacy device that made cutting the umbilical cord a particularly hazardous operation.

Despite these bizarre precautions, the male midwife was forever under fire, and just over two hundred years ago a learned textbook on the theory and practice of midwifery was openly condemned as 'the most bawdy, indecent and shameful book which the press ever brought into the world'. Needless to say, it was usually the men who complained and always the women who suffered. For centuries the sexual nature of the intimacies involved in assisting at the birth of a child stood in the way of efficient medical attention. Usually, properly qualified men were banished altogether from the labour bed, and the duties were performed by unskilled and often highly superstitious female midwives. (The word 'midwife' means simply 'with-the-wife', and makes no reference to the sex of the person concerned, although today we automatically think of it as indicating a woman, a fact that reflects this early ban on men.) As a result of this, an enormous number of women died in childbirth and many thousands of children succumbed at birth or in the first month of life. A large number of these cases were due entirely to the anti-intimacy rules that prevented skilled aid from being provided.

Here, then, is an instance of the sexual taboos on bodily contact creating a major social disaster and influencing the whole course of history. Year after year was to pass, and countless human miseries were to be heaped one on top of the other, before sanity prevailed and science was able to sweep away the ancient prejudices. Only by obeying the strictest possible code of conduct has the profession gradually been able to eliminate these early stupidities. Even so, the echoes of ancient fears can still be felt, and the modern gynaecological examination remains comfortless in the sphere of bodily contact.

There is only one area of social activity where sexual contacts do not suffer in this way, and that is the theatrical profession. Actors and actresses, including ballet dancers, opera singers and photographic models, all enjoy a professional life where they are widely licensed to touch one another in a sexual manner. In their performances they kiss and fondle, embrace and stroke, as the director demands. If it is in the script, it is within the social 'law', and the actor or actress can, during his or her working day, enjoy many bodily-contact comforts. For such an insecure profession this is undoubtedly a major benefit, although the extremes sometimes demanded can lead to difficulties. It is hard to pretend to make love to someone, even a professional colleague, time and time again, without the basic emotional reactions beginning to creep into the relationship, and this often happens, to the detriment of other intimate relationships in the 'real' world outside. If sexual intimacies are mimicked well enough, it is not easy to suppress the true biological responses which normally accompany them.

Another hazardous contact reward for the stars who dazzle us in the entertainment world is the physical acclaim of their more ardent followers. In public places they may find themselves hemmed in by eager fans, desperate to touch their idols. At a mild level this can provide a pleasant emotional reward, but occasionally it can lead to bruises and even injury. The powerful urge to touch the bodies of certain star musicians and singers – and even some of the more glamorous politicians – has recently reached staggering proportions. For the groupie girls who follow the more famous pop stars, there are literally no holds barred. Perhaps the most intimate example is that of the 'plaster-casters', groupies who persuade their pop idols to permit them to take plaster casts of their erect genitals, so that these effigies can be touched later at leisure, after their gods have departed.

In dealing with these interactions between pop stars and their fans, we have moved away from the situation in which touching is an inherent part of the professional activity itself. A masseur or a hairdresser *has* to touch his client or he cannot perform his task, but a singer does not have to touch or be touched in order to perform his songs. The fact that his special role in society makes him more touchworthy is a secondary factor. A similar condition applies in other spheres, an obvious example being the police.

It is not the policeman's job to touch people, but he is nevertheless licensed to do so with much greater freedom than the rest of us. He can lay hands on us in a way we would resent in a member of the general public. He can take a child's hand in the street without causing comment. In a crowd he can push against us to keep us back, and we accept his contact role with equal ease. If he manhandles us when we are violent, again we are less likely to lash out wildly at him than we are at someone else who treats us in a similar manner. Only at the extremes of violence, where his own restraint breaks down and he begins to behave, under intense provocation, like a uniformed thug, do we fully disinhibit our reactions to him. Then, by contrast, our fury knows no bounds, as recent riot-scenes have testified all too often. It is as if, having given him a limited licence to touch us, we find abuse of that licence particularly unacceptable, as when a choir-master behaves improperly with a choirboy, or a schoolmaster with a pupil. The result is that, if driven to repeated breakdown in restraint, the police rapidly become dubbed as hated men and violently persecuted whenever angry crowds gather. Only in countries such as Great Britain, where the police are deliberately sent out into the streets completely unarmed, have there been any signs at all of slight restraints being applied on both sides during the worst civil rioting of recent years. It is as if the fact that both sides are forced to indulge in the greater body intimacy of hand-to-hand grappling, rather than the remoter savageries of overhead club- and stick-beating, or the totally remote brutalities of firearms, has some kind of restraining influence on the hostilities. There is nothing inherently less vicious about such encounters; even without weapons, eyes can be gouged out and genitals kicked in, but such cruelties are extremely rare. When compared with the cracked and bleeding skulls of other riot scenes, the hand-to-hand battles of London and other British cities begin to look almost civilized, and it is ironic that they do so by returning to the more intimate forms of pre-civilized, pre-weapon combat.

There is a well-known film-sequence cliché in which two tough and otherwise admirable men set about one another with their fists to settle some long-standing quarrel. A sophisticated movie audience knows full well that, if the two men both begin to lose, each beating the other into total physical exhaustion, they are about to witness the birth of a great new friendship. As the two bruised hulks sprawl weakly on the ground, sure enough, one pair of cracked and bleeding lips spits out a loose tooth, and grins admiringly at his equally beaten opponent. In no time at all, our heroes are helping one another up and crawling to the bar (there is usually one nearby) to share a reviving drink. After this we can be sure that nothing will ever separate them again, and that they will become indomitable partners in righting all wrongs, until, at the end of the film, one of them will die bravely saving the life of the other, breathing his last gasp cradled in the loving arms of the man whose face he once succeeded in beating to a pulp.

The moral of this highly coloured story is, of course, that a warm enemy is better than a cold friend, and it bears some investigation in terms of the body intimacies involved. It is almost as if any form of intimacy, even violent intimacy, providing it is performed on a sufficiently personal basis, can produce a bond of attachment between two antagonists. Needless to say, it is dangerous to generalize, and it can certainly not be offered as a general excuse for violence, but to ignore the phenomenon completely because it frightens us is equally unwise.

The difficulty is that impersonal violence has in recent times reached such a horrific scale that an almost total taboo has descended on the subject. For the sexually permissive society, violence, all violence, regardless of scale or context, has become the new philosophical restriction. In the broad context in which it is intended, the creed that we should 'make love not war' is unassailable, but the message underlying the ritualized film fight may perhaps lead us to consider a possible exception to this general rule. Clearly I am not thinking of anything as savage as the brawl I described above. I am imagining instead a situation in which certain people have so suppressed their aggressiveness that, even under intense provocation, they will 'not so much as lay a finger on' their partners' bodies. To take non-violence to such an extreme in every instance can create a new form of anti-intimacy. Let me give an example.

If, unavoidably, two individuals have grown cold towards one another, for whatever reason, the relationship can finally freeze to death in an atmosphere of hypocritical restraint. The thin hard smile of inhibited anger can cut sharp as any knife. Sometimes, under such conditions, an explosion into a flaming row, accompanied with mild but nevertheless aggressive interaction, can clear the air like a long-awaited thunderstorm, and release the damaging tension. Perhaps for the first time in months, a bickering couple actually take one another in their arms and, even though it is to shake the partner violently by the shoulders rather than to embrace lovingly, the result is the feel of the first truly meant body contact in ages. It is, of course, a desperate situation to have arrived at, when to touch at all means to touch in this hostile way, and it may very well fail. But just occasionally it may succeed, and to ignore this fact because it is out of step with the current cultural mood is to disregard another facet of the powerful emotional impact that body intimacy can have on the bonds of attachment between two human beings.

A related pattern of behaviour is the rough-and-tumble play of children, or the 'horseplay' or 'rough-housing' that can sometimes be observed between friendly adults. Again, the body contacts involved make their emotional impact, and they do so because they are accompanied by the unspoken message, 'Even though I am being aggressive, you can see that I am not *really* aggressive.' The message is a subtle one, however, and play-fighting at any age can become a delicately balanced interaction. The man who playfully slaps a companion on the back can easily reverse the signal, so that it becomes, 'Even though I am pretending to be only playfully aggressive, you can tell by the way I am doing it that I am not.' He uses the slap because it has become formalized as an accepted play-fighting pattern, but by his accompanying actions and by the hardness of the slap it becomes instantly apparent to his companion that he has twisted the message back to front.

A similar complexity exists in the case of the bickering couple mentioned above. If, under extreme provocation, the action is no more than a mild slap on the cheek, or the shaking of the partner's shoulders, then the message reads, 'Although you have made me want to kick your teeth in, this is all I am doing to you.' But if the provocation is less than extreme, then even the most moderate of aggressive contacts transmits a signal that is merely surly and unpleasant.

The subtle dangers of play-fighting can sometimes be observed very clearly when two boys start idly wrestling on a street-corner. At first they both obey the conventions of playful aggression. Each body-push and arm-lock is performed with exactly the right intensity – strong enough to make it forceful, but not so strong that it becomes truly violent. If this delicate balance is accidentally upset and one of them is hurt, the mood changes. Now he retaliates more powerfully and, if the situation is badly handled, a real fight can slowly grow out of the playful one. The changes that signal this are difficult to analyse, because even the playful wrestling may look real enough. Usually the tell-tale signs begin to show in the facial expressions, which, instead of being relaxed and smiling, or exaggeratedly mock-savage, become hard and set, often with accompanying changes in pallor and flushing.

Where professional wrestlers are concerned, a mimic of this change-over can be seen. The 'villain' deliberately fouls the 'hero', who then becomes expansively outraged, protesting to the referee and demanding sympathy from the crowd. Lunging wildly at his opponent, he appears to switch from the conventional combat techniques to uncontrolled violence, returning foul for foul, and the audience roars its approval. But here, even the 'uncontrolled' aggression is itself formalized, and the audience, joining in the game, knows this perfectly well. Should one wrestler genuinely hurt a rival, the bout is immediately called off and, instead of 'savage reprisals', ill-concealed concern is shown on all sides.

Leaving this dangerous subject, we can turn now to the safer and more tender intimacies of the dance-floor. As a sphere of activity in which there are professionals who are licensed to touch, dancing offers limited possibilities. True, the adult who is seeking some form of body contact can achieve it by using the services of a dancing instructor, and a male dancer can, in certain localities, visit a dance-hall which supplies professional dancing partners at a set fee per dance, but the world of social dancing is today largely one for the amateur. At parties, discotheques, dance-halls and ballrooms, adults who are strangers to one another can come together and move around the room in an intimate frontal embrace. Individuals who are already friendly can also use the situation to escalate a non-touching relationship into a touching one. The special role that social dancing plays in our society is that it permits, in its special context, a sudden and dramatic increase in body intimacy in a way that would be impossible elsewhere. If the same full frontal embrace were performed between strangers, or partial strangers, outside the context of the dance floor, the impact would be entirely different. Dancing, so to speak, devalues the significance of the embrace, lowering its threshold to a point where it can lightly be indulged in without fear of rebuff. Having permitted it to occur, it then gives a chance for it to work its powerful magic. If the magic fails to work, the formalities of the situation also permit retreat without ignominy.

Like so many other aspects of body intimacy, dancing has a long history stretching back into our animal past. In behaviour terms, its basic ingredient is the repeated intention movement. If we look at the dancing displays of various birds, we find that the rhythmic movements they perform are mostly made up of movements that start to go one way, then stop and go another, then stop again and repeat the first action, and so on. Turning from side to side, twisting back and forth, or bobbing up and down, the bird displays vigorously in front of its mate. It is in a state of conflict, one urge pulling it forward and another holding it back. During the course of evolution, the rhythm of these intention movements becomes fixed and the display becomes a ritual. The form the ritual takes varies from species to species and in each case becomes characteristic of their particular sexual preliminaries.

Most of our dance movements have originated in the same way, but in us they have not evolved into a fixed form. Instead, they have been culturally developed and are highly variable. Many of the actions of human dancers are no more than intention movements of going somewhere, only instead of carrying the action through, we check it, move back or round, and start again. In earlier centuries, many dances were like little parades, with the couple demurely holding hands and stepping round the floor, pausing every so often, turning round, and then going on again, to the rhythm of the music. Because the pattern was essentially one of going on a journey, it also frequently included mock-greetings towards the partner, with formal bowings and curtsyings, as though the two dancers had just met. In both folk-dancing and the courtly ballrooms, there was typically an intricate weaving round and round and in and out of other pairs on the floor, or arena. The body intimacies involved in such performances were so strictly limited that they gave rise to no sexual problems. They simply permitted a general social intermingling. The fact that the leading of the female by the male, around and around the floor, was so formalized, stopped any awkward questions about where he was really supposed to be leading her, or for what purpose.

The situation changed dramatically at the beginning of the last century, as a new dance craze swept across Europe. The waltz had arrived. For the first time the dancing couple embraced as they moved, a public intimacy that immediately created widespread scandal and concern. Such a major advance required a subterfuge, and it is one we have met before. When discussing the first way in which a simple hand-to-hand contact can be effected, I mentioned that a much-used trick is the intimacy-disguised-as-aid. The hand that reaches out does so ostensibly to support or steady the other person, to guide them or to prevent them from falling. In this way it can cross the vital threshold of making body contact without causing alarm. So it was with the waltz. At the very beginning of its history it was an incredibly rapid and athletic dance, so that the partners were forced to clutch at one another's bodies to prevent themselves from spinning apart. This was the 'supporting' device and, once it had enabled the waltz to gain an entrance into the ballroom, it was then only a matter of slowing down the speed of the performance to convert these actions of mutual physical aid into the more tender intimacies of a true frontal embrace.

The older generation, who had not known such delights, were outraged. The waltz, which today seems quaintly old-fashioned, was, in its early years, described as 'polluting' and 'the most degenerate dance that the last or present century can see'. The early Victorian author of *The Ladies' Pocket-Book of Etiquette* devoted ten pages to an all-out attack on this abominable new public intimacy. Among her comments: 'Ask any mother . . . can she consent to commit her daughter promiscuously to the arms of each waltzer? Ask the lover . . . could he endure the sight of the adopted of his heart . . . reclining in the arms of another? . . . Ask the husband . . . will you suffer your wife to be half-embraced by every puppy who turns on his heel or his toe?' The attacks persisted and, less than a hundred years ago, an American dancing-master in Philadelphia pronounced the waltz immoral because it involved the hugging of a lady by a gentleman she might not previously have met. But the battle was gradually lost, and the wicked waltz reigned supreme, bringing in its wake a whole variety of dances involving the full frontal embrace. These, in their turn, caused further scandalized mutterings.

The importation of the tango from South America in 1912 was again greeted with outrage. Because this dance included 'suggestive lateral movement of the hips', which reminded the hawk-eyed guardians of morals of the actions of copulation, it was instantly dubbed as depraved.

No sooner was that battle lost than the Jazz Age burst upon the scene, and the frantic dancing-teachers of the 1920s called urgent meetings to discuss this new threat to their respectability. They issued strong official protests about this new craze, pointing out that all the jazz dances had originated in Negro brothels.

Perhaps the most extraordinary attack on jazz dancing came in a newspaper report which claimed that 'The dance, and the music, with its abominable rhythm and copulative beat, was imported from Central Africa by a gang of Bolshevists in America, their aim being to strike at Christian civilization throughout the world.' Perhaps this puts into true perspective the recent claims that the current wave of student revolt, dropping-out and drug-taking is also a 'Red plot'.

Since its early days, jazz has given birth to several lusty offspring, and each in turn has caused the inevitable raised eyebrows as the dancers have taken to the floor with more and more variations of the public embrace. In the 'forties it was jitterbugging and in the 'fifties rock-and-roll, but then something strange happened. For some reason that it is perhaps still too early to understand, the couples separated. With the 'sixties, the dancing embrace went into a rapid decline. Now only the older, more staid couples clung to one another as they rotated around the floor. The younger dancers pulled apart from one another and danced more or less on the spot where they stood. It began with the twist and before long involved a confusingly large number of alternatives such as the hitch-hiker, the shake, the monkey and the frug. More and more styles were specified until eventually, as the decade drew to a close, the situation became so confused that they all merged into one more or less nameless amalgam, and became simply the dance that was danced to 'pop'. All had the same important feature – no touching. Presumably the significance of this change lies in the marked increase in sexual permissiveness. If young Victorian couples were not allowed to enjoy extensive private intimacies, then the embrace of the waltz had great meaning for them, but if matters are freer today, then who cares about a specially 'licensed' context for a mere standing embrace? It is as if the young dancers of today are publicly stating, 'We don't need it, we have the real thing.'

This brings us to the end of this brief survey of the way in which we, as adults, find specialized methods of indulging in bodily intimacy. All through the chapter, from doctors to dancers, there has always been something other than pure contact involved. At no point has there been touching merely for touching's sake. In every case there has been some excuse that provides us with a licence to touch or be touched. And yet, frequently, there is the distinct impression that it is the contact that is more important than the official activity. Perhaps one day, as the stresses of modern living increase, we shall see the appearance of an undisguised professional toucher, who will sell embraces like beads. Or perhaps to buy his wares will always be too great an admission of failure on our part, a failure to achieve the longed-for intimacies with a family unit of our own.

Whatever happens, we can always fall back on that perpetual substitute for body intimacy, namely verbal intimacy. Instead of exchanging embraces we can exchange comforting words. We can smile and talk about the weather. It is a poor substitute where emotional interchanges are concerned, but it is better than total emotional isolation. And if we still yearn after a more direct form of contact, there are other alternatives open to us: we can touch some non-human animal, or an inanimate object, using it as a stand-in for the human body we would really like to approach, or, if there is no other solution, we can always touch ourselves. The ways in which we employ animals, objects and our own bodies as substitutes for human intimates are discussed in the next three chapters.

6

SUBSTITUTES FOR INTIMACY

IN THE ADULT human world, a world full of stress and strangers, we reach out to our loved ones for comfort. If, through their indifference or through their preoccupation with the complexities of modern living, they fail to respond, we are in danger of becoming starved of the primary reassurance of bodily contact. If, through the moralizing of a warped minority, they have become inhibited in their intimacies and have been driven to accept the view that indulgence in the tactile pleasures of the body is somehow sinful and wicked, then, even in the midst of our nearest and dearest, we are liable to become touch-hungry and body-lonely. We are, however, an ingenious species, and if we are denied something we badly want or need, our resource-fulness soon urges us on to find a substitute to replace it.

If we cannot find love inside the family, we soon start looking for it outside. The ignored wife takes a lover; the husband, a mistress. Body intimacies reblossom. Unhappily, these particular substitutes do not always add to the surviving intimacies of family life; they compete with them, and perhaps eventually replace them altogether, causing varying degrees of social havoc as they do so. A less damaging alternative was the one discussed in the last chapter – the use of contacts with special-ists who are licensed to touch. These have the great advantage that they do not usually compete with the relationships inside the family unit. The extensive intimacies of the masseur, providing they are applied with a strict professionalism, cannot be cited as grounds for divorce. But even a professional toucher, no matter how valid his official excuse for touching may be, is still a physiologically functional adult being, and as such is inevitably seen as a potential sexual threat. The 'seeing' of this threat is seldom spoken of openly, except occasion-ally in jest. Instead, society quietly imposes more and more restrictions on the nature and context of the specialist intimacies. To begin with, they are rarely admitted to exist. One goes dancing, not to touch, but 'for fun'. One goes to the doctor because of a virus, not because one needs comfort. One goes to a hairdresser to have the hair styled, not to have the head caressed. These official functions are, of course, all perfectly valid and important. They have to be so in order to mask the fact that something else is going on at the same time, namely the

seeking of friendly body contact. The moment they cease to be important, this unfulfilled need becomes too obvious, and some basic questions about our way of life start demanding answers we would rather
not be forced to consider.

Unconsciously, however, we are all aware of the game that is being
played, and so, indirectly, we tie the hands that we would have caress
us. We do this by applying conventions and codes of conduct that
reduce our sexual fears. Usually we do not say why. We simply accept
the abstract rules of good etiquette, and tell one another that certain
things are 'not done' or 'not nice'. It is rude to point, leave alone touch.
It is impolite to show one's feelings.

So where do we turn? The answer is as soft and cuddly as the kitten
in your lap. We turn, in fact, to other species. If those humans closest
to us cannot supply us with what we want, and if it is too dangerous to
seek intimacies with strangers, then we can make tracks to the nearest
pet shop and, for a small sum, buy ourselves a piece of animal intimacy. For pets are innocent; they cause no questions and they ask no
questions. They lick our hands, they rub softly up against our legs, they
curl up to sleep on our thighs, and they nuzzle us. We can cuddle them,
stroke them, pat them, carry them like babies, tickle them behind the
ears, and even kiss them.

If this seems trivial, consider the scale of the operation. In the
United States, more than 5,000 million dollars is spent on pets every
year. In Britain the annual figure is 100 million pounds. In West
Germany it is 600 million Deutsche marks. In France, a few years ago
it was 125 million new francs, and estimates already indicate that this
figure has by now doubled. Trivial is not the word for figures such as
these.

The most important pets are cats and dogs. In the United States there
are 90 million of them. Puppies and kittens are born there at the rate
of 10,000 every hour. There are over 16 million dogs in France, eight
million in West Germany and five million in Britain. Precise information on cats is not available, but there are certainly as many cats as
dogs and probably more.

Putting these figures together, one can say that, at a rough guess,
there are approximately 150 million cats and dogs in these four countries alone. Making another rough guess, let us say that each owner of
one of these animals stroke, pats or caresses it, on the average, three
times a day – or about 1,000 times a year. This adds up to a total of
150,000 million intimate body contacts per year. What is astonishing
about this figure is that it represents for Americans, Frenchmen,
Germans and Englishmen intimacies performed not with other Americans, Frenchmen, Germans or Englishmen, but with alien species

belonging to the order Carnivora. Viewed in this way, the phenomenon looks even less trivial.

As we have already seen, we pat one another on the back when we embrace, and we stroke one another's hair and skin when we are lovers, or parents with children. But clearly we do not get enough, and those thousands of millions of animal caresses are there to prove it. Blocked in our human contacts by our cultural restrictions, we redirect our intimacies towards our adoring pets, our substitutes for love.

This situation has led to violent criticism from some quarters. Dubbed 'petishism' by one author, it has been condemned as reflecting a decadent failure of modern, civilized human beings to communicate intimately with one another. In particular it has been stressed that more money is usually forthcoming to support the prevention of cruelty to animals than to prevent cruelty to children. The answers given in support of modern pet-keeping are rejected as illogical and hypocritical. The argument that it teaches us the ways of animal life is considered nonsensical, in view of the gross anthropomorphism of the relationship in almost every case. The pets are humanized – they are seen as furry people, not real animals at all. The argument that animals are innocent and helpless and need our aid is seen as hopelessly one-sided in an era of battered babies and napalmed peasants. How is it that, in this enlightened age, we can have permitted a million children to be killed or wounded in Vietnam, while our cats and dogs have been provided with expert and immediate attention whenever they have needed it? How is it that in the twentieth century we can have licensed our adult males to murder 100 million members of their own species in warfare, while we have spent more millions on stuffing food into our luxuriating pet animals? How is it, to sum up, that we have come to be kinder to other species than to our own?

These are strong arguments and they cannot be dismissed lightly, but they contain a vital flaw. The answer, put very simply, is the old one that two wrongs do not make a right. Undeniably, it is monstrous to cuddle a pet and ignore a child, and it is true that in extreme cases this does happen. But to use this as an argument for not cuddling a pet is a folly. It is doubtful if, even in extreme cases, the pet 'steals' the caress from the child. If, for some neurotic reason, the child is not receiving love from the parent, it is doubtful whether the absence of a cuddlesome pet would help to improve the situation. In almost every case, a pet animal is being used either as an additional source of intimacy, or as a substitute for intimacies that are already lacking for some reason. To say that more caring for animals is actively causing less caring for other humans seems to be totally unjustified.

Imagine for a moment that a freak disease exterminated all pet animals tomorrow, and effectively eliminated all those millions of tender intimacies that would have occurred between them and their owners. Where would all that loving go? Would it magically be re-redirected, back on to other human companions? The answer, sadly, is that it would probably not. All that would happen is that millions of people, some of them lonely and incapable for a variety of reasons of enjoying any real human intimacies, would be robbed of a major form of tender body contact. The old lady who lived alone with her cats would hardly start stroking the postman. The man who fondly patted his dog would be unlikely to pat his teenage son more in its absence.

It is true that in an ideal society we should not need these substitutes or additional outlets for our intimacies, but to suggest banning them because of this is to attempt to cure the symptom and not the cause of the trouble. And even in the ideally loving and body-free society, we would probably always have plenty of intimacy to spare for our animal companions, not because we would then need such contacts, but simply because they would give us additional pleasures that would in no way compete with our human relationships.

A final word in defence of pets: if we are capable of tenderness towards animals it does at least reveal that we are capable of such tenderness. But, the answer comes back, even the commandants of concentration camps were kind to their dogs, so what does that prove? It proves, in short, that even the most monstrous of human beings is capable of some kind of tenderness, and the fact that its juxtaposition to callous brutality in this particular case offends us so deeply, and makes the brutality even more horrific, must not blind us to this fact. It serves as a constant reminder that the human animal, when not warped by what must paradoxically be called the savageries of civilization, is fundamentally endowed with a great potential for tenderness and intimacy. If witnessing the gentle, friendly touching that occurs between pet-owners and their pets does no more than bring home to us that man is basically a loving, intimate animal, then this alone is a valuable lesson to learn and relearn, all the more so in a world that grows yearly more impersonal and cold-hearted. When, under pressure, men become merciless, it is then that we need all the evidence we can muster to prove that this need not be so – that this is not the natural condition of man. If our capacity to love our pet animals serves to demonstrate one facet of this, then well-meaning critics must think twice about launching an attack on it, no matter how unreasonable it may seem when viewed from certain angles.

This said, what of the nature of the animal intimacies themselves? Why, for example, do we pat a dog and stroke a cat, but rarely stroke a dog or pat a cat? Why does one kind of animal draw forth one type of intimacy and another another? To answer this we have to look at the anatomy of the animals concerned. In their roles as pets they are, of course, acting as stand-ins for human companions, and their bodies are therefore substitutes for human bodies. Anatomically, however, there are striking disparities. The stiff legs of a dog cannot embrace us. We cannot fling our arms around a cat. Even the largest cat is no bigger than a human baby, and its body is soft and pliable. We therefore adjust our actions accordingly.

First, the dog. As our loving companion, we want to embrace him, but because his legs make this difficult, we isolate the patting element from the embrace-and-pat complex and apply it direct. Reaching out, we pat the animal's back, or perhaps its head or flanks. In a typical large dog, the back is broad and firm and a suitable substitute for the human back which we are patting by proxy.

The cat is a different matter. Being smaller and softer to the touch, it does not feel right as a back-substitute to be vigorously patted. Its soft, silky fur is more like human head-hair to the touch. We tend to stroke a loved one's hair, and so it follows that we tend to stroke a cat. As the dog was a back-substitute, so the cat is a hair-substitute. In fact, we often treat a cat as though its whole body is a stand-in for a silky-haired human head.

Extending this argument, it might be thought that patting is something we automatically do to all canines, and that stroking is an action for all felines, but it is not as simple as this. It has much more to do with the typical body quality of the domestic dog and the domestic cat. Anyone who has enjoyed the exotic luxuries of body intimacies with a tame cheetah, lion or tiger will know that there the pattern changes. Although they are true felines, they have broad, firm backs more reminiscent of the domestic dog than of the family cat. Like the typical dog, the hair is also coarser. The result is that they are patted, not stroked. By contrast, a tiny lap-dog with long flowing hair is stroked and caressed more like a cat.

Moving up the scale in size, the horse-lover is also a patter, but there is a subtle change. The original human back – where the patting began, so to speak – was a vertical surface, but the back of a horse is horizontal and therefore less satisfying as a site for the substitute action. The horse's neck, however, comes to the rescue, being both the right height and, what is more, providing the ideal vertical surface, and it is here that the majority of horse-pattings are delivered. In this respect, the horse goes one better than the dog, whose neck is generally too small to be of much use in this respect. Again, the horse's height makes it ideal for head contacts which, in the dog, force us to lower ourselves to his level, or raise him up in our arms. And so many a horse-lover can be seen with her head pressed to her animal's neck or face, while her arm embraces and her hand pats the firm, warm flesh.

For many people a pet is not merely a substitute companion, but more specifically a substitute for a child. Here the size of the animal becomes important. Domestic cats are no problem, but the typical dog is too large, and so certain types have been progressively reduced in size by selective breeding until they have been successfully scaled down to human baby proportions. Then they, like cats and various other creatures such as rabbits and monkeys, can be scooped up in their owner's pseudo-parental arms without undue exertion. This is by far the most popular form of body contact to occur, where pet animals are concerned. An analysis of a large number of photographs depicting owners in contact with their pets reveals that the act of holding the animal in the arms, as if it were an infant, accounts for 50 per cent of all cases. Patting is the next most common action (11 per cent), followed by the semi-embrace, in which one arm is wrapped around the animal (7 per cent), followed closely by the pressing of the cheek to the pet's body, usually in the region of its head. Another intimacy that appears with rather surprising frequency is the mouth-to-mouth kiss (5 per cent), the species involved ranging from the budgerigar to the whale. The whale, one might think, leaves something to be desired as an animal intimate. Captain Ahab would certainly have been star-tled at the idea of a girl kissing one on the mouth, but the recent trend in oceanarium displays has changed all that. Both tame whales and their smaller relatives, the dolphins, have become front-line favourites in recent years, and since their bulbous, swollen foreheads give a baby-like shape to their heads, they create a strong urge in their human companions to pat, tickle and caress them when they protrude their apparently grinning faces from the sides of their pools.

Birds that are hand-tame, such as parrots, budgerigars and doves, are frequently brought up to the face and held against the cheek, where the soft smoothness of their plumage can be felt against the skin. The intimacy is often elaborated with mouth-to-mouth feeding of morsels of food. Because of their small size, which rules out embracing and patting, the hand intimacies are limited to finger-stroking and gentle tickling 'behind the ear'.

If we move further away on the evolutionary scale, the possibilities for intimacies decline rapidly. For most people, reptiles, amphibians, fish and insects are singularly unrewarding to the touch. The tortoise, with its smooth, hard shell, rates an occasional pat on the back, but its scalier relatives lack the essential qualities for friendly body contact. Perhaps the only exceptions worth mentioning are the giant constrict-ing snakes. When suitably tamed, pythons, for example, can provide their owners with something that even the cats and dogs cannot offer – an all-enveloping embrace. Wrapping their strong coils around their human companions' bodies, tightening and relaxing their muscles, undulating their many ribs, and flickering their gentle tongues across their owners' skins, they create a sensuous impact that has to be felt to be believed. However, because of their difficult feeding habits and the bad press they have received ever since the fracas in Eden, not to mention our horror of their smaller and highly poisonous relatives, the big snakes have never enjoyed wide popularity as close intimates, even for the most embrace-hungry of humans.

Fish-touching, if we draw a discreet veil over the treacherous human intimacy of trout-tickling, is virtually non-existent. Perhaps the only exception here is the voluptuous hand-kissing sometimes performed by tame giant carp when sticking their gaping heads out of water as they beg for food. These fish can gape and gulp with such energy at the edge of a carp pond that even a passing bird may be coerced into a brief act of intimacy. There is an extraordinary photograph in existence which shows a small finch, its beak full of luscious insects for its hungry nestlings, pausing in front of the inviting gape of a tame carp's mouth and impulsively ramming its precious catch down the wide open throat of the fish. If a bird can be attracted in this way to make a thoroughly unnatural body contact, then it is little wonder that human visitors to carp-ponds react in much the same way.

Up to this point we have considered only friendly and parental intimacies, but for some humans the contacts go further and include full sexual interaction. These cases are rare, but they have a long and ancient history, references to them being scattered throughout art and literature from the earliest times. They take two main forms. Either a human male copulates with an animal, usually a domesticated farm animal, or masturbation occurs. In the latter case, a natural tendency on the part of a particular species to lick or suck is directed towards the human genitals, either male or female, as a device for producing sexual arousal. It says a lot for the degree of alienation and body-contact frustration that must exist in human societies that such aberrant intimacies occur at all. However, when we remember the millions of lesser intimacies, in the form of cuddlings, kissings and strokings, that take place in our modern cultures between pet-owners and their vast army of pets, it is not so surprising that, in a small minority of cases, greater intimacies such as these do occasionally arise.

In surveying the whole question of human-animal contacts, mention has so far been made only of pets and farm animals, but there are two other spheres of interaction that deserve some comment. Human-controlled animals exist not only in private homes and on farms; they are also found in large numbers in zoos and research laboratories. Here, too, frequent contacts occur and they are not always ones that meet with general approval.

Visitors to zoos not only want to see the captive creatures held there; they also want to hold the captive creatures seen there. The urge to touch is so strong that it constitutes a constant hazard for the zoo authorities. The first-aid department register of any major zoological gardens bears witness to this. For every sprained ankle or cut finger, there is a bitten hand or a scratched face. Sometimes the injuries sustained by the eager animal-gropers are serious, but they are seldom caused by carelessness on the part of the zoo staff. Two examples will suffice to illustrate this. The first concerns a woman who arrived at the first-aid department of a major zoo, holding her screaming child, who had a badly mauled hand. While it was being treated, it emerged that he had begged to be allowed to touch the body of the zoo's adult male gorilla. Complying with his wish, the woman had lifted him laboriously over the safety barrier, past the large warning sign indicating that the animal was extremely dangerous, and had thrust him forwards so that he could push his arms around the edge of the protective armoured glass screen and through the bars into the cage. The gorilla, misinterpreting this friendly act, had promptly sunk its teeth into the boy's hand. Unrepentant, the woman now presented herself, outraged, to the helpless zoo authorities.

The second case concerns the tragic affair of the 'tiger-toucher', an elderly gentleman who repeatedly clambered over the barrier at the same zoo's big cat house in order to caress one particular tigress. Removed, protesting, time and again by the zoo staff, he finally leapt over the barrier in such desperation that he broke a leg and was removed to a hospital bed. During his absence the tigress in question was dispatched to another zoo for breeding purposes. On his return to health, the man went straight back to the cage, only to find it occupied by a strange leopard. Furious at this, he marched across to the zoo office and demanded to know what they had done with his wife. At first the authorities were nonplussed by this extraordinary accusation, but after quiet questioning it emerged that the unhappy man had recently lost his real wife, after a lifetime of close companionship, and had since transferred all his emotional attachments to the tigress in question. Because the animal had, in his mind, become the embodiment of his late mate, it was only natural that he should want to continue to make intimate contact with her body in its new form, even at considerable risk to life and limb.

If these examples seem outlandish, it is worth remembering that they are only extremes of actions which, at a more moderate level, are occurring in zoos throughout the world in large numbers every day. When the urge to touch another human being is blocked, either by personal tragedy or by cultural taboo, it will nearly always find a way of expressing itself, no matter what the consequences. One is inescapably reminded here of the pathetic cases of child-molesters who are arrested for supposedly sexual assaults on infants. Unable to make proper contact with adults, they turn to children, who are innocent of the strictness of adult taboos. Frequently all that such men want is some kind of gentle, friendly body intimacy, but always the cry for blood goes up and the actions are interpreted as inevitably sexually motivated. This they may of course be, but it is by no means inevitable, and many a harmless old man has suffered heavily as a result. Needless to say, the children in such cases always suffer too, not from the original intimacies, which even in the specifically sexual cases they probably did not understand, but from the parental panic that follows and, above all, from the trauma of the court proceedings through which they are shamefully dragged.

Returning to the animal situation and leaving the zoo gates behind us, we come now to the fourth major category of man-animal contacts, namely those that exist in the world of science. Millions of laboratory animals are bred and killed annually in the course of medical research, and the contacts that occur between research workers and their experimental subjects have given rise to much heated debate. To the scientist, the interaction is a totally objective one. He admits to no emotional bond, either positive or negative, either loving or hating, with the animals he must handle while carrying out his investigations. The decision is simple enough: if he can reduce human suffering by sacrificing the lives of laboratory animals, he sees no other choice. He would avoid it if he could, but he cannot, and he refuses to place the lives of animals on a higher plane than the lives of fellow humans. That, briefly, is his case, but it is frequently and vociferously contested.

The opponents have been many, and their general attitude can best be summed up in the words of George Bernard Shaw, who said that 'If you cannot attain knowledge without torturing a dog, you must do without knowledge.' A more moderate view is expressed by those who feel that many animal experiments are pointless and that the results they obtain are worthless in any humanitarian sense, doing no more than satisfy the idle curiosity of the academic world. Surprisingly enough, such a view was voiced by the great Charles Darwin himself, in a letter to another famous zoologist, in which he said, 'Physiological experiment on animals is justifiable for real investigation, but not for mere damnable and detestable curiosity.' More recently, it has been pointed out by a respected experimental psychologist that 'One consequence of the obsessively behaviouristic and mechanistic approach is the apparent callousness of much of the experimental work carried out on the lower animals, often without any worthwhile aim.'

It is certainly true that the number of licensed animal experiments performed each year has risen sharply as the twentieth century has grown older. In Britain the figure for 1910 was 95,000; by 1945 it had exceeded 1,000,000; more recently, in 1969, it was in the region of 5,500,000, involving 600 separate research establishments. The vast scale of the operation has started to cause comment in political circles. One member of the British Parliament, speaking in 1971, protested: 'I know that the object is to preserve human life; but it does make me wonder whether a human race that can take such morally degrading practices in its stride is really worth preserving.'

It is important to separate two distinct elements in these and other criticisms of the large-scale use of animals for research. First, there is the extreme, anthropomorphic element, which sees the animals as symbolic people and therefore dislikes the idea of causing them pain for whatever purpose. Second, there is the humanitarian element, which sees animals as *similar* to people, in that they are capable in their own ways of feeling fear, pain and distress, and dislikes the idea that they should be caused any unnecessary suffering at human hands. This second element accepts, however, that it is necessary to cause some degree of suffering, but only if it is kept to an absolute minimum and only if the research is directly aimed at reducing a greater suffering.

The research scientist responds to these two criticisms in the following way. To the first critic he says, 'Tell that to the mother of a thalidomide baby.' If more extensive animal experimentation had been carried out, she might have had a normal child. Or he may say, 'Tell that to the mother whose child died of diphtheria.' Only a few years ago this disease killed thousands of children annually, but now, thanks to the development of a vaccine developed entirely by experiments on living animals, it has practically disappeared. Or he may say, 'Ask the mother of a polio child how she feels about the fact that it costs the life of an experimental monkey for every three doses of the polio vaccine that could have saved her child.'

In other words, what the out-and-out anti-vivisectionist is proclaiming is that it is better for a child to die or suffer agonies than for living animals to be used for experimental research. Whilst this may reflect an admirable concern for the welfare of animals, it also reveals a startlingly callous attitude towards human children. This putting of animals before people takes us back again to the pet-keeping situation, but here there is an important difference. Where pets were concerned, it was perfectly possible to be kind both to pets and to people. One did not automatically exclude the other, and the anti-pet argument to the contrary was shown to be false. But here the situation does demand that in order to be kind to the child it is unhappily necessary to be unkind to the experimental animal. We simply cannot have it both ways. An unpleasant choice has to be made.

To the second and more moderate critic, the research scientist says, 'I agree; animal suffering must be kept to a minimum, but there are problems.' A great deal of detailed study has been made in recent years of ways in which experimental procedures can be made less painful for the animal subjects, and everything is done to devise tests which use fewer animals, which cause them the minimum of distress and, where possible, replace them altogether. On this basis we might expect to see the number of laboratory animals killed annually declining steadily. As the figures I quoted show, however, this is not the case. The research scientist's answer is that this does not mean that more wasteful methods are being used, but rather that research programmes are becoming increasingly extensive and investigating more and more ways of alleviating human suffering. Furthermore, he will point out that one of the great problems with research is that it is impossible to limit it to areas which are directly and obviously connected with specific forms of suffering. Many of the greatest and ultimately most beneficial discoveries are made as a result of animal experiments in 'pure' rather than 'applied' research. To say that an animal experiment must not be done because, at the moment, it has no obvious application in such spheres as medicine or psychiatry is to stifle the whole progress of scientific understanding.

This is the point at which some of the least emotional and most educated critics begin to get worried. How far, to use Darwin's words, does 'real investigation' have to go before it becomes 'mere damnable and detestable curiosity'? This involves a much more difficult and delicate argument. Reading some of the scientific journals, especially those concerned with experimental psychology, it is difficult to escape the conclusion that many research workers in recent times have, by any reasonable standards, gone too far. By so doing they are endangering the public acceptance of scientific endeavour as a whole, and many authorities believe that it is high time that a drastic revision was made of the direction that many research projects are taking. If this is not done there may be a large-scale public backlash that will, in the long run, do untold harm to scientific progress.

Having made these general points, it now remains to ask why the man/animal contacts that occur in the laboratory should cause so much heated debate and concern. The obvious – too obvious – answer is that, even when we accept that it is justified and necessary, we do not like the idea of a man causing pain to the animal he handles. But what about the man who finds mice in his kitchen, or the slum-dweller who finds rats in his bedroom, and beats them to death with a stick or condemns them to a slow and painful death by putting down poison? He does not receive our criticism, only our sympathy. There are no protection societies formed to protect the wild rats and mice that infest our dwelling-places, and yet these are the same species that are used in the laboratory experiments which cause so much comment. Killing a wild rat is approved of because it may spread disease, but killing a laboratory rat is disapproved of, even though its death may also help to prevent the spread of disease, via the agency of scientific discovery.

How can we explain this inconsistency? Clearly it has little to do – whatever we may say – with our objective concern for the welfare of rats, wild or tame. If we really cared about the laboratory rat for its own sake, as an exciting form of animal life, we would not treat its wild counterpart so brutally. No, what is happening is that we are responding in a much more complex and subtle way than we imagine. We respond to the wild rat in a very basic way as an invader of our private territory, and we feel justified in defending that territory with any means at our disposal. No treatment is too harsh for a dangerous intruder. But what of the tame white rat in the laboratory? Is this not the creature whose ancestors brought the great plague into our midst? Certainly it is, but now it appears in a new role, and we must understand what that role is if we are to understand the strong emotions its experimental death causes in us.

To start with, the white rat is no longer a pest, but a servant of man. He is gently handled, well fed, well housed and cared for in every possible way. The attitude of his human companion is that of a doctor tending a patient before an operation. Then he is experimentally infected with cancer. Later he is killed by the same hands that cared for him. Except for the cancer element, this sequence could also apply to the relationship between a farmer and his domestic stock. He cares for them, then kills them. Yet we do not complain about the behaviour of the average farmer towards his animals, any more than we complain about the man who poisons a wild rat in his kitchen. Where does this leave us? The laboratory sequence involves tender handling, then causing pain, then killing. The farm sequence involves tender handling, then killing. The pest sequence involves causing pain and killing. In other words, we do not object to killing after caring, or to killing after causing pain, but we do object to causing pain after caring. The symbolic role that the white rat plays in the research laboratory is that of the humble and faithful servant who is loved by his master until, one day, without warning or provocation, this loving master starts to torture his servant and does so, not for the servant's own good, but for the benefit of the master himself. This is the allegory of betrayal that causes all the trouble.

The critics of animal experiments will hotly deny this, claiming that it is the rat they are thinking about and not this symbolic relationship, but unless they are full-time vegetarians who literally would not swat a fly, they are deluding themselves. If ever they have received any kind of medical aid, they are hypocrites. If, however, they are honest, they will admit that it is the *betrayal of intimacy* inherent in the symbolic man-rat relationship that really concerns them.

Now it should become clear why I have gone at some length into a pattern of human behaviour that, at first sight, does not appear to have any close connection with the subject of this book. The whole essence of the experimentalist's dilemma is that, to allay fears, he must emphasize over and over again how well he treats his laboratory animals: how gently he handles them, how relaxed and contented they are in their hygienic cages where they await the important part they have to play in his research. It is the contrast between this tender intimacy and what he then proceeds to do to them that is the crux of the passionate antagonism he arouses in his critics. For, as we have seen throughout this book, intimacy means trust, and the symbolic rat-servant is made to trust his master totally, only then to be subjected to pain and disease at his gentle, caring hands. If this betrayal of intimacy occurs only occasionally and for very special reasons, then most of the critics can reluctantly accept it, but when it takes place millions of times every year, then they begin to get the creepy feeling that they belong to a nation of emotional traitors. If a man can inflict deliberate pain on an animal that trusted him and which, a moment before, he was handling so gently and carefully, then how can he be trusted in his human relationships? How, when in all other respects of his social life he behaves in a perfectly reasonable and friendly way, can we ever be sure again that reasonable friendliness is any true guide to the nature of the members of the society in which we live? How can he behave so well towards his real children, when he constantly double-crosses his symbolic laboratory 'children'? These are the fears that run unspoken through the minds of his critics.

There is a similarity here to the case mentioned earlier of the concentration camp commandant who was kind and gentle to his pet dogs, whilst brutally torturing his prisoners. There, the kindness to his animals reminded us that even such a monstrous human being was not totally devoid of tender feelings. Here, the position is reversed, with a man who is capable of being kind to his fellow men nevertheless being able to spend his working days inflicting pain on his experimental animals. It is the contrast that frightens us. If we see a friendly-looking soldier patting his dog on the head, we cannot help wondering whether he, too, would be capable of gassing helpless human victims. If we see a friendly-looking father playing lovingly with his children, we cannot help wondering whether, beneath the surface, he is capable of cruel experiments. We begin to lose our sense of values. Our faith in the bonding power of body intimacies begins to waver, and we rebel against what we call the callousness of science.

We know perfectly well that this rebellion is unjustified because of the immense benefits that scientific research has brought us, but it hits so hard at our basic concepts of what gentle, caring intimacy means that we cannot help it. We still rush to the pharmacy when we get sick and quickly swallow our pills and tablets, but we try not to think of the trusting, betrayed animals who suffered to give us these antibiotic blessings.

If the situation is bad for the general public, what must it be like for the experimentalist? The answer is that it is not bad at all, for the simple reason that he has trained himself specifically not to see his relationship with his animals as a symbolic one. In applying a ruthlessly objective approach to his subject, he overcomes the emotional difficulty. If he handles his animals with gentle care, he does it to make them better subjects for experimental procedure, not to satisfy his emotional needs for substitute body intimacy like the ardent pet-keeper. This often requires considerable restraint and self-discipline, because, of course, even the most intellectually controlled act of body contact can begin to work its basic magic and start to form bonds of attachment. It is not unknown for a large laboratory to house, in a corner cage, a fat, lop-eared rabbit that has become the departmental mascot, a pampered pet that no one would dream of using for an experiment, because it has slipped into a totally different role.

For the non-scientist it is difficult to make these rigid distinctions. For him, all animals belong in Disneyland. If, through the modern educational media of film and television, he broadens his horizons and begins at last to forget the toy-animal images of his childhood, he does so not at the deft hands of experimentalists, but in the company of naturalists, whose basic approach is that of the observer, rather than the manipulator, of animal life.

The plight of the serious experimentalist therefore remains unsolved. Like the surgeon who operates to save his patients' lives, he strives to improve our lot, but unlike the surgeon he gets little thanks for it. Like the surgeon, he remains strictly objective and unemotional throughout his operations. In either case, emotional involvement would be damaging. For the surgeon this is less obvious, for he must put on a bedside manner outside the operating theatre. Once inside, however, he treats the bodies of his patients as coldly and objectively as the experimentalist, carving them up like a master chef performing with a choice joint of meat. If this were not so, we would all suffer for it in the long run. If the experimentalist became emotionally involved with his animals and treated them all like beloved pets, he would soon be incapable of carrying on his arduous research projects that bring us so much widespread relief from disease and agony. He would be driven to drink by the enormity of what he was doing. Likewise, if the surgeon allowed himself to become emotionally aroused by the plight of his patients, his knife might waver as it sliced, and vital damage would be done. If hospital patients could hear the conversations that take place in many operating theatres, they would probably be horrified by the sometimes jocular, sometimes matter-of-fact tones, but their response would be misguided. The terrible intimacy of entering another person's body with a sharp instrument demands a dramatic switching-off of the emotional impact of the act. If the action is performed as a piece of desperate, loving care, then the patient's next piece of body intimacy is liable to be at the cold hands of an undertaker.

In this chapter we have been looking at the use of living substitutes for human bodies in a contact-hungry world. Where the contacts have been loving, as with cuddly pets, the intimacies involved have caused considerable pleasure; where they have been strictly non-loving, as with experimental animals, they have caused considerable displeasure. In sum total they account for a vast number of tactile interactions, and animals are clearly of great importance to us in this respect. We have been considering mostly cases of adult human activity; but pet-keeping is also a significant pattern for the older child, when it starts to imitate its parents by showing intense pseudo-parental care for small animals, cuddling them, carrying them, nursing them and caring for them as if they were totally dependent infants. Since cats and dogs are often already stamped as the pseudo-infants of the real parents in the family group, the young pseudo-parents frequently become more devoted to other species of a type that adults normally scorn, such as rabbits, guinea-pigs and tortoises. These species, being uncontaminated by parental involvements, then provide a separate and more private world for the substitute intimacies of the juvenile pseudo-parents.

For younger children the problem is solved by the use of toy animals – substitutes for the substitutes for love. These are cared for and loved exactly as though they were living beings, and the attachment to a favourite Mickey Mouse or Teddy bear is as passionate and powerful as that of any older child to a favourite rabbit or, later, to an adored pony. For many girls the attachment to a large cuddly toy animal survives right into adulthood, and a newspaper photograph of recent hijack victims shows a teenage girl returning to safety 'still clutching the Teddy bear that comforted her through her ordeal in the desert'. When we are badly in need of some kind of reassuring body contact, even an inanimate object will do, and this is the subject of the next chapter.

7

OBJECT INTIMACY

ON A HOARDING in Zurich, Switzerland, there is a large poster showing a man's head reproduced twice, one beside the other. The heads are identical except for one detail: between the lips of one there is a cigarette; between the lips of the other there is instead a baby's dummy, or pacifier. It is assumed that the message is obvious, since not a single word accompanies the picture. Without realizing it, the designers of this poster have said far more about the importance of smoking than they intended. In one simple visual statement they have explained why so many thousands of people are prepared to risk a painful death, coughing and spewing as their lungs clog with cancerous cells.

The poster is, of course, supposed to put adult smokers to shame by making them feel immature and babyish, but it can also be read backwards. If the man with the dummy in his mouth is gaining some comfort from it, just like a baby, then all that is wrong with that part of the picture is that it looks so infantile. Now switch to the other head – here the problem is solved. Like the dummy, the cigarette gives comfort, too, and in one stroke the babyish element is gone. Seen this way, it might almost be an advertisement to encourage smoking in those who have not yet discovered the basic comfort of this activity. Smoke a cigarette and you can be pacified without feeling immature!

Even if we do not perversely twist the well-intended message in this way, it nevertheless provides us with a valuable clue concerning the world-wide smoking problem that faces society today. It is a problem that has been dealt with recently for the first time. A large-scale campaign to alert smokers to the dangers of filling the lungs with carcinogenic smoke has been undertaken in many countries. Cigarette promotions have been banned on television in several major areas, and there has been endless discussion on how to discourage children from taking up the habit. Gruesome films are shown of pathetic hospital patients in the advanced stages of lung cancer. Some smokers have responded intelligently and given up, but many others have become so alarmed that instead they have been forced to light up an extra cigarette to calm their shattered nerves. In other words, although the problem is at last being dealt with, it is by no means solved. Simply to tell people they must not do something because it is harmful may be a wise step to take, but it is also a short-term one. It is like using war to solve the population problem. War kills millions, but as soon as it is over there is a post-war birth bulge and the population growth goes soaring on again. Similarly, every time there is an anti-smoking scare, thousands of people stop smoking, but after the scare is over, the shares of the cigarette companies start to soar again.

The great error of the anti-smoking campaigners is that they rarely stop and ask the basic question: why do people want to smoke in the first place? They seem to think it has something to do with drug addiction – with the habit-forming effects of nicotine. There is an element of this, certainly, but it is by no means the most important factor operating. Many people do not even inhale their smoke and can be absorbing only minute amounts of the drug, so that the causes of their addiction to cigarettes must be sought elsewhere. The answer clearly lies in the act of oral intimacy involved in holding the object between the lips, as the Zurich poster so beautifully demonstrates, and this answer almost certainly applies as the basic explanation for the full inhalers as well. Until this aspect of smoking is properly investigated, there will be little long-term hope of eliminating it from our stressed, comfort-seeking cultures.

What we are plainly dealing with here is a case of an inanimate object being used as a substitute for a real intimacy with another human being. In examining this phenomenon, we are moving one step further away from the original source – namely, intimacy with intimates. The first step away took us to intimacy with semi-strangers (the professional touchers), the second step took us to intimacy with live substitutes (pet animals), and now the third step takes us into the world of dummies – objects with a hidden intimacy factor. There are many of them besides the cigarette, but it helps to start with this one because it leads us naturally back to the start of the whole story, at the point where a distraught mother jams a rubber substitute for her nipple into the mouth of her screaming infant.

The baby's dummy, sometimes called a pacifier or a comforter, is usually described as a 'blind' nipple, since, unlike the teat of a feeding-bottle, it has no hole in it. The description is a little misleading because no mother can boast such a huge bulbous nipple as that found on the average commercial dummy. This is a super-nipple, milkless, but with a greatly magnified tactile quality. At its outer end there is a flat disc to simulate the mother's breast and to stop the rubber super-nipple being sucked right into the mouth.

Dummies of this sort have been in use for centuries, but not so long ago they fell into disrepute because they were considered a dangerous source of infection. More recently they have started to make a come-back, and today they are recommended in many cases by medical authorities. Babies given dummies in their early months are much less likely to become thumb-suckers (the obvious alternative if no nipple is available to give comfort when comfort is needed). Also, it is no longer believed that dummies deform the mouth or damage the developing teeth, and recent experiments have shown the experts what many mothers knew already, namely that dummies do have a really dramatic calming effect upon a distressed infant. 'Non-nutritive sucking', as it is officially termed, was studied carefully in a large number of babies and their responses recorded. It was found that after only thirty seconds with the dummy in the mouth, crying was reduced to one-fifth of its previous level, and agitated hand and foot movements to one-half. It was also pointed out that, even without active sucking, the presence of the super-nipple between the baby's lips has a calming effect. If a baby is half asleep and has stopped sucking, the removal of the dummy can easily start it crying again.

All this adds up to the fact that having something between the lips is a comforting experience for the human animal, since it spells reassur-ing contact with the primary protector, the mother. It is a powerful form of symbolic intimacy, and when we look at an old man sucking contentedly on the stem of his pipe, it becomes abundantly clear that it is one that stays with us all our lives.

The important thing for the adult 'sucker' is that he should not appear to be doing what he is doing; hence the message of the Zurich poster. The use of a baby's dummy by a stressed adult would probably be as calming as anything else, if only it did not carry the 'infantile' stigma. Since it does, however, he is forced to adopt disguised dummies of various kinds. The cigarette, in this respect at least, is ideal, because it is so totally adult. Being prohibited for children means that it is not only non-infantile, but also non-childlike, and therefore completely removed in context from the baby-sucking that is its true origin. The object feels soft between the lips and the smoke warms it, which makes it even more like the genuine mother's nipple than a rubber dummy. Furthermore, the sensation of something being sucked out of the end of it and drawn down the throat adds to the illusion. A new symbolic equation is set up: warm inhaled smoke equals mother's warm milk.

Many smokers, when putting a cigarette to their mouths, or when taking it away, let their fingers fall on the outsides of their lips, simulating in this way the touch of the mother's breast. Some put the cigarette between their lips and leave it there for long periods of time, only drawing on it occasionally. When they do this, the non-drawing moments are similar to those of the half-asleep baby who still held the dummy in its mouth after it had stopped sucking on it. Other smokers, when they take the cigarette from the mouth, continue to fondle it between their fingers, even though it would be easy to put it down on an ashtray or some other surface. Deeply stained 'nicotine fingers' bear silent testimony to this urge to hold on to the comforting tobacco-nipple, even when it is not in oral use.

Variations on this theme include the businessman's super-nipple, the cigar, the tip of which is suitably rounded and smooth where it touches the mouth. With quiet ceremony this smooth 'blind' nipple is pierced and snipped with special gadgets, to ease the comforting flow of warming smoke-milk. For some, the soft touch of the cigarette or cigar is sacrificed to increased smoothness in the form of a cigarette-holder, a cigar-holder or a pipe. Here the tongue can play with a surface as smooth and slippery as a nipple of flesh or a teat of rubber. It is surprising that some device has not been used that is both soft and slippery – say a rubber holder – but perhaps this would not be sufficiently disguised, and would begin to look too much like the real thing to maintain its adult respectability. It would certainly make it more difficult for pipe-smokers to employ one of their favourite devices, namely sucking an empty pipe. That is already getting dangerously close to the obvious, and a rubber pipe-stem would be the final give-away.

The enormous amount of tobacco-smoking that takes place around the world today bears evidence to the vast demand that exists for comforting acts of symbolic intimacy. If the damaging side-effects of this pattern of behaviour are to be eradicated, it will be necessary either to de-stress the population to the appropriate degree, or to provide alternatives. Since there is no sign of any great or immediate hope for the former, the solution will have to be the latter. Plastic cigarettes have been suggested and even tried, but there seems little hope for them. The suggestion is valid enough in itself, but it overlooks the important factors of the warmth and true 'suckability' of real cigarettes. Also, it fails to provide any official excuse for the action. There must be a disguise of some kind if the activity is to be readily acceptable. It is true that many people suck the ends of pencils, pens, matchsticks, and the tips of the side-pieces of spectacles, but all these objects have other 'official' functions. A plastic cigarette would fail in this respect and would therefore appear too much like the baby's dummy in the Zurich poster. Some other solution will have to be found, and it looks very much as though this will have to come from the cigarette manufacturers themselves, in the form of a synthetic or herbal tobacco that does not damage the lungs. Research is already progressing in this direction, and perhaps the most valuable contribution made by the recent lung-cancer scares and propaganda campaigns will be to force a drastic speeding up in these investigations. Bearing in mind the deep significance of smoking, as I have outlined it here, this is probably the only long-term way in which these campaigns will be able to help.

Those people who have given up smoking, or who have attempted to give it up, complain that they start getting fatter soon after they abandon their non-nutritional tobacco-nipples. This immediately gives a clue about certain kinds of feeding. A great deal of the nibbling and food-sucking we do is primarily concerned with symbolic oral intimacies, rather than true adult food-intake. The cigarette-hungry ex-smoker, when in sudden need of added comfort, grabs a sweet food morsel and stuffs it into his nipple-empty mouth. Sucking candies and sweets is yet another of our disguised breast-feeding substitutes. For most of us it is the pattern of behaviour that fills the gap between the dummies of infancy and the cigarettes of adulthood. The confectionery shop is the world of the child. Too old for rubber pacifiers, he takes to sucking gob-stoppers and bull's-eyes, lollipops and sticks of seaside rock. They may rot the teeth, but they help to replace the lost comfort. As adults we largely turn our backs on these delights, but many a young lover still brings to his 'sweetheart' the comforting gift of a box of assorted chocolate nipples. And many a bored housewife dips into a box of soothing candies. A trick sometimes used to give these objects adult respectability is to fill them with anti-childish alcohol and pop them into our mouths in the form of 'liqueur chocolates'.

Although these food objects do not last as long as nipples, they do have the important qualities of softness and sweetness to help them in their symbolic role. One special form overcomes this drawback of being short-lived, and that is chewing-gum. Chewing-gum consists of an elastic substance known as chicle gum, sweetened and flavoured. (One part of chicle gum to three parts of sugar, warmed and kneaded together and flavoured with cloves and cinnamon, wintergreen or peppermint.) It can be chewed for hour after hour and is advertised as something to 'calm your nerves and help you concentrate'. Symbolically, it is nothing short of a rubbery, detachable nipple. Because of its special properties it should enjoy enormous success, but it is badly hindered by the conspicuous jaw movements that accompany its use. These cause no problem for the chewer, but for those near him they create the impression that he is incessantly eating. Since he never swallows the 'food' in his mouth, this conveys the feeling that the object in his mouth is in some way unpleasant, like a tough piece of gristle, and as he becomes soothed, his observers become irritated. The result has been that, in many social settings, chewing a piece of gum has become looked upon as a 'dirty habit', and the activity has there-fore remained a restricted one.

Since mother's milk is a warm, sweet liquid, it is not surprising that adults employ a variety of warm, sweet drinks to soothe themselves at moments of tension or boredom. The millions of gallons of tea, coffee, drinking-chocolate and cocoa that are consumed annually have little to do with the real demands of thirst, but again thirst is there to provide the vital official excuse. The cups and mugs from which we sip these milk-substitutes so eagerly also provide a pleasantly smooth, slippery surface to press against our comfort-seeking lips, and the outcry when modern 'disposability' demands the use of non-smooth, non-slippery paper cups is easy to understand.

Once more, it is interesting to see the way we avoid the obvious: we drink tea hot, but milk cold. To drink hot milk is too overtly babyish. Only invalids drink hot milk, but that is permissible because, as we have already seen, the invalid has given up the adult struggle and has become an 'instant-baby' in other ways as well, so for him one more baby pattern makes no difference.

Apart from cold milk, or milk-shakes, which significantly are usually sucked through a straw, there are many other types of cold, sweet drink that we employ as comforters. They are nearly always advertised as thirst-quenchers, but in this respect they always fall far short of simple, plain water. They do, however, provide that vital sweet taste, and the increasingly acceptable habit of drinking them straight from the bottle helps to improve their symbolic value. The bottles in ques-tion have therefore shrunk in size from the traditional dimensions down to something remarkably close to that of the baby's bottle. In fact, if someone were to emulate the Zurich cigarette poster and show a man drinking from a cola or lemonade bottle with a rubber teat on it, the game would be up.

Many other objects, such as the stems of plants, or the beads of a necklace worn around the neck, are often brought up to the lips in fleeting moments of self-comfort, but enough has been said now to show that the oral intimacies of infancy remain an important part of our adult lives, even outside the more obvious spheres of friendly or sexual kissing, and we can move on to consider other parts of the adult body.

Another basic form of contact in babyhood is the pressing of the cheek against the mother's body when resting. Cheek-pressing with soft substitute objects is rare amongst adult males, but it remains fairly common with females. Many advertisements for soft bedding, blankets and linen display a serenely smiling female hugging to her body the cuddly product, her head tilted to one side and her cheek pressed lovingly to the smooth surface of the cloth. This is particularly common with blanket advertisements, to the extent that it is almost the only pose employed, despite the obvious fact that, once on the bed, the blanket will be prevented from making such a contact by the intervention of the inevitable bed-sheets.

Fur coat advertisements follow a similar course, frequently showing the fur collar turned up, or pushed up with the hands, so that its ultra-soft surface caresses the wearer's cheeks. Fur rugs offer a more extensive contact surface, like a giant maternal body flattened out on the floor or bed.

Perhaps the most widespread and common form of soft cheek contact, and one that is employed by both males and females, is the use of a down-filled pillow when sleeping at night. The caress of this tender pillow-breast provides a major soothing element at the end of the day, helping to calm us into a condition where we are prepared at last to sink into a deep slumber and give up the adult battle of the day. Great subtlety has been employed by pillow manufacturers in producing exactly the right balance between springiness and softness, and at any large bedding store it is possible to select a new pillow from a wide range of slightly varying tactile qualities. For many adults, one particular pillow, or pillow 'strength', becomes extremely important as an aid to falling asleep, and if they are faced (in both senses of the word) with the wrong type of pillow when trying to sleep in a strange bed, at a hotel or in a friend's house, they may find it difficult to drift off into a peaceful sleep as quickly as they can at home. This phenomenon is much more pronounced in the case of 'home-lovers' who travel little and who, over a period of years, become fixated on a specific pillow quality, such as resilience, thickness or sagginess.

A similar development occurs with the rest of the bed. In addition to sensitive pillow responses, adults come to prefer a particular softness or hardness in the mattress beneath them, and a certain lightness or heaviness, a looseness or a tight 'tucked-in-ness' in the bedding that covers them, as they settle down to the vitally important nocturnal bed-embrace that will envelop them for a total of one-third of their entire lives.

In 1970 a new type of bed appeared on the market in America – the 'waterbed'. Essentially, this is a vinyl mattress filled with water. Lying on it, the sleeper sinks gently into its liquid embrace as if returning to a semi-womb. A thermostat and heating element inside it keeps the water at an appropriately soothing temperature. In the second half of 1970, over 15,000 of these beds were sold, and the demand soon outstripped the supply. The advertisers encouraged their prospective buyers with significant phrases such as 'Live and love in liquid luxury' and 'You can make them rock-and-roll you to sleep'. The only hazard, to use a gynaecological expression, is rupturing the membrane. Accidentally puncturing a waterbed is almost as messy and comfortless as being born. Perhaps this slight but nagging fear will, in the end, keep most of us wrapped in the safer embrace of our old-fashioned cloth-beds.

Examined objectively, our sleeping habits, with our soft pillows, bedding and mattresses, begin to take on a special significance. They are more than a device for acquiring dream-time, in order for our computer-brains to sort out and file away the confusing bombardment of new thoughts from the past day, and much more than the mere act of obtaining physical rest for the exertions of the new day that is coming. They represent also a massive, world-wide indulgence in abandoning ourselves to the comforting intimacy of an inanimate envelopment that is part cloth-womb and part cloth-mother embrace.

Even during our waking hours we do not entirely reject these primal delights, as the modern furniture industry can demonstrate so clearly. 'Easy' chairs and sofas, of a voluptuous softness and bed-like snugness unknown in previous centuries, have become the almost inevitable centre-piece of every drawing-room, sitting-room and lounge. There, at the end of the hard working day, we sink gratefully into the soft intimacy of our favourite piece of soft furniture, whose 'arms' may not actually embrace us, but whose yielding surfaces nevertheless provide great body comfort. Cuddled snugly on the symbolic laps of our chair-mothers, we then settle down with childlike security to view at a safe distance the chaos of the harsh adult world outside, as symbolically portrayed on our television screens or between the covers of our novels.

If, in describing the act of watching television from the soft comfort of an easy chair in this way, as an infantile act similar to looking out of a window while safely held on a mother's lap, I appear to be condemning it, I hasten to add that this is not my intention. On the contrary, it is an added advantage to this now world-wide pattern of behaviour. In addition to providing entertainment and education, the act of watching television can, as I have shown, provide a vitally important soothing element in our stressful adult world. The glass screen that covers the pictures we watch traps them safely inside the television box where they cannot harm us. This helps to compensate for the fact that our chair-mothers only provide one of the two vital security factors that the real mother gives to her infant. The real mother provides both the intimacy of soft body contact *and* protection against the outside world. Our chair-mothers provide only the soft contact – they cannot protect us. But this is where the inpenetrable glass surface of the television screen comes to our rescue, compensating for the missing protection by safely walling us off from the adult dramas unfolding inside the box. The symbolic equation is therefore simple enough: real mother that protects and comforts = screen that protects + chair-mother that comforts.

Viewing our home life in this way, it is not surprising to find that when we travel or go on a vacation most of us prefer to stay in hotels which simulate in almost every way the conditions we once knew in the nursery. As in our infancy, everything is done for us and we do not need to lift a finger. Our food is prepared by the chef-mother, served to us by the waitress-mother, and our beds are made and our rooms cleaned by the maid-mother. At the best hotels, the use of room service can take us back virtually to the cradle, with our baby-crying replaced by the simple act of pressing the wall-button or lifting the house-phone. And frequently one of the first things that people do when they become rich is to introduce these nursery conditions into their own homes by employing personal staff-mothers. Also, as I pointed out in an earlier chapter, the sick-bed and the hospital provide a similar condition for the invalid who has temporarily given up the adult struggle completely.

Sometimes we indulge ourselves briefly in the even more basic luxury of a return to a womb-like condition by the act of taking a hot bath. It is no accident that nearly all of us prefer to do this at womb-temperature, floating blissfully in our mock-amniotic fluid and feeling beautifully secure inside the curving walls of our bath-wombs, with the bathroom door safely locked against the adult world outside. Sooner or later, however, we are forced to remove the cervical bath-plug and reluctantly face the trauma of a new birth. As if knowing of our fears at this dreaded moment, the towel manufacturers compete to provide us with the softest embrace they can produce. As one towelling advertisement puts it, 'Our towels just hug you dry!'; and the girl in the accompanying picture is shown clutching the object in question ecstatically to her body and face, as if her very life depended on it.

When this towel-hugging girl finally gets dressed, she need not fear that these tender intimacies will cease. Advertisers of body clothing – underwear, sweaters, skirts, and the rest – all promise her similar rewards. Those brief panties are much more, it seems, than a matter of mere modesty, for we learn that they also offer you a 'bare hug' which 'gently, caressingly . . . stretches to follow every curve of your body'. And those tights – they are 'silky soft and sensual' and 'hug you snugly from top to toe', not to mention those stockings which 'embrace your legs in a gentle, lingering caress', or those jersey dresses with that 'clinging feeling'. The lucky girl can therefore walk about fully dressed and apparently alone, but symbolically clad in a whole mass of caressing, embracing, hugging cloth-lover intimacies. If all the clothing advertisements put together had a cumulative effect, it would be surprising if she could perform the simple act of crossing a room without experiencing a multiple orgasm. Luckily for her real lovers, however, the impact of her assorted cloth-lovers is much milder, in most cases, than the advertisers would have us believe. Nevertheless, mild though it may be, it is still a genuine and important part of the body reward of wearing the soft and comfortable clothing of today.

This intimacy between clothing and wearer is more than a one-way affair. Not only do the clothes embrace the wearer – in addition, the wearer also embraces the clothes. This is, after all, a fair return for all that snug hugging and gentle caressing. The favourite way of repaying the compliment is to thrust one or both hands deep into some suitable fold of cloth. Napoleon's characteristic pose, with one hand inserted into his jacket, immediately comes to mind, but today the most widespread version is the hands-in-pockets action. Pockets are officially there to carry small objects, and if we put a hand into one, it is supposedly in the process of taking something out. But the vast majority of hand-in-pocket poses have nothing to do with object-retrieving. Instead, they are prolonged contact actions in which we, so to speak, hold hands with our pockets. Schoolboys and soldiers are often ordered to 'take your hands out of your pockets', with no explanation given other than that it is slovenly or untidy. The truth is, of course, that the posture indicates that they have relaxed into a symbolic act of intimacy, and this conflicts with their official roles as attentive, subordinate males. For males who are not restricted in this way, there are several alternatives open, and the choice made in any particular instance follows a rather curious rule. It is this: the higher up the body the hand/clothing contact is made, the more assertive it is. The most assertive of all is the grasping of the lapels. Coming a close second is the thumbs-in-waistcoat action. Next is the Napoleonic hand-in-jacket-front posture. Further down there is the hands-in-jacket-side-pockets pose, and further still the very common hands-in-trouser-pockets. Going any lower, with a hands-grasp-trouser-legs action, becomes appropriately low on the assertiveness scale.

The reason for this rule seems to be that the higher up the hand goes in making the contact, the more it adopts what amounts to an intention movement of striking a blow. Whenever a real blow is struck, it must be preceded by a raising of the hitting arm, prior to lashing out at the opponent. As we have already seen, this action becomes frozen as a formal signal in the case of the raised fist of the communist salute. The firm grasping of the lapels goes a long way towards this – as far, in fact, as a hand-to-clothing contact can go – and it is natural therefore that this should be the most truculent message of the various alternatives. Along with the thumbs-in-waistcoat posture, it has become almost a caricature of assertiveness, and a serious dominant male today is much more likely to adopt the lower-level hands-in-jacket-side-pockets posture when he finds himself displayed in a public place. This latter action is particularly favoured by tycoons, generals, admirals and political leaders, and it also became a cliché posture for the big-time gangsters of the roaring 'twenties. Such men are much more reluctant to adopt the lower-level posture of hands-in-trouser-pockets, at least when they are in a context that demands the assertion of their dominant rights.

An intriguing exception to the above rule is the thumbs-hooked-in-belt posture. Although this occurs rather low down on the body, it conveys a decidedly truculent flavour. It is popular with 'he-men', cowboys, pseudo-cowboys and mock-aggressive girls. Its assertive qualities appear to stem not only from a quick-on-the-draw intention movement, but also from the fact that it has become the modern, waistcoatless version of the old-fashioned thumbs-in-waistcoat posture. Sometimes the whole hand is slipped beneath the belt, or into the top of the trousers, but it then immediately loses much of its aggressiveness and fits more suitably on to the scale outlined above.

In addition to these actions, there are many small ways in which the hands perform minor intimacies with different parts of the clothing. All of them appear under stress, and many of them seem to represent symbolic versions of grooming acts that we would like someone else to be applying to our bodies to soothe us. Men can frequently be seen adjusting their cuff-links or smoothing their ties. President Kennedy frequently fingered his central jacket button at moments of public stress. Winston Churchill was often pictured at times of tension with his hand pressed flatly against the lower part of his jacket, as if engaged in a fragmentary self-embrace.

With the female sex, bracelets and necklaces come in for a great deal of handling and fingering at moments of strain, just as nuns no doubt obtain soothing comfort from the physical actions of telling their beads. At other times, the smooth caress of lipstick on lips, or powder-puff on cheeks, will provide a reassuring sensation of touch for a nervous female taking time off from a stressful social engagement. At more private moments, the repeated combing or brushing of hair, far beyond the demands of 'hair adjustment', can also have a marked calming effect, playing the role of a self-directed lover's caress.

In some instances, the act of making contact with a companion is performed indirectly, through an intermediate object of some kind, as when we clink our glasses together in making a toast, instead of making direct skin contact. A classic example can be found in any old Victorian scrap-book where a family group photograph has been taken. Typically, the mother sits in a central chair, with the latest addition to the large family nestled on her lap. Her husband, whose natural inclination is to put his arm around her shoulder, is too inhibited to do this in so public a situation and instead embraces the back of the chair in which she is sitting. The modern version of this can often be seen when two friends sit together informally and one stretches his arm along the back of the sofa they are sharing, aiming it in the direction of the other's back. Similarly, if a person is sitting alone in a chair, he may embrace the arms of it lovingly while talking animatedly to a companion in a chair opposite. Occasionally, additional chair-comfort is obtained by the use of a rocking-chair – another favourite action of President Kennedy's when under stress. The rocking comfort, needless to say, relates directly back to the rocking of the cradle or the mother's arms.

Finally we come to those objects that provide, quite specifically, substitutions for sexual intimacies. At the mildest level there are the photographs of loved ones, or the 'pin-ups' of those we would like to make love to, which can be kissed and touched in the unavoidable absence of the real thing. The use of life-size 'pillow-posters' is a new trend in the direction, it now being possible to buy a pillow-slip imprinted with a picture of the face of a favourite film star, to fit over the pillow that graces your bed. Then, at bedtime, you can lie cheek to cheek with the adored one, and fall gently asleep in this surrogate, cloth embrace.

Moving on to copulation itself, it was claimed during World War II that enemy soldiers (it is always enemy soldiers) at the front line were supplied with inflatable rubber dummy-females, complete with sexual orifices, for purposes of relieving sexual tensions. Whether this was merely propaganda to show how sex-starved and badly off the enemy were, or whether it actually happened, I have not been able to ascertain.

Inanimate substitutes for the male penis have, by contrast, a long and factual history, and even rate a mention in the Old Testament. Usually called dildoes, but also referred to as godemiches, consolateurs, bijoux indiscrets and dil-dols, they were known even before biblical times, and appear in ancient Babylonian sculptures dating from hundreds of years before Christ. In ancient Greece they had the name 'olisbos', meaning 'slippery bull', and they were apparently particularly popular amongst the Turkish harems. As the centuries passed, their use spread to practically every land in the world. Their popularity waxed and waned, apparently reaching a peak in the eighteenth century, when they were sold openly in London, a phenomenon not encountered again until the second half of the present century. Great care and skill is said to have been put into their manufacture 'so as to enhance realism in a coition fantasy'. In the 1970s they are now on sale in several versions in the 'sex shops' of a number of Western countries. Those that are not purchased by males out of sheer curiosity are used by lesbians or by solitary females for the purpose of masturbation.

Two forms of mechanically operated dildo have also appeared in recent times. One sort was of a highly technical nature, designed especially for an American research investigation into the nature of human copulation. Devised by radio-physicists, electrically powered, made of plastic with the optical properties of plate glass, equipped with cold-light illumination to facilitate intra-vaginal film-making, and fitted with masturbator-operated controls for varying both speed and depth of thrust, this was, by any standards, a truly remarkable instrument, a sensitive and tireless dummy lover, the sex-surrogate to end all sex-surrogates. A less ambitious and much less expensive mechanical device, and one that has gained widespread popularity in recent years, is the comparatively simple 'vibrator', or 'vibro massager'. It is a small, smooth-surfaced, battery-operated, plastic object, long and thin, with a rounded tip, the original and official function of which was localized muscle massage. It soon found a new and more sexual function as a gently vibrating masturbation dildo, and since it could be bought in its official role as a simple massage machine, had the added advantage that it could be obtained without undue embarrassment by purchasers who would have hesitated to acquire the more obvious forms of sexual equipment. Even in the outspoken underground press, the charade is maintained. A typical advertisement reads: 'Personal Massagers – penetrating, stimulating massager. Throbs away aches and frustrating pains. 7 in. long by $1\frac{1}{2}$ in. Standard batteries included.' The coyness employed in this advertisement is completely out of step with the rest of the text of underground papers, where the bluntest and most uninhibited forms of sexual comment are to be found. Once again we see the operation of the rule we have met so many times before, namely that adult intimacies require some sort of disguise, either to ourselves or to others, to obscure the real purpose of what is going on.

A most unusual and ingenious form of sex-surrogate sometimes used by females in Japan is the rin-no-tama, also known as the watama or ben-wa. This consists of two hollow balls, roughly the size of pigeon's eggs, which are inserted into the vagina. Originally made of leaf brass, but nowadays probably of plastic, one ball is empty and the other contains a small quantity of quicksilver (mercury). The empty ball is inserted first and pushed up to the end of the vagina, next to the cervix. The second is then placed against it and the opening of the vagina plugged with paper or cotton. Thus equipped, and with no embarrassing outward signs of her peculiar condition, the female proceeds to amuse herself with apparent innocence by swinging on a swing or rocking in a rocking-chair. The rhythmic back-and-forth motions then produce shifting pressures inside the vagina which simulate the thrusting of a male penis. Although, as a sexual 'toy', the rin-no-tama has the great advantage of permitting private enjoyment in public, it has not achieved the widespread popularity of the ubiquitous vibrator, presumably because, unlike the latter, it lacks a non-sexual, 'official' function.

Toys themselves, of completely non-sexual types, should also, incidentally, be a medium where tactile rewards are made available to us through inanimate objects. The possibilities are enormous, but few are attempted and even fewer succeed. When they do appear, they are usually presented as some kind of athletic exercise. The trampoline was one. The main reward here was the strange sensation of being embraced by the springy surface, flung into the air and then embraced again in a new posture. But the whole process had to be carried out under the cover of a highly muscular and sporting atmosphere, which ruled it out for many people. The short-lived hula-hoop was another case, combining the rotating embrace of the hoop around the performer's waist with an undulating action of the hips. Its appeal, however, was strictly limited and did not survive past the novelty stage.

The art world has several times attempted to present intimacy objects to an intimacy-hungry world, but with little success. In 1942 the Museum of Modern Art in New York exhibited for the first time a new type of sculpture called 'handies' or hand sculpture. They consisted of small, smoothly rounded pieces of polished wood in abstract shapes that would fit comfortably into the human hand and could be squeezed and turned this way and that to vary the tactile sensation. The artist who created them stressed that they were meant to be felt rather than looked at, and suggested that they would make an excellent substitute for cigarettes, chewing-gum or doodling for those who tended to be fidgety in committee meetings. Unhappily, it was not to be, and handies have hardly been heard of since. Once again, the message was too obvious, no committee member wishing it to be known that he was so obviously in need of a little comforting pseudo-body contact.

More recently, in the 1960s, certain artists have attempted more ambitious assaults on the bodies of art-lovers by creating 'environmental sculptures'. These have taken many forms, some of which have included a kind of play-space into which the visitor walks, there to be assailed by a series of varying tactile impressions as he passes through tubes, tunnels and passageways, walled and hung with a wide variety of textures and substances. Again, success has been short-lived and great possibilities have been wasted.

A final example neatly sums up the whole situation. One particular artist devised a simulated copulation capsule, into which the 'art-lover' was placed and wired up in various ways. The capsule was then shut and the machine switched on, with the idea of producing a massive sensory experience. At an art institute, the creator of this machine delivered a lecture about his concepts to an engrossed audience, who listened with interest as he explained that, owing to technical hitches, he had now devised a much simpler version of the machine, in which he had great faith. The modified device consisted basically of a large vertical sheet of rubber, or some such material, with a small hole at genital height, through which a male art-lover could insert his penis. For the female art-lover there was a similar vertical sheet, with a penis-shape protruding through it. In serious tones he then explained that, in addition to its simplicity, this new exhibit had the advantage that it could be used simultaneously by a male and female art-lover, with one standing on each side of it.

The absurdity of this story brings us back inevitably to the inherent absurdity of many of the examples given in the present chapter. It *is* absurd that an adult human being should have to coat his lungs with carcinogens in order to enjoy a crude substitute of the pleasures he once knew at his mother's breast, or when a baby's bottle was held to his lips. And it *is* absurd that grown men should endlessly have to mouth a disembodied, rubberized nipple in the form of chewing-gum, or that adult females should have to use a plastic massager instead of a live penis to achieve sexual gratification. But although these actions may seem absurd, pathetic or even downright repulsive to some, to many they are the only solution that seems to be available, and it must always be borne in mind that any intimacy, no matter how far removed it becomes from the real thing, is still better than the frightening loneliness of no intimacy at all. In other words, we must stop attacking the symptoms and take a closer look instead at the causes of the problem. If only we can become more intimate with our 'intimates', then we should need less and less in the way of substitutes for intimacy. In the meantime, almost any dummy-touch is better than no touch.

8

SELF-INTIMACY

THE WOMAN standing on the railway platform, about to board
a train, is horrified. Her husband has just asked her if she remembered to lock the kitchen door and she has realized that she did not.
What does she do? Before uttering a word, her mouth drops open and
she clasps a palm to one of her cheeks. Even as she starts to speak, the
hand stays there, pressed against the side of her face. Then, after a few
moments, it descends, and the next phase of the behaviour sequence
begins. We will not follow it further, however; instead we will concentrate on that hand, for it is the clue to a whole new world of body
intimacies – intimacies with oneself.

In her fleeting moment of horror, the woman on the platform gave
herself the instantaneous self-comfort of a swift caress – the clasping
of a cheek. Her sudden feeling of emotional distress drove her, unconsciously, to provide the soothing contact which, under other circumstances, a loved one's hand might have offered, or which, long ago, her
parents would have provided when she was a tiny, hurt child. Now, in
place of a lover's or a mother's hand reaching out to touch her cheek,
it is her own hand that flies up to make the contact. It does so
automatically, unthinkingly, and without hesitation. In performing this
act, her cheek has remained *her* cheek, but her hand has symbolically
become someone else's – the lover's or the mother's.

Self-contacts of this kind are a form of body intimacy that we hardly
recognize as such, and yet they are fundamentally the same as the others
discussed in previous chapters. They may appear to be 'one-person'
acts, but in truth they are unconscious mimes of two-person acts, with
part of the body being used to perform the contact movement of the
imaginary companion. They are, in other words, pseudo-interpersonal.

In this respect they provide the fifth, and final, major source of body
intimacies. The five can be illustrated as follows. (1) When we are
feeling nervous or depressed, a loved one may attempt to reassure us
by giving us a comforting hug or a squeeze of the hand. (2) In the
absence of a loved one, it may have to be one of the specialist touchers,
such as a doctor, who pats our arm and tells us not to worry. (3) If our
only company is our pet dog or cat, we may take it in our arms and
press our cheek to its furry body to feel the comfort of its warm touch.

(4) If we are completely alone and some sinister noise startles us in the night, we may hug the bedclothes tightly around us to feel secure in their soft embrace. (5) If all else fails, we still have our own bodies, and we can hug, embrace, clasp and touch ourselves in a great variety of ways to help soothe away our fears.

If you spend some time as an observer, simply watching the way people behave, you will soon discover that acts of self-contact, or auto-contact, are extremely common, much more common than you might at first suppose. It would be wrong, however, to consider all these contacts as substitutes for interpersonal intimacies. Some of them have other functions. A man scratching an itch on his leg, for instance, is not doing it as a mime of someone else doing it to him. He is doing it as a simple act of self-cleaning, performed in its own right, with no hidden intimacy factor. It is important, therefore, not to overstate the case for self-intimacies. In order to put them into their true perspective, it is best to start out with a basic question, namely how and why do we touch our own bodies?

With this question in mind, I analysed several thousand examples of human actions involving self-contact. The first fact to emerge was that the head region was the most important area for *receiving* these contacts, and the hand the most important organ for *giving* them. Although the head is only a small part of the total surface area of the human body, it nevertheless received approximately half the total number of self-contacts.

Surveying these head contacts first, it was possible to identify 650 different types of action. This was done by recording which part of the hand was used, how it made the contact, and which part of the head was involved. It soon became clear that there were four major categories. (The first three, although interesting in their own right, do not directly concern us here and will only be mentioned briefly. Their inclusion is important, however, in order to make it clear that they must be kept separate and not confused with the true self-intimacies.) The four categories are as follows.

1. *Shielding actions.* The hand is brought up to the head to cut off or reduce input to the sense organs. The man who wants to hear less puts his hands over his ears. The man who wants to smell less holds his nose. If the light is too bright, he shields his eyes, and if he cannot bear the sight in front of him, he covers them completely. Similar actions are used to reduce output, as when a hand comes up to cover the mouth and thereby conceal a facial expression.

2. *Cleaning actions.* The hand is brought up to the head to perform a scratch, rub, pick, wipe, or some similar action. A variety of hair-tidying actions also come under this general heading. Some of these

movements are genuine attempts to clean and tidy the head region, but in many instances they are 'nervous' actions, caused by emotional tensions, and are similar to the 'displacement activities' described by ethologists in other species.

3. *Specialized signals.* The hand is brought up to the head to perform a symbolic gesture of some kind. A man who says 'I am fed up to here' holds the back of his hand to the underside of his chin, indicating that he is so full of symbolic 'food' that he cannot take any more. A boy who 'thumbs his nose' presses his thumb to his nose and spreads his fingers out in the shape of a vertical fan. This insult originates from the symbolic act of imitating the comb of a fighting cock, which is how it has become a threatening gesture. This is also the reason why it is sometimes called 'cocking a snook'. Another animal symbolism used as an insult in certain countries is making a pair of horns by holding the hands to the temples, with the forefingers raised and slightly curled. A common form of self-insult is aiming a forefinger at the temple and firing an imaginary gun.

4. *Self-intimacies.* The hand is brought up to the head to perform some action that copies or imitates an interpersonal intimacy. Surprisingly, no fewer than four-fifths of the different hand-to-head actions fall into this self-intimacy category. It seems as if the main reason we have for touching our heads is one of obtaining comfort from unconsciously mimed acts of *being touched by someone else.*

The most common form that this takes is resting part of the head on the hand, with the elbow of the arm in question in contact with a supporting surface, and the forearm acting as a prop to take the weight of the head. It could, of course, be argued that this simply indicates that the neck muscles are tired. A little close observation of these actions soon reveals, however, that physical exhaustion cannot account for the majority of cases.

In this action, the hand is being used as more than a hand. By providing it with support from the elbow, it has become something more solid, and seems instead to be acting as a substitute for the shoulder or chest of the imaginary 'embracing companion'. When, as a child or a lover, we are held in another's arms, we frequently rest the side of our face against their body and feel their soft warmth through the skin of our cheek. By resting the side of our face on our supported hand, we are able to re-create that feeling in their absence, and thereby give ourselves a welcome sensation of comfort and intimacy. Furthermore, since the origin of the act is suitably obscure, we can do this in public without any fear of being thought of as infantile. Sucking a thumb in imitation of childhood breast-feeding might do as well, but there the disguise would wear too thin, and so we tend to avoid it.

Clasping an unsupported hand to the head is also a common act, like the one in the example given of the woman on the railway platform. When this is done, the head cannot lean so heavily, and it appears as if this type of action is more related to the caressing or clasping of the face or hair that is often performed by the embracing companion as an embellishment of the general intimacy. Here, the hand is acting as a symbolic companion's hand, rather than as a symbolic chest or shoulder.

The mouth is a region that receives a great deal of attention, but here the most common action is to touch it in some way with the fingers or thumb, rather than with the whole hand. When making oral contact, the fingers or thumb are being used as substitutes for the mother's breast and nipple. Full thumb-sucking, as I have said, is rare; but modified, less obvious versions of it are common. The simplest modification, and one that is frequently seen, is the pressing of the tip of the thumb between the lips. It is not inserted into the mouth or sucked, but the comforting contact is there nevertheless. The tip, side or back of the forefinger is also used a great deal in this way, and it is often held in contact with the lips for a considerable period of time, while its worried owner gains reassurance from its presence, as dim, unconscious echoes from the infantile past make themselves felt in the brain.

As an elaboration of this form of mouth contact, the finger or thumb is sometimes gently and slowly rubbed across the surface of the lips, re-creating the movements of the baby's mouth on the mother's breast. At more intense moments of anxiety, knuckle-biting and nail-biting put in an appearance. When frustrated aggression is added to the act in this way, it can, in the case of nail-biting, become so persistent as to cause near-mutilation, with the nails bitten down to tiny stumps and the nearby skin chewed raw.

Of all the many different kinds of hand-to-head contact, the most common, in order of frequency, are: (1) jaw rest, (2) chin rest, (3) hair clasp, (4) cheek rest, (5) mouth touch and (6) temple rest. All are performed by both adult males and females, but in two cases there is a strong sexual bias. Hair-clasping is three times as common in women as men, and temple-resting is twice as common in men as in women.

If we leave the head and move down the body, we soon find other forms of self-intimacy. We are all familiar with the tragic newsreel scenes of the aftermath of a disaster such as an earthquake or a mine collapse. A distraught woman in such a situation does not merely clasp one hand to her cheek. The act would be inadequate under the circumstances. Instead she goes much further, hugging her body with her arms and rocking herself pathetically from side to side, as she sits outside the ruins of her home or waits desperately at the pit-head. If she and another sufferer do not find comfort in a mutual embrace, she responds by embracing herself and by rocking herself gently back and forth as her mother once would have done when she was a frightened infant.

This is an extreme case, but we all use a similar device almost every day of our lives, when we fold our arms across our chest. Because the situation is less intense, so is the action, and the folding of arms on the chest is a much weaker form of self-embrace than the full self-hug of misery. It nevertheless provides a mildly comforting sensation of self-intimacy and is typically seen in moments when we are slightly on the defensive. If we are talking in a group of semi-strangers, for example, at a party or some other social gathering, and one of them is coming rather 'too close for comfort', we regain some of our lost comfort by bringing up our arms and folding them across our chest. Usually we are hardly aware that we have performed the act, or that it has any relation to the movements around us, but the fact that it operates in this way has led to its use as an unconscious social signal. For instance, a man who wishes to block a doorway against intruders may stand in front of it, fold his arms across his chest and say, 'No one is allowed inside.' The act of arm-folding in this case, which comforts the man in question, begins to look positively threatening to those in front of him. It signals the fact that he is shutting them out of his embrace, and that he finds self-sufficient strength in his own private act of self-embrace.

Another act of intimacy we all indulge in daily is what can be described as 'holding hands with ourself'. One hand acts as our own, while the other, which clasps or grasps it, acts as the hand of an imaginary companion. We do this in several ways, some more intense than others. When, for example, we are in a particularly strong hand-holding mood with a real companion, we often interlock our fingers with theirs, making the interaction somehow more binding and complex. Similarly, in the absence of such a companion, we can re-create this sensation by interlocking the fingers of our left hand with those of our right. At moments of tension this is sometimes done with such force that the flesh shows white with the great pressure we are unconsciously exerting.

Similar pressures are exerted lower down on the body when we sit with one leg twined tightly around the other. Leg-crossing again seems to provide us with a remarkable degree of self-comfort, providing, as it does, the reassuring pressure of one part of the body against another, and reminding us perhaps of the comforting pressure we felt on our legs when, in a clinging embrace, we straddled the bodies of our parents.

In Victorian times, ladies were expressly forbidden – by the official rules of etiquette then applying – to cross their legs in public or social situations. Victorian males were less restricted in this way, but they were, nevertheless, requested not to hug their knees or feet when performing the act. Today there are no such restrictions, and a random count of a large number of leg-crossings revealed that 53 per cent were female and 47 per cent male, so that no sex difference has survived the passage of time from the last century into this. Two sex differences do exist, however, in the form which the act takes. If it is done by placing the ankle of one leg on the knee or thigh of the other, then it is nearly always a male performance, presumably because for the female this means an undue amount of crotch-exposure. It is intriguing that this applies even where women are wearing trousers, so that it would appear that a trousered woman is still mentally wearing her skirt. The second difference concerns the position of the feet of the crossed legs. If the foot of the 'upper' leg remains in contact with the surface of the 'lower' leg after the legs have been crossed, then the performance is almost always a female one. (The exception to this rule is the low-level ankle/ankle cross, where there are no sex differences, and where the feet are almost bound to be touching one another because of the nature of the act.)

A more intimate form of leg contact is leg-hugging. At its highest intensity this involves bringing the thighs up and the chest down until the two meet. The pressure is increased by embracing the knees or lower legs with the arms. As an addition, the head is lowered on to the knees and the chin or side of the face rested there. In such cases, the bent-up legs are being used as the substitute for the trunk of the imaginary companion, with the knees acting as the chest or shoulders. This is predominantly a female act – of a number of cases recorded at random, 95 per cent were female and only 5 per cent male.

Another typically female action is the clasping of the thigh with the hand, a survey of a large number of such contacts revealing that 91 per cent were female and only 9 per cent male. There appears to be an erotic element present here, the female hand acting as if it were a man's hand placed on her thigh in a sexual context, an act more typical of a courting male than a courting female.

In this survey of self-intimacies, it has nearly always been the hands and arms, and sometimes the legs, that have performed as the active organs, the ones making the contact, but there are a few exceptions to this rule. Sometimes, and again this is typically a female movement, the head is actively lowered on to one shoulder and pressed or rested there, the contact being made by the cheek, jaw or chin. Here it is the shoulder that is being used as the symbolic chest or shoulder of the imaginary companion. Another example concerns the tongue, which may be used to caress the lips, or some other part of the body, certain females even being capable of making contact with their own nipples in this way.

In addition to all these varied methods of making bodily contact with oneself, there is one important aspect of self-intimacy that remains to

496

be discussed, and that is the auto-erotic stimulation usually referred to as masturbation. The word itself appears to be a corruption of *manustuprare*, 'to defile with the hand', and reflects the fact that the most common method of sexual self-stimulation involves a hand-genital contact. For males, this usually means grasping the penis with one hand and rhythmically raising and lowering the arm. The hand then takes on two simultaneous symbolic roles. Its movements up and down the penis mimic the male's own pelvic thrusts, whilst with its grip it acts as a pseudo-vagina. For females, the equivalent action is the stroking of the clitoris with the fingers. Here the fingers are acting as substitutes for the rhythmic pressure applied indirectly to the clitoris by the pelvic thrusting of the male during copulation. Alternative methods for the female are the stroking of the labia or the rhythmic insertion of the fingers into the vagina, with the fingers then acting as a substitute penis. Another technique is thigh-rubbing, in which the thighs are squeezed together, with an alternating tightening and relaxing of the inner muscles to produce a rhythmic pressure on the compressed genitals.

Surveys carried out in the middle of the present century revealed that masturbation is an extremely common form of self-intimacy and is indulged in by the vast majority of individuals at some time in their lives. Although it has always been little more than a harmless substitute for the interpersonal act of copulation, social attitudes towards it have varied considerably at different times. It appears to be widely practised amongst so-called 'primitive tribes', but is usually referred to as something of a joke, indicating that the masturbator is a failed copulator.

An entirely different and much less healthy view was prevalent in our own cultures in earlier centuries, when serious attempts were made to suppress the activity completely. In the eighteenth century, masturbation was denounced as 'the heinous sin of self-pollution'. In the nineteenth century, it became 'the horrid and exhausting vice of self-abuse', and young Victorian ladies were forbidden to wash their genitals in case the gentle friction of such an act, when regularly performed, 'might induce impure thoughts'. The wicked French bidet was not allowed to cross the English Channel. In the early part of the twentieth century, masturbation declined in horror to the level of a 'nasty habit', but religious authorities were still seriously concerned that it might actually give some sensual reward to the masturbator. They did, however, allow that 'the effusion of semen would be legitimate for medical purposes if only it could be achieved without causing pleasure'. By the middle of the twentieth century, attitudes had undergone a dramatic change, and it was at last boldly announced that masturbation is 'a normal and healthy act for a person of any age'. In the past two decades this new approach has continued to gain ground to the point where, in 1971, a respectable women's magazine was able to publish the following words of advice, words which would have astonished a Victorian reader: 'Masturbation . . . is wholesome, normal and sound . . . you are training your body to become a superb instrument of love. Masturbate to your heart's content.'

Today's adolescent who, in the absence of copulatory opportunities, feels inclined to indulge in this form of sexual self-intimacy is lucky indeed. The adolescent of yesterday, far from being freely permitted to perform this activity, was often severely punished for doing so. During the past two centuries all kinds of harsh restraints have been applied, some of which we now find hard to believe. In some cases the young male offender was fitted with a silver ring that was slotted through holes pierced in his foreskin. Alternatively, he might have been equipped with a small penis-belt armed with spikes that automatically pricked the penis if it started to become erect. Blistering the penis with red mercury ointment was another 'remedy' that was sometimes recommended. Both sexes of maturing children were occasionally forced to sleep with their hands tied together or to the bed-posts, to prevent them from 'playing with themselves' at night, or were equipped with modern versions of chastity-belts. Young females might be forced to suffer the mutilation of having the clitoris cauterized or completely removed by surgery, and circumcision for males was advised by some medical authorities as an imagined aid to stamping out the 'evil act' of self-stimulation.

Happily, with the single exception of male circumcision, none of these painful hazards has survived to the present day as a common practice. Society's age-old urge to mutilate its growing juveniles seems, at long last, to be under control. Bearing this in mind, it is worth digressing for a moment to consider why the curious ritual of circumcision should have escaped this general change of attitude. Today the anti-masturbatory excuse is no longer given. Instead, the foreskin of the male infant is amputated for 'religious, medical or hygienic' reasons. The frequency of the operation varies from country to country; in Britain it is thought to be performed on fewer than half the male babies born, whereas in the United States a figure as high as 85 per cent has been quoted.

The medical reason given in favour of foreskin-removal is that it eliminates certain (extremely rare) disease dangers. These only occur, however, if the unmutilated adult male fails to keep his penis reasonably clean by the simple act of pulling back the foreskin and washing the tip of the organ. If this is done regularly, there is, according to medical authorities, no more of a health risk for an uncircumcised male than for a circumcised one. Since the vast majority of foreskin removals are not performed for religious reasons, and since the medical grounds are hardly worth considering, the true reason for the thousands of sexual mutilations carried out on male babies each year remains something of a mystery. Referred to recently by one doctor in America as 'the rape of the phallus', it appears to be a hang-over from our distant cultural past. It has, since early times, been a common practice in most African tribes and was adopted by the ancient Egyptians, whose priest-doctors made sure that no self-respecting male retained his foreskin. Because of the social stigma attached to an attached foreskin, the Jews borrowed the circumcision ritual from the Egyptians and made it even more obligatory for the male members of their religion. In becoming a social or religious 'law', the original significance of the operation had already been forgotten, and it is not easy today to trace it back to its source. Even amongst the African tribes where it is part of elaborate initiation ceremonies, it is usually merely referred to as being 'the custom', but a number of explanations have been forthcoming from modern investigators. One suggestion is that the male foreskin was considered to be a feminine attribute, presumably because it covered up the head of the male organ in the way that the female labia cover the female genital opening. By the same argument, the female's clitoris was considered to be a masculine organ, so that when boys and girls reached sexual maturity, they were both made more true to their sex by having the offending opposite-sex attributes removed. Another suggestion is that the shedding of the foreskin was a symbolic imitation of the shedding of a snake's skin, an action that was widely thought to endow that reptile with immortality, since it reappeared so shining and bright after each shedding. The symbolic equation was straightforward enough: snake = phallus, therefore snakeskin = foreskin.

These and many other ingenious explanations have been put forward, but all seem to be inadequate when the phenomenon of sexual mutilation is viewed as a whole. It has occurred at some time or another in almost every corner of the globe, in literally hundreds of different cultures, and the precise form it has taken has varied considerably. It does not always involve simple foreskin or clitoris removal. In certain cases the parts removed are more extensive, or the mutilations are slits and cuts rather than amputations. In some tribes the female may be stripped of her labia as well as her clitoris, and in others the male may suffer the painful loss of the entire skin surface covering the lower belly, pelvis, scrotum and inner legs, or he may be subjected to the ordeal of having his penis split in two, down its whole length. The only overall common factor seems to be the act by human adults of perpetrating mechanical damage to the genitals of their juniors.

That this ancient form of adult aggression should have survived into present times in the form of male circumcision is something that might bear closer examination by the modern medical profession. Not since the anti-masturbatory assaults of the last century have young females been attacked in this way, presumably because, unlike the males, there was no hygienic justification left for the removal of any part of their genitals. It is fortunate that the situation was not reversed, for if the clitoris could have been proved to be unhygienic and a suitable medical excuse therefore found for its removal, the female would have suffered a considerable loss of sexual responsiveness. Recent careful tests have shown that the penis, by contrast, suffers little or no loss of sensitivity as a result of foreskin-removal, so that males who are mutilated in this way by the respectable modern equivalents of the ancient witch-doctors do not, at least, experience any reduction of sexual performance. These modern tests do, of course, make complete nonsense of the earlier, anti-masturbatory reason for surgically removing the foreskin. Mutilated or unmutilated, the adult male is still going to be able to obtain an unhindered sexual reward from his solitary indulgences in genital self-intimacy.

Summing up, then, it can be said that the reason why male circumcision has survived so widely, when virtually all other forms of archaic genital dismembering have been abandoned in 'civilized' communities, is that it is the only one which does not impair sexual activity and which, at the same time, has been able to acquire a respectable white-wash of medical rationalization.

Returning to masturbation itself, there only remains the question of whether, in the new-found self-stimulation freedom of the latter half of the twentieth century, there are any future hazards waiting in store for us. If we are all advised to 'masturbate to our heart's content' by popular magazine articles, has the pendulum of sexual opinion swung too far? Clearly the earlier rubbish about masturbation causing untold misery and sickness had to be thrown out with a vigorous propaganda campaign, and this has now successfully been done; but is there perhaps a danger that, in sweeping the ridiculous old ideas away, we may go too far in the opposite direction? Masturbation is, after all, a second-rate form of intimacy, like all the substitute social activities discussed in previous chapters. Anything done alone that is a mimic of something done with a companion must, of necessity, be inferior to the genuine act of body intimacy, and this rule must apply to masturbation as much as to any other form of self-intimacy. When there is nothing better available, then of course no justifiable argument can be brought against these substitute activities; but supposing something better is hoped for in the near future, is there not a danger of developing a fixation on the inferior substitute acts which later makes it more difficult to effect a transfer to the real thing?

Contemporary words of advice to a masturbating female stress that every woman should develop her own individual masturbation style, and that it is important to set aside several hours a week so that the new response pattern will become a stable one. She is informed that when she has educated her body in this way she will be able to guide the male, when making love, to positions that give her the maximum sensations. At least this approach is honest: the female works out and stabilizes her self-reward pattern and then it is up to the male partner to service her accordingly. This is recommended as a method of training the female body to become 'a superb instrument of love'. As a system for providing considerable sexual reward for a lonely or frustrated female it may be excellent, but as a system for enhancing love it perhaps leaves something to be desired. It overlooks completely the fact that human copulation is much more than an act of mutual sexual servicing. To approach a moment of intense, reciprocal body intimacy with a previously fixed pattern of reward-demand is to put the cart before the horse. It is no better than using the male's actions as substitutes for masturbation, rather than the other way around. Similarly, if a male has become too heavily fixated on a particular kind of manual masturbation, he may end up using the female vagina as a substitute for his hand, instead of the reverse. To approach copulation in this way is to reduce the partner to a small stimulation device, instead of a complete, intimate and loving person. Over-emphasis on the importance of advanced masturbatory techniques is therefore not perhaps as entirely innocent as the 'new liberalism' would have us believe.

This said, however, it cannot be stressed too strongly that such a warning must in no way be taken as an excuse for a return to the guilt-ridden restrictions of yesterday's forbidden self-intimacies. If the pendulum has perhaps swung a little too far, we are still in a much better position than our immediate ancestors, and we should be grateful for the sexual reformers of the twentieth century who have made this possible. In all probability the dangers of self-intimate fixations will not usually be too serious. If two people come to love one another sufficiently, the emotional intensity of their relationship stands a good chance of sweeping away the rigidity of their previous, solitary patterns of self-gratification, and allowing an increasingly free growth of the sexual interactions that occur between them. If their relationship is less intense and this does not happen, then they will at least be able to enjoy a mutual exchange of their stylized erotic stimulations, which is a good deal better than the Victorian situation, where the marriage partners felt themselves obliged to 'get the nasty business over' as quickly as possible before falling gratefully asleep.

9

RETURN TO INTIMACY

W E ARE BORN into an intimate relationship of close bodily contact with our mothers. As we grow, we strike out into the world and explore, returning from time to time to the protection and security of the maternal embrace. At last we break free and stand alone in the adult world. Soon we start to seek a new bond and return again to a condition of intimacy with a lover who becomes a mate. Once again we have a secure base from which to continue our explorations.

If, at any stage in this sequence, we are poorly served by our intimate relationships, we find it hard to deal with the pressures of life. We solve the problem by searching for substitutes for intimacy. We indulge in social activities that conveniently provide us with the missing body contacts, or we use a pet animal as a stand-in for a human partner. Inanimate objects are enlisted to play the vacant role of the intimate companion, and we are even driven to the extreme of becoming intimate with our own bodies, caressing and hugging ourselves as if we were two people.

These alternatives to true intimacy may, of course, be used as pleasant additions to our tactile lives, but for many they become sadly necessary replacements. The solution seems obvious enough. If there is such a strong demand for intimate contact on the part of the typical human adult, then he must relax his guard and open himself more easily to the friendly approaches of others. He must ignore the rules that say, 'Keep yourself to yourself, keep your distance, don't touch, don't let go, and never show your feelings.' Unfortunately, there are several powerful factors working against this simple solution. Most important of these is the unnaturally enlarged and overcrowded society in which he lives. He is surrounded by strangers and semi-strangers whom he cannot trust, and there are so many of them that he cannot possibly establish emotional bonds with more than a minute fraction of them. With the rest, he must restrict his intimacies to a minimum. Since they are so close to him physically, as he moves about in his day-to-day affairs, this requires an unnatural degree of restraint. If he becomes good at it, he is likely to become increasingly inhibited in *all* his intimacies, even those with his loved ones.

In this body-remote, anti-intimate condition the modern urbanite is in danger of becoming a bad parent. If he applies his contact restraint to his offspring during the first years of their life, then he may cause irreversible damage to their ability to form strong bonds of attachment later on. If, in seeking justification for his inhibited parental behaviour, he (or she) can find some official blessing for such restraint, then it will, of course, help to ease the parental conscience. Unhappily, such blessings have occasionally been forthcoming and have contributed harmfully to the growth of personal relationships within the family.

One example of this type of advice is so extreme that it deserves special mention. The Watsonian method of child-rearing, named after its perpetrator, an eminent American psychologist, was widely followed earlier in this century. In order to get the full flavour of his advice to parents, it is worth quoting him at some length. Here are some of the things he said:

> Mothers just don't know, when they kiss their children and pick them up and rock them, caress them and jiggle them upon their knee, that they are slowly building up a human being totally unable to cope with the world it must later live in . . . There is a sensible way of treating children. Treat them as though they were young adults . . . Never hug or kiss them, never let them sit on your lap. If you must, kiss them once on the forehead when they say goodnight . . . Can't a mother train herself to substitute a kindly word, a smile, in all of her dealings with the child, for the kiss and the hug, the pickup and the coddling? . . . If you haven't a nurse and cannot leave the child, put it out in the backyard a large part of the day. Build a fence around the yard so that you are sure no harm can come to it. Do this from the time it is born . . . If your heart is too tender and you must watch the child, make yourself a peephole so that you can see it without being seen, or use a periscope . . . Finally, learn not to talk in endearing and coddling terms.

Since this was described as treating a child like a young adult, the obvious implication is that the typical Watsonian adults never kiss or hug one another either, and spend their time viewing one another through metaphorical peepholes. This is, of course, precisely what we are all driven to do with the *strangers* who surround us in our daily lives, but to find such conduct seriously recommended as the correct procedure between parents and their babies is, to say the least, remarkable.

The Watsonian approach to child-rearing was based on the behaviourist view, to quote him again, that in man 'There are no instincts.

We build in at an early age everything that is later to appear . . . there is nothing from within to develop.' It therefore followed that to produce a well-disciplined adult it was necessary to start with a well-disciplined baby. If the process was delayed, then 'bad habits' might start to form which would be difficult to eradicate later.

This attitude, based on a totally false premise concerning the natural development of human behaviour in infancy and childhood, would merely be a grotesque historical curiosity were it not for the fact that it is still occasionally encountered at the present day. But because the doctrine lingers on, it requires closer examination. The main reason for its persistence is that it is, in a way, self-perpetuating. If a tiny baby is treated in this unnatural manner it becomes basically insecure. Its high demand for bodily intimacy is repeatedly frustrated and punished. Its crying goes unheeded. But it adapts, it learns – there is no choice. It becomes trained and it grows. The only snag is that it will find it hard ever to trust anyone again, in its entire life. Because its urge to love and be loved was blocked at such a primary stage, the mechanism of loving will be permanently damaged. Because its relationship with its parents was carried on like a business deal, all its later personal involvements will proceed along similar lines. It will not even enjoy the advantage of being able to behave like a cold automaton, because it will still feel the basic biological urge to love welling inside it, but will be unable to find a way of letting it out. Like a withered limb that could not be fully amputated, it will go on aching. If, for conventional reasons, such an individual then marries and produces offspring, the latter will stand a high chance of being treated in the same way, since true parental loving will now, in its turn, have become virtually impossible. This is borne out by experiments with monkeys. If an infant monkey is reared without loving intimacies with its mother, it later becomes a bad parent.

For many human parents the Watsonian regime appeared attractive, but far too extreme. They therefore employed a softened, modified version of it. They would be stern with their baby one moment, then give in the next. In some ways they applied rigid discipline, in others they coddled it. They left it to cry in its cot, but they gave it lots of expensive toys and cooed over it at other times. They forced it into early toilet training, but they kissed and cuddled it. The result, of course, was a totally confused baby which grew into what was called a 'spoilt child'. The fundamental error was then made of ascribing the 'spoiltness', not to the confusion, or to the early baby-stage disciplinary elements, but entirely to the moments of 'softness'. If only they had stuck to the rigid regime and not given in so often, the parents told themselves, then all would have been well. The growing child, now being awkward and demanding, was therefore told to 'behave itself', and discipline was strengthened. The result, at this stage and later, was tantrum and rebellion.

Such a child had seen what love was, in those early 'softer' moments, but, having been shown the entrance, had then had the gate repeatedly slammed in its face. It knew how to love, but it had not been loved enough, and in its later rebellions it repeatedly tested its parents, hoping to prove at last that they loved it no matter what it did – that they loved it for itself and not for its 'good behaviour'. All too often it got the wrong answer.

Even when it got the right answer, and the parents forgave its latest outrage, it still could not believe that all was well. The early imprints were too deeply engraved, the early, intermittent disciplines too unloving for a baby's mind. So it tested them again, going further and further in its desperate attempt to prove that, after all, they really did love it. Then the parents, faced with chaos, either finally applied strict discipline and confirmed the child's darkest fears, or they gave in over and over again, condoning increasingly anti-social acts out of a sense of dawning guilt – 'Where did we go wrong, how have we failed? We have given you everything.'

All this could have been avoided if only the baby had been treated as a baby in the first place, instead of a 'young adult'. During the first years of life, an infant requires total love, nothing less. It is not 'trying to get the better of you', but it does need the best of you. If the mother is unstressed, and has not herself been warped in infancy, she will have a natural urge to give her best, which is why, of course, the disciplinarian has to repeatedly warn mothers against giving in to those tender 'weaknesses' that 'tug at their heart-strings', to use a favourite Watsonian phrase. If the mother is under pressure, as a result of our modern way of life, it will not be so easy; but even so, without an artificially imposed regime, it is still not impossible to come close enough to the ideal to produce a happy and well-loved baby.

Far from growing into a 'spoilt child', such an infant will then be able to mature into an increasingly independent individual, remaining loving, but with no inhibitions about investigating the exciting world around it. The early months gave it the assurance that there is a truly safe and secure base from which to venture forth and explore. Again, experiments with monkeys bear this out. The infant of a loving monkey mother readily moves off to play and test the environment. The offspring of a non-loving mother is shy and nervous. This is the exact opposite of the Watsonian prediction, which expects that an 'excess' of early loving, in the intimate, bodily sense, will make for a soft, dependent creature in later years. The lie to this can even be seen by the time the human child has reached the third year of life. The infant that was lavished with love during its first two years already begins to show its paces, launching out into the world with great, if unsteady, vigour. If it falls flat on its face it is not more, but less, likely to cry. The infant that was less loved and more disciplined as a tiny baby is already less adventurous, less curious about what it sees, and less inclined to start making the first fumbling attempts at independent action.

In other words, once a totally loving relationship has been established in the first two years of life, the infant can readily move on to the next stage in its development. As it grows, however, its headlong rush to explore the world *will*, at this later phase, require some discipline from the parents. What was wrong at the baby stage now becomes right. The Watsonian distaste for the doting, over-protective parents of *older* children has some justification, but the irony is that where protection of this type occurs to excess, it is probably a reaction against the damage caused by earlier Watsonian baby-training. The child that was a fully loved baby is less likely to provoke such behaviour.

During later life the adult who, as a baby, formed a strong bond of attachment with its parents in the primary phase of total love will also be better equipped to make a strong sexual bond of attachment as a young adult and, from this new 'safe base', to continue to explore and lead an active, outgoing, social life. It is true that, in the stage before an adult bond of attachment has formed, he or she will be much more sexually exploratory as well. All exploring will have been accentuated, and the sexual sphere will be no exception. But if the individual's early life has been allowed to pass naturally from stage to stage, then the sexual explorations will soon lead to pair-formation and the growth of a powerful emotional bond, with a full return to the extensive body intimacies typical of the loving baby phase.

Young adults who establish new family units and enjoy uninhibited intimacies within them will be in a much better position to face the harsh, impersonal world outside. Being in a 'bond-ful', rather than a bond-starved, condition, they will be able to approach each type of social encounter on its own terms and not make inappropriate, bond-hungry demands in situations which, inevitably, will so often require emotional restraint.

One aspect of family life that cannot be overlooked is the need for privacy. It is necessary to have private space in order to enjoy intimate contacts to the full. Severe overcrowding in the home makes it difficult to develop any kind of personal relationship except a violent one. Bumping into one another is not the same as performing a loving embrace. Forced intimacy becomes anti-intimate in the true sense, so that, paradoxically, we need more space to give body contact greater meaning. Tight architectural planning that ignores this fact creates unavoidable emotional tension. For personal body intimacy cannot be a permanent condition, like the persistent impersonal crowding of the urban world outside the home. The human need for close bodily contact is spasmodic, intermittent, and only requires occasional expression. To cramp the home-space is to convert the loving touch into a suffocating body proximity. If this seems obvious enough, then it is hard to understand the lack of attention that has been given to private home-space by the planners of recent years.

In painting this picture of the 'intimate young adults', I may have given the impression that, if only they can acquire an adequate private home-space, have a loving infancy behind them, and have formed strong new bonds of attachment to one another, then all will be well. Sadly, this is not the case. The crowded modern world can still encroach on their relationship and inhibit their intimacies. There are two powerful social attitudes that may influence them. The first is the one that uses the word 'infantile' as an insult. Extensive body intimacies are criticized as regressive, soft or babyish. This is something that can easily deter a potentially loving young adult. The suggestion that to be too intimate constitutes a threat to his independent spirit, summed up in such sayings as 'the strongest man is the man who stands alone', begins to make an impact. Needless to say, there is no evidence that for an adult to indulge in body contacts typical of the infant stage of life necessarily means he will find his independence impaired at other times. If anything, the contrary is the case. The soothing and calming effects of gentle intimacies leave the individual freer and better equipped emotionally to deal with the more remote, impersonal moments of life. They do not soften him, as has so often been claimed; they strengthen him, as they do with the loved child who explores more readily.

The second social attitude that tends to inhibit intimacies is the one which says that bodily contact implies sexual interest. This error has been the cause of much of the intimacy restraint that has been needlessly applied in the past. There is nothing implicitly sexual about the intimacies between parent and child. Parental love and infantile love are not sexual love, nor need the love between two men, two women, or even between a particular man and a particular woman be sexual. Love is love – an emotional bond of attachment – and whether sexual feelings enter into it or not is a secondary matter. In recent times we have somehow come to overstress the sexual element in all such bonds. If a strong, primarily non-sexual bond exists, but with minor sexual feelings accompanying it, the latter are automatically seized upon and enlarged out of all proportion in our thinking. The result has been a massive inhibition of our non-sexual body intimacies, and this has applied to relationships with our parents and offspring (beware, Oedipus!), our siblings (beware, incest!), our close same-sex friends (beware, homosexuality!), our close opposite-sex friends (beware, adultery!), and our many casual friends (beware, promiscuity!). All of this is understandable, but totally unnecessary. What it indicates is that in our true sexual relationships we are, perhaps, not enjoying a sufficiently erotically exhausting degree of body intimacy. If our pair-bond sexual intimacies were intensive and extensive enough, then there should be none left over to invade the other types of bond relationships, and we could all relax and enjoy them more than we seem to dare to do at present. If we remain sexually inhibited or frustrated with our mates, then of course the situation is quite different.

The general restraint that is applied to non-sexual body contacts in modern life has led to some curious anomalies. For example, recent American studies have revealed that in certain instances women are driven to use random sex simply for the purpose of being held in someone's arms. When questioned closely, the women admitted that this was sometimes their sole purpose in offering themselves sexually to a man, there being no other way in which they could satisfy their craving for a close embrace. This illustrates with pathetic clarity the distinction between sexual and non-sexual intimacy. Here there is no question of body intimacy leading to sex, but of sex leading to body intimacy, and this complete reversal leaves no doubt about the separation between the two.

These, then, are some of the hazards facing the modern intimate adult. To complete this survey of intimate human behaviour, it remains to ask what signs of change there are in the attitudes of contemporary society.

At the infant level, thanks to much painstaking work by child psychologists, a greatly improved approach to the problems of child-rearing is being developed. A much better understanding now exists of the nature of parent/offspring attachments, and of the essential role that warm loving takes in the production of a healthy growing child. The rigid, ruthless disciplines of yesterday are on the wane. However, in our more overcrowded urban centres, the ugly phenomenon of the 'battered baby syndrome' remains with us to remind us that we still have a long way to go.

At the level of the older child, constant gradual reforms are taking place in educational methods, and a more sensitive appreciation is growing of the need for social as well as technical education. The demands for technological learning are, however, heavier than ever, and there is still a danger that the average schoolchild will be better trained to cope with facts than with people.

Amongst young adults, the problem of handling social encounters seems, happily, to be solving itself. It is doubtful whether there has ever before been a period of such openness and frankness in dealing with the intricacies of personal interaction. Much of the criticism of the conduct of young adults, on the part of the older generation, stems from a heavily disguised envy. It remains to be seen, however, how well the new-found freedom of expression, sexual honesty and disinhibited intimacies of present-day youth survive the passage of time and approaching parenthood. The increasingly impersonal stresses of later adult life may yet take their toll.

Amongst older adults there is clearly a growing concern about the survival of resolved personal life inside the ever-expanding urban communities. As public stress encroaches more and more on private living, a mounting alarm can be felt concerning the nature of the modern human condition. In personal relationships, the word 'alienation' is constantly heard, as the heavy suits of emotional armour, put on for social battle in the streets and offices, become increasingly difficult to remove at night.

In North America, the sounds of a new rebellion against this situation can now be heard. A new movement is afoot, and it provides an eloquent proof of the burning need that exists in our modern society for a revision of our ideas concerning body contact and intimacy. Known in general terms as 'Encounter Group Therapy', it has appeared only in the last decade, beginning largely in California and spreading rapidly to many centres in the United States and Canada. Referred to in American slang as 'Bod Biz' (for 'show business' read 'body business'), it goes under a number of official titles, such as 'Transpersonal Psychology', 'Multiple Psychotherapy' and 'Social Dynamics'.

The principal common factor is the bringing together of a group of adults for sessions lasting from roughly one day to one week, in which they indulge in a wide variety of personal and group interactions. Although some of these are largely verbal, many are non-verbal and concentrate instead on body contacts, ritual touchings, mutual massage, and games. The aim is to break down the façade of civilized adult conduct, and to remind people that they 'do not *have* bodies; they *are* bodies'.

The essential feature of these courses is that inhibited adults are encouraged to play like children again. The avant-garde scientific atmosphere licenses them to behave in an infantile manner without embarrassment or fear of ridicule. They rub, stroke and tap one another's bodies; they carry one another around in their arms and anoint one another with oil; they play childlike games and they expose themselves naked to one another, sometimes literally, but usually metaphorically.

This deliberate return to childhood is explicitly expressed in the following words, in connection with a four-day course entitled 'Become as You Were':

> The adjusted American achieves a dubious state of 'maturity' by burying many child parts under layers of shame and ridicule. Relearning how to be a child may enrich the man's experience of being masculine and the woman's experience of being feminine. Re-experiencing being child with mother may shed light on one's approaches to loving, love-making and love-seeking. Paradoxically, making contact with childish helplessness releases surges of power and contacting childish tears opens the channels for expression and joy.

Other similar courses called 'Becoming More Alive through Play' and 'Sensory Reawakening: Rebirth' also emphasize the need to return to the intimacies of childhood. In some cases the process is taken even further with the use of 'womb-pools' kept at precisely uterine temperature.

The organizers of these courses refer to them as 'therapy for normal people'. The visitors are not patients; they are group members. They go there because they are urgently seeking some way of finding a return to intimacy. If it is sad to think that modern, civilized adults should need official sanction to touch one another's bodies, then it is at least reassuring that they are sufficiently aware that something has gone wrong to actively do something about it. Many of the people who have been through such sessions repeatedly return for more, since they find themselves loosening up emotionally and unwinding in the course of the ritual body contacts. They report a sense of release and a growing feeling of warmth in connection with their personal interactions at home.

Is this a valuable new social movement, a passing fad, or a dangerous, new, drugless addiction?* With dozens of new centres opening up every month, expert opinions are varied. Some psychologists and psychiatrists vigorously support the encounter-group phenomenon, others do not. One argues that group members 'don't improve – they just get a maintenance dose of intimacy'. If this is true, then even so the

---

* The encounter group movements of the 1970s have since gone into decline, partly because of the successful encouragement of more extensive sexual intimacy within the private world of the paired couple, and partly because Aids has inhibited more widespread body intimacies.

courses may at least see certain individuals through a difficult phase in their social lives. This puts group attendance at the intimacy level of going dancing, or going to bed with a cold and being comforted there, but there is nothing wrong with that. It merely adds one more string to the bow of a person seeking a 'licensed to touch' context. Other criticisms, however, are more severe. 'The techniques that are supposed to foster real intimacy sometimes destroy it,' says one. A theologist, no doubt sensing a new form of serious competition, comments that all that people learn in encounter groups is 'new ways to be impersonal – a new bag of tricks, new ways to be hostile and yet appear friendly'.

It is certainly true that, listening to the leaders of the movement talking to the general public about their methods and their philosophy, there is sometimes an unmistakable air of smug condescension. They give the impression of having discovered the secret of the universe, which they are gracious enough to impart to other, lesser mortals. This point has been stressed as a serious criticism by some, but it is probably no more than a defence against anticipated ridicule. It is reminiscent of the tactics of the world of psychoanalysis in earlier days. Like encounter-group veterans, those who had been through analysis could not help smiling smugly down at those who had not. But analysis is past this stage now, and if encounter groups survive the novelty phase, the attitude will no doubt change, as the new cult matures to become an accepted pattern.

The more severe criticism that the group sessions actually do serious harm has yet to be proved. 'Instant intimacy', as it has been called, does, however, have its hazards for the returning devotee when he steps back, fully or partially 'reawakened', into his old environment. He has been changed, but his home companions have not, and there is a danger that he may make insufficient allowance for this difference. The problem is essentially one of competing relationships. If an individual visits an encounter group, has himself massaged and stroked by total strangers, plays intimate games with them, and indulges in a wide variety of body contacts, then he is doing more with them than he will have been doing with his true 'intimates' in his home setting. (If he is not, then he had no problem in the first place.) If – as will inevitably happen – he later describes his experiences in glowing detail, he is automatically going to arouse feelings of jealousy. Why was he prepared to act like that at the encounter centre, when he was so remote and untouching at home? The answer, of course, was the official, scientific sanction for such acts in the special atmosphere of the centre, but that is no comfort to his 'real life' intimates. Where couples attend intimacy sessions together, the problem is greatly reduced, but the 'back home' situation still requires careful handling.

Some have argued that the most distasteful aspect of the encounter groups is the way in which they are converting something which should be an unconscious part of everyday life into a self-conscious, highly organized, professional pursuit, with the act of intimacy in danger of becoming an end in itself, rather than as one of the basic means by which we can intuitively help ourselves to face the outside world.

Despite all these understandable fears and criticisms, it would be wrong to scorn this intriguing new trend. Essentially, its leaders have seen an increasing and damaging shift towards impersonality in our personal relationships and have done their best to reverse this process. If, as so often happens, by the 'law of reciprocal errors', they are swinging the pendulum rather wildly in the opposite direction, then this is a minor fault. If the movement spreads and grows to a point where it becomes a matter of common knowledge, then, even for the non-enthusiasts, it will exist as a constant reminder that something is wrong with the way in which we are using – or, rather, not using – our bodies. If it does no more than make us aware of this, it will be serving its purpose. Again, the comparison with psychoanalysis is relevant. Only a small proportion of the general population have ever been directly involved in analysis, and yet the basic idea that our deepest, darkest thoughts are not shameful or abnormal, but are probably shared by most others, has permeated healthily throughout our culture. In part, it is responsible for the more honest and frank approach to mutual personal problems in young adults today. If the encounter-group movement can provide the same indirect release for our inhibited feelings concerning intimate bodily contact, then it will ultimately have proved to have made a valuable social contribution.

The human animal is a social species, capable of loving and greatly in need of being loved. A simple tribal hunter by evolution, he finds himself now in a bewilderingly inflated communal world. Hemmed in on all sides, he defensively turns in on himself. In his emotional retreat, he starts to shut off even those who are nearest and dearest to him, until he finds himself alone in a dense crowd. Unable to reach out for emotional support, he becomes tense and strained and possibly, in the end, violent. Lost for comfort, he turns to harmless substitutes for love that ask no questions. But loving is a two-way process, and in the end the substitutes are not enough. In this condition, if he does not find true intimacy – even if it is only with one single person – he will suffer. Driven to armour himself against attack and betrayal, he may have arrived at a state in which all contact seems repellent, where to touch or to be touched means to hurt or be hurt. This, in a sense, has become one of the greatest ailments of our time, a major social disease of modern society that we would do well to cure before it is too late. If the danger remains unheeded, then – like poisonous chemicals in our food – it may increase from generation to generation until the damage has gone beyond repair.

In a way, our ingenious adaptability can be our social undoing. We are capable of living and surviving in such appallingly unnatural conditions that, instead of calling a halt and returning to a saner system, we adjust and struggle on. In our crowded urban world, we have battled on in this way, further and further from a state of loving, personal intimacy, until the cracks have begun to show. Then, sucking our metaphorical thumbs and mouthing sophisticated philosophies to convince ourselves that all is well, we try to sit it out. We laugh at educated adults who pay large sums to go and play childish games of touch and hug in scientific institutes, and we fail to see the signs. How much easier it would all be if we could accept the fact that tender loving is not a weakly thing, only for infants and young lovers, if we could release our feelings, and indulge ourselves in an occasional, and magical, return to intimacy.

CHAPTER REFERENCES

It is impossible to list all the many works that have been of assistance in writing *Intimate Behaviour*. I have therefore included only those which either have been important in providing information on a specific point, or are of particular interest for further reading. They are arranged below on a chapter-by-chapter and topic-by-topic basis. From the names and the dates given, it is possible to trace the full references in the bibliography that follows. The bibliography also contains the titles of several broader works, not listed below against specific topics, which have proved of value in relation to the general subject.

Most of the statements concerning body contacts are, however, based on my own personal observations. Where specific statements are made, such as that a particular action is, say, three times as common in females as in males, they are based on a sample of 10,000 randomly selected units of behaviour. An archive of human contact actions, based on this study, was compiled and used as the basis for future publications, such as *Manwatching* and *Bodywatching*.

1 THE ROOTS OF INTIMACY

Foetal behaviour: Munn, 1965; Tanner and Taylor, 1966.
Behaviour at birth: Prechtl, 1965; Smith, 1968.
Rocking: Ambrose, in Bowlby, 1969; Bowlby, 1969; Morris, 1967.
Heartbeat: Morris, 1967; Salk, 1966.
Swaddling: Smith, 1968; Spock, 1946.
Discipline: Watson, 1928.
Crying and smiling: Ambrose, 1960; Bowlby, 1969.
Transitional objects: Spock, 1963; Vosper, 1969.
Adolescence: Cohen, 1964; Freud, 1946.

2 INVITATIONS TO SEXUAL INTIMACY

Crotch: Morris, 1967.
Codpiece: Broby-Johansen, 1968; Fryer, 1963; Rabelais, 1653.
Human self-mimicry: Morris, 1967; Morris and Morris, 1966a; Wickler, 1967.

Prehistoric buttocks: Ucko, 1968.
Bustle: Broby-Johansen, 1968; Laver, 1963.
Breast variations: Ford and Beach, 1952; Levy, 1962.
Depilation: Gould and Pyle, 1896.
Chin: Hershkovitz, 1970.
Blush: Darwin, 1873.
Belladonna: Wedeck, 1962.
Pupil dilation: Coss, 1965; Hess, 1965.
Hair in Egypt: Murray, 1949.
Dancing: Bloch, 1958; Fryer, 1963; Licht, 1932.

3 SEXUAL INTIMACY

General: Kinsey *et al.*, 1948, 1953; Masters and Johnson, 1966; Morris, 1967, 1969.
Oral-genital contacts: Legman, 1969.
Sexual prudery: Fryer, 1963.
Primate copulation: Carpenter, 1934, 1942; Goodall, 1965; Hall and DeVore, 1965; Morris and Morris, 1966; Reynolds and Reynolds, 1965; Simonds, 1965; Southwick, Beg and Siddiqui, 1965; Yerkes, 1943.
Cross-cultural variations: Ford and Beach, 1952; Rachewiltz, 1964.

4 SOCIAL INTIMACY

Baby clapping: Ainsworth, 1964.
Hand-waving nationalities: Brun, 1969.
Men holding hands: Froissart, 1940.
Non-nutritional sucking: Bowlby, 1969; Wolff, 1969.
Bow and curtsy: Wildeblood and Brinson, 1965.
Religious kiss: Beadnell, 1942; Wildeblood and Brinson, 1965.
Handshake: Sorell, 1968; Wildeblood and Brinson, 1965.
Modern etiquette: Lyons, 1967; Page, 1961; Sara, 1963; Vanderbilt, 1952.

5 SPECIALIZED INTIMACY

Microbes on man: Rosebury, 1969.
Physical illness: Morris, 1971.
Mental illness: Szasz, 1961.

Hairdressers: Williams, 1957.
Dancing: Bloch, 1958; Brend, 1936; Fryer, 1963; Lewinsohn, 1958; Licht, 1932.
Midwives: Forbes, 1966; Fryer, 1963.
Groupie girls: Fabian and Byrne, 1969.

6 SUBSTITUTES FOR INTIMACY

Man/animal relationships: Morris, 1967, 1969; Morris and Morris, 1965, 1966a, 1966b.
Petishism: Szasz, 1969.
Animal experiments: Heim, 1971; Matthews, 1964; Russell and Burch, 1959.

7 OBJECT INTIMACY

Dummies and pacifiers: Bowlby, 1969; Spock, 1946.
Hand sculpture: Miller, 1942.
Dildoes: Bauer, 1926; Dearborn, 1961; Henriques, 1963; Masters and Johnson, 1966.
Copulation machine: Lacey, 1967.

8 SELF-INTIMACY

Female leg-crossing: Birdwhistell, 1970.
Masturbation: Comfort, 1967; Dearborn, 1961; 'J', 1970; Kinsey *et al.*, 1948, 1953; Lewinsohn, 1958; Malinowski, 1929; Masters and Johnson, 1966; Morris, 1969.
Circumcision: Comfort, 1967; Lewinsohn, 1958; Mollon, 1965; Morris and Morris, 1965; Rachewiltz, 1964; Smith, 1968; Spock, 1946; West, 1966.

9 RETURN TO INTIMACY

Child discipline: Watson, 1928.
Monkey rearing: Harlow, 1958.
Overcrowding: Morris, 1969; Russell and Russell, 1968.
Sex as comfort: Hollender *et al.*, 1969; Hollender, 1970.
Encounter-group therapy: Gunther, 1969; Howard, 1970.

BIBLIOGRAPHY

Adams, A. L., *Notes of a Naturalist in the Nile Valley and Malta* (Edmonston and Douglas, 1870)

Ainsworth, M. D. 'Patterns of infantile attachment to the mother', *Merrill-Palmer Quart.* 10 (1964), pp. 51–8

Ambrose, J. A., 'The smiling response in early human infancy' (Ph.D. thesis, London University, 1960), pp. 1–660

Appelman, F. J., 'Feeding of zoo animals by the public', in *Internat. Zoo Yearbook* 2 (1960), pp. 94–5

Ardrey, R., *African Genesis* (Atheneum, 1961)

—— *The Territorial Imperative* (Collins, 1967)

Argyle, M. *Social Interaction* (Methuen, London, 1969)

Bastock, M., D. Morris and M. Moynihan, 'Some comments on conflict and thwarting in animals', in *Behaviour* 6 (1953), pp. 66–84

Bataille, G., *Eroticism* (Calder, 1962)

Bauer, B. A. *Woman* (Cape, London, 1926)

Beach, F. A., *Sex and Behaviour* (Wiley, 1965)

Beadnell, C. M. *The Origin of the Kiss* (Watts, London, 1942)

Berelson, B., and G. A. Steiner, *Human Behavior* (Harcourt, Brace and World, 1964)

Berkowitz, L., *Aggression* (McGraw-Hill, 1962)

Berlyne, D. E., *Conflict, Arousal and Curiosity* (McGraw-Hill, 1960)

Birdwhistell, R. L. *Kinesics and Context* (University of Pennsylvania Press, Philadelphia, 1970)

Bloch, I. *Sexual Life in England Past and Present* (Arco, London, 1958)

Boule, M., and H. V. Vallois, *Fossil Men* (Thames & Hudson, 1957)

Boullet, J., *Symbolisme Sexuel* (Pauvert, 1961)

Bowlby, J. *Attachment and Loss* (Hogarth Press, London, 1969)

Brackbill, Y., and G. G. Thompson, *Behavior in Infancy and Early Childhood* (Free Press, 1967)

Brend, W. A. *Sacrifice to Attis* (Heinemann, London, 1936)

Broby-Johansen, R. *Body and Clothes* (Faber, London, 1968)

Broca, P., *On the Phenomena of Hybridity in the Genus Homo* (Longman, Green, Longman & Roberts, 1864)

Brun, T. *The International Dictionary of Sign Language* (Wolfe, London, 1969)

Caine, M., *The S-Man* (Hutchinson, 1960)

Calhoun, J. B., 'A "behavioral sink",' in *Roots of Behaviour*, (ed. E. L. Bliss) (Harper and Brothers, New York, 1962), pp. 295–315

Cannon, W. B., *Bodily Changes in Pain, Hunger, Fear and Rage* (Appleton-Century, New York, 1929)

Carpenter, C. R. 'A field study of the behavior and social relations of Howling Monkeys', *Comp. Psychol. Monogr.* 10 (1934), pp. 1–168

—— 'Sexual behaviour of free ranging Rhesus Monkeys', *J. Comp. Psychol.* 33 (1942), pp. 113–62

Carthy, J. D., and F. J. Ebling, *The Natural History of Aggression* (Academic Press, 1964)

Chance, M. R. A., 'An interpretation of some agonistic postures; the role of cut-off acts and postures', in *Symp. Zool. Soc. London* 8 (1962), pp. 71–89

Clark, G., and S. Piggott, *Prehistoric Societies* (Hutchinson, 1965)

Clark, W. E. Le Gros, *The Antecedents of Man* (Edinburgh University Press, 1959)

Cohen, Y. A., *The Transition from Childhood to Adolescence* (Aldine, 1964)

Colbert, E. H., *Evolution of the Vertebrates* (Wiley, New York, 1955)

Cole, S., *The Neolithic Revolution* (British Museum, 1959)

Comfort, A., *Nature and Human Nature* (Weidenfeld and Nicolson, 1966)

—— *The Anxiety Makers* (Nelson, London, 1967)

Coon, C. S., *The Origin of Races* (Cape, 1963)

—— *The Living Races of Man* (Cape, 1963)

Coss, R. G., *Mood Provoking Visual Stimuli* (University of California, 1965)

Crawley, E., *Dress, Drinks and Drums* (Methuen, 1931)

Dart, R. A. and D. Craig, *Adventures with the Missing Link* (Hamish Hamilton, 1959)

Darwin, C. *The Expression of the Emotions in Man and Animals* (Murray, London, 1873)

Dearborn, L. W. 'Autoerotism', in *The Encyclopedia of Sexual Behavior* (Hawthorn, New York, 1961)

Eimerl, S. and I. DeVore, *The Primates* (Time Life, New York, 1965)

Fabian, J., and J. Byrne. *Groupie* (New English Library, London, 1969)

Fast, J. *Body Language* (Evans, New York, 1970)

Forbes, T. R. *The Midwife and the Witch* (Yale University Press, New Haven, 1966)

Ford, C. S., and F. A. Beach, *Patterns of Sexual Behaviour* (Eyre & Spottiswoode, 1952)

Frank, L. K. 'Tactile Communication', in Carpenter and McLuhan (eds), *Explorations in Communication* (Cape, London, 1970), pp. 4–11

Freeman, G., *The Undergrowth of Literature* (Nelson, 1967)

Fremlin, J. H., 'How many people can the world support?' in *New Scientist* 24 (1965), pp. 285–7

Freud, A. *The Ego and Mechanisms of Defense* (International Universities Press, New York, 1946)

Froissart, J. *The Chronicles of England, France and Spain* (Everyman Library, London, 1940)

Fryer, P. *Mrs Grundy* (Dobson, London, 1963)

Goodall, J. 'Chimpanzees of the Gombe Stream Reserve', in DeVore (ed.), *Primate Behavior* (Holt, Rinehart and Winston, New York, 1965)

Gould, G. M. and W. L. Pyle, *Anomalies and Curiosities of Medicine* (Saunders, Philadelphia, 1896)

Guggisberg, C. A. W., *Simba. The Life of the Lion* (Bailey Bros. and Swinfen, 1961)

Gunther, M., 'Instinct and the nursing couple'. *Lancet* (1955), pp. 575–8

Gunther, B. *Sense Relaxation* (Macdonald, London, 1969)

Hall, K. R. L., and I. DeVore, 'Baboon social behaviour', in *Primate Behavior* (Editor: I. DeVore), (Holt, Rinehart, Winston, 1965)

Hardy, A. C., 'Was man more aquatic in the past?' *New Scientist* 7 (1960), pp. 642–5

Harlow, H. F., 'The nature of love'. *Amer. Psychol.* 13 (1958), pp. 673–85

Harlow, H. H., and M. K. Harlow, 'Social deprivation in monkeys', in *Sci. Amer.* 207 (1962) pp. 136–46

—— —— 'The effect of rearing conditions on behaviour', in *Bull. Menninger Clin.* 26 (1962), pp. 213–24

Harrison, G. A., J. S. Weiner, J. M. Tanner and N. A. Barnicott, *Human Biology* (Oxford University Press, 1964)

Hartwich, A., *Aberrations of Sexual Life* (After the *Psychopathia Sexualis* of Kraft-Ebing), (Staples Press, 1959)

Hass, H. *The Human Animal* (Putnam, New York, 1970)

Hayes, C., *The Ape in our House* (Gollancz, 1952)

Hediger, H., *Wild Animals in Captivity* (Butterworth, 1950)

—— 'Environmental factors influencing the reproduction of zoo animals', in *Sex and Behaviour* (Editor: F. A. Beach), (Wiley, 1965)

Heim, A. *Intelligence and Personality* (Pelican, London, 1971)

Henriques, F. *Prostitution in Europe and the New World* (MacGibbon & Kee, London, 1963)

Hershkovitz, P. 'The decorative chin', *Bull. Field Mus. Nat. Hist.* 41 (1970), pp. 6–10

Hess, E. H. 'Attitude and pupil size', *Sci. Amer.* 212 (1965), pp. 46–54

Hollender, M. H. 'The need or wish to be held', *Arch. Gen. Psychiat.* 22 (1970), pp. 445–53

—— L. Luborsky and T. J. Scaramella. 'Body contact and sexual excitement', *Arch. Gen. Psychiat.* 20 (1969), pp. 188–91

Hooton, E. A., *Up from the Ape* (Macmillan, New York, 1947)

Howard, J. *Please Touch* (McGraw-Hill, New York, 1970)

Howells, W., *Mankind in the Making* (Secker and Warburg, 1960)

Hutt, C. and M. J. Vaizey, 'Differential effects of group density on social behaviour'. *Nature* 209 (1966), pp. 1371–2

Inhelder, E., 'Skizzen zu einer Verhaltenspathologie reactiver Störungen bei Tieren', in *Schweiz. Arch. Neurol. Psychiat.* 89 (1962), pp. 276–326

'J'. *The Sensuous Woman* (Lyle Stuart, New York, 1970)

Jennison, G., *Animals for Show and Pleasure in Ancient Rome* (Manchester University Press, 1937)

Jourard, S. M. 'An exploratory study of body accessibility', *Brit. J. soc. clin. Psychol.* 5 (1966), pp. 221–31

Kellogg, R., *What Children Scribble and Why* (Author's edition, San Francisco, 1955)

Kinsey, A. C., W. B. Pomeroy and C. E. Martin, *Sexual Behavior in the Human Male* (Saunders, 1948)

—— —— —— and P. H. Gebhard, *Sexual Behavior in the Human Female* (Saunders, 1953)

Kleiman, D., 'Scent marking in the Canidae'. *Symp. Zool. Soc.* 18 (1966), pp. 167–77

Kleitman, N., *Sleep and Wakefulness* (Chicago University Press, 1963)

Knight, R. P., and T. Wright, *Sexual Symbolism* (Julian Press, 1957)

Kruuk, H., 'Clan-system and feeding habits of Spotted Hyenas'. *Nature* 209 (1966), pp. 1257–8

Lacey, B. 'An evening with Bruce Lacey' (Lecture-demonstration at the Institute of Contemporary Arts, London, 1967)

Lang, E. M., 'Eine ungewöhnliche Stereotype bei einem Lippenbären', in *Schweiz. Arch. Tierheilk.* 85 (1943), pp. 477–81

Laver, J., *Dress* (John Murray, 1950)
—— *Clothes* (Burke, 1952)
—— *Costume* (Cassell, 1963)
—— *Modesty in Dress* (Heinemann, London, 1969)
Legman, G. *Rationale of the Dirty Joke* (Cape, London, 1969)
Levy, M. *The Moons of Paradise* (Barker, London, 1962)
Lewinsohn, R. *A History of Sexual Customs* (Longmans, Green, London, 1958)
Leyhausen, P., *Verhaltensstudien an Katzen* (Parey, 1956)
Licht, H. *Sexual Life in Ancient Greece* (Routledge & Kegan Paul, London, 1932)
Lipsitt, L., 'Learning processes of human newborns'. *Merril-Palmer Quart. Behav. Devel.* 12 (1966), pp. 45–71
Lorenz, K., 'Der Kumpan in der Umwelt des Vogels', in *J. f. Ornith.* 83 (1935), pp. 137–213, 289–413
—— *King Solomon's Ring* (Methuen, 1952)
—— *Man Meets Dog* (Methuen, 1954)
—— *On Aggression* (Methuen, 1966)
Lowen, A. *Physical Dynamics of Character Structure* (Grune & Stratton, New York, 1958)
Lyall-Watson, M., 'A critical re-examination of food "washing" behaviour in the raccoon', in *Proc. Zool. Soc. London* 141 (1963), pp. 371–94
Lyons, P. *Today's Etiquette* (Bancroft, London, 1967)
Malinowski, B. *The Sexual Life of Savages* (Routledge & Kegan Paul, London, 1929)
Marks, I. M. and M. G. Gelder, 'Different onset ages in varieties of phobias'. *Amer. J. Psychiat.* (July 1966)
Masters, W. H. and V. E. Johnson, *Human Sexual Response* (Churchill, 1966)
Matthews, L. H. 'Animal Relationships', *Med. Sci. and Law* (1964), pp. 4–14
Miles, W. R., 'Chimpanzee behaviour: removal of foreign body from companion's eye'. *Proc. Nat. Acad. Sci.* 49 (1963), pp. 840–3
Miller, D. C. (ed.) *Americans 1942* (Museum of Modern Art, New York, 1942)
Mollon, R. *The Nursery Book* (Pan, London, 1965)
Monicreff, R. W., 'Changes in olfactory preferences with age'. *Rev. Laryngol.* (1965), pp. 895–904
Montagna, W., *The Structure and Function of Skin* (Academic Press, London, 1956)

Montagu, M. F. A., *An Introduction to Physical Anthropology* (Thomas, Springfield, 1945)

Morris, D., 'Homosexuality in the ten-spined stickleback', in *Behaviour* 4 (1952), pp. 233–61

—— 'The reproductive behaviour of the zebra finch, with special reference to pseudofemale behaviour and displacement activities', in *Behaviour* 6 (1954), pp. 271–322

—— 'The causation of pseudofemale and pseudomale behaviour', in *Behaviour* 8 (1955), pp. 46–57

—— 'The function and causation of courtship ceremonies', in *Fondation Singer Polignac Colloque Internat. Sur L'Instinct, June, 1954* (1956), pp. 261–86

—— 'The feather postures of birds and the problem of the origin of social signals', in *Behaviour* 9 (1956), pp. 75–113

—— ' "Typical Intensity" and its relation to the problem of ritualization'. *Behaviour* 11 (1957), pp. 1–12

—— *The Biology of Art* (Methuen, 1962)

—— 'Occupational therapy for captive animals', in *Coll. Pap. Lab. Anim. Cent.* 11 (1962), pp. 37–42

—— 'The response of animals to a restricted environment', in *Symp. Zool. Soc. London* 13 (1964), pp. 99–118

—— *The Mammals: a Guide to the Living Species* (Hodder and Stoughton, 1965)

—— 'The rigidification of behaviour', in *Phil. Trans. Roy. Soc. London B.* 251 (1966), pp. 327–30

—— (Editor), *Primate Ethology* (Weidenfeld & Nicolson, 1967)

Morris, R., and D. Morris, *Men and Snakes* (Hutchinson, 1965)

—— —— *Men and Apes* (Hutchinson, 1966)

—— —— *Men and Pandas* (Hutchinson, 1966)

Moulton, D. G., E. H. Ashton and J. T. Eayrs, 'Studies in olfactory acuity. 4. Relative detectability of n-Aliphatic acids by dogs'. *Anim. Behav.* 8 (1960), pp. 117–28

Munn, N. L. *The Evolution and Growth of Human Behavior* (Mifflin, Boston, 1965)

Murray, M. A. *The Splendour that was Egypt* (Sidgwick & Jackson, London, 1949)

Napier, J. and P. Napier, *Primate Biology* (Academic Press, 1967)

Neuhaus, W., 'Über die Riechschärfe der Hunden für Fettsäuren'. *Z. vergl. Physiol.* 35 (1953), pp. 527–52

Oakley, K. P., *Man the Toolmaker*. Brit. Mus. (Nat. Hist.)

Opie, I., and P. Opie, *The Lore and Language of School-children* (Oxford University Press, 1959)

Packard, V., *The Status Seekers* (Longmans, 1960)

Page, A. *Etiquette for Gentlemen* (Ward, Lock, London, 1961)

Pickering, C., *The Races of Man* (Bohn, 1850)

Piggott, S. (Editor), *The Dawn of Civilization* (Thames and Hudson, 1961)

—— *Ancient Europe* (Edinburgh University Press, 1965)

Prechtl, H. F. R. 'Problems of behavioral studies in the newborn infant', in Lehrman, Hinde and Shaw (eds), *Advances in the Study of Behavior* (Academic Press, New York, 1965)

Rabelais, F. *The Works of Mr Francis Rabelais* (Navarre Society, London, 1931)

Rachewiltz, B. de *Black Eros* (Allen & Unwin, London, 1964)

Read, C., *The Origin of Man* (Cambridge University Press, 1925)

Reynolds, V., and F. Reynolds. 'Chimpanzees of the Budongo Forest', in DeVore (ed.), *Primate Behavior* (Holt, Rinehart and Winston, New York, 1965)

Richardson, L. F., *Statistics of Deadly Quarrels* (Stevens, 1960)

Romer, A. S., *The Vertebrate Story* (Chicago University Press, 1958)

Rosebury, T. *Life on Man* (Secker & Warburg, London, 1969)

Russell, C., and W. M. S. Russell, *Human Behaviour* (André Deutsch, 1961)

—— —— *Violence, Monkeys and Man* (Macmillan, London, 1968)

—— and R. L. Burch. *The Principles of Humane Experimental Technique* (Methuen, London, 1959)

Salk, L., 'Thoughts on the concept of imprinting and its place in early human development'. *Canad. Psychiat. Assoc. J.* II (1966), pp. 295–305

Sara, D. *Good Manners and Hospitality* (Collier, New York, 1963)

Schaller, G., *The Mountain Gorilla* (Chicago University Press, 1963)

Schutz, F., 'Homosexualität und Prägung', *Psychol. Forschung* 28 (1965), pp. 439–63

Scott, J. P., 'Critical periods in the development of social behaviour in puppies', *Psychosom. Med.* 20 (1958), pp. 45–54

—— J. L. Fuller, *Genetics and the Social Behaviour of the Dog* (Chicago University Press, 1965)

Segal, R., *The Race War* (Cape, 1966)

Shirley, M. M., 'The first two years, a study of twenty-five babies'. Vol. 2, *Intellectual development. Inst. Child Welf. Mongr.*, Serial No. 8 (University of Minnesota Press, Minneapolis, 1933)

Simon, W., and J. H. Gagnon. 'Pornography – Raging menace or paper tiger?', in Gagnon and Simon (eds), *The Sexual Scene* (Aldine, New York, 1970)

Simonds, P. E. 'The bonnet macaque in South India', In DeVore (ed.), *Primate Behavior* (Holt, Rinehart and Winston, New York, 1965), pp. 175–96

Sluckin, W., *Imprinting and Early Learning* (Aldine, 1965)

Smailes, A. E., *The Geography of Towns* (Hutchinson, 1953)

Smith, A., *The Body* (Allen & Unwin, 1968)

Smith, M. E., 'An investigation of the development of the sentence and the extent of the vocabulary in young children'. *Univ. Iowa Stud. Child. Welf.* 3, No. 5 (1926)

Sorell, W. *The Story of the Human Hand* (Weidenfeld & Nicolson, London, 1968)

Southwick, C. H. (editor), *Primate Social Behaviour* (van Nostrand, Princeton, 1963).

—— M. A. Beg and M. R. Siddiqi. 'Rhesus monkeys in North India', in DeVore (ed.), *Primate Behavior* (Holt, Rinehart and Winston, New York, 1965), pp. 111–74

Sparks, J., 'Social grooming in animals'. *New Scientist* 19(1963), pp. 235–7

Spock, B. *Baby and Child Care* (Giant Cardinal, New York, 1946)

—— 'The striving for autonomy and regressive object relationships', *Psychoan. Study Child* 18 (1963), pp. 361–4

Stengel, E., *Suicide and Attempted Suicide* (Penguin, 1964)

Storr, A., *Human Aggression* (Penguin Press, 1968)

Szasz, K. *Fetishism* (Holt, Rinehart and Winston, New York, 1969)

Szasz, T. S. *The Myth of Mental Illness* (Hoebar-Harper, New York, 1961)

Tanner, J. M. and G. R. Taylor. *Growth* (Time-Life, New York, 1966)

Tax, S. (Editor), *The Evolution of Man* (Chicago University Press, 1960)

Tiger, L., Research report: Patterns of male association. *Current Anthropology* (vol. VIII, No. 3, June 1967)

Tinbergen, N., *The Study of Instinct* (Oxford University Press, 1951)

—— *The Herring Gull's World* (Collins, 1953)

Tomkins, S. S. *Affect, Imagery, Consciousness* (Springer, New York, 1962–3)

Turner, E. S., *All Heaven in a Rage* (Michael Joseph, 1964)

Ucko, P. J. *Anthropomorphic Figurines* (Szmidla, London, 1968)

Vanderbilt, A. *Complete Book of Etiquette* (Doubleday, New York, 1952)

Van Hooff, J., 'Facial expression in higher primates'. *Symp. Zool. Soc. Lond.* 8(1962), pp. 97–125

Vosper, J. *Baby Book* (Ebury, London, 1969)

Washburn, S. L. (Editor), *Social Life of Early Man* (Methuen, 1962)

—— (editor), *Classification and Human Evolution* (Methuen, 1964)

—— and I. DeVore, 'Social behaviour of baboons and early man', in *Social Life of Early Man* (Editor: S. L. Washburn), (Methuen, 1962)

Watson, J. B. *Psychological Care of Infant and Child* (Norton, New York, 1928)

Wedeck, H. E. *Dictionary of Aphrodisiacs* (Peter Owen, London, 1962)

West, D. J., *Homosexuality* (Aldine, 1968)

West, J. *Parent's Baby Book* (Parrish, London, 1966)

Whitman, C. O., *The Behaviour of Pigeons* (Carnegie Institution, 1919)

Wickler, W., 'Die biologische Bedeutung auffallend farbiger, nackter Hautstellen und innerartliche Mimikry der Primaten'. *Die Naturwissenschaften* 50 (13) (1963), pp. 481–2

—— 'Social-sexual signals and their intra-specific imitation among primates', in *Primate Ethology* (Editor: D. Morris), (Weidenfeld & Nicolson, 1967)

—— *Mimicry in Plants and Animals* (World University Library, 1968)

Wildeblood, J., and P. Brinson. *The Polite World* (Oxford University Press, London, 1965)

Williams, N. *Powder and Paint* (Longmans, Green, London, 1957)

Woddis, G. M., 'Depression and crime', in *Brit. J. Delinquency* (1957), pp. 85–94

Wolff, C. *A Psychology of Gesture* (Methuen, London, 1945)

Wolff, P. H. 'The natural history of crying and other vocalizations in early infancy', in Foss (ed.), *Determinants of Infant Behaviour*, vol. 4 (Methuen, London, 1969)

Wyburn, G. M., R. W. Pickford and R. J. Hirst, *Human Senses and Perception* (Oliver and Boyd, 1964)

Yerkes, R. M. *Chimpanzees. A Laboratory Colony* (Yale University Press, New Haven, 1943)

—— and A. W. Yerkes, *The Great Apes* (Yale University Press, 1929)

Young, P. and E. A. Goldman, *The Wolves of North America* (Constable, 1944)

Zeuner, F. E., *A History of Domesticated Animals* (Hutchinson, 1963)

Zuckerman, S., *The Social Life of Monkeys and Apes* (Kegan Paul, 1932)

INDEX